AF841758

METHODS IN MOLECULAR BIOLOGY™

Series Editor
John M. Walker
School of Life Sciences
University of Hertfordshire
Hatfield, Hertfordshire, AL10 9AB, UK

For further volumes:
http://www.springer.com/series/7651

Mouse Molecular Embryology

Methods and Protocols

Edited by

Mark Lewandoski

Frederick Cancer Research & Development Center, National Cancer Institute, Frederick, MD, USA

Humana Press

Editor
Mark Lewandoski
Frederick Cancer Research & Development Center
National Cancer Institute
Frederick, MD, USA

ISSN 1064-3745 ISSN 1940-6029 (electronic)
ISBN 978-1-60327-290-2 ISBN 978-1-60327-292-6 (eBook)
DOI 10.1007/978-1-60327-292-6
Springer New York Heidelberg Dordrecht London

Library of Congress Control Number: 2013951232

© Springer Science+Business Media New York 2014
This work is subject to copyright. All rights are reserved by the Publisher, whether the whole or part of the material is concerned, specifically the rights of translation, reprinting, reuse of illustrations, recitation, broadcasting, reproduction on microfilms or in any other physical way, and transmission or information storage and retrieval, electronic adaptation, computer software, or by similar or dissimilar methodology now known or hereafter developed. Exempted from this legal reservation are brief excerpts in connection with reviews or scholarly analysis or material supplied specifically for the purpose of being entered and executed on a computer system, for exclusive use by the purchaser of the work. Duplication of this publication or parts thereof is permitted only under the provisions of the Copyright Law of the Publisher's location, in its current version, and permission for use must always be obtained from Springer. Permissions for use may be obtained through RightsLink at the Copyright Clearance Center. Violations are liable to prosecution under the respective Copyright Law.
The use of general descriptive names, registered names, trademarks, service marks, etc. in this publication does not imply, even in the absence of a specific statement, that such names are exempt from the relevant protective laws and regulations and therefore free for general use.
While the advice and information in this book are believed to be true and accurate at the date of publication, neither the authors nor the editors nor the publisher can accept any legal responsibility for any errors or omissions that may be made. The publisher makes no warranty, express or implied, with respect to the material contained herein.

Printed on acid-free paper

Humana Press is a brand of Springer
Springer is part of Springer Science+Business Media (www.springer.com)

Preface

These days, mouse embryology sits at the crossroads of many interests and activities. Those interested in generating a mouse model of human disease must have a working knowledge of mouse development to do their work. Also they are usually using genetic technologies, such as conditional gene inactivation, which were developed either by mouse embryologists or with problems of mouse embryology in mind. The annotation of the human genome relies heavily on determining the requirements of the mouse gene ortholog of the human version, and many of these requirements often impact the mouse embryo. The field of stem cell biology that is currently blossoming naturally emerges from mouse development studies. Those concerned with regulatory safety testing often use the mouse in their teratological studies. And lastly, let us not forget those fascinated by the development of the mouse embryo for its own sake!

This volume provides detailed protocols for the novice with a basic education in laboratory techniques. Most of the protocols focus on techniques that are "close to the embryo": for example, manipulating embryonic gene expression, culturing explanted embryonic tissue and harvesting embryonic RNA. It also contains several reviews of important areas in the field, such as fluorescence imaging, lineage tracing, and genetic ablation. Given the importance of the mouse embryo to other aspects of biological research, the reader will find many valuable protocols throughout other volumes in this excellent series, such as Mammalian Oocyte Regulation (# 957), Neural Development # 1018), Embryo Culture (912), Transgenic Mouse Methods and Protocols (# 693) and Gene Knockout Protocols (# 530).

Finally, I would like to sincerely thank the authors for their contribution and patience during the process of completing this volume.

Frederick, MD, USA *Mark Lewandoski*

Contents

Contributors

RICHARD A. BALDOCK • *Human Genetics Unit, Institute of Genetics and Molecular Medicine, Edinburgh, UK*
KRISTIN K. BIRIS • *Frederick National Lab of Cancer Research, NCI, NIH, Frederick, MD, USA*
C. CLARE BLACKBURN • *MRC Centre for Regenerative Medicine, Institute for Stem Cell Research, University of Edinburgh, Edinburgh, Scotland, UK*
NICK BURTON • *MRC Human Genetics Unit, Institute of Genetics and Molecular Medicine, Edinburgh, UK*
JEFFREY H. CHRISTIANSEN • *Australian National Data Service, Monash University, Melbourne, Australia*
YOANNE M. CLOVIS • *Department of Neuroscience and Brain Technologies, Istituto Italiano di Tecnologia, Genoa, Italy*
BRIAN G. CONDIE • *Department of Genetics, University of Georgia, Athens, GA, USA*
LUIS COVARRUBIAS • *Departmento de Genética del Desarrollo y Fisiología Molecular, Instituto de Biotecnología/UNAM, Cuernavaca, Morelos, Mexico*
RODRIGO CUERVO • *Departmento de Genética del Desarrollo y Fisiología Molecular, Instituto de Biotecnología/UNAM, Cuernavaca, Morelos, Mexico*
DUNCAN R. DAVIDSON • *Human Genetics Unit, Institute of Genetics and Molecular Medicine, Edinburgh, UK*
ROBERT MICHAEL DORAN • *Department of Biomedical Sciences, College of Veterinary Medicine, Cornell University, Ithaca, NY, USA*
CATHERINE DUMUR • *Departments of Pathology, Virginia Commonwealth University, Richmond, VA, USA*
SUSAN M. DYMECKI • *Department of Genetics, Harvard Medical School, Boston, MA, USA*
JONATHAN EGGENSCHWILER • *Department of Molecular Biology, Princeton University, Princeton, NJ, USA*
DAVID K. GARDNER • *Department of Zoology, The University of Melbourne, Melbourne, VIC, Australia*
YELENA GOLUBEVA • *Histotechnology Laboratory, SAIC, Frederick National Lab of Cancer Research, NCI, NIH, Frederick, MD, USA*
JULIE GORDON • *Department of Genetics, University of Georgia, Athens, GA, USA*
KAI SU GREENE • *Department of Biomedical Sciences, College of Veterinary Medicine, Cornell University, Ithaca, NY, USA*
DAMIEN GRÉGOIRE • *Laboratory of Genetics and Development, Institut de Recherches Cliniques de Montréal (IRCM), Université de Montréal, Montréal, Québec, Canada*
PRABHAKAR R. GUDLA • *Optical Microscopy and Analysis Laboratory, SAIC, Frederick National Lab of Cancer Research, NCI, NIH, Frederick, MD, USA*
JACK L. HAAR • *Departments of Anatomy and Neurobiology, Virginia Commonwealth University, Richmond, VA, USA*

ANNA-KATERINA HADJANTONAKIS • *Developmental Biology Program, Sloan-Kettering Institute, New York, NY, USA*
ROCÍO HERNÁNDEZ-MARTÍNEZ • *Departmento de Genética del Desarrollo y Fisiología Molecular, Instituto de Biotecnología/UNAM, Cuernavaca, Morelos, Mexico*
WIELAND B. HUTTNER • *Max Planck Institute of Molecular Cell Biology and Genetics, Dresden, Germany*
PATRICIA JENSEN • *Laboratory of Neurobiology, Department of Health and Human Services, National Institute of Environmental Health Sciences, National Institute of Health, Research Triangle Park, NC, USA*
VANESSA J. JONES • *Children's Medical Research Institute, Sydney Medical School, The University of Sydney, Sydney, Australia*
TAKEHITO KANEKO • *Kyoto University, Institute of Laboratory Animals Graduate School of Medicine, Kyoto, Japan*
CHARLES KELLER • *Departments of Cellular & Structural Biology and Pediatrics, The University of Texas Health Science Center, San Antonio, TX, USA*
JUSUB KIM • *Rhythm & Hues Studios, Los Angeles, CA, USA*
SATOSHI KISHIGAMI • *School of Biology-Oriented Science and Technology, Kinki University, Kinki, Japan*
MARIE KMITA • *Laboratory of Genetics and Development, Institut de Recherches Cliniques de Montréal (IRCM), Université de Montréal, Montréal, Québec, Canada*
YOSHIHIRO KOMATSU • *School of Biology-Oriented Science and Technology, Kinki University, Kinki, Japan*
MICHAEL I. KOTLIKOFF • *Department of Biomedical Sciences, College of Veterinary Medicine, Cornell University, Ithaca, NY, USA*
MICHELLE LANE • *Research Centre for Reproductive Health, University of Adelaide, South Australia, Australia*
MARK G. LARMAN • *Department of Zoology, The University of Melbourne, Melbourne, VIC, Australia*
JANE CONSTANCE LEE • *Department of Biomedical Sciences, College of Veterinary Medicine, Cornell University, Ithaca, NY, USA*
AIMIN LIU • *Department of Biology, The Penn State University, University Park, PA, USA*
JOYCE A. LLOYD • *Department of Human and Molecular Genetics, Virginia Commonwealth University, Richmond, VA, USA*
STEPHEN J. LOCKETT • *Optical Microscopy and Analysis Laboratory, SAIC, Frederick National Lab of Cancer Research, NCI, NIH, Frederick, MD, USA*
DAVID A.F. LOEBEL • *Children's Medical Research Institute, Sydney Medical School, The University of Sydney, Sydney, Australia*
NANCY R. MANLEY • *Department of Genetics, University of Georgia, Athens, GA, USA*
MATTHEW MCAULIFFE • *Center for Information Technology, NIH, Bethesda, MD, USA*
DEAN P. MCCULLOUGH • *High Performance Technologies, Inc.,, Reston, VA, USA*
KAREN J. MEABURN • *Cell Biology of Genomes Group, NCI, NIH, Bethesda, MD, USA*
YUJI MISHINA • *Department of Biologic and Materials Sciences, School of Dentistry, University of Michigan, Ann Arbor, MI, USA*
BILLIE A. MOORE • *Department of Genetics, University of Georgia, Athens, GA, USA*
JUNICHI NAKAI • *Laboratory for Memory and Learning, RIKEN Brain Science Institute, Wako-shi, Saitama, Japan*
KAUSTAV NANDY • *Optical Microscopy and Analysis Laboratory, SAIC, Frederick National Lab of Cancer Research, NCI, NIH, Frederick, MD, USA*

LISA NEVELL • *Department of Anthropology Center for Advanced Studies in Hominid Paleobiology, The George Washington University, Washington, DC, USA*
SONJA NOWOTSCHIN • *Developmental Biology Program, Sloan-Kettering Institute, New York, NY, USA*
CHRISTOPHER J. PANG • *Departments of Human and Molecular Genetics, Virginia Commonwealth University, Richmond, VA, USA*
ALAN O. PERANTONI • *Frederick National Lab of Cancer Research, NCI, NIH, Frederick, MD, USA*
SERGEY PLISOV • *Frederick National Lab of Cancer Research, NCI, NIH, Frederick, MD, USA*
MELINDA POWER • *Children's Medical Research Institute, Sydney Medical School, The University of Sydney, Sydney, Australia*
SURESH I. PRAJAPATI • *Greehey Children's Cancer Research Institute, The University of Texas Health Science Center, San Antonio, TX, USA*
TANIA RADZIEWIC • *Children's Medical Research Institute, Sydney Medical School, The University of Sydney, Sydney, Australia*
JIANGUO RAO • *MRC Human Genetics Unit, Institute of Genetics and Molecular Medicine, Edinburgh, UK*
LATASHA C. REDMOND • *Departments of Human and Molecular Genetics, Virginia Commonwealth University, Richmond, VA, USA*
LORNA RICHARDSON • *MRC Human Genetics Unit, Institute of Genetics and Molecular Medicine, University of Edinburgh, Edinburgh, UK*
DAVID R. RODRIGUEZ • *Greehey Children's Cancer Research Institute, The University of Texas Health Science Center, San Antonio, TX, USA*
CALUM WILLIAM RUSSELL • *Department of Biomedical Sciences, College of Veterinary Medicine, Cornell University, Ithaca, NY, USA*
DAISUKE SAKAI • *Stowers Institute for Medical Research, Kansas City, MO, USA*
DAVID J. SCHULZ • *Division of Biological Sciences, University of Missouri, Columbia, MO, USA*
NIRMALA SHARMA • *Frederick National Lab of Cancer Research, NCI, NIH, Frederick, MD, USA*
JULIE M. SHERIDAN • *Walter and Eliza Hall Institute of Medical Research, Parkville, VIC, Australia*
BO SHUI • *Department of Biomedical Sciences, College of Veterinary Medicine, Cornell University, Ithaca, NY, USA*
PETER STEVENSON • *MRC Human Genetics Unit, Institute of Genetics and Molecular Medicine, Edinburgh, UK*
JOSHUA B. STUDDERT • *Children's Medical Research Institute, Sydney Medical School, The University of Sydney, Sydney, Australia*
DAVID SYMER • *Departments of Molecular Virology, Immunology and Medical Genetics, Internal Medicine, and Biomedical Informatics, Ohio State University Comprehensive Cancer Center, Columbus, OH, USA*
YVONNE NORINE TALLINI • *Department of Biomedical Sciences, College of Veterinary Medicine, Cornell University, Ithaca, NY, USA*
PATRICK P.L. TAM • *Children's Medical Research Institute, Sydney Medical School, The University of Sydney, Sydney, Australia*
SATOMI S. TANAKA • *Department of Kidney Development, Institute of Molecular Embryology and Genetics, Kumamoto University, Kumamoto, Japan*

NADYA TARASOVA • *Frederick National Lab of Cancer Research, NCI, NIH, Frederick, MD, USA*
DAVIDE DE PIETRI TONELLI • *Department of Neuroscience and Brain Technologies, Istituto Italiano di Tecnologia, Genoa, Italy*
PAUL A. TRAINOR • *Stowers Institute for Medical Research, Kansas City, MO, USA*
WILLIAM M. URBANSKI • *Department of Genetics, University of Georgia, Athens, GA, USA*
SHANMUGASUNDARAM VENKATARAMAN • *Human Genetics Unit, Institute of Genetics and Molecular Medicine, Edinburgh, UK*
ERIC VILLALON • *Christopher S. Bond Life Sciences Center, University of Missouri, Columbia, MO, USA*
HONGHE WANG • *Frederick National Lab of Cancer Research, NCI, NIH, Frederick, MD, USA*
SAMUEL T. WATERS • *Christopher S. Bond Life Sciences Center, University of Missouri, Columbia, MO, USA*
TERRY P. YAMAGUCHI • *Frederick National Lab of Cancer Research, NCI, NIH, Frederick, MD, USA*
YASUKA L. YAMAGUCHI • *Department of Kidney Development, Institute of Molecular Embryology and Genetics, Kumamoto University, Kumamoto, Japan*
YIYA YANG • *Human Genetics Unit, Institute of Genetics and Molecular Medicine, Edinburgh, UK*

Chapter 1

In Situ Hybridization Methods for Mouse Whole Mounts and Tissue Sections with and Without Additional β-Galactosidase Staining

Yoshihiro Komatsu, Satoshi Kishigami, and Yuji Mishina

Abstract

In situ hybridization is a powerful method for detecting endogenous mRNA sequences in morphologically preserved samples. We provide in situ hybridization methods, which are specifically optimized for mouse embryonic samples as whole mounts and section tissues. Additionally, β-Galactosidase (β-gal) is a popular reporter for detecting the expression of endogenous or exogenous genes. We reveal that 6-chloro-3-indoxyl-β-D-galactopyranoside (S-gal) is a more sensitive substrate for β-gal activity than 5-bromo-4-chloro-3-indolyl-β-D-galactoside (X-gal). S-gal is advantageous where β-gal activity is limited including early stage mouse embryos. As a result of the increased sensitivity as well as the color compatibility of S-gal, we successfully combined β-gal staining using S-gal with in situ hybridization using DIG-labeled probes in both whole mounts and sections.

Keywords β-Galactosidase, In situ hybridization, Mouse, S-gal (6-chloro-3-indoxyl-β-D-galactopyranoside), X-gal (5-bromo-4-chloro-3-indolyl-β-D-galactoside)

1 Introduction

Expression patterns of endogenous genes change drastically during embryogenesis. Knowledge of tissue- and stage-specific gene expression patterns is essential to speculate about genetic function. In addition, expression of particular genes can be used as markers to identify the presence or absence of a particular lineage, tissue, or cell type. Therefore, methods of mRNA detection in morphologically preserved samples serve as principal tools for analyses of mutations that cause abnormalities during embryogenesis [1, 2]. Generally, in situ hybridization can be performed using radiolabeled nucleic acid and non-radiolabeled (digoxigenin label) nucleic acid as a probe. There has been some debate over the sensitivity between radiolabeled and non-radiolabeled probes. Radiolabeling techniques require special instruments and it takes longer for detection than non-radiolabeled probes. In addition, radiolabeled

Mark Lewandoski (ed.), *Mouse Molecular Embryology*, Methods in Molecular Biology, vol. 1092,
DOI 10.1007/978-1-60327-292-6_1, © Springer Science+Business Media New York 2014

methods work best for sectioned tissues, and they are difficult to apply to whole mount tissues. Use of digoxigenin-labeled probes can overcome these problems. Here we describe in situ hybridization methods using non-radiolabeled probes for both whole mount and sectioned mouse embryonic tissues. By following our protocol, it is easy to assess intriguing gene expression patterns during all stages of mouse embryogenesis.

The *LacZ* gene of *Escherichia coli*, which encodes β-gal, has been applied as a reporter gene in the developmental biology field because of its high sensitivity and ease of detection. In the mouse, *lacZ* has been used by a variety of ways: e.g., by knocking it into the targeted gene locus for analysis of gene expression, as a transgenic reporter of tissue specificity, for detection of gene recombination by Cre-mediated recombination, and for detection of mutant ES cell derivatives in chimeric analyses. In these contexts, double staining to detect β-gal activity and mRNA presence via in situ hybridization techniques can provide a powerful strategy to examine the expression pattern of endogenous genes while monitoring β-gal activity in the same sample. However, a similar color is produced by β-gal activity using 5-bromo-4-chloro-3-indolyl-β-D-galactoside (X-gal) and in the detection of mRNA using in situ hybridization by DIG-labeled probes using BM purple (AP substrate). Therefore, it is often difficult to distinguish between these two stains in the same sample.

To overcome this difficulty and allow for double staining of β-gal activity and mRNA detection by in situ hybridization techniques, we use a different β-gal substrate, 6-Chloro-3-indoxyl-β-D-galactopyranoside (S-gal), instead of X-gal, to create a different color outcome between the two techniques. S-gal creates a pink/magenta color after the reaction and has been used for detection of β-gal activity [3, 4]. There are several options depending on materials and purposes for detecting the expression patterns of endogenous genes as well as the monitoring β-gal activity in both whole mount tissue and sectioned samples during multiple stages of mouse embryonic development [5]. S-gal is more advantageous than X-gal due to its higher sensitivity and color compatibility as exemplified in Figs. 1 and 2.

2 Materials

2.1 S-gal (6-Chloro-3-Indoxyl-β-D-Galactopyranoside) Staining

1. Stock solution for 25 mg/mL S-gal (catalog number: #C-5000, BIOSYNTH INTERNATIONAL, Inc.) in Dimethyl Sulfoxide (DMSO) HYBRI-MAX (catalog number: #D2650, Sigma) and stored at −20 °C. It is vital to protect the stock solution from the light.
2. Rinse buffer for S-gal staining: 0.1 % Sodium Deoxycholate, 0.2 % IGEPAL CA-630 (catalog number: #I3021, Sigma),

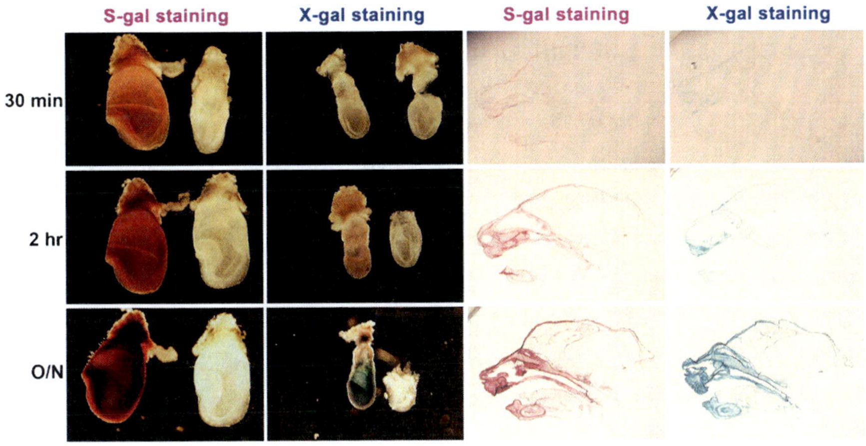

Fig. 1 Comparison with X-gal and S-gal staining on time course dependency. S-gal and X-gal staining pattern at each time point. *Upper panels* are 30 min, *middle panels* are 2 h, and *lower panels* are overnight (O/N) color development. *Two left panels* are day 7.5 embryos each set has one embryo carrying *ROSA26* (*left*) and one embryo as a negative control (*right*). The *two right panels* are sagittal sections of the mouse head for day 16.5 embryos carrying *P0-Cre* and *ROSA26*-reporter

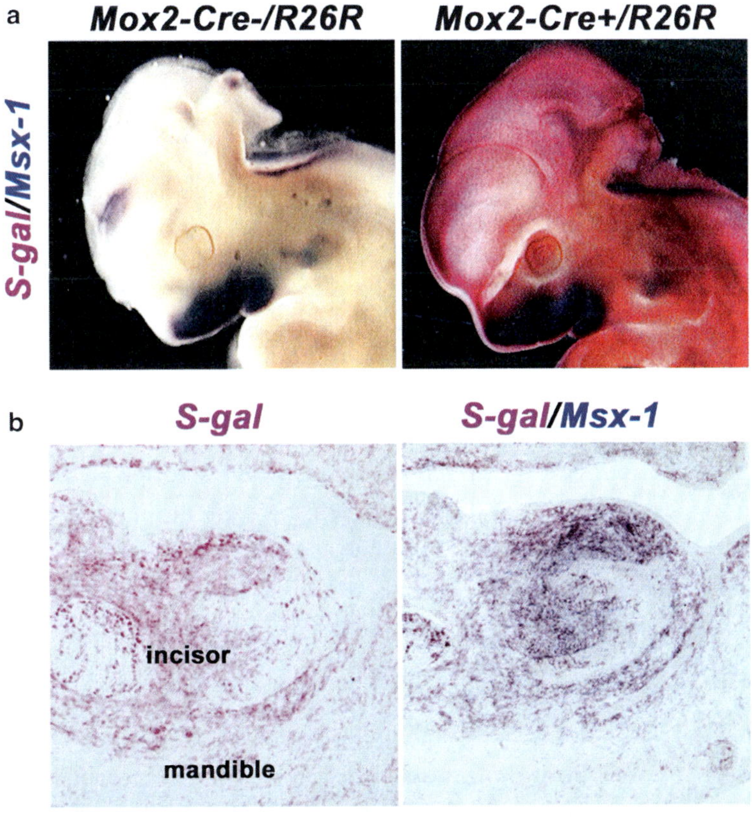

Fig. 2 Dual detection of β-gal activity and gene expression on whole mount and cryo-sectioned materials. (**a**) S-gal staining (*magenta*) and in situ hybridization signals (*purple*) were monitored for E11.5 mouse embryo. (**b**) S-gal staining and in situ hybridization signals were monitored by *magenta* and *purple* color in surrounding mesenchyme of incisor region at E14.5. *Left panel* is S-gal staining and *right panel* is double staining with S-gal and *Msx1* in situ hybridization

2 mM $MgCl_2$ in 0.1 M Na Phosphate buffer (pH 7.3). 1,000 mL of 0.5 M Na Phosphate buffer (pH 7.3) is prepared by mixing with 158 mL of 1 M NaH_2PO_4, 342 mL of 1 M Na_2HPO_4, and 500 mL water.

3. Substrate solution for S-gal needs to be made fresh and should contain 1 mg/mL S-gal from the stock, 5 mM Potassium Ferricyanide, and 5 mM Potassium Ferrocyanide in the rinse buffer.

2.2 Mouse Embryo Fixation for Whole Mount In Situ Hybridization

1. 4 % paraformaldehyde (PFA) with Phosphate Buffered Saline (PBS) in Diethylpyrocarbonate (DEPC)-treated water.
2. 25, 50, and 75 % methyl alcohol (MeOH) in PBT.

   PBT: PBS with 0.1 % Tween 20.

2.3 Embedding Procedure of Mouse Embryo for Section In Situ Hybridization

1. 4 % PFA–DEPC-treated PBS
2. 20 % sucrose ultrapure (catalog number: #802536, MP Biomedicals, LLC.) in PBS. It is necessary to make freshly on the day of use.
3. Optimal Cutting Temperature (O.C.T.) compound (catalog number: #4583, Sakura Finetek USA, Inc.).
4. Liquid nitrogen for making cryo-blocks.

2.4 Labeling of Probe for In Situ Hybridization

The following labeling kit is used for synthesizing RNA probe. DIG RNA Labeling Mix (catalog number: #11277073910, Roche), T3 RNA Polymerase (catalog number: #11031163001, Roche), T7 RNA Polymerase (catalog number: #10881767001, Roche), and SP6 (catalog number: #10810274001, Roche). RNase inhibitor (catalog number: #03335399001, Roche). DNase I (catalog number: #04716728001, Roche).

2.5 Whole Mount In Situ Hybridization for Mouse Embryo

All solutions need to be made fresh on the day of use.

1. 6 % Hydrogen peroxide in PBT solution.
2. 10 μg/mL proteinase K (catalog number: #P2308, Sigma) in PBT solution.
3. 2 mg/mL glycine in PBT solution.
4. 0.2 % glutaraldehyde, 0.1 % Tween 20 in 4 % PFA–DEPC-treated PBS solution.
5. Hybridization mix solution (final concentration): 50 % Formamide, 5× SSC (pH 4.5, use citric acid to adjust pH), 50 μg/mL yeast RNA, 50 μg/mL Heparin, and 1 % SDS in DEPC-treated water. Stored at –20 °C.
6. Solution I (final concentration): 50 % Formamide, 5× SSC, and 1 % SDS in water.

7. Solution II (final concentration): 0.5 M NaCl, 10 mM Tris–HCl (pH 7.5), and 0.1 % Tween 20 in water.
8. RNase A 100 μg/mL (catalog number: #10109142001, Roche) in Solution II.
9. Solution III (final concentration): 50 % Formamide and 2× SSC in water.
10. 10× TBST (100 mL): 8 g of NaCl, 0.2 g of KCl, 25 mL of 1 M Tris–HCl (pH 7.5), and 10 mL of 10 % Tween 20 in water. Levamisole (catalog number: #L9756, Sigma, final 2 mM) will be added to 1× TBST.
11. Sheep serum (catalog number: #S22-100 mL, Chemicon).
12. Mouse embryonic powder. Preparation: Dissect E14.5 mouse embryos from 4 to 6 wild-type litters. Fix embryos with acetone, then store at −20 °C for a couple of days. Next, homogenize these mouse embryos completely by using mortar and pestle until their form becomes a dried brown powder. Store at −20 °C.
13. Anti-Digoxigenin-AP, Fab fragments (catalog number: #11093274910, Roche).
14. NTMT buffer (final concentration): 100 mM NaCl, 100 mM Tris–HCl (pH 9.5), 50 mM $MgCl_2$, and 0.1 % Tween 20 in water. Levamisole (final 2 mM) will be added to NTMT.
15. BM Purple for AP substrate precipitating, ready-to-use solution for color development (catalog number: #11442074001, Roche).

2.6 Section In Situ Hybridization for Mouse Embryo

All solutions need to be made fresh on the day of use.

1. 20 μg/mL proteinase K (catalog number: #P2308, Sigma) in PBS solution.
2. 2 mg/mL glycine in PBS solution.
3. 1,000 mL of 0.1 M TEA buffer (pH 8.0): 18.57 g of Triethanolamine hydrochloride (catalog number: #T9534, Sigma) and 2.5 g NaOH in DEPC-treated water.
4. Acetic anhydride (catalog number: #A6404, Sigma).
5. Hybridization mix solution (final concentration): 50 % Formamide, 5× SSC (pH 4.5, use citric acid to adjust pH), 1 mg/mg Ribonucleic acid, transfer (catalog number: #R4018, Sigma), 0.1 mg/mL Heparin, 1 % Blocking reagent (catalog number: #11096176001, Roche), 0.1 % Tween 20, 0.1 % Chaps, 5 mM EDTA (pH 8.0) in DEPC-treated water. Stored at −20 °C.
6. TNE buffer (final concentration): 10 mM Tris–HCl (pH 7.6), 500 mM NaCl, and 1 mM EDTA in water.

7. RNase A 100 μg/mL (catalog number: #10109142001, Roche) in TNE buffer.
8. Buffer I (final concentration): 100 mM Tris–HCl (pH 7.5) and 150 mM NaCl in water.
9. Buffer II (final concentration): 2 % Blocking reagent (catalog number: #11096176001, Roche) in Buffer I.
10. Buffer III (final concentration): 100 mM Tris–HCl (pH 9.5), 100 mM NaCl, and 50 mM $MgCl_2$ in water.
11. Anti-Digoxigenin-AP, Fab fragments (catalog number: #11093274910, Roche).
12. BM Purple for AP substrate precipitating, ready-to-use solution for color development (catalog number: #11442074001, Roche).

3 Methods

For double staining of whole mount/section in situ hybridization and β-galactosidase, mouse embryos need to be stained with S-gal first. Therefore, we first describe the experimental procedures for LacZ staining for whole mount (Subheading 3.1) and sectioned samples (Subheading 3.2). We describe the preparation of the digoxigenin labeled RNA probes for in situ hybridization (Subheading 3.3) and their use in whole-mount (Subheading 3.4) and section in situ hybridization (Subheading 3.5).

The most important step for performing in situ hybridization is the collection of mouse embryos efficiently in order to avoid degradation of endogenous mRNA. In addition, appropriate fixation and dehydration of the mouse embryos is critical. It is also necessary to treat the embryos with an adequate amount of solution during every procedure. If at any point the tissue dries, it will cause undesirable increase in background during color development.

In regard to later stage embryos, proper penetration of substrates within the embryo is crucial for whole mount in situ hybridization. Based on our experience, we achieve good signal(s) up to stage E10.5. However, past stage E11.5, optimization is necessary by the individual researcher. Some tissues tolerate whole mount staining such as limb bud and lung tissue even after E11.5 stage if they are dissected from the rest of the embryo.

3.1 LacZ Staining of S-gal for Whole Mount Embryo (See Note 1)

1. Dissect mouse embryos in cold DEPC-treated PBS.
2. Fix with 4 % PFA–DEPC-treated PBS at 4 °C as follows:
   - E7.5 or younger—2 min
   - E8.5—5 min
   - E9.5—10 min
   - E10.5 or older—20 min

3. Quickly wash with PBS twice.
4. Wash with rinse buffer for 10 min × 3.
5. Stain with substrate solution at 37 °C for as long as needed to see stain, up to an overnight time period. Keep covered and in the dark during color development.
6. Wash with PBS for 5 min × 2.
7. The embryo can now be fixed in 4 % PFA, unless one intends to continue with in situ hybridizations. If this is the case, continue to Subheading 3.4.

3.2 LacZ Staining of S-gal for Cryo-section (See Note 1)

An example of the result is shown in Fig. 1.

1. Dissect mouse embryos in cold DEPC-treated PBS.
2. Fix with 4 % PFA–DEPC-treated PBS at 4 °C overnight.
3. Replace with 20 % sucrose in PBS at 4 °C overnight. Embryos will float in this solution. They are considered saturated by sucrose when the samples sink down completely.
4. Embed the samples in optimum cutting temperature (O.C.T.) compound on liquid nitrogen, then store at −80 °C (*see* **Note 2**).
5. Cut 14 μm cryo-sections. We use Superfrost Plus slide glass (catalog number: #12-550-15, Fisher Scientific) for tight attachment of sections to the glass. Then dry the sections completely by using a hair dryer (cooling mode) for 2 min (*see* **Note 3**).
6. Wash with PBS for 5 min × 3.
7. Wash with rinse buffer for 10 min × 3.
8. Stain with substrate solution at 37 °C overnight. Keep in the dark during color development.
9. Wash with PBS for 5 min × 2.
10. Mount the stained section using glycerol, unless one intends to continue with in situ hybridizations. If this is the case, proceed from **step 9** to continue to **step 1** of "Section in situ hybridization: day 1," below, in Subheading 3.5.

3.3 Synthesizing Digoxigenin-Labeled RNA Probes for In Situ Hybridization (See Note 4)

1. Template DNA is prepared by linearizing a plasmid, which contains the cDNA of interest and the appropriate restriction enzyme site. The authors typically digest 10 μg of plasmid in 200 μL as a final volume and incubate at 37 °C for overnight.
2. Purify the plasmid by phenol–chloroform extraction three times.
3. Precipitate the digested plasmid by adding 2.5 volume of 100 % ethanol and 1/10 volume of 3 M sodium acetate. Leave at −20 °C for 1 h.
4. Centrifuge (15,000 rpm in a microfuge, 10–15 min) and wash pellet with 70 % ethanol.

5. Add 20 μL of TE (10 mM Tris–HCl, pH 8.0 and 1 mM EDTA).
6. Take 1 μL of the DNA suspension and confirm that it is linearized with agarose gel electrophoresis. Store remaining DNA solution at −80 °C.
7. In vitro transcription is performed by using DIG RNA Labeling Mix, RNase inhibitor, DNase I, and T3/T7/SP6 RNA polymerase.
   - 10× transcription buffer—2 μL
   - Template DNA—2 μL
   - Digoxigenin labeling mix—2 μL
   - RNase inhibitor—1 μL
   - RNA polymerase—2 μL
   - DEPC-treated water—11 μL
8. Incubate at 37 °C for 2 h.
9. Add 2 μL of DNase I and incubate at 37 °C for 30 min.
10. Add 200 μL of TE, 600 μL of ethanol, and 10 μL of 4 M LiCl.
11. Leave at −20 °C for 2 h.
12. Centrifuge (15,000 rpm) and wash with 70 % ethanol.
13. Add 20 μL of TE and store at −80 °C.
14. Check 1 μL of aliquot via electrophoresis to confirm that digoxigenin-labeled RNA is successfully synthesized.

3.4 Whole Mount In Situ Hybridization

Dissection of mouse embryos for whole mount in situ hybridization

1. Dissect mouse embryos in cold DEPC-treated PBS, or use S-Gal-stained embryos from Subheading 3.1.
2. Fix with 4 % PFA–DEPC-treated PBS at 4 °C overnight.
3. Dehydrate gradually with 25 % MeOH–PBT—50 % MeOH–PBT—75 % MeOH–PBT—100 % MeOH for 10 min each on ice. Larger samples require longer time, around 30 min for each step instead of 10 min.
4. Wash with 100 % MeOH twice then store at −20 °C.

Whole mount in situ hybridization: *Day 1* (*see* **Note 4**)

1. Rehydrate the mouse embryos with 75—50—25 % MeOH–PBT solutions.
2. Wash with PBT twice.
3. Bleach with 6 % hydrogen peroxide in PBT for 1 h.
4. Wash with PBT (5 min × 3).
5. Incubate with 10 μg/mL proteinase K in PBS for 10–20 min depending on the stage and size. Incubation time needs to be

optimized by each researcher. Usually, the authors use the proteinase K condition for 5 min (mouse embryonic stage at E6.5–E7.5), 10 min (E8.5–E9.5), and 20 min (E10.5–E11.5).

6. Wash with 2 mg/mL glycine in PBT for 5 min.
7. Wash with PBT twice.
8. Re-fix the samples with 0.2 % glutaraldehyde and 0.1 % Tween 20 in 4 % PFA–DEPC-treated PBS solution for 20 min.
9. Wash with PBT twice.
10. Incubate with hybridization mix at 70 °C for 1 h.
11. Replace the hybridization mix containing 1 μg/mL digoxigenin labeled RNA probe.
12. Incubate at 70 °C overnight (*see* **Note 5**).

Whole mount in situ hybridization: Day 2

1. Wash with Solution I at 70 °C for 30 min × 2, changing solution at the end of each timepoint.
2. Wash with 1:1 mixture of Solution I and Solution II at 70 °C for 10 min.
3. Wash with Solution II 5 min × 3.
4. Incubate with 100 μg/mL RNase A in Solution II at 37 °C for 30 min × 2.
5. Rinse with Solution II briefly.
6. Wash with Solution III for 5 min.
7. Wash with Solution III at 70 °C for 30 min × 2.
8. Wash with TBST for 5 min × 3.
9. Block the samples with TBST containing 10 % sheep serum at 4 °C for 2 h.
10. During **step 9**, it is necessary to prepare the antibody solution. First, incubate 0.5 mL of TBST containing 3 mg of mouse embryonic powder at 70 °C for 30 min. Then cool the solution on ice for 30 min. Add 5 μL of sheep serum and 0.7 μL of anti-digoxigenin-AP antibody. Shake gently at 4 °C for 1 h. Spin down briefly and collect the supernatant into a new tube. Dilute this supernatant solution with TBST containing 1 % sheep serum to 2 mL as a final volume.
11. Replace the blocking solution with the antibody solution.
12. Incubate at 4 °C overnight.

Whole mount in situ hybridization: Day 3

1. Wash with TBST (5 min × 3).
2. Wash with TBST (1 h × 5).
3. Wash with NTMT (10 min × 3).

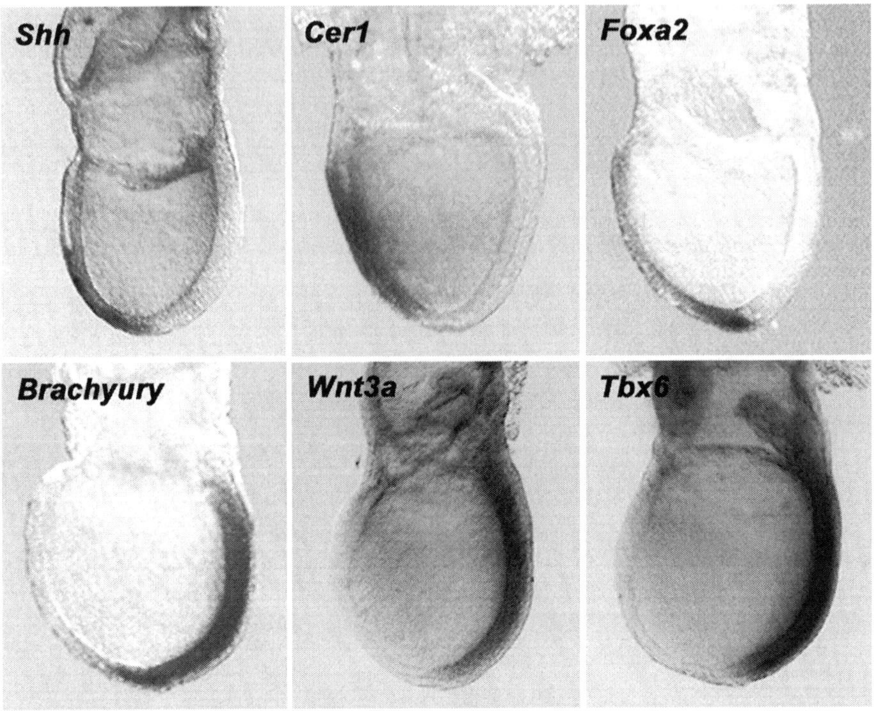

Fig. 3 Example for whole mount in situ hybridization at early stage of mouse embryonic development. Anterior and posterior marker gene expression analysis of *mouse* embryos at E7.5. Marker gene expression pattern of *Shh*, *Cer1*, *Foxa2* (*upper panels*) and *Brachyury*, *Wnt3a*, *Tbx6* (*lower panels*) respectively

4. Replace with BM purple AP substrate solution.
5. Incubate the samples until the color has developed to the desired extent. Keep in the dark during color development (*see* **Note 6**).
6. Wash with PBT (5 min × 2).

Examples of the results are shown in Figs. 3 and 4 for mRNA staining alone and Fig. 2a for lacZ and mRNA staining.

3.5 Section In Situ Hybridization Using Cryo-section

The following method is also possible with paraffin tissue section (i.e., without lacZ stained samples). In the case of paraffin-embedded section, it is necessary to perform deparaffinization first. The authors usually perform the deparaffinization in the following manner. Xylene (5 min × 3 times)—100 % EtOH (5 min × 2 times)—70 % EtOH (5 min × 2 times)—Water (5 min × 3 times). Then move to **step 1** of "Section in situ hybridization: day 1," below.

If one is not using LacZ-stained samples, then the following sections must be prepared. If one is using LacZ-stained samples, proceed to **step 1** of "Section in situ hybridization: day 1," below.

1. Dissect mouse embryos in cold DEPC-treated PBS.
2. Fix with 4 % PFA–DEPC-treated PBS at 4 °C overnight.

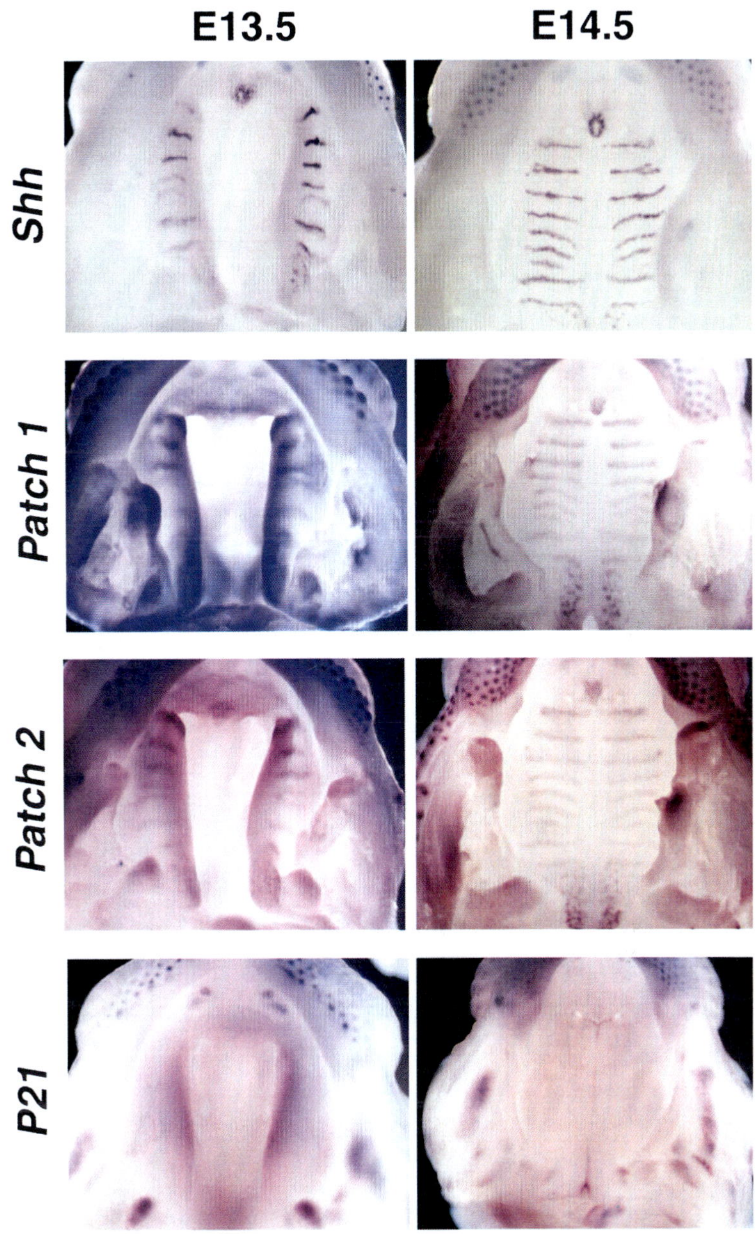

Fig. 4 Example for whole mount in situ hybridization at late stage of mouse embryonic development. Whole mount in situ hybridization was performed using *Shh*, *Patch1*, *Patch2*, and *P21* at E13.5 and E14.5. *Shh*, *Patch1*, and *Patch2* start to express in the prospective palatal rugae at E13.5 but there is no expression of *P21* on the palatal rugae at E13.5 and E14.5. *Patch1*, *Patch2*, and *P21* are strongly expressed in the edge of the palate at E13.5

3. Replace with 20 % sucrose in PBS at 4 °C overnight.
4. Embryos will float in this solution. They are considered saturated by sucrose when the samples sink down completely.

5. Embed the samples in optimum cutting temperature (O.C.T.) compound on liquid nitrogen, then store at −80 °C (*see* **Note 2**).
6. Cut 14 μm cryo-sections.
7. Start to section in situ hybridization (Day 1).

Section in situ hybridization: Day 1 (*see* **Note 4**).
It is critical to perform the experiment completely RNase free.

1. Wash the section with PBS for 10 min.
2. Incubate with 10 μg/mL proteinase K in PBS for 15 min.
3. Re-fix with 4 % PFA solution for 10 min.
4. Wash with 2 mg/mL glycine in PBS for 5 min.
5. Wash with PBS for 5 min.
6. Wash with TEA for 5 min.
7. Incubate with TEA containing 0.25 % acetic anhydride for 10 min (*see* **Note 7**).
8. Wash with 4× SSC for 10 min.
9. Incubate with 2× SSC with 50 % formamide for 30 min at 65 °C.
10. Add 100 μL of hybridization mix containing 1 μg/mL digoxigenin labeled RNA probe (*see* **Note 8**).
11. Cover the section with cover glass or parafilm.
12. Incubate the section at 65 °C for overnight (*see* **Note 9**).

Section in situ hybridization: Day 2

1. Wash with 2× SSC with 50 % formamide at 65 °C for 20 min × 3 (*see* **Note 10**).
2. Wash with TNE buffer at 37 °C for 5 min.
3. Incubate with 20 μg/mL RNase A in TNE buffer at 37 °C for 30 min.
4. Wash with TNE buffer at 37 °C for 5 min.
5. Wash with 0.2× SSC at 65 °C for 20 min × 3.
6. Wash with Buffer I for 5 min.
7. Blocking with Buffer II for 1 h.
8. Incubate with 1:1,000 dilution of anti-digoxigenin-AP antibody in Buffer II at 4 °C overnight.

Section in situ hybridization: Day 3

1. Wash with Buffer I for 15 min × 2.
2. Wash with Buffer III for 5 min.
3. Add 200–300 μL of BM purple AP substrate solution to the section.

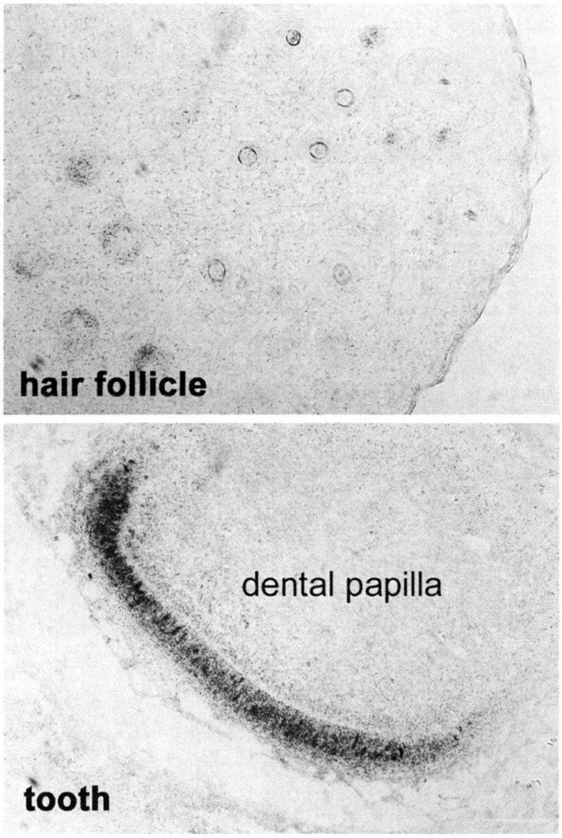

Fig. 5 Example for section in situ hybridization by paraffin section. Section in situ hybridization was performed using *Shh* specific RNA probe. Sagittal paraffin section of craniofacial region at E18.5. Note that *Shh* expression is specifically detected in hair follicle (*upper panel*) and pre-ameloblasts (*lower panel*)

4. Incubate the section until color has developed to the desired extent. Keep in the dark during color development (*see* **Note 6**).
5. Wash with PBS for 5 min × 2.

Examples of the results are shown in Fig. 5 for mRNA staining alone and Fig. 2b for lacZ and mRNA staining.

4 Notes

1. If the investigator is only staining for LacZ activity and therefore not also performing in situ hybridizations, this protocol can be applied for general LacZ staining by using X-gal instead of S-gal.
2. It is necessary to freeze the O.C.T. block slowly. The authors put the O.C.T. embedded samples on the 10 cm cell culture

dish, floating them on liquid nitrogen. The color of the O.C.T. compound will be changed from clear to white after 5 min. Then, collect these cryo-blocks and store in 50 mL tube at −80 °C.

3. For consistent results, the authors prepare cryo-sections on the day of use.
4. It is critical to perform this section completely RNase free.
5. The probe can be recycled once hybridization is done. The authors usually use RNA probes at least three times to obtain the same levels of in situ signals. Store the recycled RNA probes at −80 °C.
6. After color begins to appear, the color development can be confirmed under the microscope. If the color of the BM purple AP substrate solution changes from yellow to dark purple, you can replace the new substrate solution to enhance further color development.
7. Acetic anhydride is unstable in aqueous solution. The authors usually prepare the TEA containing 0.25 % acetic anhydride solution just before starting this process.
8. It is necessary to wipe the excess amount of 2× SSC with 50 % formamide solution from around the section, before adding the hybridization mix containing RNA labeling probe. Without this step, final concentration of the RNA probe is reduced and it may lower the detection of signals.
9. The authors use a plastic container with an airtight seal and add paper towels containing 2× SSC with 50 % formamide on the bottom. This plastic box needs to be sealed completely during hybridization to prevent evaporation.
10. In order to avoid any damage to the section, it is necessary to remove the cover glass or parafilm as follows: the authors carefully sink the slide glass into a beaker containing enough pre-warmed 2× SSC with 50 % formamide solution (65 °C) to cover the sample. After 1–2 min, the cover glass or parafilm will spontaneously detach from the slide glass.

Acknowledgments

We thank Drs. Ken-ichi Yamamura for *P0-Cre* mice, and Philippe Soriano for *Mox2-Cre* and ROSA26 reporter mice; Drs. Andrew P. McMahon, Michael Rauchman, William Shawlot, Hiroshi Sasaki, Bernhard G. Herrmann, Terry P. Yamaguchi, Deborah L. Chapman, and Hiroshi Hamada for in situ probes; Dr. Gen Yamada for sharing with us the unpublished results; Ms. Trisha Castranio and Sudha Rajderkar for critical reading; Dr. Mark Lewandoski for generous editing of this manuscript. This work was supported by the National Institutes of Health (K99DE021054 to Y.K., R01DE020843 and ES071003-11 to Y.M.), a fellowship from the Japan Society for the Promotion of Science (Y.K.), and the conditional gift from RIKEN-Brain Science Institute (Y.M).

References

1. Wilkinson DG (1992) In situ hybridization. A practical approach. Oxford University Press, New York, NY
2. Wilkinson DG, Nieto MA (1993) Detection of messenger RNA by *in situ* hybridization to tissue sections and whole mounts. Methods Enzymol 225:361–373
3. Varlet I, Collignon J, Robertson EJ (1997) Nodal expression in the primitive endoderm is required for specification of the anterior axis during mouse gastrulation. Development 124:1033–1044
4. Baker JC, Beddington RS, Harland RM (1999) Wnt signaling in Xenopus embryos inhibits bmp4 expression and activates neural development. Genes Dev 13:3149–3159
5. Kishigami S, Komatsu Y, Takeda H, Nomura-Kitabayashi A, Yamauchi Y, Abe K, Yamamura K, Mishina Y (2006) Optimized beta-galactosidase staining method for simultaneous detection of endogenous gene expression in early mouse embryos. Genesis 44:57–65

Chapter 2

Two-Color In Situ Hybridization of Whole-Mount Mouse Embryos

Kristin K. Biris and Terry P. Yamaguchi

Abstract

RNA in situ hybridization is a powerful technique used to identify the spatial localization of a specific RNA in a tissue section or whole tissue. In this protocol, we describe a reliable method for two-color in situ hybridization that can be used to accurately assess the expression of multiple genes with contrasting or overlapping expression patterns in whole mouse embryos.

Key words In situ hybridization, Development, Embryo, RNA probe, Color detection, Antibody

1 Introduction

In situ hybridization is a method of detecting specific mRNA sequences in fixed cell populations and whole tissue. The basic technique involves the use of an enzymatically labeled probe that anneals to complementary sequences in the target tissue, and is visualized with a chromogenic substrate. This technique can be routinely used to visualize the expression patterns of multiple genes in whole mammalian embryos.

The following whole-mount in situ hybridization (WISH) protocol is modified from the protocols of Wilkinson et al. [1], Conlon et al. [2], and Parr et al. [3]. The procedure described here provides a detailed step-by-step method for two-color analysis using Digoxigenin (DIG)-labeled and Fluorescein (FLU)-labeled probes. Simultaneous detection of two or more probes is achieved with two chromogenic substrates, producing beautiful contrasting patterns of gene expression. This protocol has been optimized for the analysis of gene expression in the gastrulating embryo [4–6], but has been readily applied to fetal stages, as well as isolated organs and cultured tissue explants (Fig. 1).

Mark Lewandoski (ed.), *Mouse Molecular Embryology*, Methods in Molecular Biology, vol. 1092,
DOI 10.1007/978-1-60327-292-6_2, © Springer Science+Business Media New York 2014

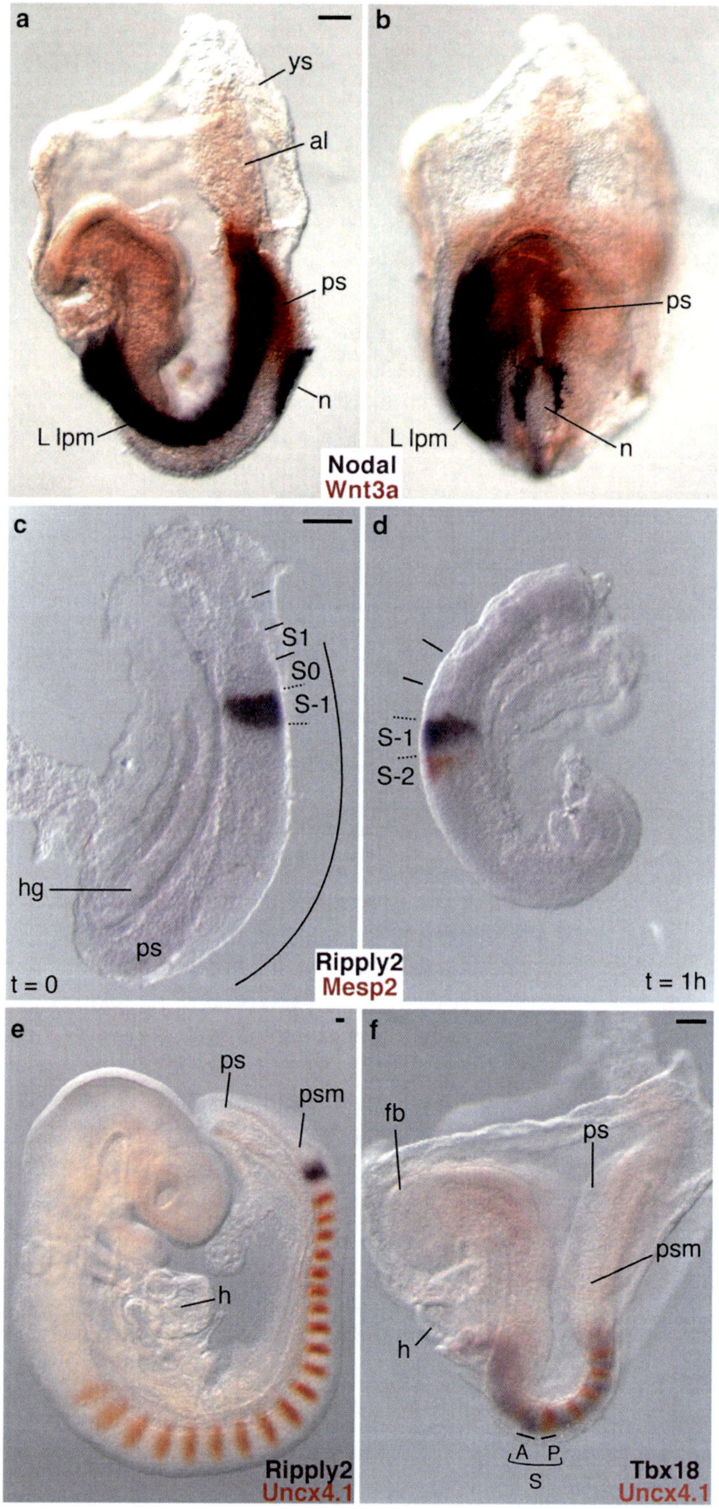
a
ys
al
ps
n
L lpm
b
ps
L lpm
n
Nodal
Wnt3a
c
S1
S0
S-1
hg
ps
t = 0
d
S-1
S-2
t = 1h
Ripply2
Mesp2
e
ps
psm
h
Ripply2
Uncx4.1
f
fb
ps
psm
h
A
P
S
Tbx18
Uncx4.1

2 Materials

Fixatives, hybridization solution, and reagents may be prepared in advance and stored in aliquots, while remaining wash and incubation solutions are prepared fresh on the day of use. All procedures up to the hybridization step must be carried out under RNase-free conditions to preserve the integrity of the RNA. Prepare solutions with RNase-free water purchased from the manufacturer. Autoclave all glassware, plastic ware, and pipette tips, and clean all equipment and work area with RNaseZAP® decontamination solution or similar solution (*see* **Note 1**). It is recommended that sterile plastic transfer pipettes are used for changing solutions, and that they are replaced frequently.

2.1 Preparation of Embryos

1. Phosphate-buffered saline (PBS), pH 7.4.
2. Micro dissecting tweezers pattern #5 and #55, light superfine points, tip 0.05×0.01 mm, material: Dumostar alloy (Roboz RS4978 and RS4984; *see* **Note 2**).
3. VWR® 2 mL screw-cap microcentrifuge tubes (VWR 89004-302; *see* **Note 3**).
4. 4 % paraformaldehyde (PFA) in PBS. Dissolve in PBS and stir at 65 °C on a hot plate for 1 h. Filter and aliquot into 15 mL and 50 mL tubes and store at −20 °C until use.

Fig. 1 Examples of two-color WISH performed on mouse embryos and cultured embryo explants. In all cases, DIG-labeled RNA probes were detected with BM purple, and FLU-labeled RNA probes were visualized with INT/BCIP. (**a**, **b**) Two-color WISH analysis of a 4 somite stage (E8.2) embryo demonstrating the complementary expression of genes encoding the secreted signaling molecules *Nodal* (*purple*) and *Wnt3a* (*orange*). *Nodal* is asymmetrically expressed in the node periphery and in the (*left*) lateral plate mesoderm [9, 10], while *Wnt3a* mRNA is restricted to the midline primitive streak [11] (Reproduced from ref. 5 with permission from Development). (**c**, **d**) Embryo-half culture experiments illustrate how dynamic, oscillating gene expression patterns can be visualized with two-color WISH. The bHLH transcription factor, *Mesp2* (*orange*), and the transcriptional corepressor *Ripply2* (*purple*), are important for segment boundary formation [4, 5, 12, 13]. In the uncultured half-embryo explant ($t = 0$), both genes are coexpressed in S-1 in the anterior presomitic mesoderm (**c**). After culturing the complementary half-embryo explant for 1 h, *Ripply2* expression remains in S-1 while *Mesp2* expression is activated in S-2 (**d**). (**e**) Analysis of E9.5 embryos shows that the paired type homeodomain transcription factor *Uncx4.1* (*orange*) is expressed in the caudal half of segmented somites [14–16], and in the posterior half of S0 immediately adjacent to the broad somite-wide stripe of *Ripply2* expression (*purple*) in S-1 [4]. (**f**) Two-color WISH clearly illustrates the complementary expression patterns of segment polarity markers. Expression of the anterior half-somite marker *Tbx18* (*purple*) complements *Uncx4.1* expression in the caudal half [5, 17]. All embryo views are lateral with the exception of (**b**), which is ventroposterior. *ys* yolk sac, *al* allantois, *ps* primitive streak, *n* node, *L lpm* left lateral plate mesoderm, *hg* hindgut, *S* somite, *S-1* presumptive somite, *S0* forming somite, *S1* first newly formed somite, *curved line or psm* presomitic mesoderm, *h* heart, *fb* forebrain, *A* anterior half somite, *P* posterior half somite. Scale bars: 100 μm

5. PBT: PBS with 0.1 % SigmaUltra Tween® 20 (Sigma P7949).
6. Methanol–PBT series (25, 50, 75 % methanol in PBT). Methanol is considered hazardous and should be handled and disposed of using established laboratory procedures. Store in a sealed flammable storage cabinet.
7. RNaseZap® RNase decontamination solution (Sigma R2020).

2.2 Embryo Powder Preparation and Storage

1. Mortar and pestle.
2. Tissue homogenizer.
3. Acetone. Follow established laboratory procedures when handling and disposing of this chemical. Store in a sealed flammable storage cabinet.
4. Whatman filter paper.

2.3 Generation of Template

1. Purified plasmid DNA.
2. Restriction enzymes.
3. Phenol–Chloroform–Isoamyl Alcohol 25:24:1 saturated with 10 mM Tris–HCl, pH 8.0, 1 mM EDTA. Phenol can cause severe burns. Handle in a fume hood and wear gloves.
4. Phase Lock Gel® Heavy, 2 mL tubes (PLG tubes) (Eppendorf 0032005.152).
5. RNase-free absolute ethanol.
6. RNase-free 3 M Sodium Acetate (NaOAc), pH 5.2.
7. RNase-free 70 % ethanol.
8. 10 mM Tris–HCl, pH 7.6.

2.4 Synthesis of Digoxigenin or Fluorescein Labeled RNA Probes

1. Sterile Dnase, RNase, Protease-free water.
2. 10× transcription buffer (Roche 1277073).
3. 0.1 M dithiothreitol (DTT). Store in aliquots at −20 °C.
4. 10× DIG RNA labeling mix 10 mM ATP, 10 mM CTP, 10 mM GTP, 6.5 mM TTP, 3.5 mM DIG-11-UTP, pH 7.5 (Roche 1277073).
5. 10× Fluorescein RNA labeling mix 10 mM ATP, 10 mM CTP, 10 mM GTP, 6.5 mM UTP, 3.5 mM FLU-12-UTP, pH 7.5 (Roche 1685619).
6. Linearized DNA template (1 μg/μL).
7. Protector RNase inhibitor (40 U/μL) (Roche 03335399001).
8. Sp6, T3, or T7 RNA polymerase (20 U/μL) (Roche 10810274001, 11031163001, 10881767001).
9. RNase-free DNaseI recombinant (10 U/μL) (Roche 04716728001).
10. TE: 50 mM Tris–HCl, 1 mM EDTA, pH 8.0.

11. 7.5 M LiCl precipitation solution (Ambion AM9480).
12. Prehybridization/hybridization solution: 50 % formamide (Ambion AM9342), 5× standard saline sodium citrate (SSC), pH 4.5 (Ambion AM9763), 1 % sodium dodecyl sulfate (SDS) (Ambion AM9823), 100 μg/mL tRNA (Sigma R8505), 100 μg/mL heparin (Sigma H3149; *see* **Note 4**). Formamide is a hazardous material and should be handled using established laboratory procedures.

2.5 Prehybridization and Hybridization of Embryos

1. Methanol–PBT series (25, 50, 75 % methanol in PBT).
2. 100 % Methanol.
3. PBT: PBS with 0.1 % Tween-20.
4. 30 % hydrogen peroxide solution (Sigma H1009).
5. 4 % PFA in PBS.
6. 25 % glutaraldehyde solution (Sigma G5882). Glutaraldehyde is packaged in 10 mL glass ampoules. Upon opening, aliquot into tubes and store at −20 °C.
7. 10 mg/mL proteinase K (Roche 03115836001) in RNase-free water. Store in 25 μL aliquots at −80 °C.
8. Glycine, OmniPur* (VWR EM4810).
9. Prehybridization/hybridization solution.
10. DIG-labeled RNA probe and FLU-labeled RNA probes. Store at −80 °C.
11. Hybaid Shake “n” Stack Hybridization Oven (Thermo Scientific 6241).

2.6 Posthybridization Washes and Antibody Incubation

1. Posthybridization wash solution I: 50 % formamide, 4× SSC, 1 % SDS.
2. Posthybridization wash solution II: 0.5 M NaCl, 10 mM Tris–HCl, pH 7.5, 0.1 % Tween-20.
3. Posthybridization wash solution III: 50 % formamide, 2× SSC.
4. RNase A (Sigma R6513).
5. MABTL: 0.1 M Maleic acid pH 7.5, 0.15 M NaCl, 0.1 % Tween-20, and 2 mM levamisole, adjust pH with NaOH. Filter before use (*see* **Note 5**).
6. Blocking reagent solution: 2 % w/v blocking reagent (Roche 11096176001) in MABTL.
7. Sheep serum (Sigma S2263; *see* **Note 6**).

2.7 Preadsorption of Antibody

1. Sheep anti-Digoxigenin (Roche 1093274), and anti-Fluorescein (Roche 1426338) Fab fragments conjugated to alkaline phosphatase (AP). Store at 4 °C.

2.8 Antibody Washes and Color Detection

1. NTMT: 100 mM NaCl, 100 mM Tris–HCl pH 9.5, 50 mM MgCl2, 0.1 % Tween-20.
2. NTMTL: NTMT with 2 mM levamisole.
3. BM Purple: AP substrate precipitating (NBT/BCIP ready-to-use solution) (Roche 11442074001). Store at 4 °C protected from light.
4. INT/BCIP Stock Solution (Roche 11681460001). Store at 4 °C protected from light.

2.9 Embryo Storage and Photography

1. Glycerol SigmaUltra (Sigma G6279).
2. Leica MZFLIII high-performance stereomicroscope.
3. Zeiss AxioCam high-resolution digital camera.
4. NCL150 Fiber Optic Illuminator.

3 Methods

Subheadings 3.1–3.4 should be performed in advance. Embryos should be gently rocked on a single speed orbital mixer or in a hybridization oven equipped with a variable speed rocking platform during the prehybridization, hybridization, and washing steps. At the beginning of each day of the protocol, prepare and warm solutions in advance. All steps should be performed at room temperature unless otherwise stated.

3.1 Preparation of Embryos

1. Dissect embryos in ice-cold PBS. Reflect extraembryonic membranes and remove the amnion. Puncture the fourth ventricle in the hindbrain to prevent trapping of reagents and consequent background in the neural tube (*see* **Note** 7).
2. Place the embryos in a 2 mL screw-cap tube, and fix in 4 % PFA in PBS overnight at 4 °C.
3. Wash three times with PBT for 5 min each.
4. Dehydrate the embryos in a Methanol series (25, 50, 75 % Methanol–PBT) for 5 min each. Wash twice with 100 % Methanol for 10 min each, and store indefinitely in 100 % Methanol at −20 °C.

3.2 Embryo Powder Preparation and Storage

1. Dissect wild-type embryos (embryonic day (E) 12.5–14.5 mouse embryos) in cold PBS.
2. Homogenize embryos in a 50 mL conical tube on ice in a minimum volume of PBS.
3. Add 4 volumes of ice-cold acetone, gently mix, and then incubate on ice for 30 min.
4. Centrifuge at 10,000 × g for 10 min.

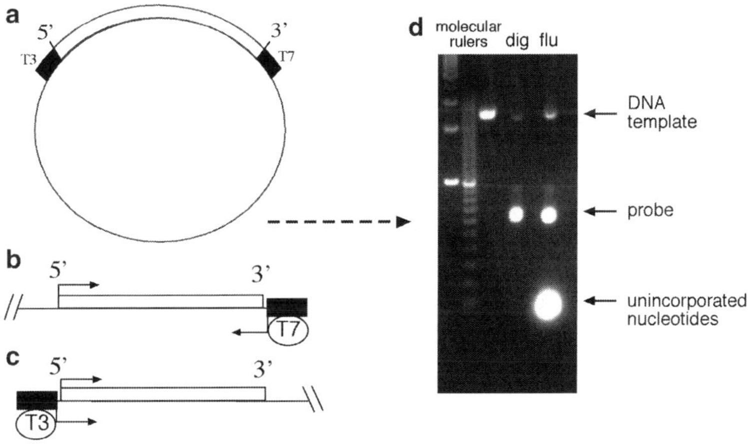

Fig. 2 Schematic illustrating the generation of the linear cDNA template and RNA probes. (**a**) The plasmid must contain unique restriction enzyme sites and RNA polymerase initiation sites for T7, T3, or SP6, flanking the cDNA of interest. (**b**) In this example, the 5′ end of the insert is digested with an appropriate restriction enzyme and then transcribed with T7 RNA polymerase to generate an antisense probe. (**c**) Similarly, a sense probe is generated by digesting the 3′ end with an appropriate enzyme, and transcribing with T3 polymerase. (**d**) Gel electrophoresis of freshly synthesized RNA probes. RNA should migrate to a position that roughly correlates with the size of the cloned insert, while fainter, slowly migrating DNA template bands should also be apparent

5. Wash the pellet with ice-cold acetone and centrifuge again.
6. Spread pellet out on a piece of Whatman filter paper. Then, cover with a second piece of filter paper and let dry overnight at room temperature.
7. Grind into a fine powder with a newly autoclaved mortar and pestle.
8. Store dry in an eppendorf tube at 4 °C.

3.3 Generation of Template

1. Quantify circular plasmid DNA with a standard spectrophotometer.
2. Prepare template DNA by linearizing 6 μg of DNA with the appropriate restriction enzymes (*see* **Note 8** and Fig. 2). Digest at 37 °C for 3 h in a 100 μL volume. Remove a 5 μL aliquot and run out on a 1 % agarose gel to ensure that the digestion reaction is complete and the plasmid is completely linearized.
3. Dilute the remaining 95 μL digest with 105 μL RNase-free water, then phenol–chloroform extract the DNA using PLG tubes following manufacturer's recommendations (*see* **Note 9**).
4. Add 2.5 volumes of 100 % ethanol and 1/10 volume of 3 M NaOAc to precipitate the DNA. Mix well and incubate at −80 °C for a minimum of 1 h.

5. Spin samples at 4 °C for 15 min, then wash pellet with 70 % ethanol and air-dry.
6. Resuspend pellet in 5 μL 10 mM Tris–HCl pH 7.6. Use 1 μL (1 μg) as the template for RNA probe synthesis and store remaining template at –20 °C.

3.4 Synthesis of Digoxigenin or Fluorescein Labeled RNA Probes

1. Mix these reagents in the following order at room temperature:

Sterile Dnase, RNase, Protease-free water	11.5 μL
10× transcription buffer	2.0 μL
0.1 M DTT	2.0 μL
10× DIG or FLU RNA labeling mix	2.0 μL
Linearized template (1 μg)	1.0 μL
RNase inhibitor (40 U/μL)	0.5 μL
Sp6, T7, or T3 RNA polymerase (20 U/μL)	1.0 μL

2. Incubate at 37 °C for 2 h.
3. To estimate the transcript amount, a 1 μL aliquot is removed and electrophoresed on a 1 % agarose gel containing 0.5 μg/mL ethidium bromide (*see* **Note 10**). A faint template band and an RNA band ten times more intense than the plasmid band should be observed, indicating that approximately 10 μg of probe has been synthesized. Two or more RNA bands are commonly observed.
4. To the transcription reaction, add 2 μL Dnase I (10 U/μL), and incubate at 37 °C for 15 min.
5. Add 100 μL TE, 5.3 μL of 7.5 M LiCl, and 300 μL 100 % ethanol, mix, and precipitate RNA at –80 °C for 1 h.
6. Spin in a microcentrifuge at 4 °C for 30 min.
7. Wash the pellet twice with 70 % ethanol and air-dry.
8. Dissolve pellet in 200 μL of warm hybridization solution at approximately 25–50 ng/μL, and store at –80 °C. Probes can withstand multiple freeze–thaw cycles and are stable for years.

3.5 Prehybridization and Hybridization of Embryos

1. Rehydrate the embryos by washing through descending Methanol–PBT series (75, 50, 25 % Methanol–PBT) followed by washes in PBT for 5 min each.
2. Bleach embryos with 6 % hydrogen peroxide solution in PBT for 1 h.
3. Wash three times with PBT for 5 min each.
4. Treat embryos with 10 μg/mL proteinase K in PBT. Timing is important and depends upon tissue size and the batch of proteinase K used. We recommend the following digestion

times as a guideline: digest embryos younger than E7.5 for 1 min or less, E7.5 for 3 min, E8.5 for 7.5 min, E9.5 for 15–20 min if internal tissues are to be probed, or 5 min if gene expression is assessed in superficial ectoderm, and E10.5 for at least 25–30 min (*see* **Note 11**).

5. Inactivate proteinase K with freshly prepared 2 mg/mL glycine in PBT for 5 min.
6. Wash twice with PBT for 5 min each.
7. Refix the embryos with 4 % PFA/0.2 % glutaraldehyde in PBS for 20 min.
8. Wash twice with PBT for 5 min each. Remove as much of the last wash as possible, before adding the prehybridization solution.
9. Add prehybridization solution to each tube and hybridize at 70 °C for 1 h. Incubate with the tubes vertical in a microcentrifuge tube rack.
10. Preheat the RNA probes to 70 °C for 10 min prior to the hybridization step. Resuspend with gentle agitation.
11. Replace the prehybridization solution with fresh hybridization solution, and add 5 μL of 25–50 ng/μL of DIG-labeled or FLU-labeled RNA probe for each 1 mL of hybridization solution for a total of 125–250 ng of RNA probe per 1 mL of hybridization solution. For two-color in situ hybridization, add 125–250 ng of each labeled RNA probe per 1 mL hybridization solution. Make sure that the probe(s) is/are in solution by gently flicking the tube.
12. Place tubes vertically in a tray and incubate at 70 °C overnight.

3.6 Posthybridization Washes and Antibody Incubation

1. Prepare posthybridization wash solutions I–III. Filter solution 2 through a 0.45 μm filter or equivalent, then pre-warm solutions 1 and 3 at 65 °C.
2. Wash embryos twice with solution I for 30 min each at 70 °C (*see* **Note 12**).
3. Wash embryos with a 1:1 mix of with solution I: solution II for 10 min at 70 °C.
4. Wash three times with solution II for 5 min each.
5. Incubate with 100 μg/mL RNase A in solution II for 1 h at 37 °C (*see* **Note 13**).
6. Prepare the MABTL and blocking reagent solution during the 1 h RNase step. Heat the MABTL in a microwave until the solution becomes cloudy, about 30–45 s. Avoid boiling the solution. Dissolve the blocking reagent by vortexing until completely in solution. Cool on ice before using.

7. Wash one time with solution II, followed by one wash with solution III for 5 min each.
8. Wash twice with solution III for 30 min each at 65 °C. Let embryos settle by standing tubes vertically in the Hybaid oven before changing wash solutions. Near the end of the second wash, begin the antibody preadsorption with the embryo powder (*see* Subheading 3.7).
9. Wash embryos three times with MABTL for 5 min each.
10. Block the embryos with 10 % sheep serum in blocking reagent solution (2 % w/v blocking reagent in MABTL) for 1–3 h. Longer blocking times are acceptable.
11. Replace blocking solution with the preadsorbed antibody (*see* Subheading 3.7, **step 5**). Rock tubes on an orbital mixer overnight at 4 °C.

3.7 Preadsorption of Antibody

1. Place 3 mg of embryo powder into a 1.5 mL tube containing 1 mL block reagent stock (*see* **Note 14**). Briefly vortex, and heat inactivate for 30 min at 70 °C.
2. Cool on ice, and add 10 μL sheep serum and appropriate amount of anti-DIG-AP or anti-FLU-AP antibody for a final dilution of 1:4,000 dilution of anti-DIG-AP Fab fragment, or a 1:2,000 dilution of anti-FLU-AP Fab fragment.
3. Rock gently at 4 °C for the duration of the blocking procedure. This should be no less than 2 h.
4. Centrifuge for 10 min at 4 °C.
5. Recover the supernatant containing the antibody and dilute to desired final concentration with chilled 1 % sheep serum in blocking solution. Keep the antibody mixture at 4 °C until ready to use.

3.8 Antibody Washes

1. Remove the antibody and wash embryos three times with MABTL for 5 min each.
2. Wash embryos eight to ten times for 1 h each with MABTL, and continue overnight for a total of 24 h. Decreased background may be obtained by washing for a second or third day especially if large tissues or embryos are being analyzed.

3.9 Color Detection for Single and Two-Color WISH

1. Wash three times with NTMTL for 10 min each.
2. Remove the NTMTL and incubate with 1–2 mL of chromogenic substrate of your choice. If using BM Purple, add substrate to the embryos directly. Alternatively, if INT/BCIP is desired, dilute substrate in NTMT (7.5 μL INT/BCIP per 1 mL of NTMT). Cover in foil, then stand the tubes in a rack and develop protected from light (*see* **Note 15**).
3. Monitor the reaction frequently until the desired signal intensity is produced, or background is observed in sense strand controls.

The signal usually becomes visible within 1–2 h, but the color reaction may need to proceed overnight at room temperature or 4 °C for low-abundance transcripts.

4. Stop the color reaction with two rinses in PBT, followed by three washes in PBT for 5 min each. Proceed directly to Subheading 3.11 if only a single color is desired. For two-color analysis, photograph embryos in PBT to document the first color singly, or proceed directly to Subheading 3.10.

3.10 Antibody Incubation and Washes for Two-Color WISH

Briefly post-fix embryos in 4 % PFA in PBS for 20 min.

1. Wash three times with PBT for 5 min each.
2. Block embryos with 10 % sheep serum in blocking reagent solution (2 % w/v blocking reagent in MABTL) for 1–3 h.
3. Begin the preadsorption of the second antibody (*see* Subheading 3.7).
4. After the blocking step, remove the blocking solution and replace with the preadsorbed antibody. Rock tubes on an orbital mixer overnight at 4 °C.
5. Remove the antibody and wash embryos three times with MABTL for 5 min each followed by a minimum of eight to ten washes over a 24 h period. Incubate with the desired AP color reagent, usually INT/BCIP, until the desired signal intensity is produced.

3.11 Embryo Storage

1. Wash three times with PBT for 5 min each.
2. Fix embryos in 4 % PFA/0.2 % glutaraldehyde in PBS for 1 h.
3. Wash three times with PBT for 5 min each.
4. Wash embryos through a Glycerol/PBT series (50 and 80 % glycerol PBT), and store in 80 % glycerol/PBT. Photograph embryos as soon as possible if using INT/BCIP as a chromogenic substrate as the orange signal fades rapidly. Note also that this precipitate is soluble in alcohols.

3.12 Photography

Photograph embryos in 80 % glycerol using a stereo microscope equipped with a high quality digital camera. We have attained excellent results with the Zeiss Axiocam and Axiovision Imaging software.

4 Notes

1. Caution should be taken to minimize the introduction of contaminating RNases which can result in RNA degradation and loss of signal. Always wear gloves when working with RNA or when handling RNA probes as human skin contains copious amounts of RNases.

2. A "biologie" tip style is recommended for dissecting mouse embryos as it is finer then standard tips. The soft alloy material used in the construction of these tools gives the tweezer flexibility making it an excellent choice for dissecting mouse embryos. Be warned however, that the tip is easily damaged by inadvertent contact with hard surfaces. For guidelines on how to stage embryos refer to ref. 7.
3. Clear, screw-cap, 2 mL tubes with conical bottoms work best for embryos younger than E10.5. Glass scintillation vials work well for larger samples.
4. Hybridization solution can be made in advance and stored at –20 °C for up to 1 year. This solution becomes cloudy and viscous when stored at –20 °C. Pre-heat at 65 °C prior to the prehybridization step to clear.
5. Levamisole is an inhibitor of endogenous AP activity and is used to reduce background staining.
6. Heat-inactivate sheep serum for 30 min at 70 °C. Aliquot into small volumes (1–3 mL) and store at –20 °C for up to 1 year. Keep in mind that the serum should always match the species from which the antibody is made.
7. Background frequently arises when the probe or other reagents become trapped in the ventricles of the forebrain and hindbrain of E9.5 or older embryos. Membranes should be punctured before fixation to ensure thorough and even fixation of all tissues. *See* ref. 8 for practical tips on embryo dissection, and removal of extraembryonic tissue from E6.5 to E13.5 embryos.
8. Templates generated with a restriction enzyme that leaves a 5′ overhang or blunt end give the best results. Avoid using enzymes that generate 3′ overhangs as they may give rise to unusual transcripts. When making complementary antisense strand probes, choose a unique restriction site at the 5′ end of the insert. The antisense transcript is synthesized using a promoter located at the 3′ end of the insert (Fig. 2b). Conversely, for the sense strand probe, choose a unique restriction site at the 3′ end, and transcribe with the promoter found at the 5′ end (Fig. 2c). A sense strand probe serves as a negative control that should not hybridize with the target mRNA and can be used to monitor background activity.
9. Phase Lock Gel columns efficiently separate the water soluble (aqueous) phase from the organic phase preventing contamination while increasing recovery. The procedure is faster than traditional phenol–chloroform extraction methods, and offers added protection from exposure to hazardous compounds (*see* Subheading 2.3, **item 4**).
10. When analyzing the amount of RNA, a clean dedicated gel apparatus is recommended to avoid RNA degradation during

electrophoresis. Ethidium bromide should always be disposed of safely, in a manner according to regulations set forth by your Office of Environmental Health and Safety.

11. Permeabilization of the tissues is necessary to increase the accessibility of the target RNA. Proteinase treatment times vary depending on the stage and type of tissue, and should be empirically determined. It should be emphasized that digestion times are important as over-digestion may result in loss of tissue integrity and decreased signal, whilst under-digestion may result in high background.
12. Embryos become translucent in formamide solution, so allow embryos to settle to the bottom of the tube before changing the solution.
13. The RNase A digestion step reduces background, but decreases signal as well. This step can sometimes be omitted altogether, but this is probe-dependent. Two-color in situ hybridizations generally benefit from RNase A treatments as the reduced background makes it easier to visualize overlapping gene expression patterns.
14. 1 mL of embryo powder containing blocking solution is sufficient to preadsorb up to 10 μL of anti-DIG antibody or 20 μL anti-FLU antibody.
15. Development of chromogenic substrate can be performed in any order, however, BM purple is recommended first since the signal develops more quickly. The overall color of the substrate solution should be monitored regularly and replaced if a color change is observed.

Acknowledgements

This research was supported by the Intramural Research Program of the NIH, National Cancer Institute, Center for Cancer Research.

References

1. Wilkinson DG (1992) Whole mount in situ hybridization of vertebrate embryos, in situ hybridization: a practical approach. IRL, Oxford
2. Conlon RA, Hermann BG (1993) Detection of messenger RNA by in situ hybridization to postimplantation embryo whole mounts. Methods Enzymol 225:373–783
3. Parr BA, Shea MJ, Vassileva G, McMahon AP (1993) Mouse Wnt genes exhibit discrete domains of expression in the early embryonic CNS and limb buds. Development 119: 247–261
4. Biris KK, Dunty WC Jr, Yamaguchi TP (2007) Mouse Ripply2 is downstream of Wnt3a and is dynamically expressed during somitogenesis. Dev Dyn 236:3167–3172
5. Dunty WC Jr, Biris KK, Chalamalasetty RB, Taketo MM, Lewandoski M, Yamaguchi TP (2008) Wnt3a/beta-catenin signaling controls posterior body development by coordinating mesoderm formation and segmentation. Development 135:85–94
6. Nakaya MA, Biris K, Tsukiyama T, Jaime S, Rawls JA, Yamaguchi TP (2005) Wnt3a links

left-right determination with segmentation and anteroposterior axis elongation. Development 132:5425–5436
7. Kaufman MH (1999) The atlas of mouse development. Academic, London
8. Hogan B, Beddington R, Costantini F, Lacy E (1994) Manipulating the mouse embryo: a laboratory manual. Cold Spring Harbor Laboratory Press, New York
9. Collignon J, Varlet I, Robertson EJ (1996) Relationship between asymmetric nodal expression and the direction of embryonic turning. Nature 381:155–158
10. Lowe LA, Supp DM, Sampath K, Yokoyama T, Wright CV, Potter SS, Overbeek P, Kuehn MR (1996) Conserved left-right asymmetry of nodal expression and alterations in murine situs inversus. Nature 381:158–161
11. Takada S, Stark KL, Shea MJ, Vassileva G, McMahon JA, McMahon AP (1994) Wnt-3a regulates somite and tailbud formation in the mouse embryo. Genes Dev 8:174–189
12. Morimoto M, Sasaki N, Oginuma M, Kiso M, Igarashi K, Aizaki K, Kanno J, Saga Y (2007) The negative regulation of Mesp2 by mouse Ripply2 is required to establish the rostro-caudal patterning within a somite. Development 134: 1561–1569
13. Saga Y, Hata N, Koseki H, Taketo MM (1997) Mesp2: a novel mouse gene expressed in the presegmented mesoderm and essential for segmentation initiation. Genes Dev 11: 1827–1839
14. Mansouri A, Yokota Y, Wehr R, Copeland NG, Jenkins NA, Gruss P (1997) Paired-related murine homeobox gene expressed in the developing sclerotome, kidney, and nervous system. Dev Dyn 210:53–65
15. Neidhardt LM, Kispert A, Herrmann BG (1997) A mouse gene of the paired-related homeobox class expressed in the caudal somite compartment and in the developing vertebral column, kidney and nervous system. Dev Genes Evol 207:330–339
16. Leitges M, Neidhardt L, Haenig B, Herrmann BG, Kispert A (2000) The paired homeobox gene Uncx4.1 specifies pedicles, transverse processes and proximal ribs of the vertebral column. Development 127:2259–2267
17. Kraus F, Haenig B, Kispert A (2001) Cloning and expression analysis of the mouse T-box gene Tbx18. Mech Dev 100:83–86

Chapter 3

Detection and Monitoring of MicroRNA Expression in Developing Mouse Brain and Fixed Brain Cryosections

Davide De Pietri Tonelli, Yoanne M. Clovis, and Wieland B. Huttner

Abstract

MicroRNAs (miRNAs) are 20–25 nucleotide long, noncoding, and single-strand RNAs that have been found in almost all organisms and shown to exert essential roles by regulating the stability and translation of target mRNAs. In mammals most miRNAs show tissue specific and developmentally regulated expression. Approximately 70 % of all miRNAs are expressed in the brain and a growing number of studies have shown that miRNAs can modulate both brain development function and dysfunction. Moreover, miRNAs have been involved in a variety of human pathologies, including cancer and diabetes and are rapidly emerging as new potential drug targets. In order to further characterize miRNA functions, it is therefore crucial to develop techniques enabling their detection in tissues (both fixed and in vivo) with single-cell resolution. Here, we describe methods for the detection/monitoring of miRNA expression, that can be applied in both developing embryos and fixed samples, which we and others have applied to the investigation of both embryonal and postnatal neurogenesis in mice, but also in zebrafish, and cell cultures.

Key words Development, Neurogenesis, MicroRNAs, In utero electroporation, In situ hybridization

1 Introduction

Most neurons of the mouse brain are generated during prenatal development, and almost all the neurons of the neocortex originate from differentiation of neural stem cells located in the developing telencephalon (i.e., the foremost region of the neural tube). During the early stages of telencephalic development, neural stem cells undergo proliferative types of divisions, but later (around embryonic day 10, E10) neural stem (and progenitor) cells switch to neurogenic-type of divisions (i.e., at the onset of neurogenesis). The onset of neurogenesis is not synchronous in all the cells and therefore these two types of divisions (i.e., proliferative and neurogenic) coexist within the same telencephalic regions. Moreover

Mark Lewandoski (ed.), *Mouse Molecular Embryology*, Methods in Molecular Biology, vol. 1092,
DOI 10.1007/978-1-60327-292-6_3, © Springer Science+Business Media New York 2014

different subclasses of neural stem (progenitor) cells, such as neuroepithelial cells (NE), radial glia (RG), and basal (or intermediate) progenitors (BP) exist in the developing telencephalon [1]. Interestingly, the order of appearance of these different subclasses of neural stem (progenitor) cells follows the onset of neurogenesis. Therefore, these different subclasses of neural stem (progenitor) cells also coexist within the same regions of the developing telencephalon. Taken together, these notions highlight the complexity of cell biology at the basis of embryonal neurogenesis [1].

MiRNAs are 20–25 nucleotide long, noncoding, single-stranded RNAs that are present in virtually all the organisms (including viruses, fungi, animals, and plants) and modulate gene expression, mostly at the posttranscriptional level [2]. MiRNA action occurs through the nearly perfect, base-pairing with target messenger RNAs (mRNAs) and results in the repression of mRNA translation and/or stability [2]. In mammals, many miRNAs show tissue-specific and developmentally regulated expression and a growing number of studies (including ours) have shown that miRNAs are crucially involved in the control of several aspects of neurogenesis and proper neocortical development [3–7]. In order to further characterize miRNA functions, it is therefore crucial to develop techniques enabling the simultaneous detection of miRNA expression and of lineage markers in tissues, with single-cell resolution. Furthermore, given the complex cell biology at the basis of embryonal neurogenesis (especially in telencephalon), the establishment of techniques allowing the detection and monitoring of miRNA expression during cell-fate change in vivo is crucial to investigate the role of miRNAs in neurogenesis.

Here, we have optimized a robust method to detect simultaneously miRNAs and lineage markers in fixed brain cryosections. One important technical improvement of our method is in the use of conventional NBT/BCIP chromogenic stain and laser scanning confocal imaging to acquire a fluorescent in situ hybridization signal. Using this approach we can detect, within the same optical section, miRNAs and lineage markers (stained with conventional immunofluorescence), thereby achieving a much greater cellular resolution compared to conventional bright-field based in situ hybridization. Furthermore, we also describe a fluorescence reporter-based assay (Dual Fluorescence Reporter Sensor, DFRS) that allows the detection of miRNA activity [8]. One innovative aspect of this reporter-based strategy is in the possibility of achieving greater cellular resolution even when compared to our in situ hybridization method. Importantly, in our reporter-based assay, we acutely deliver the DFRS plasmid in developing mouse telencephalon by means of in utero electroporation, thereby enabling the detection of miRNA activity, in vivo.

2 Materials

2.1 Tissue Fixation and Sectioning

1. Phosphate buffered saline 1×: 37 mM NaCl, 2.7 mM KCl, 10 mM sodium phosphate dibasic, 2 mM potassium phosphate monobasic and a pH of 7.4. The stock solution is prepared at a 10× concentration (*see* **Notes 1–3**).
2. Fixation: 4 % w/v paraformaldehyde (PFA) in 1× PBS (final volume 500 ml): to 300 ml of bidistilled autoclaved water add 50 ml of 10× PBS, add 16 g of PFA and let PFA dissolve (under fume hood) by stirring overnight at 45 °C until dissolved, chill, aliquot, and store at −20 °C for up to 3 months (*see* **Note 4**).
3. Embedding/sectioning: 30 % w/v sucrose in 1× PBS, store at +4 °C.
4. Tissue-Tek OCT compound (Sakura, Zoeterwoude, The Netherlands).
5. Truncated shape embedding molds (Electron Microscopy Sciences, Warrington, USA).
6. Cryostat blades (Leica Microsystems S.P.A, Milano, Italy).
7. SuperFrost Slides (Menzel-Gläser GmbH, Braunschweig, Germany).

2.2 In Situ Hybridization

1. NaCl 5 M (final volume 1 l). This is an almost saturated solution so it might take a long time for the NaCl to be completely dissolved (*see* **Notes 2**, **3**, and **5**).
2. Tris–HCl (Trizma® base, powder, Sigma, Milano Italy) 1 M pH 8.0 (and pH 9.5) (final volume 500 ml): take to pH 8.0 (or 9.5) by adding dropwise 5 N HCl solution and monitor the pH, while stirring, using a pH-meter (*see* **Notes 2** and **3**).
3. EDTA 0.5 M (final volume 1 l): add EDTA to 800 ml of bidistilled water take to pH with about 20 NaOH pellets, monitor the pH using a pH-meter while stirring, and bring the pH to 8.0 (*see* **Notes 2**, **3**, **6**, and **7**).
4. RIPA buffer, prepare 1 l of 150 mM NaCl; NP-40 1 % v/v; 0.5 % w/v sodium deoxycholate ; 0.1 % w/v SDS, 1 mM EDTA (using stock 0.5 M pH 8.0), and 50 mM Tris–HCl (using stock 1 M pH8.0) (*see* **Notes 2** and **3**).
5. Triethanolamine buffer (final volume 1 l): 100 mM triethanolamine; 0.002 % acetic acid, check that pH is around 8.0 (*see* **Notes 2**, **3**, and **8**).
6. SSC 20×, prepare 1 l of 0.15 M NaCl and 0.015 M Sodium Citrate (*see* **Notes 2** and **3**).
7. Hybridization solution, prepare 20 ml of 50 % formamide; 5× SSC; 5× Denhardt's (Sigma, Milano, Italy); 250 μg/ml of Yeast RNA; 500 μg/ml salmon sperm DNA (*see* **Note 2**),

aliquot to 5 ml. This solution can be stored at −20 °C for up to 3 months (*see* **Note 9**).

8. Post-hybridization solution: prepare 200 ml of 50 % v/v formamide; 2× SSC; and 0.1 % Tween (*see* **Notes 2**, **3**, and **9**).
9. Maleic acid–NaCl pH 7.5 10× stock: prepare 1 l of 1 M maleic acid; 1 M NaCl, take to pH 7.5 with NaOH tablets (*see* **Notes 2**, **3**, and 7).
10. Buffer B1 (final volume 1 l): to 1× maleic acid–NaCl pH 7.5, add 0.1 % (final v/v) Tween (*see* **Notes 2** and **3**).
11. Buffer B2 (blocking buffer): prepare 100 ml of 10 % v/v fetal calf serum (FCS) in B1. This solution can be stored at +4 °C up to 1 week.
12. Buffer B3 (reaction buffer): prepare 100 ml of 100 mM Tris–HCl pH 9.5; 50 mM $MgCl_2$; 100 mM NaCl; 0.1 % v/v Tween (*see* **Note 2**). This solution cannot be stored, so prepare it fresh.
13. Anti-digoxigenin (DIG) antibody, Fab fragment, alkaline phosphatase conjugated (Roche, Milano, Italy).
14. NBT/BCIP (Sigma, Milano, Italy) substrate solution for alkaline phosphatase. This solution needs to be stored in a dark container.
15. LNA probes for miRNAs (Exiqon, Milano, Italy): prepare a stock of 25 μM LNA-modified ribonucleotide oligo, Store at −20 °C (*see* **Note 4**).
16. 16 DIG 3′ end labeling kit (Roche, Milano, Italy).
17. Pap Pen.

2.3 Preparation of DFRS Plasmids

1. Synthetic oligonucleotides, purified by PAGE; restriction/modification enzymes: PacI, FseI, EcoRI, NotI; Polynucleotide Kinase (10 U/μl); T4 DNA Ligase.

2.4 Immunofluorescence and Imaging

1. Blocking and incubation solution: 10 % v/v FCS in 1× PBS, prepare 250 ml (*see* **Note 2**). Store at +4 °C for up to a week.
2. Nuclear DNA staining: BisBenzimide H 33342 trihydrochloride (Hoechst 33342, Sigma, Milano, Italy), stock 1 mg/ml in DMSO.
3. Mounting resins: Anti-fade reagent (Invitrogen, Milano, Italy); or Mowiol. To prepare stock of Mowiol, mix 6 g of Mowiol 4-88 (Polysciences, Warrington, USA) and 15 g Glycerol (ReagentPlus grade, Sigma, Milan, Italy) in 15 ml bidistilled water and stir overnight; add 30 ml of 0.2 M Tris–HCl (pH 8.5) heat to 50 °C in a water bath, stir. Clear from residual precipitates by centrifugation at 5,000×*g* for 15 min. Store at −20 °C (*see* **Note 4**).
4. Coverslips (Menzel-Gläser GmbH, Braunschweig, Germany).
5. Imaging systems used: for bright-field microscopy, upright Olympus BX51 (Olympus Italia S.r.l., Milano, Italy) equipped

with motorized stage (Neurolucida, MicroBrightField Europe, E.K., Magdeburg, Germany); for laser scanning confocal microscopy, inverted confocal microscope Leica TCS AOBS SP5 (Leica, Mannheim, Germany), equipped with lasers: 405 nm laser diode; an argon laser with five lines: 458, 476, 488, 496, and 514 nm (RYBplus LSOS, Salt-Lake city, USA); DPSS 561 nm; HeNe 633 nm. Objectives used were HC PL FLUOTAR 20×/0.50; HCX PL APO 40×/1.25–0.75 Oil CS; and HCX PL APO 63×/1.40–0.60 Oil (Leica, Mannheim, Germany).

3 Methods

3.1 Detection of miRNAs on Fixed Brain Cryosections by Non-radioactive In Situ Hybridization of LNA-Modified Riboprobes

Locked Nucleic Acid (LNA) is a conformational restricted nucleic acid analogue, in which the ribose ring is "locked" with a methylene bridge connecting the 2′-O atom and the 4′-C atom. LNA-modified nucleotides can form base pairs with their complementary nucleotides according to standard Watson-Crick base pairing rules. Incorporation of one LNA-modified nucleotide in one strand of a nucleic acid duplex will increase the melting temperature of the duplex with 2–8 °C per LNA monomer. LNA vastly improves the thermal stability and specificity of duplexes formed by complementary DNA or RNA. Therefore, LNA-modified probes are widely used for miRNA detection (e.g., Northern blotting and microarray) [9].

We have used LNA-modified probes to detect miRNAs in mouse embryo cryosections, with in situ hybridization [3, 8].

1. Tissue preparation: embryos/brains are fixed in 4 % w/v PFA in 1× PBS overnight at +4 °C. Tissues are washed several times in 1× PBS (to remove the excess PFA), prior to de-hydration with about 3 ml/embryo (or brain) of 30 % w/v sucrose in 1× PBS (e.g., in a 14 ml Falcon 2059 tube, Becton Dickinson, Le Pont de Claix, France) overnight at +4 °C (or until they sink). Embryos/tissues are washed in 1× PBS (to remove the excess sucrose), then embedded in OCT in a disposable embedding mold, and immediately frozen on dry ice (*see* **Note 10**). Prepare 8–12 μm-thick cryosections using a cryostat, attaching tissue sections on glass slides (*see* **Note 10**).
2. Probe labeling: use 3′ DIG labeling kit (Roche, Milano, Italy) following the manufacturer's instructions. Label 4 μl of LNA stock in a 20 μl final reaction volume. The reaction is performed in a standard thermal cycler. Terminate the reaction by adding 2 μl of 0.2 M EDTA pH 8.0, mix (*see* **Note 11**).
3. Probe purification: add bidistilled autoclaved water to a final volume of 25 μl, purify LNA probe from unincorporated DIG-UTP with a microspin G-25 column (GE Healthcare, Milan,

Italy) following the manufacturer's instructions (*see* **Notes 4, 10**, and **11**).

4. (*Day one*) Probe hybridization: let slides dry on the bench at RT for about an hour, then make a square around the tissue area on the slides with a Pap pen (to minimize the use of reagents in the subsequent reactions). Dissolve the OCT with three brief washes in 1× PBS (*see* **Notes 12, 13**, and **14**). Permeabilize the tissue by treating twice for 10 min with RIPA buffer, then postfix for 10 min with 4 % PFA in 1× PBS and remove the excess of PFA with three brief washes with 1× PBS. Block the positive charges in the tissue by treating slides for 15 min with acetic anhydride (0.25 % v/v final) in triethanolamine buffer, followed by three brief washes with 1× PBS.

   Pre-hybridize (directly on slides) with 200–250 μl of hybridization solution at the appropriate temperature in hybridization oven (or water bath) for 1 h (*see* **Notes 15** and **16**). Change the pre-hybridization solution, and hybridize (directly on slides) overnight in hybridization oven with hybridization solution containing 20–40 μl/ml of DIG-labeled LNA probe previously prepared (*see* **steps 2** and **3** above).

5. (*Day two*) Post-hybridization/DIG immunodetection: washes are performed at the same temperature as the hybridization (to remove the excess of unbound probe). Wash with (pre-warmed) post-hybridization solution briefly, and then wash twice for 1 h. DIG immunodetection: wash briefly twice with B1 buffer, and then treat once for 20 min. Block tissue with B2 buffer for 1 h. Dilute the anti-DIG antibody (conjugated with alkaline phosphatase) 1:2,000 in an appropriate volume of B2 buffer (*see* **Note 2**) and incubate (directly on slides) overnight at +4 °C.

6. (*Day three*) Alkaline phosphatase detection: briefly wash the slides twice with buffer B1 and then once for 30 min. Incubate (directly on slides) with NBT/BCIP (*see* **Note 2**), let the reaction develop slowly in the dark, either at +4 °C (high expressed miRNAs), or for 2 h at 37 °C, and then at RT (low expressed miRNAs) from one to several days. Terminate the reaction by several washes with 0.1 % Tween 20 in 1× PBS. After this step, sections can be mounted using mounting resins and imaged using conventional bright-field microscopy (*see* Fig. 1).

3.2 Immunofluorescence-Mediated Detection of Proteins in Fixed Brain Cryosections Following Non-radioactive In Situ Hybridization

Immunofluorescence-mediated detection of proteins following in situ hybridization is challenging. Several antigens/proteins are lost/misfolded during the above-described treatments. Nonetheless, some antigens remain and can be revealed by conventional immunofluorescence protocols, except that permeabilization and quenching are not necessary.

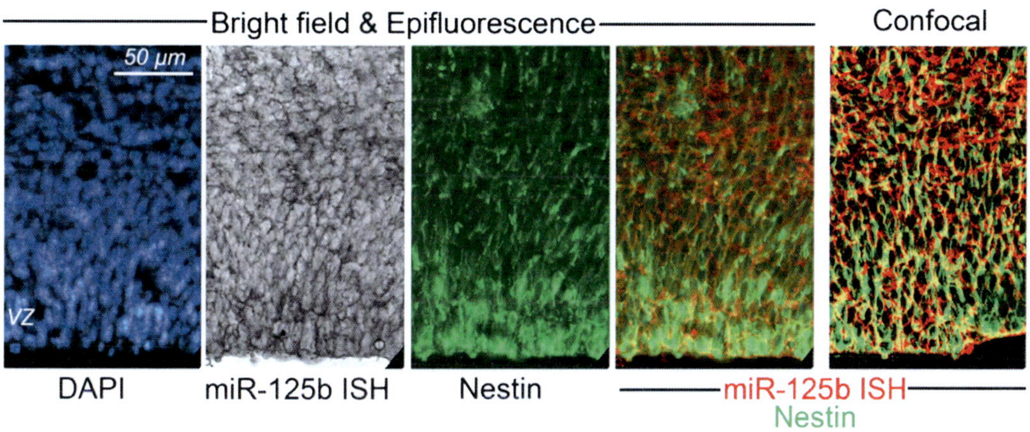

Fig. 1 In situ hybridization with LNA-modified oligonucleotides for miR-125b combined with immunofluorescence staining with anti-nestin primary antibody, on 12 μm-thick coronal cryosections through the cortex of E 15.5 mouse brain, the sections were also co-stained for nuclear DNA with DAPI. In situ staining (miR-125b, *red*) and Nestin (*green*) were imaged with a conventional microscope (bright-field and epifluorescence, respectively; red is a false color), or with a laser scanning confocal microscope (far red fluorescence emitted by NBT/BCIP chromogenic stain of LNA probe hybridized in situ). Note that in the same section, a much greater colocalization of miR-125b and Nestin in neural stem (and progenitor) cells is revealed by confocal microscopy compared to standard microscopy. Images were processed with Adobe Photoshop. *VZ* ventricular zone

3.3 Detection of In Situ Hybridization by Laser Scanning Confocal Microscopy

A combination of BCIP (5-Bromo-4-Chloro-3′-Indolyphosphate *p*-Toluidine Salt) and NBT (Nitro-Blue Tetrazolium Chloride) react with alkaline phosphatase, yielding a black-purple precipitate. Alkaline phosphatase-NBT/BCIP staining has greater sensitivity, precision and is often the preferred method over fluorescent in situ techniques, which in the case of miRNAs, is particularly challenging [9]. For this reasons, when NBT/BCIP chromogenic staining is used, in situ hybridization is normally documented by bright-field microscopy. Confocal imaging of NBT/BCIP-stained samples was recently documented for mRNAs in Vibratome sections and brings several advantages that facilitate gene expression analysis [10]. We applied a similar protocol for the detection of miRNAs in mouse brain cryosections. This imaging method, in combination with immunofluorescence, allows a greater tissue resolution compared to standard bright-field imaging (*see* Fig. 1).

1. Brain cryosections (12 μm-thick) subject to in situ hybridization with DIG-labeled LNA probes, staining with NBT/BCIP and immunostaining for neural progenitor lineage markers (e.g., Nestin), are subject to laser scanning confocal microscopy using the following parameters: Laser HeNe, excitation 633 nm; emission is recorded with photomultipliers between 707.3 and 795 nm.

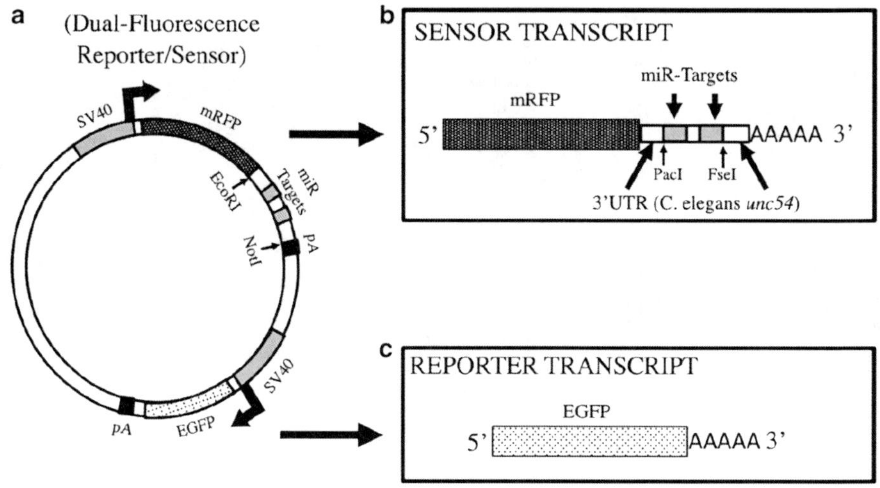

Fig. 2 (**a**) Schematic map of the Dual-Florescence Reporter/Sensor plasmid. DFRS express two independent transcripts, driven by identical SV40 promoters, encoding a sensor (mRFP) and reporter (GFP). pA, polyadenylation sites. EcoRI, NotI, restriction sites for the extraction of 3′UTR and miR-targets. (**b**) Sensor transcript encoding mRFP, contains a 3′-UTR derived from *C. elegans Unc54* gene. *Light grey boxes* indicate the two sequences (miR-targets) complementary to the miRNA of interest. PacI, FseI, restriction site to exchange miR-targets. (**c**) Reporter transcript encoding for cytosolic GFP (or Gap43-GFP, plasma membrane targeted-GFP, not shown)

3.4 Detection of miRNA Activity in Developing Mouse Brains with Fluorescence-Based Reporter Assay: Dual Fluorescence Reporter Sensor (DFRS) Plasmid

Despite the above-described advantages of the combination of NBT/BCIP in situ hybridization, immunofluorescence and laser scanning confocal microscopy (*see* Subheadings 3.1–3.3), this approach also has some limitations, especially in the developing brain. Indeed, given the nature of NE cells (the primary neural stem cells of the developing brain) which are highly elongated, with most of their cell volume occupied by the nucleus (*see* Fig. 1 and ref. 1) and given that mature miRNAs (as well as mRNAs) are expressed only in the cytoplasm (which in NE cells is very small, *see* Fig. 1), it is extremely difficult to achieve single-cell resolution in this particular tissue, using in situ hybridization. To overcome these limitations we developed a fluorescent reporter-assay. This assay is based on a *D*ual *F*luorescence *R*eporter *S*ensor (DFRS) plasmid (*see* Fig. 2 and ref. 8), which is delivered in developing mouse telencephalon by in utero electroporation, allowing to monitor the presence (and activity) of an endogenous miRNA by the silencing of a fluorescent sensor, thus achieving single-cell resolution (*see* Fig. 4 and ref. 8).

1. Details for the cloning of the DFRS plasmid were previously described [8]. The DFRS plasmid (*see* Fig. 2) encodes for (a) a fluorescent sensor (monomeric-RFP or mRFP), containing in its 3′-UTR (obtained by PCR cloning of *C. elegans Unc54* 3′UTR) two sequences complementary to the miRNA of interest (*see* Figs. 2 and 3) and (b) a fluorescent reporter (GFP, cytosolic; or Gap43-GFP, plasma membrane targeted)

```
a
Mmu-miR-125b      3' AGUGUUCAAUCCCAGAGUCCCU 5'      3' AGUGUUCAAUCCCAGAGUCCCU 5'
                     ||||||||||||||||||||||           ||||||||||||||||||||||
S125F          5'-TAA TCACAAGTTAGGGTCTCAGGGA[GGCGCGCCC]TCACAAGTTAGGGTCTCAGGGA GGCCGG-3'
                  ||| |||||||||||||||||||||| ||||||||| |||||||||||||||||||||| ||
S125R        3'-TAATT AGTGTTCAATCCCAGAGTCCCT[CCGCGCGGG]AGTGTTCAATCCCAGAGTCCCT CC-5'

b
Mmu-miR-125b      3' AGUGUUCAAUCCCAGAGUCCCU 5'      3' AGUGUUCAAUCCCAGAGUCCCU 5'
                     ||||||||||||||||||•••|           ||||||||||||||||||•••|
S125mut-F      5'-TAA TCACAAGTTAGGGTCTCACCCA[GGCGCGCCC]TCACAAGTTAGGGTCTCACCCA GGCCGG-3'
```

Fig. 3 Example of annealed oligos for the generation of mmu-miR-125b sensor cassette, and mutated control. Mature mmu-miR-125b sequences are written above the annealed sensors. (**a**) Regions complementary to miR-125b are shown as *underlined* text; overhangs for cloning into modified *C. elegans Unc54* 3′UTR sequence are shown in *bold* (left PacI, right FseI site overhang); a spacer region (AscI restriction site) is delimited by *square parentheses*. (**b**) Example of mutated (forward) oligo used as control for miR-125b sensor; mutated nucleotides are shown in *bold*

used to identify the targeted cells after electroporation. Both reporter and sensor are under the transcriptional control of identical SV40 promoters, thus to minimize differences in the expression of the two transcripts (i.e., GFP, reporter; and mRFP, sensor). The DFRS plasmid (*see* Fig. 2) is designed to allow rapid exchange of miRNA sensor cassette. This can be obtained by ligating the sensor cassette (obtained by annealing synthetic oligonucleotides, *see* Fig. 3) into PacI and FseI-digested DFRS plasmid (*see* Fig. 2).

2. Annealing of synthetic oligonucleotides to prepare the sensor cassette for a given miRNA. In an appropriate PCR tube, set up a 100 μl reaction as follows: add 76.8 μl of H_2O; 10 μl of 10× PNK buffer; 5 μl of forward oligo (from 100 μM stock); 5 μl of reverse oligo (from 100 μM stock) (*see* Fig. 3); 1 μl of ATP (from 100 mM stock); 0.5 μl of PNK (stock 10 U/μl). Incubate the reaction in a conventional thermal cycler programmed as follows: temperature (Time—*Reaction Step*): 37 °C (1 h—*oligo phosphorylation*); 94 °C (5 min—*PNK inactivation*); followed by a controlled oligonucleotide annealing: decrease by 0.1 °C/s to 90 °C; incubate at 90 °C (3 min); decrease by 0.1 °C/s to 70 °C; incubate at 70 °C (3 min); decrease by 0.1 °C/s to 50 °C; incubate at 50 °C (3 min); decrease by 0.1 °C/s to 25 °C; incubate at 25 °C (3 min). Alternatively, phosphorylation and inactivation may be performed in two heated blocks. Annealing can be performed after inactivation by immediately incubating the reaction in a water bath at 70 °C for 5 min, the water bath is then switched off, and the reaction left to cool down slowly to RT. At this stage the synthetic DNA fragment is phosphorylated, PNK inactivated, and oligos annealed. Annealing of oligos can be controlled on a conventional 4 % agarose gel for DNA: load 2–5 μl of the annealing reaction on the gel; as control load

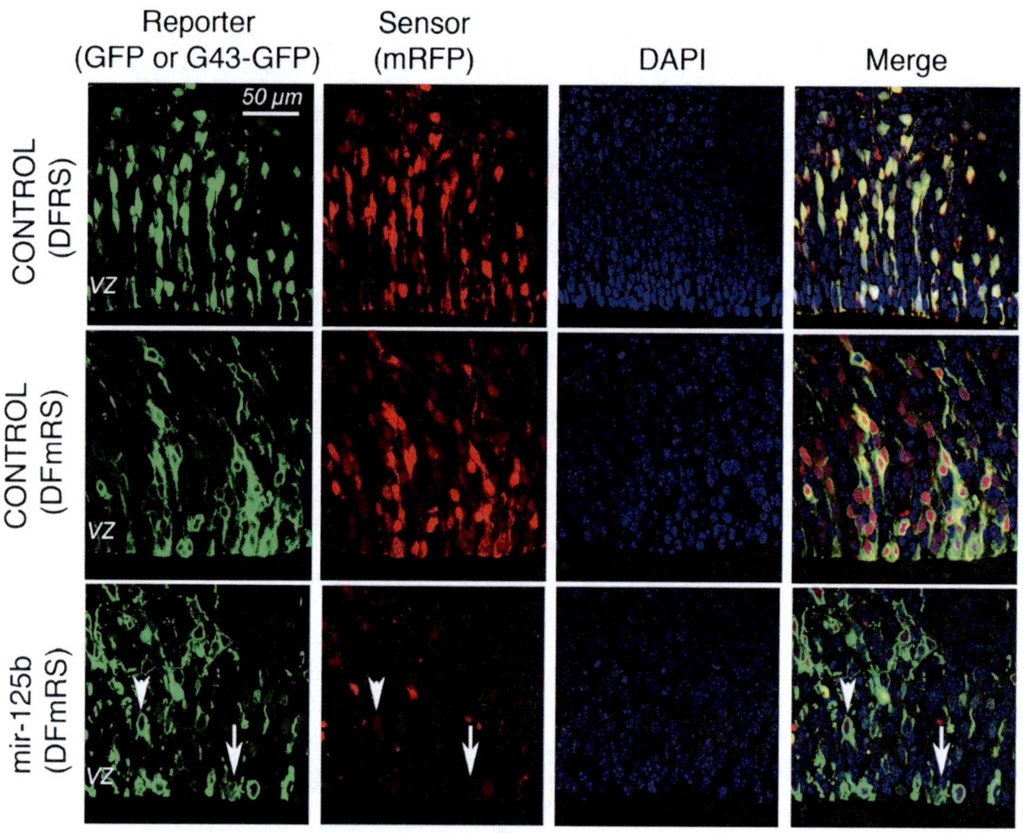

Fig. 4 DFRS-mediated detection of miR-125b in the neuroepithelium of E15.5 mouse embryos. DFRS plasmid (encoding a cytosolic GFP reporter) or DFmRS (encoding a plasma membrane-targeted GFP reporter), with control sequences (Control) or antisense sequence for miR-125b (miR-125b) was electroporated into a telencephalic vesicle of E13.5 developing mouse embryo, followed by 48 h of in utero development. Brains were fixed and the electroporated region of the telencephalon was examined for the presence of reporter GFP and sensor mRFP with laser scanning confocal microscope. Note that in control brains, most if not all the cells positive for GFP were also positive for mRFP. In contrast, most of the cells targeted with DFmRS plasmid for miR-125b (GFP positive cells), expressed mRFP at low (*arrowhead*) or not detectable (*arrow*) level. *VZ* ventricular zone

0.5–1 μl of a corresponding single-strand (either forward or reverse) oligo from stock 100 μM, which runs faster than the double-strand annealed DNA.

3. The fragment is ligated (without purification) into the DFRS vector cut with PacI and FseI and dephosphorylated (*see* Fig. 2). To set up the ligation reaction add 5 μl of annealed fragment in a 10 μl final ligation reaction, incubate the reaction +4 °C overnight (*see* **Notes 17** and **18**).
4. Electroporation of developing mouse embryos. DFRS plasmid is delivered in the neuroepithelium of developing mouse embryos by directed electroporation either ex utero or in utero (*see* Fig. 4). These two methods were previously described in refs. 8, 11.

4 Notes

1. Unless otherwise stated all solutions should be prepared in bidistilled autoclaved water.
2. Filter with 0.2 μm filter for bottle (or syringe).
3. Store at RT.
4. Avoid repeated freezing–thawing.
5. Heating the solution moderately might help to dissolve the salt.
6. EDTA will not completely dissolve until the pH is around 8.
7. Allow NaOH pellets to dissolve completely before adding some more.
8. 0.25 % acetic anhydride must be added just before use.
9. Add more SSC in the final solution to increase stringency. Stringency is a critical parameter and needs to be optimized.
10. Can be stored for several weeks at −20 °C.
11. Probe labeling can be performed during the early steps of the hybridization procedure.
12. Washes/treatments are performed at RT, unless a different temperature is indicated.
13. Unless stated otherwise, wash slides in a slide container.
14. A "brief" wash/treatment means one for 5 min.
15. Pre-hybridization, hybridization, and post-hybridization reactions are typically performed at temperature of 20 °C less than the calculated melting temperature of the LNA probe. This is a critical parameter and needs to be optimized.
16. When incubation is performed directly on slides, use a closed humidified chamber. To minimize the amount of reaction buffers, slides can be covered with Parafilm of appropriate size. To make the humidified chamber, use a Pyrex box with a tight cover; dispose two bars (e.g., plastic disposable pipettes) in the bottom as slide support. Before use, place on the bottom of the Pyrex box two layers of 2 mm Whatman paper soaked with bidistilled water. Make sure that slides are in a horizontal position and that they do not touch the Whatman paper.
17. Dephosphorylation of the plasmid is not strictly necessary.
18. In case of low ligation efficiency, it might be convenient to extract from DFRS the *C. elegans Unc54* 3′UTR with EcoRI–NotI digestion (*see* Fig.1) and sub-clone into a cloning vector by (e.g., pBluescript or pGEM), ligate the annealed oligos with PacI–FseI, and reinsert the modified *C. elegans Unc54* 3′UTR into DFRS.

Acknowledgements

We thank Dr. Laura Cancedda for critical reading of the manuscript; Dr. Mattia Pesce for technical help in the optimization of confocal microscopy and the staff of IIT mouse facility, especially Dr. Federica Piccardi and Daniela Cantatore, for excellent support. Y.C. was supported by a PhD fellowship from University of Genova, Italy. W.B.H. was supported by grants from the DFG (SFB 655, A2; TRR 83, Tp6) and the ERC (250197), by the DFG-funded Center for Regenerative Therapies Dresden, and by the Fonds der Chemischen Industrie. This work was supported by Fondazione Istituto Italiano di Tecnologia.

References

1. Fietz SA, Huttner WB (2011) Cortical progenitor expansion, self-renewal and neurogenesis—a polarized perspective. Curr Opin Neurobiol 21:23–35
2. Pasquinelli AE (2012) MicroRNAs and their targets: recognition, regulation and an emerging reciprocal relationship. Nat Rev. Genet 13(4): 271–282
3. De Pietri Tonelli D, Pulvers JN, Haffner C, Murchison EP, Hannon GJ, Huttner WB (2008) miRNAs are essential for survival and differentiation of newborn neurons but not for expansion of neural progenitors during early neurogenesis in the mouse embryonic neocortex. Development 135:3911–3921
4. Clovis YM, Enard W, Marinaro F, Huttner WB, De Pietri Tonelli D (2012) Convergent repression of Foxp2 3'UTR by miR-9 and miR-132 in embryonic mouse neocortex: implications for radial migration of neurons. Development 139(18):3332–3342
5. Nigro A, Menon R, Bergamaschi A, Clovis YM, Baldi A, Ehrmann M, Comi G, De Pietri Tonelli D, Farina C, Martino G, Muzio L (2012) MiR-30e and miR-181d control radial glia cell proliferation via HtrA1 modulation. Cell Death Dis. 3:e360. doi:10.1038/cddis.2012.98
6. McNeill E, Van Vactor D (2012) MicroRNAs shape the neuronal landscape. Neuron. 75(3): 363–379
7. Li X, Jin P (2010) Roles of small regulatory RNAs in determining neuronal identity. Nat Rev Neurosci 11:329–338
8. De Pietri Tonelli D, Calegari F, Fei JF, Nomura T, Osumi N, Heisenberg CP, Huttner WB (2006) Single-cell detection of microRNAs in developing vertebrate embryos after acute administration of a dual-fluorescence reporter/sensor plasmid. Biotechniques 41:727–732
9. Obernosterer G, Martinez J, Alenius M (2007) Locked nucleic acid-based in situ detection of microRNAs in mouse tissue sections. Nat Protoc 2:1508–1514
10. Trinh LA, McCutchen MD, Bonner-Fraser M, Fraser SE, Bumm LA, McCauley DW (2007) Fluorescent in situ hybridization employing the conventional NBT/BCIP chromogenic stain. Biotechniques 42:756–759
11. Osumi N, Inoue T (2001) Gene transfer into cultured mammalian embryos by electroporation. Methods 24:35–42

Chapter 4

Laser Capture Microdissection of Embryonic Cells and Preparation of RNA for Microarray Assays

Latasha C. Redmond, Christopher J. Pang, Catherine Dumur, Jack L. Haar, and Joyce A. Lloyd

Abstract

In order to compare the global gene expression profiles of different embryonic cell types, it is first necessary to isolate the specific cells of interest. The purpose of this chapter is to provide a step-by-step protocol to perform laser capture microdissection (LCM) on embryo samples and obtain sufficient amounts of high-quality RNA for microarray hybridizations. Using the LCM/microarray strategy on mouse embryo samples has some challenges, because the cells of interest are available in limited quantities. The first step in the protocol is to obtain embryonic tissue, and immediately cryoprotect and freeze it in a cryomold containing Optimal Cutting Temperature freezing media (Sakura Finetek), using a dry ice–isopentane bath. The tissue is then cryosectioned, and the microscope slides are processed to fix, stain, and dehydrate the cells. LCM is employed to isolate specific cell types from the slides, identified under the microscope by virtue of their morphology. Detailed protocols are provided for using the currently available ArcturusXT LCM instrument and CapSure® LCM Caps, to which the selected cells adhere upon laser capture. To maintain RNA integrity, upon removing a slide from the final processing step, or attaching the first cells on the LCM cap, LCM is completed within 20 min. The cells are then immediately recovered from the LCM cap using a denaturing solution that stabilizes RNA integrity. RNA is prepared using standard methods, modified for working with small samples. To ensure the validity of the microarray data, the quality of the RNA is assessed using the Agilent bioanalyzer. Only RNA that is of sufficient integrity and quantity is used to perform microarray assays. This chapter provides guidance regarding troubleshooting and optimization to obtain high-quality RNA from cells of limited availability, obtained from embryo samples by LCM.

Key words Laser capture microdissection, RNA, Microarray, Infrared laser

1 Introduction

As a prerequisite to comparing the global gene expression profiles of different embryonic cell types, it is necessary to purify the cells of interest. Laser capture microdissection (LCM) is a technique that allows the precise identification and isolation of homogeneous populations of cells from complex heterogeneous tissues. The principles and applications of the technique have recently been reviewed [1, 2]. In short, a microscope slide with a tissue cryosection is

Mark Lewandoski (ed.), *Mouse Molecular Embryology*, Methods in Molecular Biology, vol. 1092,
DOI 10.1007/978-1-60327-292-6_4, © Springer Science+Business Media New York 2014

placed on the platform of the LCM instrument, and the specific cells of interest are identified by morphology using the inverted microscope. An LCM cap covered by a thermoplastic film is placed over the section. An infrared laser is directed through the cap to the cells of interest, and these cells are attached to the thermoplastic film. The cells can then be removed from the cap using a standard RNA preparation denaturing solution, and RNA can be isolated. This chapter provides a step-by-step protocol to perform LCM on embryo samples and obtain sufficient amounts of high-quality RNA for microarray hybridizations. With some modifications, the protocol described here could be applied in any case where limited tissue samples are available for LCM and subsequent microarray analyses, for example, for patient samples in research settings.

Recently, LCM followed by quantitative reverse transcriptase-PCR (qRT-PCR) has been employed to compare gene expression patterns in different embryonic cell types [3–5] or to compare mutant and normal embryonic cells [6]. In these cases, specific candidate genes were selected for qRT-PCR, to determine if they are differentially expressed. In order to obtain a more global profile of differential gene expression in embryos, however, an LCM/microarray strategy can be employed. This strategy requires more RNA than does qRT-PCR, but it is becoming increasingly more popular. In our experience, high-quality qRT-PCR data can be obtained with RNA prepared from hundreds of LCM-isolated cells (or less depending on the abundance of the mRNA), whereas high-quality Affymetrix GeneChip microarray assays require at least a few thousand cells [7].

In the past 3 years, the LCM/microarray approach has been used for several applications in developmental biology [7–13]. For example, Redmond et al. identified erythroid-enriched genes expressed in the mouse embryonic day 9.5 (E9.5) yolk sac, using Affymetrix GeneChips [7]. Xiao et al. identified genes that are differentially expressed in E10 and E12 mouse lens cells [10]. Williams et al. identified genes that are differentially expressed along the dorsal–ventral, medial–lateral, and anterior–posterior axes of the E17.5 mouse olfactory bulb. Bhattacherjee et al. isolated neural crest- and mesoderm-derived mesenchymal cells from the first brain branchial arch of E9.5 mouse embryos, and used Affymetrix GeneChip microarrays to determine which genes are differentially expressed in the two distinct lineages [9]. Brunskill et al. generated an atlas of gene expression in the developing mouse kidney at microanatomic resolution. They used LCM to isolate some kidney cell populations. To isolate other populations, the study employed transgenic mice expressing GFP in specific cells, followed by fluorescence-activated cell sorting (FACS) [13]. Our laboratory has used the LCM/microarray strategy to compare the genetic profiles of normal E9.5 yolk sac erythroid cells to cells from embryos with a null mutation in the gene encoding Krüppel-like factor 2 (KLF2−/−), embryos that are KLF1 null, and double null mutants [14, 15]. Clearly, the results from these LCM/microarray experiments can be used to develop hypotheses about the genes controlling development in specific tissues.

The LCM part of the protocol described here is designed for the ArcturusXT microdissection instrument, but could easily be adapted for use with other Arcturus instruments. Microdissection instruments from other manufacturers differ in that they employ specialized platforms for tissue attachment, rather than regular microscope slides. In any case, for microdissecting cells from mouse embryonic tissue, infrared (IR) laser capture as opposed to UV laser cutting is the best option, because a limited number of cells within a small area on the slide are being isolated. UV laser cutting is designed to capture larger numbers of cells than are available in mouse embryos. The tissue and RNA preparation methods and troubleshooting guides described here should be applicable with any LCM instrument. The methods described here do not duplicate a manufacturer's protocol; rather they combine various manufacturer recommended kits, other kits and reagents found to be both reliable and economical. The protocol is written specifically for mouse embryonic yolk sac, but notes indicate how it can be adapted for use with other embryonic tissues, or any other tissue that has a limited number of cells of interest.

2 Materials

2.1 Preparation and Freezing of Tissue

1. Aluminum foil.
2. RNase AWAY (ISC BioExpress, Kaysville, UT, USA).
3. 2-Methylbutane (isopentane).
4. Freshly made 20 % (w/v) sucrose cryoprotection buffer (2 g of sucrose per 10 ml of 1× PBS).
5. Optimal Cutting Temperature freezing media (OCT, Sakura Finetek USA, Inc., Torrance, CA, USA).
6. Watchmaker's forceps.
7. Metal crucible.
8. Intermediate 15 × 15 × 5 mm cryomolds.
9. Plastic disposable 1-ml sterile Pasteur pipettes.
10. 60 × 15 mm plastic Petri dishes.
11. Dissecting microscope.
12. Zip-lock sandwich bags.

2.2 Cryosectioning of Tissue

1. RNase AWAY (ISC BioExpress).
2. Low-Profile Accu-Edge Disposable Microtome Blades (Sakura Finetek).

3. Vibratome UltraPro 5000 (Global Medical Instrumentation, Inc., Ramsey, Minnesota, USA).
4. Paintbrushes with 0.5–1 cm bristles.
5. Silane-Prep microscope slides (Sigma-Aldrich).
6. Rapid Stain (American Master Tech Scientific, Inc., Lodi, CA).
7. Uni-Cassette® (Sakura Finetek).
8. Desiccant.
9. Plastic 25-slide boxes.

2.3 Dehydration and Staining of Microscope Slides for LCM

1. HistoGene™ LCM Frozen Section Staining Kit (MDS Analytical Technologies, Sunnyvale, CA, USA). This kit is purchased mainly for the stain, **items 2–4** are purchased separately to extend the kit.
2. Anhydrous ethanol (VWR, West Chester, PA, USA, catalog number IB15720).
3. Xylene.
4. Nuclease-free and proteinase-free H_2O.
5. SUPERaseIn™ (20 U/μl) (Ambion, Foster City, CA, USA).
6. 75 % EtOH and 95 % EtOH dilutions (made with anhydrous ethanol and nuclease-free H_2O).
7. 50 ml sterile conical polypropylene screw cap tubes.
8. RNase AWAY (ISC BioExpress).
9. Glass Staining Dishes for 20 slides, including two slide-staining trays (Fisher).
10. Molecular sieves, 4 Å (Sigma-Aldrich).

2.4 Laser Capture Microdissection Using the ArcturusXT

1. ArcturusXT Microdissection Instrument (MDS Analytical Devices).
2. CapSure® HS LCM Caps (MDS Analytical Technologies).
3. GeneAmp™ 500 μl Thin-walled PCR Reaction Tubes (Applied Biosystems, Foster City, CA, USA).
4. Post-it® Notes (3 M). Keep them in their original packaging or a clean zip-lock bag to avoid RNase contamination.
5. Denaturing solution from ToTALLY RNA™ kit (Ambion).

2.5 RNA Preparation

1. 1.5 ml round bottom screw cap microcentrifuge tubes with O-rings (Fisher Scientific).
2. 3 M sodium acetate and Acid Phenol from the Totally RNA™ kit (Ambion).
3. Glycogen (5 mg/ml, nucleic acid- and nuclease-free).
4. Isopropanol.

5. 75 % EtOH made with anhydrous EtOH and nuclease-free H_2O.
6. SUPERaseIn™ (20 U/μl) (Ambion).
7. Nuclease-free and proteinase-free H_2O (USB).

2.6 Assessing RNA Quality

1. Agilent 2100 bioanalyzer with chip priming station (Agilent Biotechnologies, Santa Clara, California, USA).
2. Pico RNA Chip, reagents and electrode cleaner (Agilent).
3. RNA 6000 Pico ladder (Ambion).
4. RNase AWAY (ISC BioExpress).
5. Nuclease-free H_2O.
6. Pipette accurate for measuring 1 μl volume, such as Rainin P10.
7. RNase-free 1.5 ml snap cap microcentrifuge tubes.
8. Vortex mixer with Agilent adapter.

3 Methods

3.1 Preparation and Freezing of Tissue

1. Test the OCT media to make sure that is in good condition for freezing specimens (*see* **Note 1**).
2. Ensure that all reagents, dissection equipment, instruments, pipettes, slides, and storage vessels that will contact samples are RNase free (e.g., handle only with gloved hands, treat with RNase AWAY) to maintain RNA quality. Use sterile pipette tips designated for RNA use only.
3. Chill the isopentane in a metal crucible surrounded by dry ice in a Styrofoam box.
4. The uterine horns from staged pregnant mice are removed and placed in a Petri dish containing 1× PBS at 4 °C.
5. Individual E9.5 yolk sacs are dissected from the embryo and immediately placed in 1.5 ml microcentrifuge tubes containing 1 ml of cryoprotection buffer at 4 °C. Ensure yolk sac is not floating at the top of the cryoprotection buffer by gently tapping the tube on a tabletop.
6. After 25–30 min, when the yolk sac sinks to the bottom of the tube, use a sterile 1 ml plastic disposable Pasteur pipette to transfer yolk sac into a Petri dish containing 1:1 20 % sucrose PBS:Optimal Cutting Temperature freezing media. Rinse well in the solution. It is helpful to cut off the tip of the pipette with a clean razor blade to avoid shearing the tissue.
7. Carefully transfer the yolk sac into another Petri dish containing OCT. A watchmaker's forceps can be used to scoop and transfer the yolk sac to prevent bubbles from forming in the OCT.

Rinse yolk sac in OCT and examine Petri dish under the dissecting scope to ensure that there is no sucrose residue, which will negatively affect the integrity of the yolk sac upon freezing.

8. Gently place the yolk sac in the middle of a cryomold containing OCT. Examine under a dissecting microscope, and remove or move away any bubbles near the yolk sac, using a forceps or Pasteur pipette. With the edge of dry forceps, manually sink the yolk sac close to the bottom of the cryomold (*see* **Note 2**).
9. Lower cryomold into the metal crucible containing isopentane chilled with dry ice until cryomold floats on the surface of the isopentane. Aluminum foil or gauze strips can be used as a sling to position the cryomold while slowly lowering it. Allow OCT media to completely cover the cryomold over the course of 2 min, and then drop cryomold into the crucible to completely submerge in cold isopentane, and freeze for one additional minute (*see* **Note 3**).
10. Remove cryomold from isopentane and wrap with chilled aluminum foil, storing in a chilled zip-lock sandwich bag. Store temporarily on dry ice until transferring to −80 °C for storage.

3.2 Cryosectioning of Tissue

1. Clean the knife, brushes and slide box with RNase AWAY. Equilibrate the cryostat, metal chuck, knife, slide box, and brushes to −23 °C (*see* **Note 4**). The cryostat should be completely defrosted and thoroughly cleaned with RNase AWAY at least once per month.
2. Mount OCT block onto metal chuck using OCT and let freeze completely. The block should be mounted so the yolk sac is facing away from the metal chuck.
3. Ensure that knife and OCT block are firmly in place. Adjust relative knife-to-block angle. Lower angles can cause sections to curl more severely while steeper angles can cause chattering of sections.
4. Orient the OCT block so that the knife cuts evenly on both the left and right sides of the block.
5. Place 8-μm sections on room temperature Silane-Prep microscope slides, number slides with a pencil, and store slides in the chilled 25-slide box in the cryostat. Place tissue in the middle (at least 0.5 cm away from the edges) of slide for effective LCM (*see* **Notes 5** and **6**).
6. Stain periodic landmark slides (every tenth section) with Rapid Stain to observe cellular morphology.
7. Place a cassette containing dust-free desiccant behind the 23rd slide of a 25-slide box.
8. Wrap slide box with chilled aluminum foil and store at −80 °C.

3.3 Dehydration and Staining of Microscope Slides for LCM

1. Place molecular sieves into stock solution of 100 % EtOH to ensure that the 100 % EtOH bath will completely dehydrate the slides. Incomplete dehydration will result in poor transfer from slide to cap during LCM.
2. Aliquot and prepare Histogene LCM stain by adding 1 μl of SuperaseIn™ for every of 20 μl of stain to be used in the session, keep on ice.
3. Frozen sections are processed for LCM using reagents from the HistoGene LCM frozen staining kit, supplemented with anhydrous ethanol, xylene, and nuclease-free water purchased separately. All instructions assume that slides are blotted on absorbent paper between solutions to avoid carryover of solution. To maintain good RNA quality, there should be as little delay as possible in placing slides in subsequent solutions.
4. Remove two slides at a time from the slide box and place back-to-back into staining dish filled with 75 % EtOH. Transfer all slides into dish within 2 min of removing slide box from −80 °C. Alternatively, keep the foil-wrapped slide box on dry ice while transferring slides. If processing only 2–4 slides at a time, this can be done back-to-back in one or two 50 ml tubes, rather than in slide dishes.
5. One slide at a time is vigorously dipped 8–10 times in a 50 ml conical tube containing nuclease-free H_2O for about 10–15 s to remove OCT residue. Use fresh tubes with fresh H_2O after every ten slides that are processed.
6. Pipette 20 μl of Histogene LCM stain directly onto the yolk sac tissue on the slide, leave for 10 s, and wash off stain by pipetting H_2O on the slide and discarding.
7. The slide is dipped in H_2O several times to remove stain residue and placed into another staining dish with 75 % EtOH. Repeat H_2O and staining steps with the rest of the slides. All slides should remain in the second 75 % EtOH wash for at least 30 s.
8. The slide-staining rack is transferred to a staining dish containing 95 % EtOH for 30 s and then to a dish with 100 % EtOH for 3 min.
9. The rack is dipped twice in a dish containing xylene, and then placed in another staining dish with xylene for 10 min.
10. To maintain RNA integrity, stained slides should remain in xylene no longer than 2 h total time prior to LCM.

3.4 Laser Capture Microdissection Using the ArcturusXT

1. Wipe the cap loading dock, slide holders, and the QC station with Kimwipes sprayed with RNase AWAY before using the LCM instrument. One to two slides may be removed from the xylene and allowed to air-dry for 2 min by standing an edge on absorbent paper. Since xylene dissolves the LCM cap polymer,

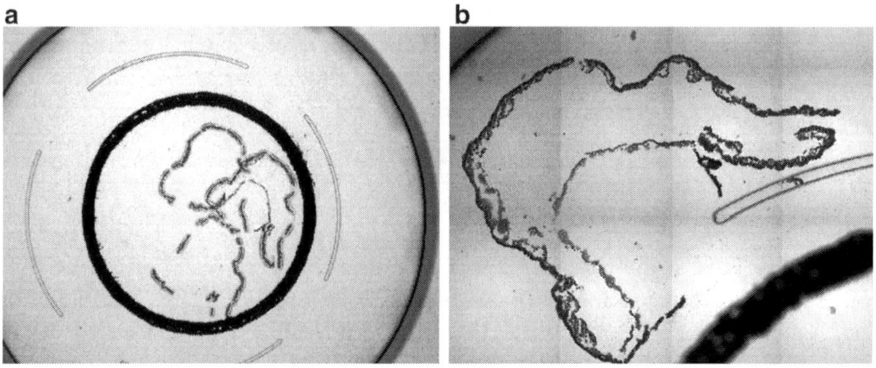

Fig. 1 Positioning of the LCM cap on the yolk sac A: The yolk sac can be placed within the black ring on the LCM cap (**a**, 20× magnification) or outside of it (**b**, 100× magnification). Avoid overlapping the yolk sac with previously collected cells, with the black ring, and with the clear railing. Because our RNA extraction method exposes the entire cap to the denaturing solution, RNA from all captured cells inside and outside of the black ring will be obtained

the slide should be completely dry. To preserve RNA integrity, duration of LCM session should be no longer than 20 min per LCM cap, and no longer than 20 min per slide.

2. Selecting "macro" as the cap type will allow cell capture outside of the cap black ring. At 2× setting, position the yolk sac under the cap. Avoid overlapping the black ring, another captured region, or the 12-μm railing circumscribing the black ring. Examples of suitable cap placements are illustrated in Fig. 1. When the "place cap" icon is clicked, the automated robot arm will place the cap in the center of the current view.
3. Set the power and duration of laser pulses as low as possible, ensuring however that there is good contact between the thermoplastic cover on the cap and the slide after the pulse (Fig. 2). The following parameters are used for LCM of yolk sac erythroid cells on the ArcturusXT Microdissection Instrument: 10 μm spot size, 70 power setting (coarse adjustment), 30 duration (fine adjustment), 80 IR (infrared) spot spacing. In 20× magnification, fire several test pulses by double clicking on a blank slide region to check that the laser melts the cap polymer to the slide making a well-defined black ring as illustrated in Fig. 2. For the Arcturus PixCell II Laser Capture Microdissection system, the parameters are as follows: 7.5 μm laser spot size, power setting at 80 mW, and laser pulse duration of 1.2 ms.
4. The blue cross marks where the IR beam is focused. After firing a test pulse, check that the blue cross is centered within the pulse region. If not, right-click in the center and select "locate IR Beam" from the drop-down menu to align the laser and the blue cross.
5. Right-click on the viewing screen and select "Draw line of IR pulses." Laser pulses are fired in the same order that they are

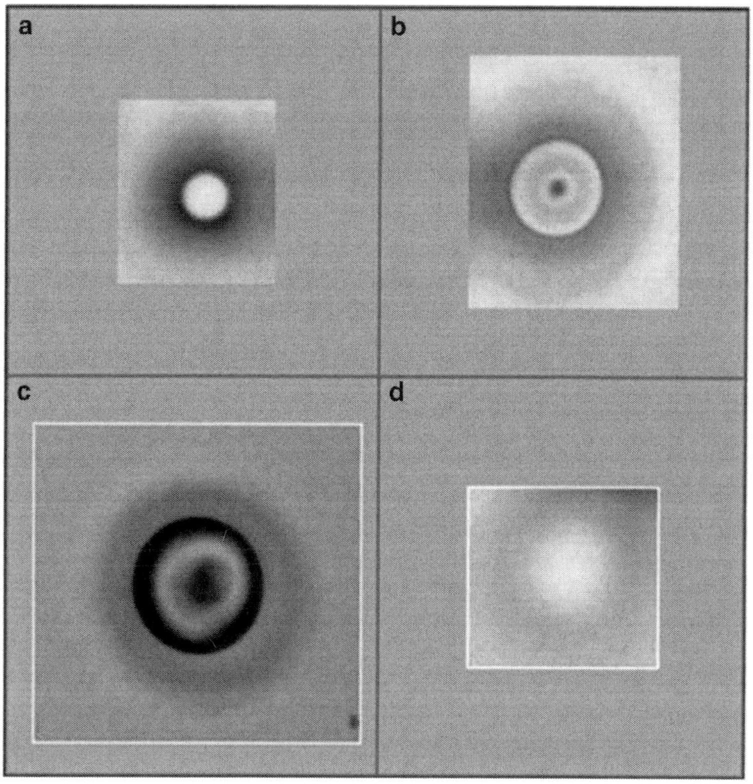

Fig. 2 Various laser pulse intensities and melting of the LCM cap polymer. The cap polymer contains a dye that allows visualization of the melted polymer. In (**a**, **b**), the appropriate laser pulse power, duration, and size is indicated, and the polymer is making good contact with the slide. In (**c**), the laser pulse was fired with too much power, resulting in blackening inside the ring. (**d**) Depicts inadequate power of the laser, and the cap film is not firmly contacting the slide. If this occurs, increase the power and/or duration, or adjust the positioning of the cap. If good contact is still not achievable, the LCM cap may be defective

drawn (*see* **Note** 7). An example of the steps in cell collection is illustrated in Fig. 3.

6. After marking the laser pulses on the viewing screen, click the "IR laser pulse" icon to proceed with firing the laser.
7. The ArcturusXT Microdissection Instrument tracks the number of pulses fired as well as surface area melted in the "I" menu of the "Select" toolbar. At least 300–400 erythroid cells were collected from three slides on a single LCM cap within 20 min.
8. After the 20 min of LCM for the cap, click the "Move to QC" icon to inspect for purity of collected cell population. Click "present stage" icon and remove cap.
9. Place cap directly on three different clean sticky regions of a 3 M Post-it® to remove nonspecific binding to the cap (Fig. 4).

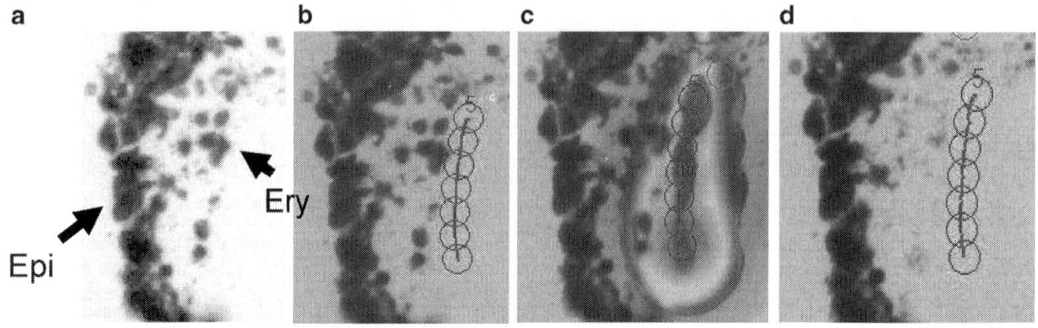

Fig. 3 Embryonic yolk sac tissue sample before and after laser capture microdissection. (**a**) Stained slide photographed at 200× magnification with erythroid (Ery) and epithelial cells (Epi) indicated. (**b**) The seven concentric circles represent the position of the intended line of IR pulses in relation to the erythroid cells (cells of interest). (**c**) The actual line of IR pulses, showing the width and intensity of the melted polymer. The single *circle* at the *top right* marks the center of the microscope field. (**d**) Erythroid cells have been removed with the LCM cap, leaving the epithelial cells remaining on the slide. In this specialized case, the position of the laser pulses was offset from the erythroid cells of interest, to allow an increase of power to remove these very adherent cells from the slide, without collecting contaminating epithelial cells

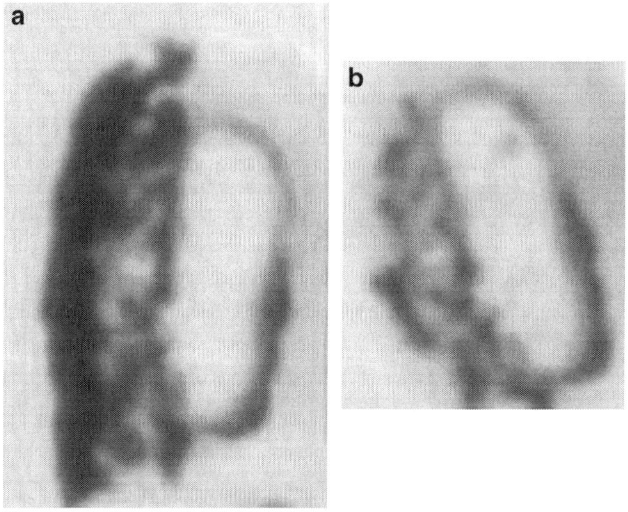

Fig. 4 Removing nonspecifically bound cells contaminating the LCM cap. (**a**) Both erythroid and epithelial cells on the LCM cap at the QC station, prior to placing the cap on the 3 M Post-It adhesive region. (**b**) Captured erythroid cells remain on the cap and nonspecifically bound epithelial cells have been removed by pressing on a Post-It adhesive region. It is difficult to accurately focus images on the QC station

Place cap back on the LCM QC and right-click the respective cap and select "view cap." Check for purity of cell population. Place cap on the 3 M as necessary until nonspecific adherents are removed.

10. Immediately place the cap on top of a GeneAmp™ 500 μl Thin-walled PCR Reaction Tube containing 200 μl of 4 °C denaturing solution. Invert tube and vortex for 3 min, place tube on ice.

11. Carefully zip spin the tube with attached LCM cap at a low speed to recover solution from the cap (make sure that the LCM cap on top of the tube does not obstruct the centrifuge rotor from spinning normally; it will be taller than a tube alone). Remove LCM cap from the GeneAmp™ tube and store the tube at −80 °C.

3.5 RNA Preparation

1. Make sure that all reagents, working environment and pipettes are RNase free. Allow the 0.5 ml tubes containing the LCM material (200 μl) to thaw on ice. Once thawed, mix the cellular extract by pipetting twice and transfer to a 1.5 ml microcentrifuge tube with O-ring screw cap. If there are any bubbles, zip spin tubes again. (Please note that material of the same cell type collected on three LCM caps can be combined for a total of 600 μl, and RNA can be extracted in a single tube.)
2. Add 1/10 volume of 3 M NaAC (sodium acetate) pH 4.50 (ToTALLY RNA isolation kit; Ambion) and mix by pipetting up and down (for every 200 μl of denaturing solution, use 20 μl of 3 M sodium acetate pH 4.50).
3. Add 200 μl of acid-phenol:chloroform (5:1, pH 4.5, ToTALLY RNA isolation kit; Ambion) per LCM cap to the tube (i.e., 600 μl if three caps are combined per tube).
4. Vortex the tubes for 2 min and then centrifuge at 12,000 × g (11,400 rpm) at 4 °C for 5 min.
5. Transfer the aqueous phase to a fresh 1.5 ml screw cap tube leaving behind 20 μl at the interphase, to avoid DNA contamination (for three LCM caps combined, leave behind at least 60 μl).
6. To 180 μl of the aqueous phase, add 1.8 μl of 5 mg/ml glycogen as a carrier, and 180 μl of chilled isopropanol.
7. Vortex the tube and store overnight at −80 °C.
8. Thaw the tube on ice, vortex, and centrifuge at 12,000 × g (11,400 rpm) at 4 °C for 30 min. During this time, prepare the nuclease-free water containing 1/20 volume of SuperaseIn™.
9. Discard the isopropanol supernatant without disturbing the RNA pellet. The pellet should be readily visible because of the use of glycogen as a carrier. Wash the pellet with 1 ml of cold 75 % ethanol (made with nuclease-free H_2O). Add the ethanol down the sides of the tubes, washing to remove any residual salt. Briefly vortex, zip spin and remove all of the ethanol. Wash the pellet again with 1 ml of cold 75 % ethanol, using the same technique as above. Vortex, zip spin the tube, and remove 900 μl of the ethanol. Zip spin again and remove the remaining ethanol. Allow the pellet to air-dry for 1 min. RNA can become difficult to resuspend if pellet is left to dry longer.

10. Redissolve the RNA by adding 12 μl of nuclease-free water with 1/20 volume of SuperaseIn™ to the same side of the tube as the pellet, zip spin, and mix by pipetting up and down ten times. Keep tubes on ice or store at –80 °C. If combining RNA from several tubes for a single microarray assay or for qRT-PCR, *see* **Note 8**. RNA collected for qRT-PCR can be used for assays at this step; typically not enough cells are collected to assess RNA quality using the Agilent bioanalyzer.

3.6 Assessing RNA Quality

1. Allow the reagents from the Pico kit to equilibrate to room temperature for at least 30 min.
2. Prepare dilutions of RNA and of the RNA 6000 ladder (Ambion). For RNA samples, prepare a 1:4 and a 1:8 dilution using nuclease-free H_2O (*see* **Note 9**). Undiluted RNA can contain too much salt to run true to size on the Agilent chip; the dilution amount can be adjusted depending on the number of cells collected. Transfer 5 μl (150 ng/μl) of the RNA ladder into a 1.5 ml microcentrifuge tube and heat at 70 °C for 2 min. Place the tube on ice and dilute the ladder to 1 ng/μl (1,000 pg/μl) using nuclease-free H_2O. Use the ladder immediately, or aliquot and store the prepared ladder at –80 °C for later use.
3. Prepare the gel mix by filtering 550 μl in the supplied filter column. Centrifuge at 1,500 × *g* for 10 min. Aliquot 65 μl of the filtered gel into RNase-free microcentrifuge tubes. The filtered gel can be stored and used for up to 2 months.
4. Decontaminate the electrodes on the bioanalyzer before each run. First slowly pipette 350 μl of RNase AWAY into any well of the electrode cleaner chip, so that it covers the bottom of all of the wells. Place the electrode cleaner in the bioanalyzer for 5 min. Open the lid, remove the chip, and empty the RNase AWAY. Add 350 μl of RNase-free H_2O to the electrode cleaner and place in the bioanalyzer for about 1 min but no more than 5 min. Allow the bioanalyzer to dry for about 1 min with the lid open.
5. Vortex the dye concentrate for 10 s and zip spin. Add 1 μl of dye to 65 μl of the filtered gel-matrix aliquot. Vortex thoroughly and centrifuge at 13,000 × *g* for 10 min. The dye is light sensitive, so care should be taken to return it quickly to the kit box and close the lid securely.
6. Take a Pico chip out of the bag and place in the chip priming station. Carefully pipette 9 μl of the gel-dye mix without introducing any bubbles into the well marked G with a black dot. Make sure that the syringe on the chip priming station is set to the 1 ml mark. Close the chip priming station so that the latch audibly clicks. Depress the plunger so that it is tucked under the clip for 30 s. Release the plunger using the clip release and

wait about 15 s before pulling the plunger to the 1 ml mark. Slowly pull the plunger to the 1 ml mark and open the chip priming station. If the syringe gasket is properly sealed, the plunger should quickly be at the 0.7 ml mark before time expires.

7. Pipette 9 µl of the gel-dye matrix to the other two wells marked G and discard the remaining mixture. Pipette 9 µl of the conditional solution to the well marked CS.
8. Pipette 5 µl of Pico marker into the well marked ladder and to each of the sample wells to be used for the experimental run. Place 6 µl of marker in the unused sample wells or the chip will not run properly.
9. Pipette 1 µl of the prepared RNA ladder into the designated well and 1 µl of the diluted RNA samples to each of the test wells. A pipette that is accurate for measuring 1 µl of volume should be used, to obtain results that are as accurate as possible with respect to the concentration of the RNA.
10. Vortex the chip for 1 min. It is recommended that the chip is held down with a finger or with tape, so it does not become dislodged from the adapter. Make certain that the chip is flat and stable, or samples in different wells may become mixed together. The chip run should begin within 5 min after mixing, to avoid evaporation.
11. Activate the software before loading the chip. In the instrument context, select Assay → Electrophoresis → RNA → Eukaryote Total RNA and press start to initiate the run. Sample names can be added while the chip is running. At the end of the run, discard the chip, clean the electrodes as above, and return the reagents to 4 °C.
12. After the run is completed, select results to examine the electropherograms and virtual agarose gel for the RNA samples and the RNA ladder. For interpreting the electropherograms, refer to **Note 10** and Fig. 5. Only RNA of high quality is processed for subsequent microarray hybridizations. For processing LCM-prepared RNA for Affymetrix microarray hybridizations, *see* **Note 11**. For troubleshooting RNA samples that are degraded, *see* **Note 12**.

4 Notes

1. Test the OCT media to verify that it has not undergone a freeze/thaw cycle prior to arrival. Gently pour the OCT into an intermediate cryomold, and freeze the cryomold in a dry ice–isopentane bath. Cut 8-µm sections of the OCT media and place on a Silane-Prep slide; use a permanent marker to identify the area. If the media has a stringy appearance during cryosec-

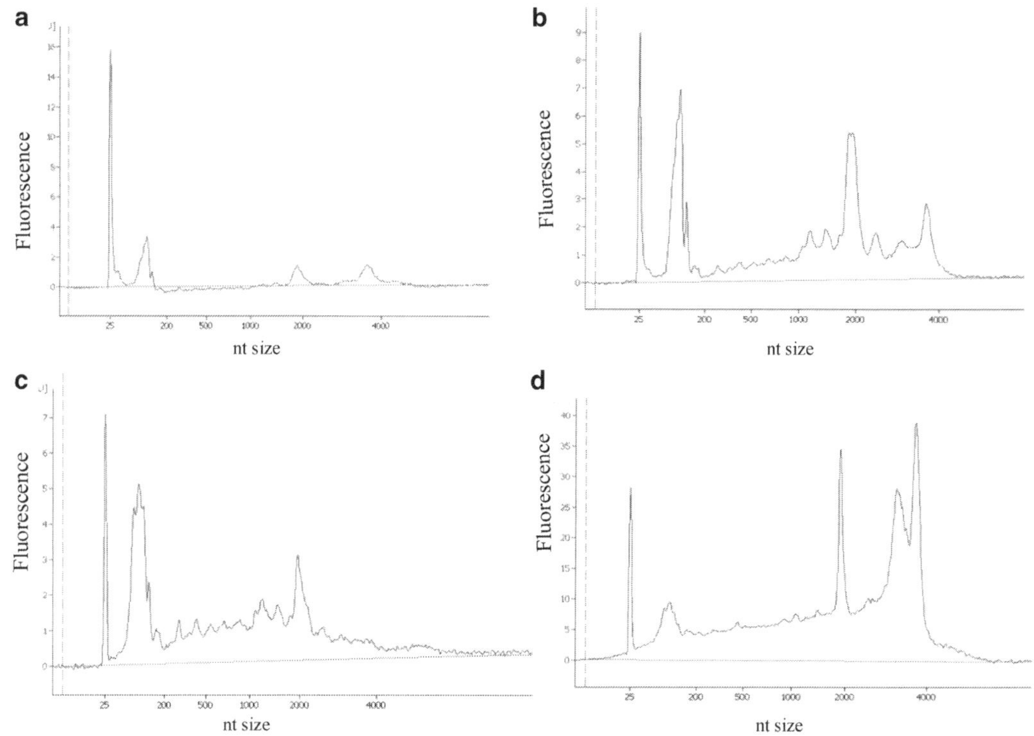

Fig. 5 Electropherograms to assess quality of RNA prepared from microdissected cells. RNA integrity is tested using the Agilent bioanalyzer, and only intact RNA is used for performing microarray assays. (**a**) Intact RNA that is properly diluted, though of a low concentration (*see* **Note 8**). The 18S rRNA fluorescent peak appears at about 2,000 nucleotides, and the 28S rRNA peak is just under 4,000 nucleotides. (**b**) Intact RNA that is not diluted can appear as though it is partially degraded, with peaks between the 18S and 28S rRNA peaks. (**c**) Partially degraded RNA may contain an apparent 18S rRNA peak, but no 28S rRNA peak. (**d**) RNA that is contaminated with DNA will have a broad 28S rRNA peak, or two peaks near the 28S rRNA peak

tioning, or appears dirty or contains holes when slides are examined on the microscope, replace it with a new batch.

2. Correct placement of the tissue in the cassette prior to freezing is important. Make certain that the tissue is completely surrounded by OCT media on all sides. It may be necessary to place the tissue/embryo at the correct angle to obtain optimal sections. To obtain the maximum amount of tissue from each cryosection, unfold and flatten out yolk sac tissue in the OCT medium prior to freezing in isopentane.
3. Cryoprotection with sucrose ensures that no ice crystal frozen artifacts form while freezing the tissue. This is essential for yolk sac which is a very thin tissue, and easily subject to freezer burn. Cryoprotection is not required for freezing some other tissues for LCM. Some tissues can simply be dried by blotting on Whatman 3M filter paper, to remove moisture and allow

the OCT to directly contact tissue, prior to freezing in OCT in a dry ice–isopentane bath. Some tissues can be dissected without rinsing in 1× PBS, and immediately frozen in OCT in a dry ice–isopentane bath.

4. Using the appropriate cryostat temperature can influence the quality of the sections, and the optimal temperature is specific to the tissue type and may also depend on the cryostat. For example, bone marrow frozen cryosections are typically prepared at −25 °C, and adult brain cryosections are prepared at −18 °C. Using a temperature as cold as possible helps maintain RNA integrity. If the temperature is too cold, however, there will be shattering throughout the section.
5. To keep cryosections flat, cut 8-μm sections until only one edge is still attached to the block. Use paintbrushes to unravel section and push the free edges inwards opposing the axis of curling.
6. Positioning multiple sections on the slide during cryosectioning can lead to further efficiency in the LCM staining steps as well as the actual LCM steps. Since the first cryosection needs to remain inside the cryostat, place other sections on the same slide within 15–20 s to ensure that they will adhere firmly to the slide.
7. Drawing a line of pulses is the preferred method for marking IR spots for collecting erythroid cells from embryonic yolk sac blood islands, about two cells are collected per pulse. Use the interactive pen to mark the touch-screen for further precision when drawing lines of pulses. For LCM collection of other cell types, other options for marking IR spots are available, including single pulses, defined circles, and rectangles. Optimal collection method, spot size, and power and duration settings need to be empirically determined for different cell types. Some cell types (erythroid precursors, for example) adhere more steadfastly to the slide and require higher power to remove than others (epithelial cells, for example). One method to avoid contamination from other nearby cells, when collecting very adherent cells by LCM, is to offset the laser pulses from the cells of interest, as shown in Fig. 3. There are several measures that can be taken to maximize cell collection efficiency in the 20 min LCM session. First, mark large blood islands and skip blood islands containing fewer erythroid cells. Second, mark up multiple yolk sac sections on an individual slide, then position the cap prior to laser capture to collect from consecutive sections simultaneously. Third, transfer the cap with collected cells more quickly onto the GeneAmp™ tube by shifting the slide platform so that the cap is exposed from underneath the LCM machine cover, and manually transfer the

cap to the QC station using RNAse-free tweezers. The LCM software can then be restarted to "reset" the robot arm.

8. RNA prepared from the same cell type but extracted in multiple tubes can be pooled for microarray or qRT-PCR assays. Dissolve the first RNA pellet, then use the 12 μl in the first tube to resuspend the RNA pellets from subsequent tubes. Dissolve the first pellet, zip spin, and transfer to the next tube containing an RNA pellet. Resuspend the second pellet and continue this process until all pellets have been combined. This process has been used to successfully combine and obtain high-quality RNA from eight tubes. Hundreds of cells can be collected per LCM cap, and RNA from three caps can be extracted in a single tube, so RNA from 24 LCM caps can be combined in a single tube using this method to resuspend pellets. In this way, RNA from the thousands of cells necessary for a single microarray can be combined. This RNA is re-precipitated with isopropanol and used to prepare labeled cRNA for Affymetrix GeneChip hybridizations, as described in **Note 11**. At least 50 ng of RNA can typically be obtained from 10,000 yolk sac erythroid cells collected by LCM.
9. To dilute RNA from LCM samples for the Agilent bioanalyzer assay, add 2 μl of RNA to a 1.5 ml tube containing 6 μl of nuclease-free H_2O and mix by pipetting (1:4 dilution). Prepare a 1:8 dilution from the 1:4 dilution. Diluting the RNA properly is essential for the bioanalyzer assay. If the RNA is not diluted enough, samples with a higher salt content will migrate incorrectly and appear to be degraded. If the RNA is too dilute, the bioanalyzer is not able to detect the ribosomal species, and the profile may suggest that the sample has no RNA or is degraded. A 1:4 and 1:8 dilution are recommended for the bioanalyzer assays. A 1:2 dilution may be used for lower yields.
10. Because LCM RNA microdissected samples contain very little RNA, the 28S:18S ribosomal ratios may not be as high as 2.0. The 28S:18S ribosomal ratios for yolk sac microdissected cells are typically between 0.3 and 1.2 and have produced successful microarray hybridizations. To determine whether the RNA is of high quality, two important quality assessment parameters should be considered. The parameters are the detection of strong 18S and 28S rRNA peaks in the electropherogram, and the area under the 18S and 28S peaks should be greater than 15 % of the total area under the curve, as determined by the Agilent software. RNA preparations not meeting these quality control criteria should not be used for subsequent microarray hybridizations.
11. To process LCM-prepared RNA for microarray assays, re-precipitate the RNA collected from multiple tubes. After

centrifugation at 12,000 ×*g* (11,400 rpm) at 4 °C for 30 min, carefully wash the pellet with 500 μl of chilled 70 % EtOH, allowing it to air-dry for 1 min and then resuspend it in 3 μl of nuclease-free H_2O. Proceed with the Affymetrix protocol using the Two-Cycle cDNA Synthesis Kit (following the manufacturer's protocol). This protocol claims to create enough labeled cRNA from 10 ng of total RNA, but a good yield of labeled cRNA can reliably be obtained from at least 50 ng of RNA. Labeled cRNA yields exceeding 40 μg are considered to be good. As a quality control parameter, the 3′/5′ ratios for housekeeping genes such as GAPDH and β-actin are expected to be higher for the Two-Cycle than for the standard One-Cycle cDNA Synthesis protocol, which was developed to be used with at least 1 μg of RNA. This is due to a more accentuated 3′ bias introduced during the cDNA and cRNA synthesis steps of the Two-Cycle protocol.

12. Samples can be tested prior to experimental LCM, to determine whether tissues and reagents will give intact RNA preparations. These steps can be used at any time to troubleshoot if RNA preparations are degraded. To troubleshoot, RNA samples should be prepared in three ways, and then tested for RNA integrity on the Agilent bioanalyzer. First, remove a slide from the −80 °C freezer and place 200 μl of denaturing solution directly onto the tissue on the slide, allow it to thaw just enough to become liquid, pipette up and down to dissolve all of the tissue, and proceed with an RNA preparation (Subheading 3.5). Second, stain, fix, and dehydrate a slide as described in Subheading 3.3, then dissolve the tissue on the slide using denaturing solution, as described for the first slide and continue RNA preparation. Finally, stain, fix, and dehydrate a slide, and blanket the slide with laser pulses to adhere all of the tissue on the slide onto an LCM cap, then continue with the RNA preparation starting at **step 10** of Subheading 3.4. Typically, entire cryosections from slides need to be diluted more than do LCM samples for the Agilent bioanalyzer assay, therefore a 1:100 and a 1:500 dilution is recommended. If the RNA from the first slide is intact, then the frozen tissue samples are of high quality. If not, then troubleshooting of the procedures for dissection, freezing and cryosectioning needs to be performed. If the RNA from the second slide preparation is intact, then the tissue sample is good, and the stain, solutions, and methods used for slide processing are producing intact RNA. If the first slide gives intact RNA and the second does not, then the slide processing steps need troubleshooting. If the RNA preparation from the final slide is the only one that is degraded, then the LCM steps need to be improved.

References

1. Chimge NO, Ruddle F, Bayarsaihan D (2007) Laser-assisted microdissection (LAM) in developmental biology. J Exp Zool B Mol Dev Evol 308:113–118
2. Murray GI (2007) An overview of laser microdissection technologies. Acta Histochem 109:171–176
3. Nawshad A, LaGamba D, Olsen BR, Hay ED (2004) Laser capture microdissection (LCM) for analysis of gene expression in specific tissues during embryonic epithelial-mesenchymal transformation. Dev Dyn 230:529–534
4. Sainson RC, Johnston DA, Chu HC, Holderfield MT, Nakatsu MN, Crampton SP, Davis J, Conn E, Hughes CC (2008) TNF primes endothelial cells for angiogenic sprouting by inducing a tip cell phenotype. Blood 111:4997–5007
5. Canto-Soler MV, Huang H, Romero MS, Adler R (2008) Transcription factors CTCF and Pax6 are segregated to different cell types during retinal cell differentiation. Dev Dyn 237:758–767
6. Toba Y, Tiong JD, Ma Q, Wray S (2008) CXCR4/SDF-1 system modulates development of GnRH-1 neurons and the olfactory system. Dev Neurobiol 68:487–503
7. Redmond LC, Dumur CI, Archer KJ, Haar JL, Lloyd JA (2008) Identification of erythroid-enriched gene expression in the mouse embryonic yolk sac using microdissected cells. Dev Dyn 237:436–446
8. Spencer MW, Casson SA, Lindsey K (2007) Transcriptional profiling of the Arabidopsis embryo. Plant Physiol 143:924–940
9. Bhattacherjee V, Mukhopadhyay P, Singh S, Johnson C, Philipose JT, Warner CP, Greene RM, Pisano MM (2007) Neural crest and mesoderm lineage-dependent gene expression in orofacial development. Differentiation 75:463–477
10. Xiao W, Liu W, Li Z, Liang D, Li L, White LD, Fox DA, Overbeek PA, Chen Q (2006) Gene expression profiling in embryonic mouse lenses. Mol Vis 12:1692–1698
11. Plummer S, Sharpe RM, Hallmark N, Mahood IK, Elcombe C (2007) Time-dependent and compartment-specific effects of in utero exposure to Di(n-butyl) phthalate on gene/protein expression in the fetal rat testis as revealed by transcription profiling and laser capture microdissection. Toxicol Sci 97:520–532
12. Williams EO, Xiao Y, Sickles HM, Shafer P, Yona G, Yang JY, Lin DM (2007) Novel subdomains of the mouse olfactory bulb defined by molecular heterogeneity in the nascent external plexiform and glomerular layers. BMC Dev Biol 7:48
13. Brunskill EW, Aronow BJ, Georgas K, Rumballe B, Valerius MT, Aronow J, Kaimal V, Jegga AG, Yu J, Grimmond S, McMahon AP, Patterson LT, Little MH, Potter SS (2008) Atlas of gene expression in the developing kidney at microanatomic resolution. Dev Cell 15:781–791
14. Redmond LC, Dumur CI, Archer KJ, Grayson DR, Haar JL, Lloyd JA (2011) Krüppel-like factor 2 regulated gene expression in mouse embryonic yolk sac erythroid cells. Blood Cell Mol Dis 47(1):1–11
15. Pang CJ, Lemsaddek W, Alhashem YN, Bondzi C, Redmond LC, Ah-Son N, Dumur CI, Archer KJ, Haar JL, Lloyd JA, Trudel M (2012) Krüppel-like factor 1 (KLF1), KLF2, and myc control a regulatory network essential for embryonic erythropoiesis. Mol Cellular Biol 32(13):2628–2644

Chapter 5

EMAGE: Electronic Mouse Atlas of Gene Expression

Lorna Richardson, Peter Stevenson, Shanmugasundaram Venkataraman, Yiya Yang, Nick Burton, Jianguo Rao, Jeffrey H. Christiansen, Richard A. Baldock, and Duncan R. Davidson

Abstract

The EMAGE (Electronic Mouse Atlas of Gene Expression) database (http://www.emouseatlas.org/emage) allows users to perform on-line queries of mouse developmental gene expression. EMAGE data are represented spatially using a framework of 3D mouse embryo models, thus allowing uniquely spatial queries to be carried out alongside more traditional text-based queries. This spatial representation of the data also allows a comparison of spatial similarity between the expression patterns. The data are mapped to the models by a team of curators using bespoke mapping software, and the associated meta-data are curated for accuracy and completeness. The data contained in EMAGE are gathered from three main sources: from the published literature, through large-scale screens and collaborations, and via direct submissions from researchers. There are a variety of ways to query the EMAGE database via the on-line search interfaces, as well as via direct computational script-based queries. EMAGE is a free, on-line, community resource funded by the Medical Research Council, UK.

Key words Mouse, Gene expression, Database, Development, Anatomy, Embryo, Atlas, Curated, In situ hybridization, Immunohistochemistry, Knock-in, Wild-type, Resource, Transcription

1 Introduction

The EMAGE database (http://www.emouseatlas.org/emage) is an on-line resource of in situ gene expression data in the developing mouse embryo. The data are housed within the framework of the EMAP (Electronic Mouse Atlas Project) (*see* **Note 1**) Anatomy Atlas (EMA), which is a digital atlas composed of 3D mouse embryo models and an accompanying integrated, textual anatomy ontology. The expression data housed in EMAGE are mapped into the 2D and 3D space of the virtual embryo models. This allows the expression patterns to be accessed via direct spatial queries without the need to refer to named anatomical structures, as well as via text-based anatomical queries.

Mark Lewandoski (ed.), *Mouse Molecular Embryology*, Methods in Molecular Biology, vol. 1092,
DOI 10.1007/978-1-60327-292-6_5, © Springer Science+Business Media New York 2014

1.1 Background

The results of the types of experiments described in EMAGE (notably in situ hybridization, immunohistochemistry, and lacZ/GFP reporter assays) are typically recorded as 2D images or, more recently, as 3D volumes. Traditionally, the information contained in these images was annotated for computational reference and analysis by describing the expression patterns using controlled vocabularies. Each aspect of the pattern was described using anatomical structures and qualifying descriptors to build up a text-based description of the expression pattern. Generally, this results in a list of tissues in which the gene is expressed. This method, while extremely useful, suffers from the fundamental drawback that it is almost impossible (and time consuming) to fully describe a complex pattern. The more complex the spatial expression pattern, the lengthier and more complicated the text description needs to be. In particular, the text description cannot represent the complex variation in expression strength across tissues and boundaries that do not relate to any morphological or histological feature. The key concept behind the EMAGE database was to use *spatial annotations* of expression patterns (i.e., to spatially transform data into a standard reference frame) to overcome this limitation of text-based annotation. It was envisioned to develop a series of digital 3D embryo models to use as a spatiotemporal framework to index and visualize the spatial information represented in the data images from in situ gene expression experiments.

1.2 Technical Developments

EMAP is a collaboration between The University of Edinburgh and the MRC Human Genetics Unit, UK and is based partly on the "The Atlas of Mouse Development" by Matthew H. Kaufman [1]. The EMAP models were originally generated from reconstructions of histological slides, photographed digitally at high resolution. These digital images are stacked in order and aligned from one section to the next (based on image correlation) to produce a 3D digital volume image.

With the development of Optical Projection Tomography (OPT) within the EMAP group [2] a new technique for producing 3D volumetric models without sectioning was possible and enabled a rapid extension of the atlas series. However, OPT visualization did not produce a high enough resolution of internal structures to allow mapping of data. The EMAP team therefore developed a technique to generate models utilizing the benefits of both OPT and histological sectioning. The embryo chosen as a model is first OPT scanned to provide a geometrically accurate 3D image of the specimen, it is then sectioned and mounted on slides to produce conventionally stained histological sections, which are digitized and restacked as before. The OPT scan is used as a scaffold to realign the sections, producing a smoother and more geometrically faithful result, while retaining the high resolution internal detail provided by the digitized histological sections. A fuller description of the process can be found on the EMA website.

At the time of writing, there are 63 EMAP models covering Theiler Stages 7–26, available on the EMA website.

1.3 Sources of Data in EMAGE

The data available in EMAGE come from three main sources: from published journal articles, from large-scale screens and collaborations, and via direct submission from researchers working in the field. EMAGE has established copyright agreements with a number of developmental biology journals, allowing the data images published in those journals to be included in the database. The inclusion of such published data is carried out in collaboration with our colleagues in the Gene Expression Database (GXD) at the Mouse Genome Informatics (MGI) resource [3]. Curators at the GXD scan the literature and code the relevant journal articles, assigning IDs to gene names and assays, and textually annotating the assay results using the EMAP anatomy ontology. The curators at EMAGE then complement this text annotation by spatially annotating the data (*see* Subheading 1.4) to the embryo models. These peer-reviewed published data represent the highest quality data in EMAGE, as the experimental procedures are optimized and in general the images provided are of a high quality. However, the gene coverage available to EMAGE from this source is limited by what data are published with images of sufficient quality to allow spatial mapping onto the embryo models, at the developmental stages covered by EMAGE. This limitation of genome coverage is overcome through the addition of data from large-scale screens with which EMAGE collaborates [4–7]. Such data are flagged in EMAGE as "screen data" as by necessity, individual assays in screens are not optimized. As such they may suffer from high background signal, potentially producing misleading expression patterns. To further assist the user in differentiating between high quality/high confidence data and more variable quality/lower confidence data, all entries in EMAGE have a star rating assigned by the curators, describing the pattern clarity and signal extraction as either 1, 2, or 3 star, relating to low, medium, and high quality. A similar star rating is also assigned to each entry with respect to the morphological match to the model (both systems are described in more detail in Subheading 1.5).

The third main source of data in EMAGE is that provided by individual investigators/laboratories. Anyone can submit data to EMAGE for spatial mapping and inclusion in the database, regardless of whether they are published. The data can be submitted to the editorial office by email, by FTP or by post, with images accepted in most common file formats. The minimum information required is an image of the expression pattern along with a specification of the gene, or genetic sequence being assayed. However, ideally as much information as is available (about the probe used and the experimental conditions) should also be included. As a guide, the minimum meta-data information for an in situ data experiment is defined by the MISFISHIE (Minimum Information Specification For In Situ Hybridization and Immunohistochemistry Experiments) data standard [8].

1.4 Spatial Annotation of Data in EMAGE

The spatial representations of expression patterns in EMAGE are generated either by manual annotation (including full 3D mapping) or using semiautomated mapping, which has been developed for the inclusion of large-scale datasets. Both of these processes are described with examples on the EMAGE website.

1.4.1 Manual Annotation

Manual spatial annotation is carried out by a team of biocurators, using bespoke in-house software called MAPaint [9]. The manual annotation process can also be divided into two broad areas: wholemount data annotation and section data annotation, although the majority of steps in these processes are identical. In both cases the starting point is an image (either wholemount or section) of the relevant expression pattern, which must first be correctly assigned to a Theiler stage. Theiler staging is based on a series of morphological criteria as specified by Karl Theiler in "The House Mouse—Atlas of Embryonic Development" [10] (*see* **Note 2**). We use Theiler staging to define the developmental age of an embryo rather than "days *post coitum*" (dpc), as the developmental state of the embryo, and its morphological similarity to the EMAP model are extremely important in producing an accurate spatial annotation of the data, and within a litter at one specific dpc value, there can be a range of developmental stages of embryo. This staging of the data embryos defines the EMAP model to which the data will be mapped. In the case of wholemount data, the model used is a 2D lateral projection of the full 3D model, similar to a lateral view of a wholemount embryo. In the case of section data, the curator will navigate through the 3D model, altering the angle (pitch and yaw) of the viewing plane to find a plane through the model that matches, as accurately as possible, the histology of the data section. From this point, the mapping processes of both data types are the same. MAPaint is used to define points of morphological equivalence between the data image and the model (lateral projection or internal section). These points form the basis of an image transformation (warp) between the data image and the model using an underlying mesh algorithm. Once the data image has been warped to overlay the model, regions of signal are extracted from the data image using color thresholding, and applied to the correct regions of the model, defined by the warp parameters. Different apparent strengths of signal can be extracted and are presented as different false colors in the final spatial representation: red = strongest signal, yellow = moderate signal, blue = weakest signal, green = possible signal, and cyan = no apparent signal.

1.4.2 Semiautomated Annotation

The semiautomated process of spatially annotating data was developed in order to map very large datasets, for example the comprehensive data from the Eurexpress screen [4]. The volume of data from the Eurexpress project was such that it would have been unrealistic to undertake the spatial annotation by the manual process described previously. Therefore, a number of steps in the manual annotation process were automated to allow the spatial integration of this dataset. It is important to note that the automation of the process was only possible due to the extremely consistent methods by which the data were generated. Aspects of the automation will be applicable to other consistently produced screen datasets, but are unlikely to be useful in the mapping of heterogeneous data like those from the literature. The process of semiautomated spatial annotation is described on the EMAGE website [11] and is summarized here.

The Eurexpress dataset [4] is an in situ hybridization screen for every transcribed mouse gene (approx 19,000 genes) at 14.5 dpc. The data were collected as approximately 24 frozen sagittal sections, each 25 μm thick, uniformly distributed across the embryo. The work was carried out in seven different labs across Europe using the same set of protocols, thus ensuring the consistency of the results. As a first pass approach to the spatial integration of the data, it was decided to define a "pseudo-wholemount" expression pattern for each gene and spatially map these onto the EMAGE wholemount model at Theiler Stage 23. To generate this pseudo-wholemount, the constituent sections from each assay were aligned and overlaid. The resulting maximum intensity projection through the composite sections produces a single image that is representative of a wholemount expression pattern for the gene. To allow the sections to be aligned, the data images were first segmented into "tissue" and "background" using the difference in color values associated with each segment. (Although no general histological stain was applied to the sections, the application of the in situ hybridization stain resulted in a color difference between the tissue and the background.) An automated process of edge recognition was then used on the embryo tissue segments to align each section to the next in the series. By carrying out the alignment for every section from an originally sectioned embryo, the embryo can be digitally reconstructed, and then manually mapped to the EMAGE model, which defines the warp parameters to be used for each constituent assay. A color model was generated comprising all color values encompassed by the in situ hybridization signal used in the assays. This color model was used to automatically extract ten levels of signal intensity (defined by color intensity) from each section in an assay. Using a maximum intensity projection again, these levels of signal intensity were used in the representation of the pseudo-wholemount data for each assay.

1.5 Star Ratings of Data

There are a number of factors involved in the spatial mapping process, which by the nature of the data can be variable. The most important of these (in terms of data representation and subsequent analysis) are the morphological match to the model, and the quality of the data image and visible signal. While we have one model for each Theiler stage as described in Theiler (1989) [10], in reality the process of embryological development is temporally continuous. It is possible to assign a Theiler stage to any data embryo based on the morphological criteria described; however, the exact stage of the data embryo may fall part-way between two Theiler stages. This morphological variation from the standard model can result in errors in the spatial and temporal registration of the data.

Similarly, data images which are unclear, off-plane, show high background, are under- or over-developed, out of focus, have shadows, etc. make interpreting the difference between regions of signal and nonsignal more difficult and therefore less accurate. For these reasons, we have implemented a star rating for both "match to model" and "image/signal quality" for all data. These star ratings are assigned by the curators, usually at the point of mapping and reflect the ease of the mapping process. They are a means by which users can filter data to include only the highest quality data in their analysis. For similar reasons, as mentioned previously, any data in EMAGE that were generated as part of a screen are annotated as "screen data."

2 Materials

As this method describes the use of an on-line resource, there are very few materials required: namely a computer and a web connection. However, it is worth noting the information on the EMAGE website specifying Web Browser Compatibility (http://www.emouseatlas.org/emage/info/browsers_tested.html).

At the time of writing, the recommended browser for the EMAGE website is Firefox 10+, tested on the following platforms:

- MacOS 10.4 and later (Firefox 10+, Safari 5+, Google Chrome 15+).
- Windows XP (Firefox 10+, Internet Explorer 8).
- Windows 7 (Firefox 10+, Internet Explorer 8, Google Chrome 15+).
- SUSE Linux (Firefox 10+, Google Chrome 15+).

Different combinations of operating systems and browser versions, from those specified above, may not display the content or functionalities of the database as intended.

3 Methods

3.1 Searching EMAGE

3.1.1 Search by Space (Wholemount Data)

The spatial search mechanism is based on the LOSSST (LOcal Spatial Similarity Search Tool) algorithm [12]. The underlying mechanism of the LOSSST algorithm is a comparison (by Jaccard index) of two spatial domains; the query domain defined by the user, and the domain of "detected" expression annotated in the model (the "detected" domain consists of the "strongest," "moderate,"and "weakest" domains combined.) The Jaccard coefficient (V) is defined (in the case of the LOSSST algorithm) as the area of the intersection of the two domains divided by the area of the union of the two domains. For the purposes of the spatial search, the query domain specified by the user is dilated by a pre-determined distance to define the "local" area over which the Jaccard coefficient will be calculated. These numerical values of similarity allow the results to be ordered for similarity to the query region.

1. Select "search by embryo space" either by clicking on the button (Fig. 1a) on the front page or by selecting "By Embryo Space" from the "Search" menu on any page.
2. You are now presented with a page entitled "Search for gene expression in an embryonic region."
3. Select an age/Theiler stage of embryo to define the query region on, from the dropdown menu provided.
4. Select wholemount data from the dropdown menu and then click "Next Step."
5. You are now presented with a page entitled "Paint a region using your cursor."
6. By hovering the cursor over the model, it becomes a cross to indicate that it is in painting mode.

   Use the "Flip L/R" button to select the opposite view of the model if necessary.
7. A suite of tools is available for painting the query region including various line widths, undo and redo, eraser tool, and a clear input option. Use these tools as required to paint your query region on the model by clicking and holding the left mouse button to paint a region, and releasing to stop painting. This process can be repeated indefinitely (with the pencil and/or the eraser) to build up the query pattern required (*see* Fig. 1a).
8. Select the appropriate option from the dropdown menu to define whether the search is for regions where genes are detected, not detected or possibly detected.
9. Select or deselect the option as appropriate to determine whether to retrieve data from both sides of the embryo or from only the side used to define the query region—*see* **Note 3**.

10. Where appropriate, select the stage range over which the query is to be carried out—*see* **Note 4**.
11. Once the query pattern and all parameters are defined to your specification, click "Begin Search" located below the model.
12. The results are returned in the standard tabular format of all EMAGE results. *See* Subheading 3.2 for a full description of the EMAGE results format.

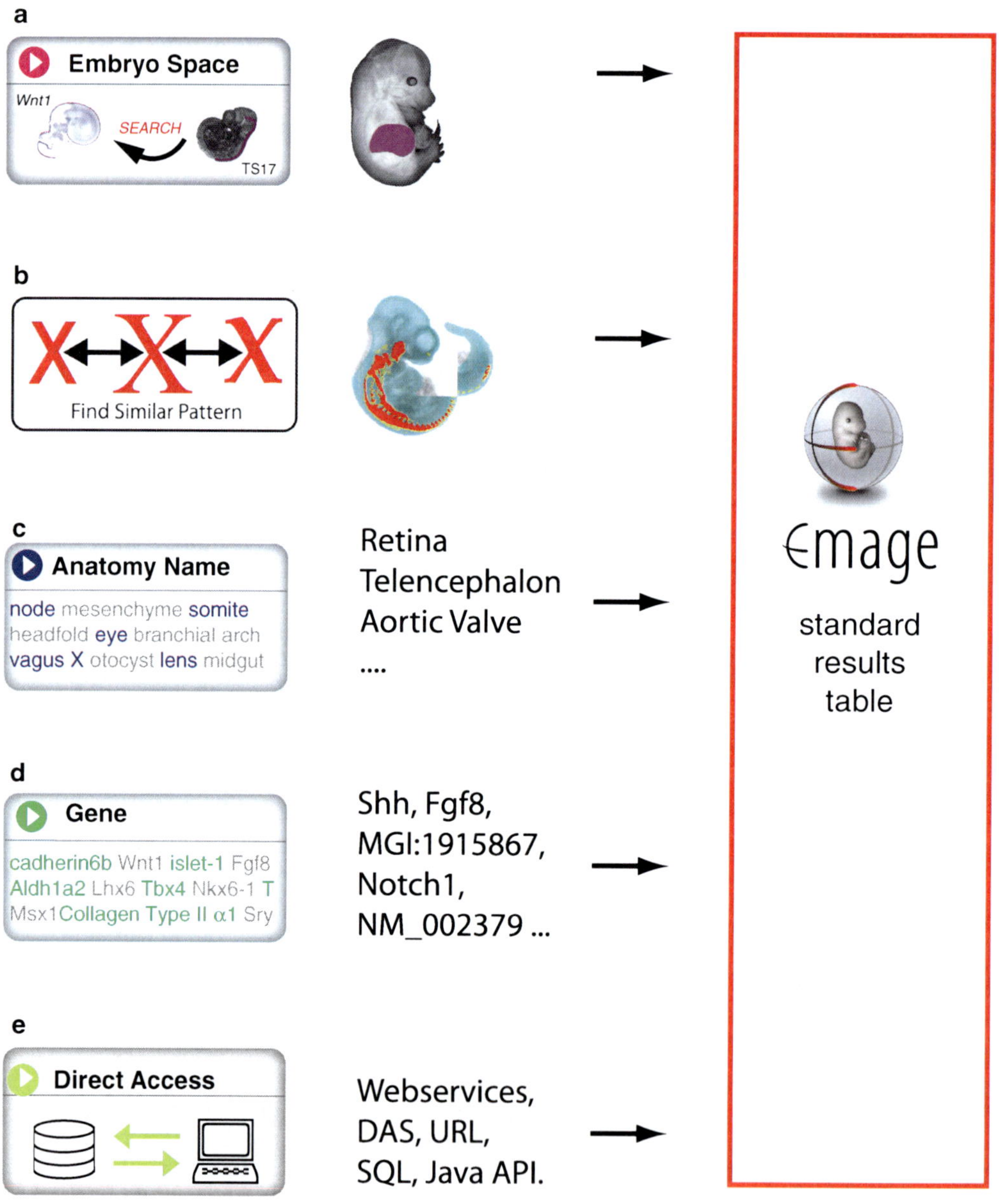

3.1.2 Search by Space (Section Data)

In the past, users were required to download a Java application to allow them to carry out a 3D spatial search in EMAGE. At the time of writing, this application has been retired, soon to be replaced by an on-line section browser tool. The new web browser-based tool will allow users to navigate through their chosen 3D model in any plane, and specify a section or sections of interest for a spatial query. When the correct section has been identified, the painting of a query region and the spatial search will be carried out in essentially the same way as for wholemount data. This new web browserbased 3D spatial query will be available from the "search by embryo space" page, alongside the wholemount spatial query.

3.1.3 Search by Pattern Similarity

Another method of spatially searching the data in EMAGE is to use the "find similar" function initiated via the "xXx" button (Fig. 1b). This button is provided in the results table, as well as with the original image, and the mapped spatial representation, within the full EMAGE entries (*see* Subheading 3.2.2 for a full description of the contents of an EMAGE entry). Clicking on the "find similar" button initiates a spatial search in a similar way to that described previously. In this case, however, the mapped expression pattern of the entry is used as the query domain in the search, and patterns are compared for similarity across the entire area of the model rather than just a local region. This allows that the most similar patterns can be returned, in order of their degree of similarity.

1. Click on any of the "find similar" buttons associated with the chosen entry.

Fig. 1 Methods for Searching EMAGE. (**a**) *Search by space*: Spatial queries are carried out in EMAGE by defining a query domain on a standard embryo model. In the example shown here, the query domain (shown in *magenta*) is defined on the TS23 wholemount model and covers the liver region. The user-defined query domain is dilated by a pre-determined distance and then compared (by Jaccard index) across the local area of the query domain to the defined expression domains held in EMAGE. (**b**) *Search by pattern similarity*: It is possible to query EMAGE for similar expression patterns using an existing expression domain as a query pattern. In the example shown here, the spatially mapped pattern of expression of an existing EMAGE entry for Sox10 is used. The *red* and *yellow* regions (constituting the "detected domain" in this mapped pattern) would be combined and used as a query domain for a spatial search, with the similarity being measured across the whole of the model, rather than just a local area (as in the case of a standard spatial search). The results are returned, ranked for similarity to the original pattern. (**c**) *Search by anatomy term*: There is an option to query EMAGE using anatomy text terms selected from the EMAP anatomy ontology. Searching by anatomy term relies on the text annotations included in EMAGE alongside the spatial annotations, with any number of terms able to be included in a query. (**d**) *Search by gene*: Gene searches allow the user to query EMAGE for all data pertaining to any gene or list of genes. The gene(s) can be specified in a number of formats including gene name, gene symbol as well as a range of common gene identifiers, including "out-of-date" or "retired" symbols and names. The option to use wildcard symbols in the query allows data relating to a family of genes to be retrieved. (**e**) *Direct access to EMAGE*: EMAGE data can be accessed and interrogated by a number of direct access methods. The available methods are listed on the EMAGE website, and some information is provided to allow users to access the data freely. In all cases, however, assistance can be provided through direct contact with EMAGE

2. The results are returned in the standard tabular format of all EMAGE results (*see* Subheading 3.2), ordered by similarity to the original pattern (which as the most similar, will appear first in the table).

3.1.4 Search by Anatomy Term

1. Select "search by anatomy name" either by clicking on the button (Fig. 1c) on the front page or by selecting "By Anatomical Name" from the "Search" menu on any page.
2. You are now presented with a page entitled "Search for gene/protein in anatomical structure(s)" and a text box to input your query term(s).
3. As you begin to type your anatomical term in the text box, a list of suggestions will appear below the text entry box. This list is refined as you continue to type more letters.
4. Select the structure(s) to be included in the query by double clicking the required term(s) to add them to the selection list. Multiple structures can be selected using the control key. Repeat this process as necessary until the list of structures to be included in the query is complete.
5. Select from the dropdown menu at the bottom of the page whether the query is for genes detected, not detected, or possibly detected in the listed structures.
6. Click "Begin SEARCH."

The results are returned in the standard tabular format of all EMAGE results. *See* Subheading 3.2 for a full description of the EMAGE results format.

3.1.5 Search by Gene

1. Select "search by gene" either by clicking on the "Gene" button (Fig. 1d) on the front page or by selecting "By Gene/Protein" from the "Search" menu on any page.
2. You are now presented with a page entitled "Search using gene/protein symbol, name, ID."
3. Specify the gene(s)/protein(s) for which you wish to search for expression data. These can be specified either by name, symbol, or any of a comprehensive list of standard bioinformatics resource unique identifiers (UID) (*see* **Note 5** for list of UIDs recognized by EMAGE). Multiple genes/proteins can be included in a query by adding them as a comma separated list.

   For information on the use of wild cards and exact matches, *see* **Note 6**.
4. Specify the search requirements: whether to search for where expression is detected, not detected, or possibly detected; and the age range over which the search is to be carried out.
5. Click "Begin SEARCH."

6. The results are returned in the standard tabular format. *See* Subheading 3.2 for a full description of the EMAGE results format.

3.2 Results Format

3.2.1 Results Table

Results in EMAGE are returned in a standard tabular format. Each row in the table refers to an individual entry or assay in EMAGE, and the columns relate to selected fields in the entry. The type of query that was carried out to generate the results dictates the columns visible by default in the table. However, columns can be added or removed to suit the user's preferences using the "add/remove columns" button above the table. Below each column title is an information link, signified by a white question mark in a blue circle. Clicking on this link takes the user to a page describing the contents of that column. Column titles that appear as hyperlinks can be used to order the returns based on the values in that column, e.g., Theiler Stage or Pattern Clarity.

The first column in the table is always the "select" column. Rather than a data value, this column contains a checkbox which allows the user to select one or more entries to be saved to a clipboard, using the "Add to Submissions clipboard" link at the bottom left hand corner of the results table page. If required, the user can continue to perform queries (of any type), each time selecting those entries of interest from the results tables to be added to the clipboard. This user-specified list of entries in the clipboard can then be viewed (as a results table) using the "View clipboard" link at the bottom left hand corner of the page.

3.2.2 Expanded EMAGE Entry

The column entitled "ID" contains the EMAGE ID for each entry in the table. Clicking on this link will open up the full EMAGE entry. The layout of an EMAGE entry is based on the MISFISHIE specifications [8] with the various components of the entry arranged according to the following subheadings: Data Images, Detection Reagent, Specimen, Expression Pattern Description, Procedures, and General Information.

Data Images

There will always be an original data image for every entry. Where our copyright permissions permit, this original data image will be displayed along with any other relevant data images from the publication (e.g., alternate views of the specimen), clicking on such a thumbnail image within the entry will open up the full-size image. In some cases, copyright legislation prevents the original image from being displayed within EMAGE; however, in all cases where the data has come from the published literature there will be a link to the original publication containing the image.

Also displayed within this section is the "Expression pattern clarity" star rating relating to the mapped image (*see* Subheading 1.5 for a description) and the "find similar" button. This button allows the mapped pattern for the entry to be used as a query domain for a spatial search, thus returning the most similar expression patterns in the database (*see* Subheading 3.1.3 for a description).

Detection Reagent

Within this section, the information relating to the in situ probe, reporter construct, or immunohistochemistry antibody is described in as much detail as is available. Where possible the sequence of the probe, insert, or antigen is given. As further data types are added to EMAGE, this section will be adapted as required, to reflect the information relating to the new detection reagents.

Specimen

Embryo developmental timing is affected by strain differences, and so when it is known, the strain is always reported in EMAGE along with the embryo given age, Theiler stage, and genotype information (in the case of transgenic reporters).

Currently, EMAGE only contains data relating to wild-type embryos. There are inherent problems with the spatial annotation of phenotypically mutant data due to differences in morphology between the data embryo and the model. Heterozygote expression patterns included in EMAGE have been defined by the authors as being representative of wild-type patterns.

Expression Pattern Description

Within this section there are potentially two descriptions of the expression pattern; a text annotation and a spatial annotation.

The text annotation utilizes the EMAP anatomy ontology, which is a comprehensive description of all the visible anatomical components existing at each Theiler stage, arranged as a "part-of" hierarchy [13]. The text annotation is compiled from the anatomical terms used by the author to describe the regions of expression, relative levels of expression, and pattern of expression of the gene in question. For those models that contain painted anatomy domains, it is also possible to generate automatic text annotation by determining which anatomy domains are overlapped by the spatial annotation.

In contrast, the spatial annotation is generated by the curators at EMAGE. The method by which the spatial annotation image is produced is described in Subheading 1.4. Clicking on the thumbnail image of the spatial annotation within an entry, opens up a full-size image of the annotation. Below the spatial annotation image is a key, indicating the colors used to denote relative levels of expression: red = strongest, yellow = moderate, blue = weakest, green = possible, cyan = not detected. There is also another "find similar" button, and the star rating describing the morphological match to model including, where appropriate, information on who approved the spatial mapping (either an EMAGE editor or the submitter).

Procedures

Further information on experimental procedures and conditions are described in this subsection, specifically relating to fixation of the specimen, and the staining procedure used to visualize the expression pattern.

General Information

In this section, details are provided of who submitted the data, who the authors are (if published), whether the experiment was part of a screen or not, as well as links to articles referred to in the entry. There are also a number of links to related databases, based either on the gene being assayed or the assay itself.

3.3 Direct Access

There are a number of ways in which the data in EMAGE can be accessed and queried directly (without the use of the user interface). These methods are available to allow bioinformations, programmers, or other interested parties, access to the data using software, in order to carry out analysis, integration with other datasets or to perform other actions not provided by the web interface.

The direct access (Fig. 1e) options available are via: Standard Query Language (SQL), Webservices, Distributed Annotation System (DAS), Formatted Uniform Resource Locator (URL), and the Java Application Programming Interface (API). In all cases (even those not requiring registration), users are encouraged to contact EMAGE staff to discuss their direct access requirements, as assistance can be provided to determine the best method to use to carry out any particular analyses. A description of the functionalities available through each of the direct access methods is given below.

3.3.1 SQL

EMAGE consists of two separate databases: the mouse anatomy ontology atlas database and the EMAGE mouse spatial gene expression database. The database schema and schema design document for each constituent database are provided on the EMAGE website. Users are required to register (using the link provided on the website SQL page) to receive a login password and access details, allowing them to query both these databases. Registration also allows assistance with querying EMAGE, as well as up-to-date documentation on the schema and content of the EMAGE databases.

3.3.2 Webservices

EMAGE provides WSDL (webservices description language) described webservices, to allow users to access the database via client software requests. The webservices are delivered by apache axis and are described on the EMAGE website through WSDL service descriptions and XML (Extensible Markup Language) documentation. The webservices provided are similar in function to the structured URL requests (*see* Subheading 3.3.4) but represent a limited subset of these requests. Users wishing to utilize the EMAGE webservices are encouraged to contact EMAGE for help and assistance and to provide feedback on the webservices provided.

A full list of the EMAGE functions currently available via webservices is provided on the EMAGE website (http://www.emouseatlas.org/axis/services).

3.3.3 DAS

EMAGE provides a Distributed Annotation Server (DAS) [14] based on Dazzle [15]. DAS defines a communication protocol used to exchange annotations on genomic or protein sequences.

In EMAGE, DAS queries are restricted to gene annotations defined by either gene symbol or Ensembl ID.

The EMAGE DAS service can be accessed at http://www.emouseatlas.org/das/emage, and a typical query would take the form of a URL, for example, http://www.emouseatlas.org/das/emagefeatures?segment=Fgf8. Results are returned as XML, formatted according to the DAS specification as described at http://www.biodas.org/documents/spec-1.53.html.

3.3.4 Formatted URL

It is possible for users to invoke an EMAGE query using a structured URL in a web browser, with the results returned as a web page. Gene/anatomy queries, spatial similarity queries, and EMAGE ID queries can all be performed by this method.

Gene/Anatomy Queries

The names/attributes available to construct a gene/anatomy query are, **all**, **genes**, **goterms**, **structures**, **stages**, **strengths**, **exactmatchgenes**, **includegenesynonyms**, **exactmatchgoterms**, **includegotermsynonyms**, **exactmatchstructures**, and **includestructuresynonyms**.

The URL is constructed using name and value pairs, concatenated using the symbol "&."

The attributes **exactmatchgenes**, **includegenesynonyms**, **exactmatchgoterms**, **includegotermsynonyms**, **exactmatchstructures**, and **includestructuresynonyms** may only have the value of "true." All other terms (except **all**) may have one or more values. The values must be comma separated and specified in the correct format as described here.

- **genes**: current MGI gene symbols.
- **goterms**: gene ontology terms from the GO project.
- **structures**: EMAP anatomy ontology terms or IDs.
- **stages**: whole numbers between 1 and 28 with or without the prefix "ts" or "TS".
- **strengths**: either "detected," "possible," or "not detected".

In the case of attributes **exactmatchgenes**, **excatmatchgoterms**, and **exactmatchstructures**, where the value of "true" is NOT applied, and implicit wild card is assumed in the search.

Where an attribute is not specified in the URL, all values of that attribute will be included in the search. The exception to this rule is the attribute **strength**. Where **strength** is not included in the URL, the default value of "detected" is assumed.

Thus, a general form of the URL for a gene/anatomy query is as follows:http://www.emouseatlas.org/emagewebapp/pages/

emage_general_query_result.jsf?genes=a,b&structures=c,d&stages=e,f&strengths=detected,possible&exactmatchgenes=true&includegenesynonyms=true&exactmatchstructures=true&includestructuresynonyms=true

which returns all entries for genes a and b (and their synonyms), expressed in structures c and d (and their synonyms), at stages e and f, with expression levels of detected and possible.

Where **all** is used as an attribute, it has no values and simply returns all possible results.

A search for all possible EMAGE results would take the form: http://www.emouseatlas.org/emagewebapp/pages/emage_general_query_result.jsf?all

Spatial Similarity Queries

The attributes available to construct a spatial similarity query are **spatialQueryID**, **spatialQueryStage**, and **spatialQueryWholemount**. Each term must be specified for every query and be given only one value. The correct format of the values is described here.

- **spatialQueryID**: EMAGE ID number with or without the prefix "EMAGE:"
- **spatialQueryStage**: whole numbers between 1 and 28 with or without the prefix "ts" or "TS"
- **spatialQueryWholemount**: either "true" or "false"

Thus, a general form of the URL for a spatial similarity query is as follows:http://www.emouseatlas.org/emagewebapp/pages/emage_spatial_query_result.jsf?spatialQueryID=a&spatialQueryStage=b&spatialQueryWholemount=true

which returns all wholemount entries (at stage b) with a spatially similar expression pattern to entry a (ranked in order of spatial similarity).

EMAGE ID Queries

The attributes available to construct an EMAGE ID query are simply, **emageIDs** and **all**. The values for **emageIDs**, which must be comma separated, are the EMAGE ID numbers with or without the prefix "EMAGE:".

Thus, a general form of the URL to carry out an EMAGE ID query is as follows:http://www.emouseatlas.org/emagewebapp/pages/emage_general_query_result.jsf?emageIDs=a,b,c,d

which returns entries a, b, c, and d.

Where **all** is used as an attribute it has no values and simply returns all possible results. The query is in effect the same as that described for a Gene/Anatomy query for **all**. As such, the format of the URL is exactly the same.

A search for all possible EMAGE results would take the form: http://www.emouseatlas.org/emagewebapp/pages/emage_general_query_result.jsf?all

3.3.5 Java API

The most comprehensive direct access to EMAGE is via the Java API.

The Java API defines an expansive set of classes and methods that allow programmable querying of the EMAGE database and customized formatting of result sets. Methods are available for all possible query types, and access can be made to the full set of data results from a query.

Users must first register to be able to use the Java API, after which they will receive instructions on how to download the EMAGE RMI (Remote Method Invocation) driver. It is also necessary to download a collection of jar files from hibernate, and the various EMAGE data classes (instructions on all aspects of setting up and using the Java API can be found either on the website or through contact with EMAGE support). It is then possible to use Java to carry out complex queries on the contents of the EMAGE database.

3.4 Data Analysis

3.4.1 Cluster Analysis of Data (Data Mining for Similar Patterns)

By virtue of the spatial annotation process, it is possible to cluster the data held within EMAGE by spatial expression pattern alone. Such spatial clustering allows for the detection of syn-expression groups without any knowledge of the specific gene functions or gene pathway activities. This also allows genes of unknown function to be clustered alongside very well characterized genes.

The data in EMAGE are clustered using a Jaccard Index and hierarchical clustering. Through the spatial annotation of the expression patterns in EMAGE, a set of expression domains are produced for each entry, mapped onto the standard framework of the EMAGE models. These expression domains can be measured for similarity using the Jaccard Index, defined as the ratio of the shared features of two expression patterns to the total number of features of the two patterns. This index provides a numerical value (V) to describe the level of similarity between two patterns. This comparison is carried out for every pair of patterns (mapped to the same model) in the database, providing a signature of similarity values for each entry. These entries can then be clustered based on the similarity of the signatures, resulting in similar expression profiles being clustered together. This cluster analysis is carried out using a version of the Cluster3.0 program [16] adapted to suit the EMAGE data, using an un-centered correlation similarity metric followed by complete linkage clustering. The output files of this program (.cdt, .gtr, and .atr files) can be visualized via the EMAGE website using an interactive tree viewer. The viewer allows the original images, as well as heat maps of the areas of expression, to be viewed in the context of the tree format. (It is worth noting that the output files can also be read into the original version of the Cluster3.0 program and viewed as a tree. However this does not include the original images and so allows little in the way of feedback on the original expression patterns. It is also possible to view the clustered patterns as an ordered list—rather than as a tree—including the original images.)

While there are instructions on the EMAGE website to carry out the cluster analysis described here, it is currently only possible to do so on a static data set, which does not represent the full data available in EMAGE. Therefore, anyone wishing to carry out a spatial cluster analysis on the EMAGE data should contact the editorial office directly (emage@emouseatlas.org) for assistance.

4 Notes

1. Originally, EMAGE and EMAP were acronyms for "Edinburgh Mouse Atlas of Gene Expression" and "Edinburgh Mouse Atlas Project," respectively. Given the nature of the projects, over time it has become more explanatory to replace Edinburgh with Electronic.
2. The full text of Karl Theiler's book is available as a pdf document from www.emouseatlas.org.
3. As default, regardless of the model view selected, the search will recover data from both sides of the embryo to increase the range of data returned. However, this option can be turned off (by deselecting the box) to allow asymmetrical expression patterns to be queried accurately.
4. For the mid-gestation stages (TS15-TS19) where the morphological differences between stages are relatively minor, a "stage-to-stage" warp has been defined from each wholemount model to the next (in time). This allows a query region defined on one particular model to be applied to the surrounding models, and so increase the volume of data that can be retrieved. The stage range available for each query depends on the model chosen to define the query region. Users can then choose the number of stages above and below the initial model to include in the search.
5. The list of recognized gene IDs is as follows:
   - MGI gene ID
   - Official mouse gene Symbol
   - Official mouse gene Name
   - Entrez mouse gene ID
   - Ensembl mouse gene ID
   - VEGA mouse gene ID
   - UniGene mouse gene cluster ID
   - RefSeq mouse gene ID
   - Genbank sequence ID

- SwissProt mouse protein ID
- Mouse gene name/symbol synonym
- Human Entrez gene ID
- Official human gene Symbol
- Official human gene Name
- RefSeq human ID
- Human gene name/symbol synonyms

6. The default settings are that the "include synonyms, common IDs and orthologues" option is checked, but "exact match" is not. Wild cards (*) can be used in the search term, and are implied at the end of the search term unless "exact match" is checked. For example, the search term "Wnt" will be read as "Wnt*" and will return data for any Wnt gene, e.g., Wnt1 Wnt2, Wnt11, etc. With the default settings activated, this means that the search term "Wnt" will also return genes whose synonym starts with "Wnt," for example Wifl (Wnt inhibitory factor 1). Both options, however, are fully user definable and it is possible to exclude synonym searches and/or to search for only an exact match. This is useful in the case of genes whose symbol is short and may be a common subset of other symbols, for example the gene Brachyury, whose official mouse gene symbol is "T."

Acknowledgments

The authors would like to gratefully acknowledge Dr. Martin Ringwald and all the curators at the GXD for a long-standing collaboration and the provision of the textual annotations used in EMAGE. We would also like to acknowledge all the staff members who have contributed to EMAP and EMAGE over the years, as well as the innumerous data providers. The work described in this chapter is funded by the Medical Research Council, UK.

References

1. Kaufman MH (1992) The atlas of mouse development. Academic Press, San Diego
2. Sharpe J et al (2002) Optical projection tomography as a tool for 3D microscopy and gene expression studies. Science 296: 541–545
3. http://www.informatics.jax.org/
4. http://www.eurexpress.org/ee/
5. Gray PA et al (2004) Mouse brain organization revealed through direct genome-scale TF expression analysis. Science 306:2255–2257
6. Vendrell V et al (2009) Gene expression analysis of canonical Wnt pathway transcriptional regulators during early morphogenesis of the facial region in the mouse embryo. Gene Expr Patterns 9:296–305
7. http://www.emouseatlas.org/emage/facebase/
8. Deutsch EW et al (2008) Minimum information specification for in situ hybridization and immunohistochemistry experiments (MISFISHIE). Nat Biotechnol 26:305–312

9. http://www.emouseatlas.org/emap/analysis_tools_resources/software/eMAP-apps.html
10. Theiler K (1989) The house mouse atlas of embryonic development. Springer, New York
11. http://www.emouseatlas.org/emage/about/data_annotation_methods.html#auto_eurexpress
12. http://www.emouseatlas.org/emage/info/lossst_method.html
13. Bard JL et al (1998) An internet-accessible database of mouse developmental anatomy based on a systematic nomenclature. Mech Dev 74:111–120
14. http://www.biodas.org
15. http://www.biojava.org/wiki/Dazzle
16. http://bonsai.hgc.jp/~mdehoon/software/cluster/software.htm#ctv

Chapter 6

Real-Time PCR Quantification of Gene Expression in Embryonic Mouse Tissue

Eric Villalon, David J. Schulz, and Samuel T. Waters

Abstract

The Gbx family of transcription factors consists of two closely related proteins GBX1 and GBX2. A defining feature of the GBX family is a highly conserved 60 amino acid DNA-binding domain, which differs by just two amino acids. *Gbx1* and *Gbx2* are co-expressed in several areas of the developing central nervous system including the forebrain, anterior hindbrain, and spinal cord, suggesting the potential for genetic redundancy. However, there is a spatiotemporal difference in expression of *Gbx1* and *Gbx2* in the forebrain and spinal cord. *Gbx2* has been shown to play a critical role in positioning the midbrain/hindbrain boundary and developing anterior hindbrain, whereas gene-targeting experiments in mice have revealed an essential function for *Gbx1* in the spinal cord for normal locomotion. To determine if *Gbx2* could potentially compensate for a loss of *Gbx1* in the developing spinal cord, we performed real-time PCR to examine levels of *Gbx2* expression in *Gbx1*$^{-/-}$ spinal cord at embryonic day (E) 13.5, a developmental stage when *Gbx2* is rapidly downregulated. We demonstrate that *Gbx2* expression is elevated in the spinal cord of *Gbx1*$^{-/-}$ embryos.

Key words Spinal cord, *Gbx2*, *Gbx1*, Real-time PCR, Mouse

1 Introduction

The selective expression of transcription factors during mouse embryogenesis provides temporal and spatial regulation of gene expression, which influences lineage specification, patterning and morphogenetic movement of cells and tissues. Interestingly, many DNA-binding transcription factors that have overlapping expression domains have been shown to functionally compensate for family members having reduced levels of expression due to hypomorphic or inactivation mutations [1–3]. Increases in gene expression levels as detected by in situ hybridization, which detects specific mRNA expression and polymerase chain reaction (PCR), which provides a quick and simple method to amplify and analyze specific nucleic acid sequences, are often correlated with the ability of a gene to functionally compensation for its mutant family member.

Mark Lewandoski (ed.), *Mouse Molecular Embryology*, Methods in Molecular Biology, vol. 1092,
DOI 10.1007/978-1-60327-292-6_6, © Springer Science+Business Media New York 2014

However, analysis of the end products of in situ hybridization and conventional PCR is limited by only allowing qualitative estimates of the relative starting concentration of the specific nucleic acid templates examined. Therefore, potential threshold requirements are unattainable by use of these methods. The advent of instrumentation providing real-time detection of PCR amplicons provides a reliable method to determine mRNA and gene copy numbers in a spatiotemporal manner during embryonic development relative to the starting concentrations of the template amplified [4].

Gbx genes encode evolutionarily conserved homeodomain-containing transcription factors that share extensive sequence similarity [5–7], are expressed in the developing neural tube, and have overlapping expression domains in the developing medial ganglionic eminence (forebrain), anterior hindbrain, and spinal cord. However temporal expression in these regions differ significantly [8]. At E 10.5, *Gbx2* expression in the spinal cord is restricted to the dorsal ventricular and mantle zones and two ventral lateral stripes. This expression persists through E 12.5. However, expression of *Gbx2* in the spinal cord is rapidly downregulated after E 12.5. In contrast, *Gbx1* is expressed throughout the dorsal and ventral ventricular zone at E 10.5 and becomes restricted to the dorsal mantle zones at E 12.5. Unlike *Gbx2*, *Gbx1* remains strongly expressed in the dorsal spinal cord through E18.5 [8, 9].

Despite partially overlapping expression domains, mutations for mouse *Gbx1* and *Gbx2* have very different consequences [10–14]. To determine whether the loss of *Gbx1* expression impacts the expression of *Gbx2* in the spinal cord, which may lead to functional compensation, we examined *Gbx2* expression in $Gbx1^{-/-}$ spinal cord. Our data shows that *Gbx2* expression persists in the spinal cord of $Gbx1^{-/-}$ embryos at E13.5, a developmental stage when *Gbx2* expression, as indicated by in situ hybridization, is considerably downregulated.

2 Materials

2.1 Real-Time Polymerase Chain Reaction (PCR) System

To generate the data used for the experiments shown in this chapter, quantitative real-time PCR was performed with an ABI 7300 Real-Time PCR system (Applied Biosystems). The results were analyzed using the 7300 System SDS software package.

2.2 PCR Reactions

1. TaqMan Universal PCR Master Mix (2×) (Applied Biosystems).
2. Power SYBR Green PCR Master Mix (Applied Biosystems).
3. Forward PCR primer.
4. Reverse PCR primer.
5. TaqMan probe (Applied Biosystems).

6. cDNA sample.
7. Nuclease-free water.
8. MicroAmp optical 96-well reaction plates (Applied Biosystems).
9. MicroAmp optical 8-tube strip (0.2 ml) (Applied Biosystems).
10. MicroAmp optical 8-cap strip (Applied Biosystems)

2.3 Tissue Collection

Mouse embryos were collected at embryonic day (E) 13.5.

1. Pregnant female mice sacrificed (13.5 days p.c.).
2. Fine scissors.
3. Two pair of fine forceps.
4. Stereoscopic dissecting microscope (Leica, Wetzlar, Germany).
5. 10× PBS.
6. Diethylpyrocarbonate (DEPC)-treated water.
7. Petri dishes.
8. Transfer pipets.
9. Cryotubes.
10. Ethanol.
11. Dry ice.

2.4 RNA Isolation

1. RNeasy (Qiagen).
2. RNase-free water.
3. 10× buffer RQ1 (Promega).
4. RQ1 RNase^{-} DNase (Promega).
5. Heated water bath.
6. 5 M ammonium acetate.
7. RNase inhibitor.
8. Microcentrifuge.

2.5 Reverse Transcription (cDNA Synthesis)

1. SuperScript III Platinum Two-Step qRT-PCR kit (Invitrogen).
2. *E. coli* RNase H.
3. Heated water bath.

3 Methods

Real-time quantitative RT-PCR overlaps extensively with the process of simple reverse transcription PCR (RT-PCR), with the simple additions of detection chemistry and real-time fluorescence data acquisition. Therefore, we will not go into all of the details of RT-PCR in this chapter. Rather, the individual investigator is encouraged to employ whatever standard techniques for tissue collection, RNA isolation, and reverse transcription are effective for their particular system in order to generate template cDNA for real-time quantitation. Subsequent to cDNA synthesis, template cDNA is added to PCR reaction mix containing the detection chemistry, and run with whatever fluorescence detection platform is available. In our experiments we will provide protocols that are standard in our system, and data are acquired using the ABI 7300 Real-Time PCR System from Applied Biosystems, including the offline data analysis software package provided by ABI.

3.1 Embryonic Spinal Cord Isolation

1. Timed-pregnant mice are sacrificed on postcoitum day 13.5 by CO_2 narcosis. Note that euthanasia procedures must be approved by your Institutional Animal Care and Use Committee.
2. To harvest the embryos, remove the uterus by making an incision into and opening the abdominal cavity. Remove the uterus by cutting across the cervix and the utero-tubal junctions. Rinse the blood off by placing the uterus in a petri dish containing 1× PBS/DEPC. Under the stereoscopic dissecting microscope, remove the muscle layer of the uterus, visceral yolk sac, and amnion.
3. Place the embryo into a fresh petri dish containing 1× PBS/DEPC. Under the stereoscopic dissecting microscope, remove the head, viscera, and limbs. Using fine scissors or forceps, trim and remove the remaining tissue surrounding the spinal cord.
4. Using a transfer pipette, place the spinal cord in a cryotube. Remove all extra 1× PBS/DEPC.
5. Snap freeze in a dry ice/ethanol bath and store at −80 °C for RNA preparation.

3.2 RNA Isolation

1. RNA was isolated using RNeasy (Qiagen) and eluted in 30 μl RNase-free water.
2. Bring the volume of each eluted RNA sample to 87 μl with DEPC-treated water.
3. Add 10 μl of 10× buffer RQ1 and 3 μl RQ1 $RNase^-$ DNase to each sample and incubate at 37 °C for 15 min.
4. Ethanol precipitate each RNA sample by adding 67 μl of 5 M ammonium acetate, 300 μl cold ethanol and incubating at −20 °C for 20 min.
5. Centrifuge the samples at 13,000 rpm at 4 °C for 10 min.
6. Wash the pellet with cold 70 % ethanol in DEPC-treated water.
7. Resuspend the pellet in 30 μl of DEPC water containing 0.5 μl RNase inhibitor (Applied Biosystems).

3.3 Reverse Transcription

Reverse transcription reactions were performed using SuperScript III Platinum Two-Step qRT-PCR kit (Invitrogen). A master mix without DNA may be prepared.

For each reaction tube add on ice:

1. 10 μl 2× RT Reaction Mix.
2. 2 μl Enzyme Mix.
3. 10 pg to 1 mg RNA.
4. DEPC-treated water for a final volume of 20 μl.
5. Incubate samples at room temperature for 10 min.
6. Incubate samples at 42 °C for 50 min.
7. To stop the reactions, incubate the samples at 85 °C for 5 min, then place on ice.
8. Add 1 μl of *E. coli* RNase H and incubate at 37 °C for 20 min.

3.4 Set Up of Absolute Quantification PCR Reactions

For absolute quantification, PCR reactions were performed in triplicate. The TaqMan probes (*see* **Note 1**) were labeled with 6-carboxyfluorescein (FAM) and quencher dye 6-carboxytetrametnylrhodamine (TAMARA) on its 5′ and 3′ ends, respectively. A master mix without DNA may be prepared.

3.5 Gbx2: Gene of Interest

The PCR primers (*see* **Note 2**) and TaqMan probe (*see* **Note 2**) used to measure the levels of *Gbx2* were designed to target the splice junction between *Gbx2* exons 1 and 2 (*see* **Note 3**).

For each reaction tube add:

1. 25 μl TaqMan Universal PCR Master Mix (2×) (1× final concentration).
2. 5 μl Forward PCR primer 5′-GGCAACTTCGACAAAGCCGAGG-3′ (50–900 nM final concentration).
3. 5 μl Reverse PCR primer 5′-CCAGGCAAATTGTCATCTGAGC-3′ (50–900 nM final concentration).
4. 5 μl TaqMan probe 5′-CAGGCGTCGCTCGTCGGGGCT-3′ (50–250 nM final concentration).
5. 5 μl DNA sample (10–100 ng final concentration).
6. 5 μl Nuclease-free water.

3.6 Gapdh-Control Gene

A 105-bp fragment of glyceraldehyde-3-phosphate dehydrogenase (*Gapdh*) was targeted for normalization of the assay.

For each reaction tube add:

1. 25 μl TaqMan Universal PCR Master Mix (2×) (1× final concentration).
2. 5 μl Forward PCR primer 5′-GCATGGCCTTCCGTGTTCCTA-3′ (50–900 nM final concentration).
3. 5 μl Reverse PCR primer 5′-CTGCTTCACCACCTTCTTGA-3′ (50–900 nM final concentration).
4. 5 μl TaqMan probe 5′-ACGTGCCGCCTGGAGAAACCTG-3′ (50–250 nM final concentration).

5. 5 μl DNA sample (10–100 ng final concentration).
6. 5 μl Nuclease-free water.

3.7 Set Up of Relative Quantification PCR Reactions

1. 25 μl Power SYBR Green PCR Master Mix.
2. 5 μl Forward PCR primer 5′-GGCAACTTCGACAAAGCCG AGG-3′ (50–900 nM final concentration).
3. 5 μl Reverse PCR primer 5′-CCAGGCAAATTGTCATCTG AGC-3′ (50–900 nM final concentration).
4. 5 μl DNA sample (10–100 ng final concentration).
5. 10 μl Nuclease-free water.

3.8 Gapdh-Control Gene

For each reaction tube add:

1. 25 μl TaqMan Universal PCR Master Mix (2×) (1× final concentration).
2. 5 μl Forward PCR primer 5′-GCATGGCCTTCCGTGTTC CTA-3′ (50–900 nM final concentration).
3. 5 μl Reverse PCR primer 5′-CTGCTTCACCACCTTCTTGA-3′ (50–900 nM final concentration).
4. 5 μl DNA sample (10–100 ng final concentration).
5. 10 μl Nuclease-free water.

3.9 PCR Cycling Conditions for Two-Step RT-PCR

1. 2 min at 50 °C.
2. 10 min at 95 °C.
3. 40 cycles of 95 °C for 15 s, 58.4 °C for 1 min.

3.10 Standard Curve Generation

For absolute quantification PCR reactions, unknown samples are compared with standard curves (*see* **Note 4**).

1. To generate the standard curves used for the experiments shown in this chapter, samples of plasmid DNA containing cDNA encoding either *Gbx2* or *Gapdh* were serially diluted from 10^7 to 10^2 copies. An example of the results produced is shown in Fig. 1.
2. After performing the PCR it is important to choose an accurate threshold for the fluorescence emission baseline of the reporter dye (FAM). The threshold should be set just above the background noise (–Rn) for maximal accuracy in threshold cycle (C_T) determination, which is inversely proportional to the original number of copies of target template. An example of the results produced is shown in Fig. 2.

3.11 Primer Validation

1. To validate the primers (*see* **Note 5**) used for the experiments shown in this chapter, the (C_T) slope method was used to determine the target amplification efficiency (E_X):

$$E_X = 10^{(-1/slope)} \quad 1 \times 100$$

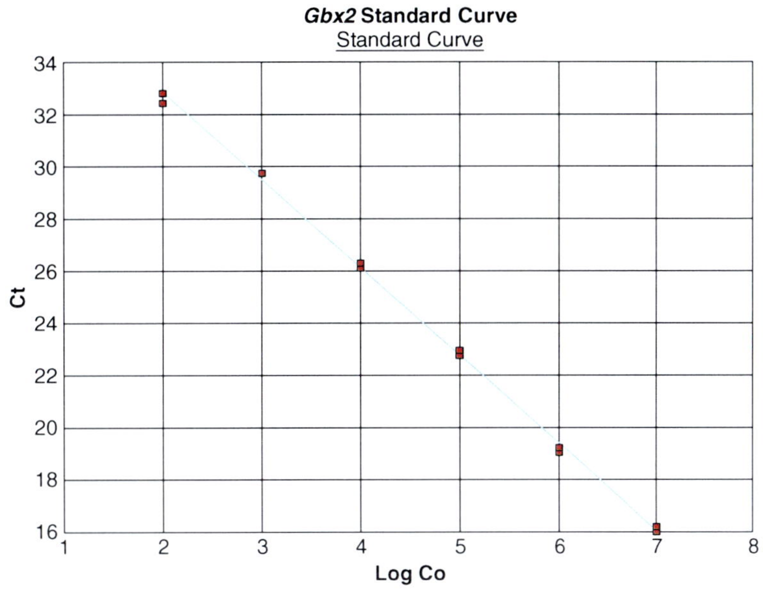

Fig. 1 The *Gbx2* target amplification standard curve with a 6 log dynamic range of template (Log C0) using an FAM-labeled probe. Ct is the cycle threshold, which corresponds to the number of PCR cycles required to reach a fixed amount of fluorescence (as set by the experimenter)

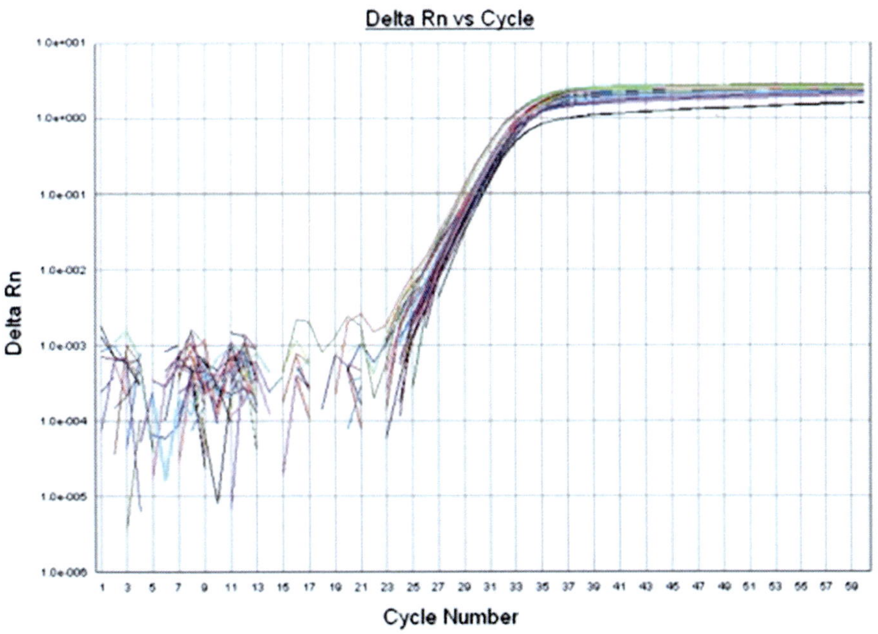

Fig. 2 Amplification plot of *Gbx2*. The emission intensity of the reporter dye (FAM), which is normalized against the emission intensity of the quencher dye (TAMRA) is used to determine the number of transcripts and is calculated by the instrument software using the equation DRn = (Rn+) – (Rn–). DRn represents the amount of cleaved probe. (Rn+) is the emission intensity of the reporter divided by the emission intensity of the quencher during amplification. (Rn–) is the emission intensity of the reporter divided by the emission intensity of the quencher prior to amplification

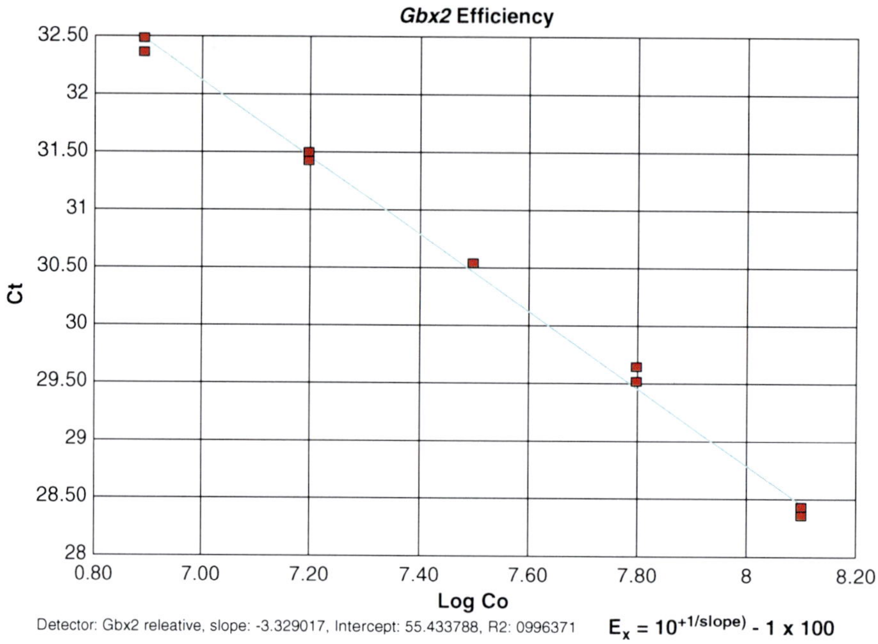

Fig. 3 Measured efficiency using the CT slope method with 5 data points generated from a serial dilution of embryonic spinal cord cDNA. The calculated efficiency for this assay is 99.7 %

2. Twofold dilutions were made of cDNA synthesized from total RNA preparations.
3. Perform amplification of the target template using the Set Up of Relative Quantification PCR reaction (Subheading 3.7).
4. Construct a plot of C_T versus log cDNA concentration. An example of the results produced is shown in Fig. 3.

3.12 Analysis of Real-Time Fluorescence Data for Absolute Quantitation

The gene copy number of *Gbx2* was determined in spinal cord of normal and *Gbx1*$^{-/-}$ embryos.

1. To control for sample variation, separate reactions were performed for *Gbx2* and *Gapdh* and standard curves prepared for both (*see* **Note 4**). In this way, *Gbx2* values were normalized to the value of the corresponding endogenous *Gapdh*. (Average number of *Gbx2* transcripts divided by the average number of *Gapdh* transcripts) for each experimental sample. An example of the results produced is shown in Fig. 4.

3.13 Analysis of Real-Time Fluorescence Data for Relative Quantitation

1. The comparative method (*see* **Note 6**) was used to determine the quantity of *Gbx2* in *Gbx1*$^{-/-}$ embryos relative to normal embryos using the formula: 2^{-DDCT}.
2. It is important that the efficiencies of the target (*Gbx2*) and the reference (*Gapdh*), as indicated by validation experiments (Subheading 3.11), are similar (Table 1).

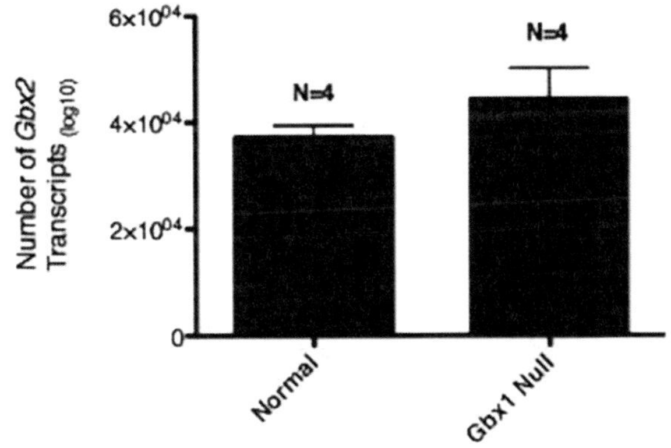

Fig. 4 Level of *Gbx2* transcripts after normalization to endogenous *Gapdh*

Table 1
Average C_T value for *Gbx2* and *Gapdh* in experimental samples

Sample $N=4$	*Gbx2* average C_T	*Gapdh* average C_T	DCT *Gbx2*—*Gapdh*	DDCT DCT—DCT, normal	*Gbx2* Rel. to normal
Normal	28.41 ± 0.58	22.13 ± 0.22	6.28 ± 0.62	0.00 ± 0.62	1.0
Gbx1$^{-/-}$	28.78 ± 0.35	22.26 ± 0.17	6.52 ± 0.38	+0.24 ± 0.38	0.84

3.14 Melt Curve Analysis to Ensure Quality of PCR Amplification (SYBR Green Only)

Since SYBR Green will detect any double-stranded DNA including primer dimmers, contaminating DNA, and PCR product from misannealed primers, it is important to run a dissociation curve following the real-time PCR. By viewing a dissociation curve, you ensure that the desired amplicon was detected and confirm the specificity of the chosen primers. A typical derivative plot of the melting curve for *Gapdh* is shown in Fig. 5. It is apparent that the melting temperature of the amplicon occurs at approximately 82 °C, with little or no contaminating products being produced in the reaction. A typical derivative plot of the melting curve for *Gbx2* is shown in Fig. 6. In contrast to the melt curve for *Gapdh*, two significant peaks are formed. The initial peak is likely representative of primer dimmers, which usually melt at lower temperatures because of their small size. The presence of primer–dimers can reduce the amplification efficiency and accuracy of the data. With a probe-based detection method, specificity of quantitation is still maintained, as long as the second product does not influence the efficiency of the PCR amplification. However, with SYBR green, quantitation would be unreliable, as signal is generated by both amplification products.

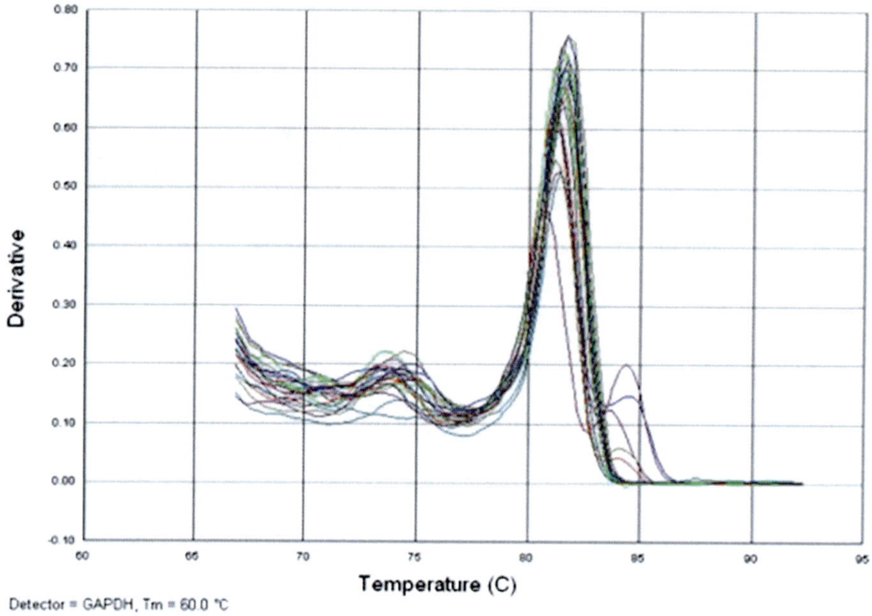

Fig. 5 Melt curve performed with SYBR green as the fluorescent detector molecule allows a post hoc analysis of the specificity of PCR amplification. Change in fluorescence is plotted over temperature to determine at what point denaturation of PCR amplicons occurs. A single peak indicates that a single PCR product was amplified in this reaction

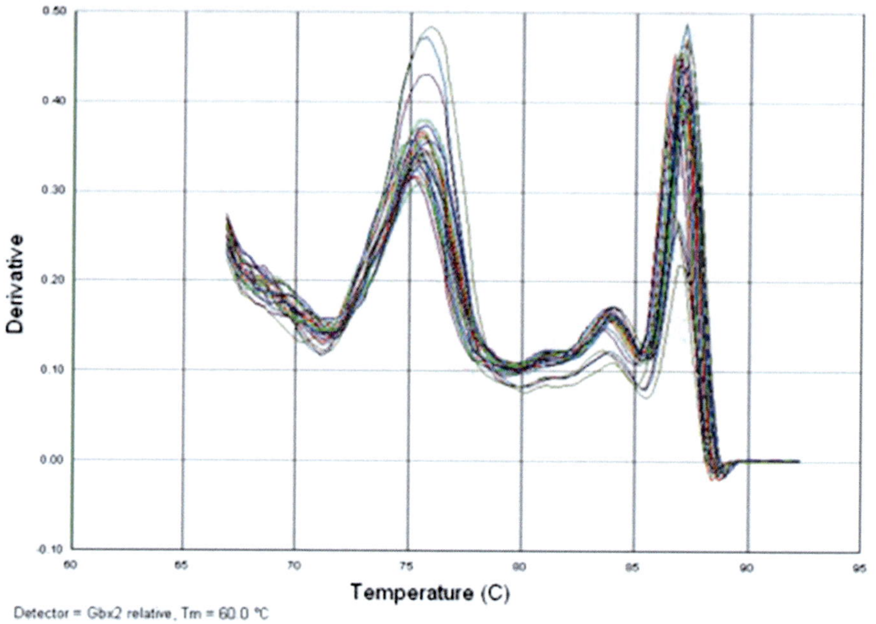

Fig. 6 A melt curve with multiple peaks indicates that multiple PCR products were generated in this reaction. However, because this is an end-point analysis, no quantitative conclusions can be made regarding the abundance of a given peak. In this situation, independent validation of PCR efficiency under experimental conditions is required to determine whether nonspecific products interfere with the quantitation of the product of interest (*see* Fig. 3)

4 Notes

1. Detection chemistry for fluorescent detection of amplified PCR products.

   Detection chemistry is highly variable. In this experiment, we employed a combination of hydrolysis probes and SYBR Green detection chemistries. Hydrolysis probes, or so-called Taqman probes, work via an interaction of fluorophore and quencher molecules covalently linked to the probe. The probe hybridizes to the PCR amplicon during the annealing phase of the PCR, and during the polymerization step the fluorophore and quencher bonds are hydrolyzed and they are released from the probe, leading to a cessation of fluorescence quenching and ultimately a detectable fluorescence signal.

   In contrast to hydrolysis probes, SYBR green is a nonspecific fluorescent dye that preferentially binds to any double-stranded DNA molecule. Binding of double-stranded DNA greatly increases the fluorescence of SYBR green, thus making it suitable as a marker for double-stranded DNA content (i.e., PCR amplicons) in a given cycle of a reaction. SYBR green has the advantages of being able to monitor the amplification of any double-stranded DNA product, thus minimizing assay optimization, and reducing costs. In addition, SYBR green incorporation is proportion to the length of the DNA amplicon, thus assay sensitivity can be increased simply by increasing the length of the PCR product being generated. Furthermore, SYBR green allows for end-point validation of PCR specificity; because double-stranded DNA binding is reversible with denaturation of the DNA, SYBR green can be used in a "melt curve" to test the specificity of amplification of a PCR reaction (see below). However, the major disadvantage to SYBR green is the generation of false positives due to its lack of specificity; fluorescence is generated from *any* double-stranded DNA generated in the reaction, including nonspecific PCR products and primer dimers and the like. This limitation puts the onus of PCR validation on the user to ensure clean amplification of specific PCR products.

2. Primer and probe design; characteristics of the primers/probe.

   Real-time PCR is a powerful and effective technique for accurate quantitation of DNA. Critical to the success of real-time PCR is the specificity and efficiency of the PCR amplification. Therefore, the design of primers for real-time PCR represents perhaps the key offline aspect of assay design that will most directly influence the success or failure of the quantitation. Two different aspects must be taken into account during the primer design phase: the characteristics of the primers themselves, as well as those of the PCR amplicon they will generate (*see* **Note 3**). The potential pitfalls of poor primer design include amplification of nontarget sequence, formation of primer dimers that can compete with specific product amplification and affect PCR efficiency, and generation of multiple PCR products that make accurate quantitation impossible.

For real-time PCR, the same strategy exists for characteristics of the primers as with any other PCR design protocol. Most primer design targets a T_m on the order of 60 °C for a primer with a length on the order of 18–22 nucleotides. The forward and reverse primers should have T_m as similar to one another as possible, and no more than 1–2 °C difference. Other characteristics that should be taken into account include a G–C content on the order of 50 %, minimal or preferably no self-dimerization, minimal repetitive nucleotides (3 or fewer), lack of hairpin formation (as a result of internal base-pair complementarity), and minimal heterodimer formation between the primer pair. It is often desirable for the terminal 3′-base to be occupied by a G or C (the so-called G–C clamp) to minimize the chances of mispriming and increase the hydrogen bonding strength at the point where extension will begin. From the point of design, probes are often short (<30–40 nucleotides), with features similar to those of the primers, including ~50 % G–C content, contain minimal nucleotide repeats and are devoid of self-complementarity as well as avoid heterodimer formation between probe molecules or probe and primer. The T_m of the probe usually is about 10 °C higher than that of the primers, on the order of 68–70 °C.

3. Primer design with regards to the PCR amplicon generated.

While target amplicon characteristics will vary greatly depending on any given experiment, some generalities can be considered when designing primers with respect to the stretch of the gene of interest to be amplified. Most PCR amplicons used in real-time quantitation are relatively small amplicons. While the approach used (SYBR Green vs. primer+probe for example) and the characteristics of the gene of interest will ultimately determine the amplicon features, the rule of thumb for real-time amplicons remains that "smaller is better." This is a good rule of thumb for many reasons. First, smaller amplicons will be made with greater efficiency in any PCR reaction. Depending on the polymerase used, shorter stretches of extension are less likely to be subject to premature termination, and quite simply smaller products are "easier" to make. Second, small amplicons are more likely to be completely denatured from cycle to cycle, allowing complete access to of the primer and probe for binding to their respective complementary sequences. Finally, keeping extension times in the PCR protocol shorter provides some protection against the generation of nonspecific products, genomic contaminants, and other opportunistic and unpredictable amplification events. For these reasons, real-time PCR amplicons are often in the range of 70–120 nucleotides long, although this again will vary based on the parameters of any given experiment. Finally, if enough genomic sequence information is available primer pairs are best designed to span multiple exons. By combining short extension times with primers that span intron–exon boundaries the likelihood of amplifying contaminating genomic DNA is greatly reduced.

In addition to the size of the amplicon, basic good PCR strategy applies with regards to the overall content of the PCR amplicon. Long repetitive stretches should be avoided. Extremely G–C rich sequences make for poor candidates for real-time PCR, as complete denaturation can be a problem with these. Conversely, A-T-rich sequences may form less stable interactions with their primer/probe complements, affecting efficiency of amplification and/or detection. As a general rule of thumb, amplicons are often targeted to have 50–60 % G–C content to avoid these issues.

4. Standard curve generation and validation of primers for use in real-time PCR.

   Following primer design, assay validation must be performed to ensure that PCR amplification proceeds in an exponential and reproducible fashion across a range of template concentrations wider than those expected from the biological sample. In addition, if absolute quantitation will be performed, a standard curve of known copy number must be generated for interpolation of acquired unknown biological samples to quantify copy number from the sample. The form of this template varies among investigators, and usually includes plasmid DNA, PCR amplicon, cDNA, or in vitro transcribed RNA. In this example, we use samples of plasmid DNA containing cDNA encoding either *Gbx2* or *Gapdh* serially diluted from 10^7 to 10^2 copies for standard curve generation and initial primer validation. It is inappropriate to extrapolate data outside the range validated by first-hand experimental data acquisition.

5. Primer validation for real-time PCR.

   Both the primer validation and the standard curve generation (*see* **Note 4** above) are complementary procedures that should be done before processing of biological samples. The overall process includes serial dilutions of a known copy number of a template containing the gene of interest across a range of concentrations, use of this serially diluted template in the real-time assay, and curve fitting of the fluorescence to copy number relationship to determine the rate of PCR amplification and efficiency of the assay. We follow this primer validation with a second method consisting of serial dilutions of biologically derived cDNA, in order to validate primers in an assay that will be consistent with PCR conditions used in actual data acquisition. PCR efficiency is usually considered acceptable between 85 and 100 % (1.7–2.0).

6. Relative quantitation of gene expression with the comparative CT method.

Relative expression is most often done using the comparative CT method outlined in User Bulletin 2, ABI PRISM 7700 Sequence Detection System, P/N 4303859. Briefly, the amount of a target gene, normalized to a reference gene, is expressed relative a calibrator sample (control) via the formula 2^{-DDCT}, where DCT represents the different in cycle threshold between a given sample and its reference gene (normalization/loading control), and DDCT represents the difference in DCT's between the gene of interest in the sample of interest relative to the calibrator sample. In this case, 2 represents the PCR efficiency, and the experimentally determined efficiency should be substituted in the formula as the base value. This analysis technique leaves a fold-difference in starting template level relative to some calibrator, be it control sample or tissue, etc.

Acknowledgment

This work was supported by NSF 1021288.

References

1. Tourtellotte WG, Nagarajan R, Bartke A, Milbrandt J (2000) Functional compensation by Egr4 in Egr1-dependent luteinizing hormone regulation and Leydig cell steroidogenesis. Mol Cell Biol 20:5261–5268
2. Relaix F, Rocancourt D, Mansouri A, Buckingham M (2004) Divergent functions of murine Pax3 and Pax7 in limb muscle development. Genes Dev 18:1088–1105
3. Hanks M, Wurst W, Anson-Cartwright L, Auerbach AB, Joyner AL (1995) Rescue of the En-1 mutant phenotype by replacement of En-1 with En-2. Science 269:679–682
4. Lie YS, Petropoulos CJ (1998) Advances in quantitative PCR technology: 5′ nuclease assays. Curr Opin Biotechnol 9:43–48
5. Murtha MT, Leckman JF, Ruddle FH (1991) Detection of homeobox genes in development and evolution. Proc Natl Acad Sci U S A 88: 10711–10715
6. Castro LF, Rasmussen SL, Holland PW, Holland ND, Holland LZ (2006) A Gbx homeobox gene in amphioxus: insights into ancestry of the ANTP class and evolution of the midbrain/hindbrain boundary. Dev Biol 295:40–51
7. Chapman G, Rathjen PD (1995) Sequence and evolutionary conservation of the murine Gbx-2 homeobox gene. FEBS Lett 364:289–292
8. Waters ST, Wilson CP, Lewandoski M (2003) Cloning and embryonic expression analysis of the mouse Gbx1 gene. Gene Expr Patterns 3: 313–317
9. John A, Wildner H, Britsch S (2005) The homeodomain transcription factor Gbx1 identifies a subpopulation of late-born GABAergic interneurons in the developing dorsal spinal cord. Dev Dyn 234:767–771
10. Waters ST, Lewandoski M (2006) A threshold requirement for Gbx2 levels in hindbrain development. Development 133: 1991–2000
11. Wassarman KM, Lewandoski M, Campbell K, Joyner AL, Rubenstein JL, Martinez S, Martin GR (1997) Specification of the anterior hindbrain and establishment of a normal mid/hindbrain organizer is dependent on Gbx2 gene function. Development 124: 2923–2934
12. Byrd NA, Meyers EN (2005) Loss of Gbx2 results in neural crest cell patterning and pharyngeal arch artery defects in the mouse embryo. Dev Biol 284:233–245
13. Lin Z, Cantos R, Patente M, Wu DK (2005) Gbx2 is required for the morphogenesis of the mouse inner ear: a downstream candidate of hindbrain signaling. Development 132: 2309–2318
14. Buckley DM, Burroughs-Garcia J, Lewandoski M, Waters ST (2013) Characterization of the Gbx1-/- mouse mutant: a requirement for Gbx1 in normal locomotion and sensorimotor circuit development. PLoS One. 2013;8(2): e56214. doi: 10.1371/journal.pone.0056214. Epub 2013 Feb 13

Chapter 7

Identifying Essential Genes in Mouse Development via an ENU-Based Forward Genetic Approach

Aimin Liu and Jonathan Eggenschwiler

Abstract

The completion of the human and mouse genome projects at the beginning of the past decade represented a very important step forward in our pursuit of a comprehensive understanding of the genetic control of mammalian development. Nevertheless, genetic analyses of mutant phenotypes are still needed to understand the function of individual genes. The genotype-based approaches, including gene-trapping and gene-targeting, promise a mutant embryonic stem (ES) cell resource for all the genes in mouse genome; however, the phenotypic consequences of these mutations will not be addressed until mutant mice are derived from these ES cells, which is not trivial. An efficient and non-biased, *N*-ethyl-*N*-nitrosourea (ENU)-based forward genetic approach in mouse provides a unique tool for the identification of genes essential for development and adult physiology. We have had great success in identifying genes essential for morphogenesis and early patterning of mouse via this approach. Combined with complete genome information and numerous genetic resources available, ENU-based mutagenesis has become a powerful tool in deciphering gene functions.

Key words ENU, Mouse, Forward genetics, Polymorphism, Linkage analysis, Genetic mapping

1 Introduction

Genetic studies in the mouse have been one of the major sources of information regarding human genetic diseases. In the post-genome era, a genome-wide investigation of the gene functions has become a top priority. In the past two decades, a prevailing approach in gene function analysis is the genotype-based, reverse genetic approach. In this approach, researchers generate pre-designed mutations in genes of interest via gene-targeting in mouse embryonic stem (ES) cells. Although it is a powerful mutagenesis and gene manipulation tool, this approach requires prior knowledge about the genes to be mutated, such as tissue-specific expression pattern, important functional domains in the protein, or homology with important genes in invertebrates. However, not all genes and signaling pathways are perfectly conserved during evolution.

Mark Lewandoski (ed.), *Mouse Molecular Embryology*, Methods in Molecular Biology, vol. 1092,
DOI 10.1007/978-1-60327-292-6_7, © Springer Science+Business Media New York 2014

In addition, many genes essential for mammalian embryonic development and adult physiology do not exhibit tissue-specific expression patterns or contain known functional motifs. Insertional mutagenesis, especially gene-trapping, has provided a helpful genetic resource for mouse geneticists [1]. Unfortunately, the biased integration of retroviral vectors carrying the gene-trap constructs into the mouse genome prevents the mutation of all the genes in the genome. A combined effort to mutate all genes in the mouse genome using both gene-trapping and gene-targeting in ES cells will eventually generate at least one mutation for every gene in the genome [2, 3]. However, two limitations of these methods exist even after the completion of this ambitious project. First, genotype-based mutagenesis in ES cells alone rarely provides information about the function of the genes and tremendous effort is still needed if mutant mice are to be generated from these cells for functional analysis. Furthermore, many human genetic diseases are the results of hypomorphic mutations in developmentally essential genes caused by single amino acid changes, which are not faithfully mimicked by direct gene-targeting or insertional mutagenesis.

The phenotype-based forward genetic approaches identify important genes based on their mutant phenotype; hence there is no bias based on protein structures, gene expression patterns, or knowledge on their homologues in other species. *N*-Ethyl-*N-Nitrosourea* (ENU), a powerful alkylating agent, typically generates single nucleotide substitutions (both transitions and transversions) in mouse spermatogonia at a frequency of one mutation per gene in ~700 genomes [4]. This high mutagenic efficiency allows successful small-scale screens for recessive mutations in average-sized labs that have yielded many important discoveries [5–8]. The single nucleotide mutations induced by ENU also mimic many human genetic conditions and can serve as better disease models than the deletions and transgene insertions as a result of reverse genetic manipulation. The main disadvantage of this approach is the time and effort that one has to invest to identify the mutation. However, the completion of mouse genome sequence and new large-scale sequence analysis techniques greatly facilitated the identification of the ENU-induced mutation. In this chapter, we describe the methods for the generation of novel recessive mutations disrupting mouse embryonic development, identification of the mutated genes involved, and genetic validation of their identity.

2 Materials

2.1 Breeding and Mutagenesis of Mice

1. Mice: C57BL/6J males (8–9 weeks old), C3H/HeJ (C3H) females and males (6 weeks old), Cast/Si females (6 weeks old), Jackson Laboratories, Bar Harbor, ME.

2. Phospho-citrate buffer: 0.1 M Na_2HPO_4/0.05 M citric acid, pH 5.0. Check pH and adjust with citric acid as needed. Sterile filter and store at 4 °C.
3. 10 % Saturated ceric ammonium nitrate (Sigma, #431338) and 0.1 N NaOH, to be used for ENU decontamination.
4. *N*-Ethyl-*N*-Nitrosourea (ENU), Sigma # N3385/1 g pack in isopac bottles (*see* **Note 1**).
5. Protective gear: lab coat, gloves, goggles, mask.
6. 26G3/8 intradermal bevel needles and 1 ml tuberculin syringes (Becton Dickinson).
7. Gross anatomy probe (Fine Science Tools, catalogue # 10088-15) for checking vaginal plugs.
8. Fume hood.

2.2 Dissection of Mouse Embryos

1. Dissection tools: student quality iris scissors (Fine Science Tools, #91461-11), Dumont forceps (Fine Science Tools, #91150-20), spoon (Fine Science Tools, #10370-18).
2. Stereo-microscope, such as Nikon SMZ645.
3. Phosphate-buffered saline (PBS): prepare 10× stock by dissolving 80 g NaCl, 2 g KCl, 14.4 g Na_2HPO_4, and 2.4 g KH_2PO_4 in 1 l of water (*see* **Note 2**), adjust pH to 7.4 and autoclave at 121 °C for 20 min. Dilute to 1× solution with water for use.
4. 4 % Paraformaldehyde (PFA): prepare 16 % stock by adding 32 g PFA powder and 100 μl 5 N NaOH in 150 ml pre-warmed water (~65 °C). Once PFA is dissolved, add 20 ml 10× PBS and adjust the volume to 200 ml with water. Filter through Whatman paper and make 10 ml aliquot in 50 ml centrifuge tubes. Store at −20 °C. Thaw one tube at ~65 °C and dilute to 4 % PFA with PBS for use. 4 % PFA can be stored at 4 °C for up to 1 week (*see* **Note 3**).

2.3 Mouse Genotyping by PCR

1. Ear punch tool (Kent Scientific, # INS500075-5) for obtaining biopsies for genotyping.
2. Tissue lysis buffer: 50 mM Tris–HCl, pH 8.8, 1 mM EDTA and 0.5 % Tween 20. Store at room temperature. Add proteinase K (Roche, #03115879001) to a final concentration of 100 μg/ml right before use.
3. 2× PCR mix: mix 2 ml GeneAmp 10× PCR buffer II, 1.2 ml 25 mM $MgCl_2$ (provided with Taq polymerase, Applied Biosystems, # N8080245), 240 μl 25 mM dNTPs (Roche, #11969064001), and 6.7 ml water. Make 500 μl aliquots and store at −20 °C.
4. Primer mix: for each of the markers to be analyzed (*see* Table 1), make a mix of forward and reverse primers at 5 μM each in water.

Table 1
Microsatellite markers that can be used for genome scanning

Chr	MIT number	Map position	C57 product length	C3H product length	Left primer	Right primer
1	231	8.7	267	219	ACCCACAATTGCCTGTGG	GTCTTTGCAAGCCACCAAAT
1	156	32.8	143	112	TCTGCTGCCACTTCTGAGAA	TGTGTGTCTATGGACATGGATG
1	26	64.5	194	222	GAGGAATCTTGAATGGGCAA	CTGACAACACCCTCTGGCTT
1	115	99.5	148	144	AAGGGAATGGAATTAGGGTCA	TAACGGACACCCATTTTAAACA
2	237	27.3	130	111	TTCCAAGTCACCTATTATCAAAAGG	TTGATACGGACACCAGCAAA
2	249	50.3	146	118	GGCCTTTCCAACTCAATCAA	CTGAGAAGGGAGTCTCGGTG
3	101	35	108	126	CCTCTAGATGCATACATGTGCC	GGTCAAGTTAAGTGTATTTTTTTCCC
3	19	66.7	160	176	CAGCCAGAGAGGAGCTGTCT	GAACATTGGGGTGTTTGCTT
4	108	16.4	133	157	TCAGCCATCTCTATCAGGTGG	CCAATGTCAGATAATGCTTATGAG
4	219	48.1	113	127	CCATGGTGTTTCATCACAGC	AGACTGCCAAACTGTGTTACTCA
4	204	61.2	104	82	CTGCTGCAGCGATTCTCTC	TCAGGCACCTAAGTACATGTGC
5	386	9.8	117	83	CTGGGTCATGGCATAGGAGT	AATGTCACATTTAAGCATAGAACACA
5	233	19.7	147	180	TCCCCTCTGATCTCCTCAGA	CCTCCTAGAATACAATTCAATGTGG
5	292	73.2	114	128	GTCTTACTCCTGCCTGAACCC	TTACCTTACACTGAGACCAGAAAGC
6	74	10.9	150	130	CATGTGCAGTGTAAGTAAGACCTC	TCTCCTCCATCCTTCTCCAT
6	249	31.7	146	162	TTGTGTATGGGACTGGGGAT	TATCATACCCAAAACACATGTGC
6	366	47	124	142	AAGGCTCTGGTTTGCTAATAACC	GACCTTTTTACAAAGTTTAGGTCCC
7	76	3.3	228	250	CATGAGCACGTGGAGAAAGA	CGTGGAAACCTGATAAACTGA
7	229	18.6	123	143	GGTTCTCTTTCCTTGTTTGCC	TACTGGTTACATCTGGTGGGTG
7	215	29.5	96	110	CATCATTTTTGTTTGTTCACTGTG	TCCTCTCTGGCCTCCATG

8	4	12	157	195	CCAACTCATCCCCAAAGGTA	GTATGTTCAAGGCTGGGCAT
8	132	35	84	110	ATTTGTTTTGCTTACTTCTCAGTGT	AGAATGAAGATATCCAGAGATGCC
8	271	59	96	120	GGCAGAACCACAGGTTGATT	GGAATGAGGTTTGGGTCAAA
9	224	13.1	100	116	AATAAATGAATGCTTACAGGAGCA	TGGGGCTCAGAAAATGATTC
9	130	21.9	121	141	TCAATTTATTGTTGCAGTTGGC	AACACAGCATGAGATGGATACG
9	104	33.9	125	117	TTAATTGAGACGCACTTTGGG	AGGGGTCATTAGAGTTGGGG
10	189	4.4	106	130	TGTGTAGGTATGTGTGTGCATAGG	ATCAGACAGCACCTGGGAAC
10	115	33.9	121	143	CCATGGAATAGAAGTCTTTAAGAAGC	ACCTGAAGGATAAAGGGTCTTACC
10	180	67.8	134	206	GACCTTCCTTTATACACAAGTCATAGC	GTGGTACAGAACTTAGGTGTTTAATTG
11	349	27.3	118	78	AGTATCAGAAGATCCAGTTGGAGG	GTAGAAAAAGATACCCAGTGTCAGC
11	61	74.3	140	184	ACCCTTTCTCAGCTTACAAGTCC	GTATCAGAATCTGGTGCTTCTGG
12	89	21.9	126	118	CTCCTTGACATCCGGGAATA	CTGTTCAAGTGCCCACCC
12	210	21.9	150	160	CTGATGTGAAATTCACAAAGAACC	TGGGGCCCACTCTACATTAG
12	263	56.8	114	126	TCAGATCTCAGCAGATAAATACTTGG	TCCCCTGGAGCATATTTGAC
13	179	15.3	96	82	GACCAATGCCCTACAATTTCA	CAGAAGCAGTTTGTCTTTGTGG
13	159	32.8	142	160	CCCATTGTCCCTGTTCAGAT	AAACCCACCATGAATTAAATGC
14	133	16.4	139	117	TTGTCAAATAATTGCATGAGGC	AACTATGACTCAGATTCCAAGTTGG
14	5	31.7	178	164	CACATGAACAGAGGGGCAG	GTCATGAAGTGCCCACCTTT
14	166	62.3	144	128	TGGGGTTAGAGTAACTAGAATATAGGG	GGGGGCATTGTATGCTTAAA
15	175	5.5	178	128	ATAGCAACTAACAAAGACATACACACA	ACCCATTGCAGTGTAAAATTCC
15	156	29.5	145	123	CCCACATTCATGCACATATAGG	AACAAATCAAGAACCAATTGGG
15	161	65.6	128	102	TCTGTTTTGTTTGTTCGTTTGC	TAAAATCTCCCTGTATACAAGTCTGTG

(continued)

Table 1
(continued)

Chr	MIT number	Map position	C57 product length	C3H product length	Left primer	Right primer
16	165	10.9	171	153	AAATCAGTTGGCTCTATTAGTTTGG	AATGTAAACCCTAACTAGGTCTCTCTC
16	13	25.1	146	114	TTAGAACTCAGTAAGCTCTC	CTTAAACAGGCACNAATCCCATT
16	152	45.9	106	84	AGAGACCTCTGGGGTGGG	TTCAAGATAGACTATTCTGGAAAAAGC
17	34	9.8	148	126	TGTTGGAGCTGAATACACGC	GGTCCTTGTTTATTCCCAGTACC
17	238	29.5	126	114	TACTCCTTCCTCAACACAATACTAATT	AATGTGCTCATGCATACATGC
17	187	41.5	247	210	TTACCTCCCAACACCTACGC	GCCATTGTTTTAAAATTCCACC
18	197	4.4	124	132	CTCCCATCTGTCTTTGTCTTCC	CCATGCTTTCTTTTTACTTGTGG
18	17	14.2	213	188	TCAGGCAGATTCCAAGCAG	CTGTGGGTAGCCCAAGTCAT
18	1	37.2	154	154	TGAGCAAAATACATTGCATG	GGGATACCAGGCCAGACATA
19	61	9.8	131	149	ATGCTGCTGCTGTTGCTG	AATCTGTGATAAGTTTTTTACACACA
19	19	26.2	142	116	CCTGTGTCCATACAGGCTCA	ACCATATCAGGAAGCACCATG
19	1	43.7	121	142	AATCCTTGTTCACTCTATCAAGGC	CATGAAGAGTCCAGTAGAAACCTC

5. Horizontal electrophoresis apparatus appropriate for large-scale screening, e.g., Thermo Scientific Owl A3-1 wide gel system.
6. Agarose: Ultrapure agarose (Invitrogen, #16500500) and Nusieve GTG agarose (Lonza, #50084).
7. TBE buffer for DNA electrophoresis: dissolve 54 g Tris base, 27.5 g boric acid in water, add 20 ml 0.5 M EDTA (pH 8.0) and adjust the total volume to 5 l with water.
8. Taq polymerase: AmpliTaq Gold polymerase (Applied Biosystems, #N8080245). Store at −20 °C (*see* **Note 4**).
9. 96-well PCR plates (BioRad, # MLL-9601) and adhesive seals for PCR plates (BioRad, #MSB-1001).

2.4 RNA Preparation and Reverse Transcription

1. Trizol reagent (Invitrogen, #15596-026), store at 4 °C.
2. Reverse transcriptase (NEB, #M0253), store at −20 °C.

2.5 Generation of a Second Allele and Complementation Test

1. Culture medium for mouse embryonic stem cells: Dulbecco's Modified Eagle's Medium (DMEM, Mediatech, #15013-CV) supplemented with 10 % fetal bovine serum (Hyclone, Embryonic Stem cell screened), 55 μM β-mercapto-ethanol (Invitrogen, #21985-023), 1 mM sodium pyruvate (Invitrogen, #11360-070), 1× nonessential amino acids (Invitrogen, 11140-050), 2 mM Glumatax (Invitrogen, #35050-079), and 1,000 U/ml mouse leukemia inhibitory factor (brand name ESGRO, Millipore, #ESG1106). Store at 4 °C.
2. Gelatin: Dilute 2 % gelatin solution (Sigma, #G1393) with cell culture-grade water (Invitrogen, 15230-162) to a final concentration of 0.1 %. Store at 4 °C (*see* **Note 5**).
3. Trypsin–EDTA (Mediatech, #25052-CI), store at 4 °C.
4. Tissue culture-grade Phosphate-Buffered Saline (PBS, Mediatech, #21040-CV), store at 4 °C.

3 Methods

The key to a successful ENU-mutagenesis project is the efficient generation and screening of the mutant mouse lines carrying single recessive mutations. Subsequently, the mutation can be located to a specific chromosome through a genome scan. Further genetic mapping narrows down the candidate region for the mutation, or the interval; and candidate gene sequencing helps to pinpoint the exact nature of the mutation underlying the mutant phenotype. Finally, the identity of the mutation can be confirmed through the generation of a second mutant allele and a genetic complementation test.

3.1 ENU Injection

1. Order 20–30 8–9 weeks old C57BL/6J males from Jackson laboratories and allow males to acclimate to new facility for 1 week (*see* **Note 6**). Weigh the males the day before injection and divide them into groups based on weight (separated by 3 g intervals). Determine the volume of a 10 mg/ml ENU solution to be injected for each group to achieve 100 mg/kg body weight.
2. Wearing protective gear and working in a fume hood, prepare fresh ENU solution by dissolving the contents of the Isopac bottle in 10 ml 95 % ethanol for 1 h. Add 90 ml phosphate citrate buffer to make a 10 mg/ml solution. Freshly prepared ENU should be kept on ice and used within 1 h after preparation.
3. Under a fume hood, inject an appropriate volume of ENU solution intraperitoneally to achieve 100 mg/kg body weight for each group. Quarantine the cages housing the injected mice for 24 h to prevent exposure to the mutagen.
4. Decontaminate all instruments, gloves, and containers that come in contact with the ENU solution by soaking in 10 % ceric ammonium nitrate. Add ceric ammonium nitrate solution to residual ENU solution and discard in biohazard containers after decontamination. Surfaces in fume hood should be cleaned with 0.1 N NaOH, followed by tap water. Twenty-four hours after injection, mice should be moved to new cages. Bedding should be safely discarded in biohazard bags and cages wiped down with 0.1 N NaOH.
5. Repeat **steps 3** and **4** once per week for the following 2 weeks (three injections of 100 mg/kg ENU in total).

3.2 Breeding Mice for the Screen

1. House males for 7 weeks and order C3H/HeJ (C3H) females around week 5. After 7 weeks, place two C3H females in a cage with one of the injected males (~25 such cages in all). Observe the females for 4–7 weeks. If the females in any of the cages became pregnant during this time, remove and euthanize the mice in this cage (*see* **Note 7**). Save those mating cages in which the male remains sterile for at least 11 weeks after the final injection.
2. After the first desired litters are born (~14 weeks after the final ENU injection), wean pups once they are 21 days old saving only F1 male pups and euthanizing the rest.
3. Continue mating, replacing older C3H females (>6 month) with younger C3H females and saving between 5 and 25 F1 males per mutagenized C57BL/6J male. In our experience, roughly 60–90 F1 lines can be effectively screened by a small laboratory per round. For this reason, approximately 100–120 F1 males in all should be saved from these crosses.

4. Once F1 males are 8 weeks or older, begin to establish F1 families. This is best done on a rolling basis by setting up 5–7 crosses between individual F1 males with two C3H/HeJ female mice each week. This is repeated until a suitable number of F1 families have been generated and screened.
5. Wean G2 pups from the F1 families once they are 21 days old. At this point, euthanize the G2 males and save the G2 females in new cages.
6. When G2 female mice are ≥7 weeks old, remove their F1 father from his mating cage and set him up with three of his G2 daughters. Check for vaginal plugs with probe (white, hardened plugs in the vagina, an indication of mating the previous night) every morning. Mark noon of the day the plugs are found as embryonic day (E) 0.5. As G2 females are plugged, remove them to new cages, notch their ears for identification, and introduce new G2 females into the mating cage with the F1 father (*see* **Note 8**).
7. If, after screening 5 litters from a given line (*see* Subheading 3.3 below), no embryos with a clear morphological mutant phenotype are observed, the F1 male and all of his remaining G2 progeny for that line should be euthanized to limit the number of cages maintained at any given time.

3.3 Screening for Recessive Mutations by Morphology

1. Euthanize a pregnant G2 female mouse (plugged by its F1 father) at E9.5 (9 days after the vaginal plug was found) or other stages of interest through regulated CO_2 inhalation. Note that euthanasia procedures must be approved by your Institutional Animal Care and Use Committee.
2. Wet the abdominal region of the mouse with 70 % ethanol. Cut the skin and open the abdominal cavity with iris scissors. Remove the two uterine horns with scissors and place them in cold PBS in 10 cm Petri dishes. Inspect the uterus and record the number of decidua (seen as bulges along the uterus). Save the liver of the mother at −20 °C for later genotyping. Additional information on mouse embryo dissection can be found in ref. [9].
3. Under a stereo-microscope, gently tear open the uterine wall and remove the embryo with its surrounding yolk sac from the decidual tissues.
4. Inspect the embryos for any morphological defects. Some common defects seen at E9.5 are neural tube defects (exencephaly and/or spinal bifida), heart-looping defects (straight or right-turning heart loops versus left-turning heart loops seen in wild-type embryos), and gastrulation defects (very small embryos which arrest prior to embryonic turning). Some examples are shown in Fig. 1b.

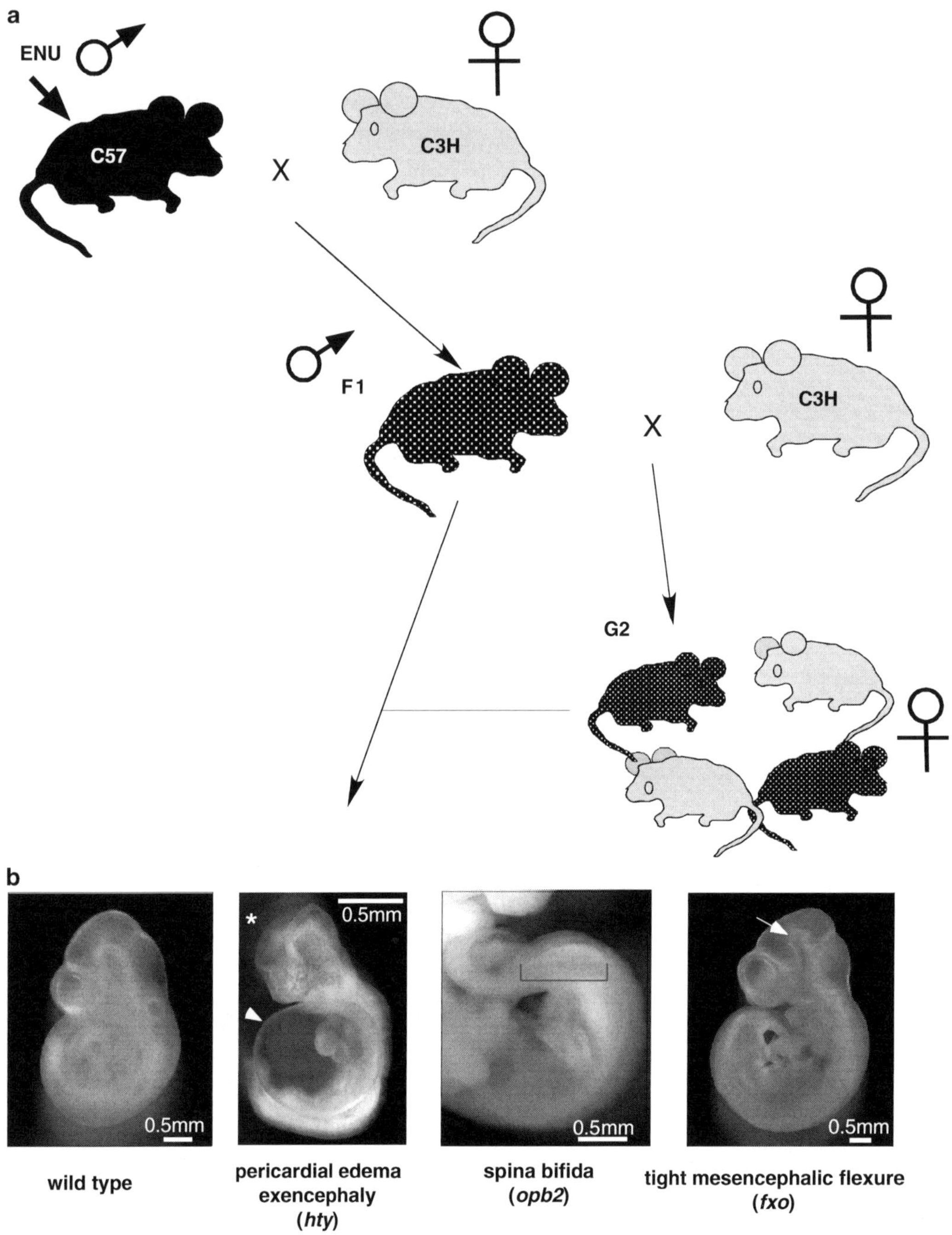

Fig. 1 A genetic screen for recessive mutations disrupting mouse embryonic development. (**a**) Male mice from the C57BL/6J (C57) strain are injected with the chemical mutagen ENU and are bred with wild-type C3H/HeJ (C3H) female mice. The resulting male pups (F1) are further bred with C3H female mice. Each F1 male is subsequently bred with his daughters (G2), from which embryos are isolated and inspected for developmental defects. (**b**) Some commonly seen morphological defects in E9.5 and E10.5 mouse embryos are shown. These include pericardial edema (*arrowhead*), exencephaly (*asterisk*), spinal bifida (*bracket*), and tight mesencephalic flexure (*arrow*). Examples shown are $C2cd3^{hty}$, $Rab23^{opb2}$, and $Ift88^{fxo}$ mutants [11–14]

5. Record the defects. Collect yolk sacs and amniotic membranes from the mutant embryos and save them at –20 °C for genotyping. Fix the mutant embryos in 4 % PFA and store them at 4 °C.

3.4 Mutation Recovery

After screening 5–7 litters for a given line, a preliminary judgment may be made as to whether that line contains a single locus recessive mutation causing developmental defects with high penetrance. Lines producing no phenotypic embryos may be euthanized. It is preferable to pursue those lines in which the morphological phenotype of the embryos is consistent and which appear to show simple Mendelian inheritance (roughly half of the litters contain mutants and within each litter, roughly 1 in 4 embryos shows the phenotype). If the inheritance patterns deviate substantially from what is expected, more G2×F1 crosses may be performed to increase statistical power.

1. Once promising lines are identified, it is important to recover the mutation in males from subsequent generations as the F1 males will not remain fertile indefinitely. This is done by returning the F1 male to the C3H mating cages and then saving both G2 males and females from these crosses.
2. When the G2 animals are old enough to mate, roughly 8 G2 bother/sister crosses should be set up. Vaginal plugs are checked and litters screened for mutants as described above in Subheading 3.3.
3. As only half of the males and half of the females should be carriers for recessive mutations, one expects to find mutant embryos in 1 in 4 of such litters on average. Identification of litters containing mutant embryos is an indication that the G2 males that sired them are carriers for the desired mutation. These males may be bred with more of their G2 sisters to obtain more mutant embryos for the genome scan experiments described below. Carrier G2 males may also then be crossed with C3H/HeJ females to generate G3 animals, while the genome scan is in progress.

3.5 Linkage Analysis/Genome Scan

As the first step toward the identification of the point mutation induced by ENU, the chromosome on which the mutation is located needs to be determined. The basis for such a linkage analysis/genome scan is the difference in the DNA sequences of the chromosomes (polymorphism) between the two mouse strains used in this experiment, C57BL/6J (C57) and C3H/HeJ (C3H). In mammalian genomes, there are numerous repetitive sequences of short units of 2, 3, or 4 nucleotides, designated as microsatellite DNAs (Fig. 2a). The lengths of these microsatellite DNAs are highly polymorphic; therefore they are commonly used to distinguish chromosomes from different origins. To detect these

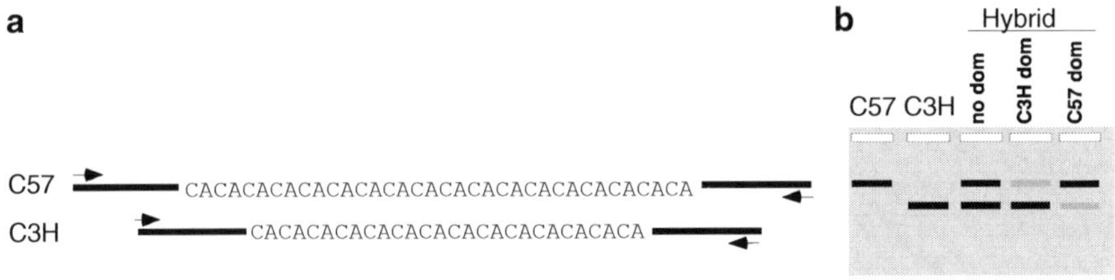

Fig. 2 Microsatellite marker-based mouse genotyping. (**a**) The lengths of microsatellite DNA, or repeat of short units (in this case, two nucleotides, CA), vary between C57BL/6J and C3H/HeJ. Primers (*arrows*) designed to amplify the microsatellites and their surrounding sequences are called microsatellite markers. (**b**) A schematic illustration showing how the PCR products appear on a 4 % agarose gel. Three potential outcomes are shown for C57/C3H hybrid. Sometimes only the C3H product is efficiently amplified (C3H dominant), whereas sometimes only the C57 product is efficiently amplified (C57 dominant). Only markers consistently showing comparable amplification of both C57 and C3H products (no dominance) should be used for mouse genotyping

polymorphisms, PCR is performed using primers flanking the microsatellites [10]. Based on the size of the PCR product, the origin of the chromosome can be determined (Fig. 2b). As the mutagenesis was initially performed on a pure C57 background, the assignment of the mutation to a specific chromosome is based on the association between the mutant phenotype and the presence of the C57 chromosome (Fig. 3a). Experimental details are as follows:

1. When at least ten mutant embryos have been recovered from one F1 family, extract DNA from the saved extra-embryonic membranes by incubating them in 100 μl tissue lysis buffer with proteinase K overnight at 55 °C. Inactivate proteinase K by boiling the lysate for 10 min. Dilute the lysate with 900 μl water.
2. Set up PCR reactions with all the genome scan primers (Table 1). Make sure to include C57, C3H, and C57/C3H heterozygote (tail clips or ear punch biopsies of the F1 males can be used) controls for each pair of primers. We set up PCR reactions in 96-well PCR plates by first making reaction mix according to the number of samples (Table 2), aliquoting 5 μl reaction mix to each well, and then adding 5 μl tissue lysate to each well. The PCR program is as follows:
   (a) 95 °C, 10 min.
   (b) 55 cycles of (95 °C, 20 s, 55 °C, 30 s, 72 °C, 45 s).
   (c) 72 °C, 7 min.
   (d) Hold at 4 °C.
3. Place a 2-l plastic beaker with 800 ml TBE on top of a stir plate. Slowly add 16 g Ultrapure agarose, then 16 g Nusieve GTG agarose into the beaker while stirring. Leave the beaker on stir plate for 10 more minutes to allow agarose to be fully

a

mutant#	D14mit133	D14mit233	D14mit142	D14mit5
1	CC	BC	BB	BB
2	BC	BB	BB	BB
3	BC	BC	BB	BC
4	BB	BB	BB	BC
5	BC	BB	BB	BB
6	BC	BC	BC	BB
7	BB	BB	BB	BB
8	BB	BB	BB	BC
3	BC	BC	BC	BB
4	BC	BB	BB	BB
F_B	60%	80%	90%	85%

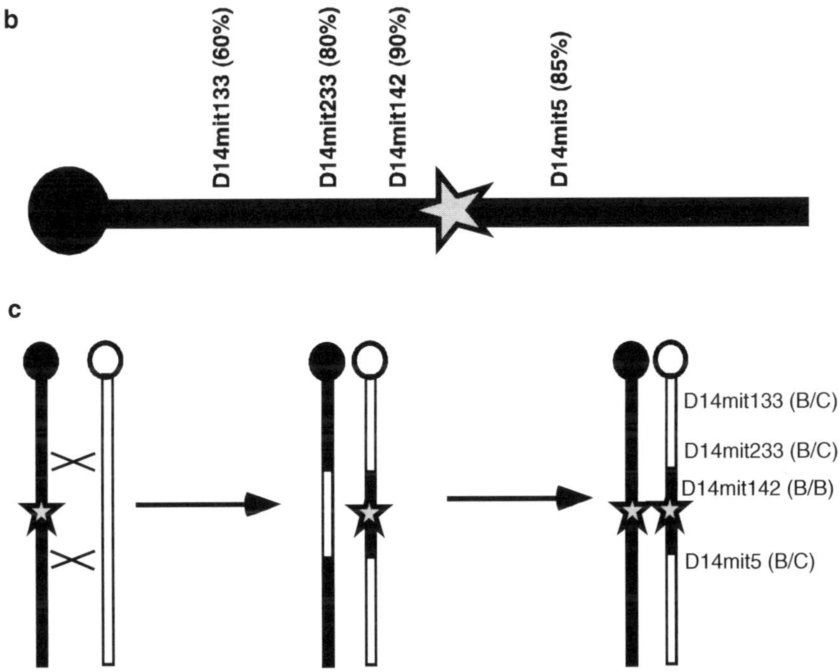

Fig. 3 Positional cloning. (**a**) To find the linkage for the mutation, a genome scan is performed and the data are summarized in a table like the one shown here. *B* stands for C57BL/6J and *C* stands for C3H/HeJ. $F_B = \#B/(\#B + \#C) \times 100\ \%$. (**b**) Increased correlation between C57 product and the mutant phenotype is observed for microsatellite markers on the chromosome near the mutation. A schematic illustration of mouse chromosome 14 is shown. The *solid circle* on the left is the centromere and the *star* denotes the mutation. (**c**) Determining the interval containing the mutation. The black chromosome is the C57 chromosome with the mutation (*star*), and the white one is the wild-type C3H chromosome (*left*). Two crossovers between the chromosomes lead to the insertion of a fragment of C57 chromosome into the otherwise C3H chromosome (*center*). When a recombinant mouse is bred with a non-recombinant mutant carrier, the genotypes for the four microsatellites are shown (*right*). If this embryo exhibits mutant phenotype (such as embryo #3), it can be concluded that the interval containing the mutation is between D14mit233 and D14mit5. By summarizing all recombinant embryos in the table shown in **a**, the mutation is further defined as between D14mit142 and D14mit5

Table 2
Setting up PCR reactions for mouse genotyping

Sample#	1	2	3	4	5	6	7	8	9	10	11	12
PCR mix	5.11	10.22	15.33	20.44	25.55	30.66	35.77	40.88	45.99	51.1	56.21	61.32
Primer mix	0.57	1.14	1.71	2.28	2.85	3.42	3.99	4.56	5.13	5.7	6.27	6.84
Taq	0.08	0.16	0.24	0.32	0.4	0.48	0.56	0.64	0.72	0.8	0.88	0.96
sample#	13	14	15	16	17	18	19	20	21	22	23	24
PCR mix	66.43	71.54	76.65	81.76	86.87	91.98	97.09	102.2	107.31	112.42	117.53	122.64
Primer mix	7.41	7.98	8.55	9.12	9.69	10.26	10.83	11.4	11.97	12.54	13.11	13.68
Taq	1.04	1.12	1.2	1.28	1.36	1.44	1.52	1.6	1.68	1.76	1.84	1.92

Based on the number of samples, make 2× PCR reaction mix by mixing PCR mix, primers and Taq polymerase in a tube. Make sure to include C57, C3H, and C57/C3H hybrid controls
For each sample, mix 5 μl of sample with 5 μl of 2× PCR reaction mix

hydrated. Microwave solution for 5 min, stir for 1 min, then microwave for 1 min. Repeat the microwaving and stirring steps until the agarose solution boils. Add 18 μl 10 mg/ml ethidium bromide and let it cool to about 60 °C while stirring (*see* **Note 9**).

4. Pour the gel and insert the combs (*see* **Note 10**).
5. When the gel sets, load the PCR samples with a 12-channel pipette and run the gel at ~200 V for ~1.5 h.
6. Compare the PCR products from each mutant embryo and compare them to the three controls (Fig. 2b). Record the results in a table. If only C57 product is present, record it as BB. If only C3H product is present, record it as CC. if both C57 and C3H products are present, record it as BC (Fig. 3a).
7. For each microsatellite marker, calculate the frequency of the C57 product (F_B):

   F_B = number of B alleles/(number of mutant embryos × 2 alleles per embryo) × 100 %.
8. If F_B is ≤50 % for all microsatellite DNAs on a given chromosome, it indicates that the mutation is not on that chromosome. If the single recessive mutation causing the mutant phenotype is in a particular region of a chromosome, the value of F_B for markers in that region should be on the order of ≥85 % such that logarithm of odds (LOD) scores are >2.0, assuming the marker and mutation are separated by ≤25 cM (*see* Fig. 3a, b). The closer the microsatellite marker is to the mutation, the higher the F_B value (*see* **Note 11**).

3.6 Mapping the ENU-Induced Mutation

1. In addition to providing the linkage information of the mutation, the result of genome scan also defines the initial candidate region in which the mutation lies. As shown in Fig. 3c, because the recessive mutation was generated on a C57 chromosome, only the regions in which both alleles are from C57 could harbor the mutation. Markers genotyping as BC that lie proximally and distally on the chromosome help set the boundaries for the candidate region, or the "interval". Therefore, in this example, the mutation can only be located between D14mit142 and D14mit5.
2. The initial candidate region for the mutation, determined in the genome scan, may span 10–50 cM and contain hundreds of genes. In order to narrow down this interval, more microsatellite makers should be assayed to define the locations of crossovers between C57 and C3H chromosomes. Additional microsatellite markers polymorphic between C57 and C3H in a given region of the genome can be found at Sloan-Kettering Mouse Project webpage (https://mouse.mskcc.org/marker/MIT/query.php). After loading this page, enter the microsatellite markers currently flanking the interval and choose the two inbred strains used in the screen (e.g., C57 and C3H) and click enter. You will see a table with all available microsatellite markers in this interval. The ones highlighted in green are those whose PCR products for C57 and C3H alleles are separated by at least 5 bp (hence can be separated on a 4 % agarose gel). Order primers for amplifying these markers and genotype the recombinants from the genome scan to further narrow the interval.
3. To identify additional recombinants, genomic DNA from carrier G2 males and females should be prepared. For G2 carrier males identified in the recovery phase, obtain an ear punch biopsy and for carrier G2 females identified in the initial screen and recovery phases, use a small (~50 mg) piece of the liver sample previously frozen at −20 °C. Prepare crude DNA samples as described above for embryos and genotype the carrier samples using the markers flanking and contained within the candidate interval. If these carrier mice are heterozygous (BC) at both ends of the interval, no recombination event has occurred. If a carrier mouse showing wild-type (CC) genotype at one end (or, in rare cases, both ends) of the interval, it indicates that the mutation and the marker were separated by a recombination event. This information may help to further narrow the candidate interval.
4. At this point, male and female carriers may be identified by PCR genotyping, rather than by breeding/dissection. Obtain ear punch biopsies from untested G2 and G3 animals and genotype with the microsatellite markers that most closely flank the candidate interval. Keep males that are heterozygous (BC) at both ends of the interval as carriers and use to propagate the

mutation through subsequent generations. Animals that are recombinant (i.e., heterozygous (BC) at one end of the interval and wild-type (CC) at the other end) should be bred with (BC, BC) carrier animals to test whether the recombinant animals are carriers. Check plugs for these crosses, dissect embryos at the appropriate stage, and determine whether phenotypically mutant embryos are present. If so, the recombinant animal is a carrier and its genomic DNA can be used to further narrow the candidate interval.

5. Record the number of generations as the mutation is bred into the C3H background. For example, the sons of F1 males are G2; the sons of G2 males are G3; etc. Note any change in the expressivity or penetrance of the mutant phenotype between generations because these can be strain background dependent.
6. Every time a recombinant embryo with the mutant phenotype is identified, further mapping of the mutation can be performed. Genotyping this embryo with additional microsatellite markers between the two markers currently defining the limits of the interval. In subsequent generations, the microsatellite markers defining the limits of the new intervals may be used for genotyping adults to immediately identify the most useful recombinants.
7. Sometimes, no polymorphic microsatellite markers are available in the minimally defined interval. This usually happens once the interval becomes relatively small. If further narrowing of the interval is desired (e.g., there are still too many genes in the interval to initiate a direct DNA sequencing analysis), PCR primers surrounding additional microsatellite DNAs in the interval can be designed. First identify the location of the microsatellite markers flanking the current interval at Ensembl (http://mouse.ensembl.org/) or Sloan-Kettering Mouse Project webpage (e.g., D14mit233 is on chromosome 14, 51,097,898 bp from the centromere).
8. Subsequently, go to the Drummond Lab mouse SSRS webpage (http://danio.mgh.harvard.edu/mouseMarkers/musssr.html), enter the locations of the flanking microsatellite markers and click "Search" (*see* **Note 12**).
9. The result of the search includes several files. One of them, "candidate SSR primer pairs for mm_chr...", recommends primers for amplifying microsatellites in the interval. Order these primers and perform PCRs with C57, C3H, and C57/C3H heterozygous DNA templates to determine whether the corresponding microsatellite repeats exhibit polymorphism between these two strains. If they are polymorphic, these primers can be used to genotype the recombinant mutant embryos to further narrow the interval for the mutation. The exact location of the microsatellite is given in the original file and can be

confirmed by performing a BLAST search against the mouse genome at Ensembl webpage using the marker sequence as a query (*see* **Note 13**).

3.7 Identifying the ENU-Induced Mutation

The ultimate goal of positional cloning is to identify the point mutation that disrupts development in the homozygous mutant embryos. A candidate gene approach, which is based on prior knowledge of the proteins encoded by the genes in the candidate interval and some understanding of the molecular mechanism underlying the mutant phenotype, can save time and effort in positional cloning. When this approach fails, especially due to lack of knowledge about the genes and the molecular mechanism underlying the mutant phenotype, sequencing of all the genes in the interval may be required.

1. When the interval is small enough such that it contains ≤20 known and predicted genes (typically <2 Mb), perform a literature search to investigate the function of known genes. Based on the literature search, the genes may be characterized as likely candidates, which exhibit functions that appear to be associated with the mutant phenotype; unlikely candidates, for which mutants with different phenotypes have been previously generated; and other genes, which include all the genes whose functions have not been previously characterized.
2. For the likely candidates, find the cDNA sequence in Ensembl webpage. Design primers for reverse transcriptase PCR (RT-PCR) reactions that amplify the entire open reading frame (ORF). Optimal length of the RT-PCR product is around 0.5–0.8 kb, which allows easy amplification and high-quality sequence.
3. Prepare RNA from homozygous mutant embryos as well as stage-matched C57 wild-type control embryos (*see* **Note 14**). We use Trizol reagent from Invitrogen for this purpose. Dissect E9.5 mouse embryos in ice-cold DEPC-treated PBS. Add 1 ml Trizol reagent for each embryo in a 1.5 ml eppendorf tube, and homogenize the embryos by passing them through hypodermal needles of decreasing gauges (gauge 18, 20, and 25). Let the homogenized tissue sit at room temperature for 5 min. Add 200 μl chloroform, shake vigorously, and centrifuge at top speed for 15 min at 4 °C in a table-top eppendorf centrifuge. Carefully transfer the top phase into a clean tube and add 500 μl isopropanol. Let it sit at room temperature for 5 min and centrifuge at top speed for 10 min at 4 °C. Decant the supernatant, rinse the pellet with 1 ml 70 % ethanol, and dry the pellet briefly. Resuspend RNA in 50 μl DEPC-H_2O. Make 5 μl aliquots and store at −80 °C.
4. Reverse transcription (RT): In a 200 μl PCR tube, add 0.5–2 μg RNA, 2 μl 40μM random hexamer, and DEPC-H_2O to a total

of 16 μl. Heat at 65 °C for 5 min and immediately place on ice (*see* **Note 15**).

5. Add 2μl 10× RT buffer; 1 μl reverse transcriptase and 1μl RNAse inhibitor, incubate at 25 °C for 5 min, then at 42 °C for 1 h. Finally inactivate the enzyme at 90 °C for 10 min.
6. PCR: Dilute the RT product by 25 times with water, and set up PCR reactions with the primers designed in **step 2**. We use the same recipe and PCR program for genotyping the embryos (*see* Subheading 3.4, **step 2**).
7. Run 5 μl of the PCR products on a 1.5 % agarose gel. If the wild-type and mutant RT-PCR products are of the same size. Purify and sequence the rest of the PCR products. We use Seqman software (DNAStar) to compare the DNA sequence between the mutant and wild-type embryos.
8. If one or more RT-PCR products in mutants are different from the ones in wild-type, such as no product is seen in the mutants, or one or more RT-PCR products of different sizes are present in the mutants, this may indicate a defect in the splicing of the mRNA (Fig. 4a, b). To confirm the presence of abnormal splicing, clone the RT-PCR products in the mutant embryos into a T-A cloning vector (such as pGEM-T from Promega) and analyze the sequence (Fig. 4d).
9. If the sequence of the RT-PCR products indicates abnormal splicing in a candidate gene, the mutation is likely in the intron sequences near the intron/exon junctions, which are known to be critical for the splicing of RNAs. To find the mutation, download the intron sequence from Ensembl webpage, PCR amplify from C57 wild-type and mutant embryo genomic DNAs and analyze the sequence of the relevant intron/exon junction regions (Fig. 4c).

3.8 Validating the Mutation

There are numerous ways a single nucleotide substitution could diminish or abolish the function of a gene and lead to abnormal development of an animal. Sometimes, a single nucleotide mutation results in severe truncation of the encoded protein by either creating a stop codon or shifting the reading frame. More frequently, a single nucleotide mutation leads to the change of a single amino acid, for which the impact on the function of the protein and the development of the animal needs further validation. A phenotype rescue experiment by reintroducing the normal gene through a bacteria artificial chromosome-based transgenic mouse approach can be performed, but it is often technically demanding. With the quick expansion of the coverage of gene-trap mouse mutant resource, another common approach for mutation validation has been the generation of a second mutant allele for the candidate gene and a complementation test between the ENU

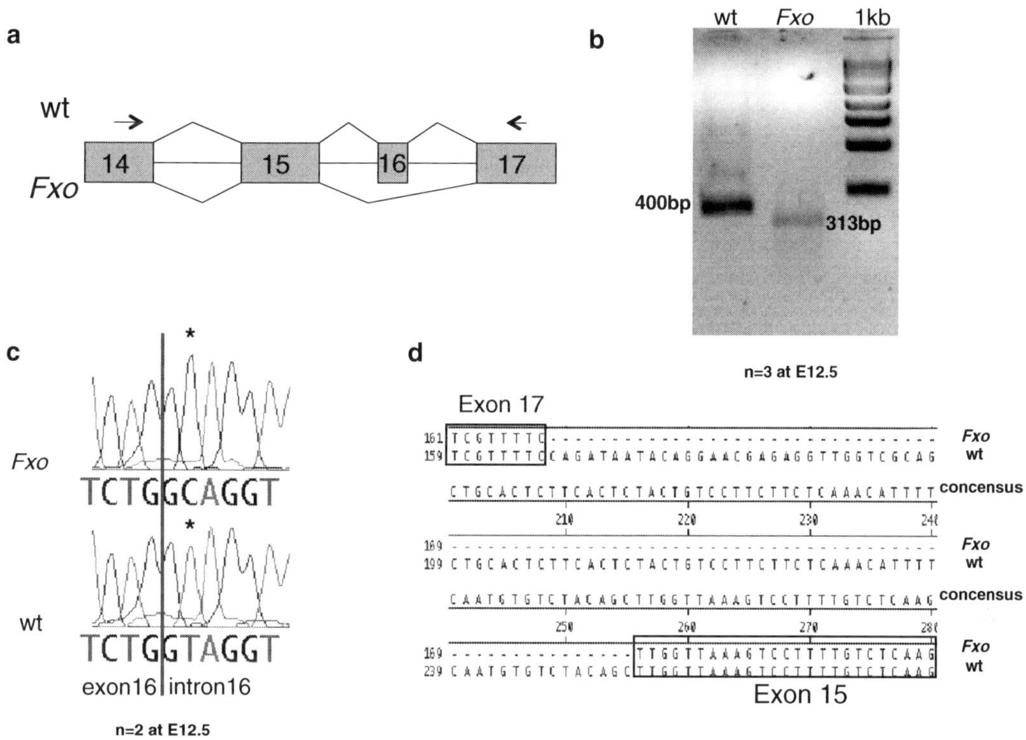

Fig. 4 Identification of the ENU-induced mutation. (**a**) Part of mouse *Ift88* gene. RT-PCR using primers in exon 14 and exon 17 amplify a predicted product of ~400 bp. In the ENU-induced *flexo* (*fxo*) mutants [12, 13], an abnormal splice event occurs between exon 15 and exon 17, skipping the 16th exon. (**b**) An agarose gel showing the RT-PCR products in wild-type and *fxo* mutant embryos. A shorter PCR product is observed in the mutants. (**c**) The DNA sequence around the junction between exon 16 and intron 16. The second nucleotide of intron 16 is mutated from T to C in *fxo* mutants. (**d**) Sequence analysis of the PCR products shown in **b**. Exon 16 is missing in the *fxo* mutant transcript

allele and gene-trap allele. Following is a detailed protocol on validating the ENU-induced mutation through complementation.

1. Search the website of International Gene-Trap Consortium (IGTC; http://www.genetrap.org) for mouse ES cell lines with gene-trap mutations in the candidate gene. We prefer to BLAST search the database using the sequence of the cDNA. Once an ES cell line is identified with a gene-trap insertion in the candidate gene, it can be obtained at low cost through IGTC.
2. IGTC ships frozen ES cells on dry ice and should be stored in a liquid N_2 tank. Thaw the ES cells in 37 °C water bath. Immediately after the cells are thawed, transfer them to a 15 ml centrifuge tube containing 10 ml ES cell medium pre-warmed at 37 °C. Centrifuge the cells in a table-top centrifuge at

1,500 rpm for 5 min. Decant supernatant and add 5 ml ES cell medium containing 125 μg/ml G418. Resuspend cells by aspiration and transfer them to a 60 mm tissue culture dish pre-coated with 0.1 % gelatin. Incubate the cells at 37 °C, 5 % CO_2 (*see* **Notes 16** and **17**).

3. Change medium the next morning and every day afterwards. When cells are 80 % confluent, split them by washing them twice with Ca/Mg-free PBS, incubating them with 0.5 ml Trypsin/EDTA at 37 °C for 10 min, inactivating the enzyme with 4.5 ml ES cell medium and dissociating the cells by aspiration. Aliquot 1 ml cell suspension into each 60 mm dish containing 4 ml ES cell medium and incubate at 37 °C, 5 % CO_2 (*see* **Notes 18** and **19**).
4. To prepare ES cells for blastocyst injection, pass cells into 100 mm tissue culture dishes with no G418 in the medium. On the day of injection, make single ES cell suspension as described in **step 3**. Centrifuge the cells at 1,500 rpm for 5 min, decant the supernatant, resuspend cells in PBS, centrifuge again, and resuspend cells in 5 ml DMEM plus 5 % FBS and keep them on ice (*see* **Note 20**).
5. Most transgenic facilities provide chimeric mice after blastocyst injection. Breed male chimeras with wild-type C57BL/6J female mice. The ES cell-derived pups are agouti, whereas the host cell-derived pups are black. The agouti pups can be genotyped by PCR using primers specific to lacZ (forward primer: 5′-CCGAACCATCCGCTGTGGTAC-3′; reverse primer: 5′-CATCCACGCGCGCGTACATC-3′; product: 345 bp). Breed the heterozygous carriers (lacZ+) with carriers of the original ENU-induced mutation and dissect embryos at appropriate stages. If the mutation identified in the ENU mutant allele is indeed responsible for the abnormal development observed in the mutant embryos, a similar mutant phenotype should be recovered in this complementation test cross (*see* **Notes 21** and **22**).

4 Notes

1. ENU is carcinogenic/genotoxic and must be handled with care in a fume hood with appropriate protective gear. Several Isopac bottles from one lot should be ordered. Isopac bottles with undissolved ENU may be stored at −20 °C for no longer than 1 year. It is advised that one bottle from the lot be prepared in advance. The solution should have a light yellow color. Ideally, UV absorption analysis and melting point (~90–95 °C) should be checked for the lot. A new Isopac bottle should be used for each day of injection.

2. Unless otherwise stated, all water (H_2O) used in this protocol should be ultra-purified water with a resistance of 18.2 MΩ.
3. PFA is highly irritant and carcinogenic. Handle PFA powder and warm solution in the fume hood and wear a mask.
4. We have obtained satisfactory results using Hot-start Taq polymerases from other vendors. However, try to avoid using non-Hot-start Taq polymerases because they frequently give unacceptable levels of nonspecific amplification.
5. 2 % Gelatin solution solidifies during storage at 4 °C, hence it needs to be warmed at 37 °C briefly prior to dilution.
6. It is advisable to order C3H/HeJ males and females at this point to set up matings for a small C3H colony to ensure a steady supply of C3H females for subsequent breeding in the following months.
7. Exposure to ENU causes short-term male sterility as it kills mature spermatozoa but generally spares the mutagenized stem cell spermatogonia. During the ~8 weeks following ENU injection, the spermatogonia begin repopulating the germ-line with mature spermatozoa and the length of the sterility period correlates with mutation frequency. It is therefore advisable to save F1 progeny from injected males that were sterile for ≥11 weeks after the final ENU injection.
8. If an F1 male is a carrier for a single locus recessive mutation, half of his G2 daughters should inherit this mutation. On average, half of the litters from G2 × F1 crosses may contain homozygous mutants (which would occur at an average frequency of 1 in 4 within the litter). For this reason, it is advisable to dissect and analyze approximately 5 litters per line before deciding whether a particular line is unlikely to be useful for further study.
9. During the cooling process, a thin film of solidified gel will form on the surface, which can be removed with a pipette tip.
10. For 4 % agarose gel, we occasionally found that the gel fails to go through the combs before it solidifies. Therefore, we always insert the combs after pouring the gel to make sure the gel is consistent.
11. Sometimes after initial linkage analysis microsatellite markers on more than one chromosome will show increased frequency of B alleles. Based on our experience, this could result from false identification of the mutants. A more careful characterization of the mutant phenotype and linkage analysis with additional mutant embryos would help. Taking expected recombination frequencies, sample sizes, LOD scores into account, we find ≥17 B (non-recombinant) chromosomes out of 20 chromosomes analyzed is a promising indicator of link-

age. In rare cases, the mutant phenotype results from more than one mutation (strong linkage to more than one locus). We do not routinely pursue such lines due to the additional effort necessary to clone the genes responsible. For these reasons, we strongly recommend that microsatellite markers on all 19 autosomes be examined even if a linkage is suspected after examining just a few chromosomes. Because of the design of the screen, recessive mutations on the X chromosome will not be identified.

12. The server can handle a region up to 3 Mb. If your interval is larger than 3 Mb, divide it into several segments and search each segments individually.
13. Sometimes, no microsatellite marker in the mutant interval exhibits polymorphism between C57 and C3H, preventing further mapping of the mutation. An out-breeding to a more distantly related strain, such as Cast/Si, may be performed to circumvent this problem. First, one should search the Sloan Kettering Website to make sure that markers in the interval are polymorphic between C57 and Cast/Si. After breeding the mutation into the new background, the mutant phenotype should be re-characterized to ensure that the expressivity and penetrance of the mutation are not affected by the new genetic background. The first generation Cast/C3H hybrid mice are hyperactive and aggressive. We always warn the animal caretakers not to open the cages and we handle them in a deep container (e.g., the sink) to prevent them from escaping. This behavior problem abates with further breeding into the Cast/Si background.
14. We recommend using pure C57BL/6J embryos for the wild-type controls. As the mutations were induced on a C57 background, this avoids potential confusion of bona fide de novo mutations with functionally irrelevant strain polymorphisms.
15. For convenience, we perform the RT reaction in a PCR machine. Instead of placing the tubes on ice, we hold the samples at 4 °C.
16. The IGTC ES cells are feeder-independent and can be maintained on gelatin-treated tissue culture dishes.
17. Detailed protocols on the freezing, thawing, and propagating mouse ES and feeder cells can also be found in ref. [9].
18. To prevent ES cell differentiation, never allow them to overgrow and always make sure cells are thoroughly dissociated when splitting the cells. Prolonged culture before blastocyst injection should be avoided.

19. It is advisable to verify the identity of the insertion site in the gene-trapped ES cell line before proceeding with generation of the mouse strain. This is typically done by RT-PCR using methods described by the gene-trap center. For example, the Sanger Institute Gene Trap Resource (SIGTR) has a protocol page (http://www.sanger.ac.uk/PostGenomics/genetrap/protocols.shtml) describing methods used for such validation.
20. The total number and density of ES cells, as well as the solution used for ES cell resuspension varies between institutions. Consult your transgenic facility about its requirement.
21. Identification of gene-trap/ENU mutant transheterozygotes form complementation crosses may be performed by genotyping. PCR for the LacZ insertion can be used to test presence of the gene-trap allele. Amplification of the region containing the ENU mutation followed by sequencing of the PCR product should reveal 2 peaks in the trace at the site of the mutation (one wild-type and one mutant) if the embryo carries the ENU-induced allele.
22. The gene-trap homozygotes or gene-trap/ENU allele transheterozygotes may display a somewhat different phenotype from the ENU allele homozygotes if one of the alleles is hypomorphic. Gene-trap alleles are frequently null, although they may be hypomorphic due to splicing around the insertion or generation of a functional fusion protein.

References

1. Nord AS, Chang PJ, Conklin BR, Cox AV, Harper CA, Hicks GG, Huang CC, Johns SJ, Kawamoto M, Liu S, Meng EC, Morris JH, Rossant J, Ruiz P, Skarnes WC, Soriano P, Stanford WL, Stryke D, von Melchner H, Wurst W, Yamamura K, Young SG, Babbitt PC, Ferrin TE (2006) The International Gene Trap Consortium Website: a portal to all publicly available gene trap cell lines in mouse. Nucleic Acids Res 34:D642–D648
2. Skarnes WC, Rosen B, West AP, Koutsourakis M, Bushell W, Iyer V, Mujica AO, Thomas M, Harrow J, Cox T, Jackson D, Severin J, Biggs P, Fu J, Nefedov M, de Jong PJ, Stewart AF, Bradley A (2011) A conditional knockout resource for the genome-wide study of mouse gene function. Nature 474:337–342
3. Collins FS, Rossant J, Wurst W (2007) A mouse for all reasons. Cell 128:9–13
4. Anderson KV (2000) Finding the genes that direct mammalian development : ENU mutagenesis in the mouse. Trends Genet 16:99–102
5. Zohn IE, Anderson KV, Niswander L (2005) Using genomewide mutagenesis screens to identify the genes required for neural tube closure in the mouse. Birth Defects Res A Clin Mol Teratol 73:583–590
6. Garcia-Garcia MJ, Eggenschwiler JT, Caspary T, Alcorn HL, Wyler MR, Huangfu D, Rakeman AS, Lee JD, Feinberg EH, Timmer JR, Anderson KV (2005) Analysis of mouse embryonic patterning and morphogenesis by forward genetics. Proc Natl Acad Sci U S A 102:5913–5919
7. Kasarskis A, Manova K, Anderson KV (1998) A phenotype-based screen for embryonic lethal mutations in the mouse. Proc Natl Acad Sci U S A 95:7485–7490
8. Herron BJ, Lu W, Rao C, Liu S, Peters H, Bronson RT, Justice MJ, McDonald JD, Beier DR (2002) Efficient generation and mapping of recessive developmental mutations using ENU mutagenesis. Nat Genet 30:185–189
9. Nagy A, Gertsenstein M, Vintersten K, Behringer R (2003) Manipulating the Mouse Embryo. Cold Spring Harbor Laboratory Press, New York, NY
10. Dietrich WF, Miller J, Steen R, Merchant MA, Damron-Boles D, Husain Z, Dredge R, Daly

MJ, Ingalls KA, O'Connor TJ (1996) A comprehensive genetic map of the mouse genome. Nature 380:149–152

11. Eggenschwiler JT, Espinoza E, Anderson KV (2001) Rab23 is an essential negative regulator of the mouse Sonic hedgehog signalling pathway. Nature 412:194–198
12. Huangfu D, Liu A, Rakeman AS, Murcia NS, Niswander L, Anderson KV (2003) Hedgehog signalling in the mouse requires intraflagellar transport proteins. Nature 426:83–87
13. Liu A, Wang B, Niswander LA (2005) Mouse intraflagellar transport proteins regulate both the activator and repressor functions of Gli transcription factors. Development 132:3103–3111
14. Hoover AN, Wynkoop A, Zeng H, Jia J, Niswander LA, Liu A (2008) C2cd3 is required for cilia formation and Hedgehog signaling in mouse. Development 134:4049–4058

Chapter 8

Generation of Mouse Embryos with Small Hairpin RNA-Mediated Knockdown of Gene Expression

David A.F. Loebel, Tania Radziewic, Melinda Power, Joshua B. Studdert, and Patrick P.L. Tam

Abstract

We are using knockdown of gene expression in mouse embryos by constitutive expression of small hairpin (sh)RNAs as a means of observing loss-of-function phenotypes more rapidly than gene targeting. Plasmid constructs that direct shRNA expression via an RNA pol III promoter are introduced into embryonic stem (ES) cells by electroporation and drug selection. Clones are propagated and the degree of knockdown assessed by quantitative protein or RNA methods. Selected ES cell clones are used to generate embryos by tetraploid complementation. Blastomeres of two cell embryos are electrofused to generate tetraploid embryos. Chimeric embryos are produced by injection of ES cells into blastocysts or aggregation with morulae. In these embryos, the tetraploid cells become excluded from the fetal tissues, resulting in ES cell-derived embryos harboring the shRNA knockdown construct. Embryos can be collected and their phenotype assessed by appropriate means.

Key words Mouse embryo, Tetraploid complementation, shRNA, RNA interference, Knockdown, Embryonic stem cell, Blastocyst injection, Aggregation

1 Introduction

Analysis of gene function in the mouse embryo has conventionally involved studying loss-of-function mutations created by gene targeting resulting in deletion or mutation of individual genes. This method has proved very valuable in efforts to understand the roles of genes in embryogenesis. However, construction of plasmid targeting vectors, selection of embryonic stem (ES) cell clones with the correctly targeted insertion, production of chimeric mice, and establishment of mouse lines are slow and laborious. In addition, mutant mice need to be kept on the shelf, with the associated maintenance and genotyping costs. The use of RNA interference (RNAi) in embryos provides a means of potentially identifying a mutant phenotype much faster than gene targeting.

Mark Lewandoski (ed.), *Mouse Molecular Embryology*, Methods in Molecular Biology, vol. 1092,
DOI 10.1007/978-1-60327-292-6_8, © Springer Science+Business Media New York 2014

1.1 RNAi and shRNA

RNAi involves introducing small double-stranded RNA molecules into cells, which then trigger the cleavage and degradation of the target transcript and consequently a reduction in the amount of gene product. Knockdown of gene function in embryos can be achieved either by directly introducing double-stranded short interfering RNAs (siRNAs) into cells or by expressing hairpin RNAs from constructs that have been integrated into the genome. Vectors have been developed that allow expression of short hairpin RNAs (shRNAs) from plasmids in mammalian cells. Expression of shRNAs is directed by RNA polymerase III promoters such as H1 [1] or U6 [2]. The shRNAs are cleaved in the cell to form double-stranded siRNAs. Expression of shRNAs in embryonic stem (ES) cells opens up the possibility for generating embryos that express shRNAs in all cells [3].

Knockdown by siRNA has been useful for studying the consequences of reducing the activity of specific genes in preimplantation and early postimplantation mouse embryos [4, 5], but analysis of embryogenesis in later postimplantation period requires that the knockdown be sustained over a longer time period. To achieve this, the shRNA construct is delivered to embryonic stem (ES) cells by electroporation and selection for drug resistance. This has the advantage of allowing the level of expression of the target gene to be tested in vitro in undifferentiated ES cells to assess the effectiveness of the knockdown. Depending on whether the gene is expressed in undifferentiated ES cells it might be necessary to allow the cells to undergo differentiation as embryoid bodies (EB). Quantitative methods are required to test the efficiency of knockdown. If an antibody is available, this can be used to measure the amount of protein produced, compared to suitable controls. Since shRNA-mediated gene silencing results in the specific degradation of RNAs, it is possible to measure the degree of knockdown by quantifying the transcript using Northern blotting or real-time RT-PCR. Cell lines exhibiting the desired degree of knockdown can then be chosen for further analysis via directed in vitro differentiation and testing effects on embryogenesis in vivo.

1.2 Design and Construction of shRNA Plasmids

The choice of target sequence and the design of the specific oligonucleotides are critical for success of knockdown experiments. There are internet resources that can assist in choosing a target sequence within your gene of interest. The shRNA oligonucleotides consist of a short sequence identical to the target sequence beginning with AA binucleotide, a loop sequence and the reverse complement of the target sequence and ending with a string of thymidines to terminate RNA polIII transcription. We have knocked down gene expression in ES cells using target sequences selected using the GeneLink shRNA Explorer (http://www.genelink.com/), but there are several other shRNA design resources provided by commercial oligonucleotide suppliers. Methods for the design and construction of shRNA vectors have recently been reviewed in this series [6]. It is also possible to purchase sets of shRNA constructs chosen for many mouse and human genes. Libraries of knockdown shRNA constructs, designed by the RNAi Consortium, a combined public–private partnership [7] are available through Sigma-Aldrich and Open Biosystems. We have successfully used constructs from this source, which are in the pLK01 vector that can be introduced into cells either in lentivirus or as a plasmid. Several sets of constructs should be tested for knockdown efficiency. Appropriate controls include (a) non-targeting shRNAs, designed to avoid homology to any mouse gene; (b) "scrambled" sequences that contain the same base composition as a gene-specific shRNA but in a jumbled order; and (c) empty vectors that do not contain any shRNA sequence.

1.3 Knockdown of Gene Expression in ES Cell-Derived Embryos

To study the effects of knockdown of gene activity in vivo, completely ES cell-derived embryos must be produced. This method takes advantage of the inability of tetraploid cells to contribute efficiently to fetal tissues. Tetraploid mouse embryos fail to develop normally and in chimeric embryos the tetraploid cells are confined to extraembryonic lineages, resulting in embryos that are composed almost entirely of ES cell-derived tissues [8, 9]. Exclusion of tetraploid cells occurs after gastrulation [8]. Creation of embryos by tetraploid complementation with ES cells has been useful for analyzing the effects of mutations on embryo development in situations where a gene has a role in both embryonic and extraembryonic tissues, such as GATA and HNF family transcription factors [10, 11]. Embryos created in this way from ES cells expressing shRNAs have been shown to exhibit comparable phenotypes to those that harbor a targeted deletion of the same gene [3, 12] and have been used to investigate mutant phenotypes of genes that are potential targets of WNT signalling in mouse development [13].

ES-embryo chimeras can be produced either by injection of ES cells into the cavity of blastocysts or aggregation with 4 or 8-cell stage embryos. We have found that production of embryos by blastocyst injection results in a higher and more consistent yield of embryos for analysis. However, this procedure requires specialized micromanipulation equipment, so aggregation with morulae may be a more amenable alternative. Here, we describe tissue culture and embryological methods for electroporation of ES cells, differentiation to EBs to assess the degree of knockdown and creation of ES cell-derived embryos by tetraploid complementation using blastocyst injection or morula aggregation.

2 Materials

2.1 Culture and Electroporation of ES Cells

1. Tissue culture plastic ware: 100 × 20 mm Tissue Culture Dish, 6-well Tissue Culture Plate, 48-well tissue culture plate, 96-well round bottom tissue culture plate (Falcon).

Table 1
ES cell media

Component	Stock	Per 500 mL	Comments
Fetal calf serum		62.5 mL	Heat inactivate for 30 min at 55 °C before use
Pen/strep (Gibco)	100×	5 mL	Store in 10 mL aliquots
Glutamax 1 (Gibco)	100×	5 mL	Add 4 weeks after making media
β-Mercaptoethanol	1 % solution in PBS−	350 μL	
Nonessential amino acids (Gibco 11140-050)	100×	5 mL	
Nucleosides		5 mL	Stock solution is Adenosine 0.16 g Cytidine 0.146 g Guanosine 0.17 g Uridine 0.146 g Thymidine 0.048 g In 200 mL
LIF (ESG1107, Millipore)	5.56×10^7 units/mL	450 μL	Supplied as 10^7 units in 1 mL. Add 800 μL and store in 100 μL aliquots at 4 °C
DMEM		Up to 500 mL	

2. 15 mL Polypropylene Tubes.
3. BioRad cuvette. 0.4 cm electrode gap—50 (BioRad. Catalogue Number 165-2088).
4. 1.8 mL Cryovials (Nunc. Catalogue Number 375418).
5. Phosphate-buffered saline (PBS−), Ca^{2+}, Mg^{2+} free.
6. 0.05 % Trypsin–EDTA.
7. ES media (*see* Table 1 for preparation).
8. 0.1 % gelatine in H_2O, autoclaved (Sigma. Catalogue Number G-9391).
9. Mouse tonicity (MT)-PBS (15 mM Na_2HPO_4, 0.4 mM $NaH_2PO_4 \cdot 2H_2O$, 0.15 M NaCl).
10. G418 geneticin (Roche. Catalogue Number 1464990).
11. Puromycin dihydrochloride, cell culture tested (Sigma. Catalogue Number P-8833).
12. 2× freeze media (50 % FCS, 30 % ES media, 20 % Dimethyl sulfoxide, DMSO).

13. Mineral oil, embryo culture tested (Sigma. Catalogue Number M-8410).
14. ES cells: R1 [14] or G4 [15].
15. shRNA and control plasmids.

2.2 Assessment of Knockdown in ES Cells or Embryoid Bodies

1. Phosphate-buffered saline, Ca^{2+}, Mg^{2+} free.
2. 0.05 % Trypsin–EDTA.
3. LIF negative ES medium (as for Table 1, but with LIF omitted).
4. 0.1 % gelatine in H_2O (Roche. Catalogue Number 1464990).
5. 65 mm bacterial dishes.

2.3 Production of ES Cell-Derived Embryos by Tetraploid Complementation

2.3.1 Preparation of ES Cells

1. Phosphate-buffered saline, Ca^{2+}, Mg^{2+} free.
2. 0.05 % trypsin–EDTA.
3. ES media.

2.3.2 Collection, Electrofusion, and Culture of Embryos

1. Folligon (Intervet)- Serum gonadotrophin 1,000 IU powder diluted in phosphate-buffered saline.
2. Chorulon (Intervet)- Chorionic Gonadotrophin 1,500 IU powder diluted in phosphate-buffered saline.
3. M16 medium (prepared from stock solutions according to Tables 2 and 3 or available from Sigma). Filter and divide into 1 mL aliquots.
4. M2 medium (prepared fresh weekly according to Tables 2 and 3 or available from Sigma). Filter and divide into 10 mL aliquots.
5. 0.3M Mannitol Solution (0.3 M mannitol, 0.3 % BSA in ultra-pure water), filtered and frozen in 1 mL or 10 mL aliquots.
6. Mineral oil, embryo culture tested (Sigma).
7. 60 × 15 mm round sterile tissue culture dish.
8. 100 × 20 mm round sterile tissue culture dish.
9. Dissection equipment: coarse forceps, fine forceps (2 pairs), fine scissors
10. 30G needle attached to 1 mL syringe.
11. Electrofusion apparatus:

    CF-150 Fusion Machine (BLS, Hungary).

    250 μm Laszlo slide (Electrofusion chamber; BLS, Hungary).

    Dissecting microscope.

    CO_2 incubator.

    Glass embryo handling pipettes.

Table 2
Stock solutions for M2 and M16 media

Component	g/100 mL
Stock A (10× conc)	
NaCl	5.534
KCl	0.356
KH_2PO_4	0.162
$MgSO_4 \cdot 7H_2O$	0.293
Sodium lactate	2.610
Glucose	1.0
Penicillin	0.060
Streptomycin	0.050
Stock B (10× conc)	
$NaHCO_3$	2.101
Phenol red	0.010
Component	g/10 mL
Stock C (100× conc)	
Sodium pyruvate	0.036
Stock D (100× conc)	
$CaCl_2 \cdot 2H_2O$	0.252
Component	g/100 mL
Stock E (10× conc)	
HEPES	5.958
Phenol red	0.010

Table 3
Components of M2 and M16 media

Stock	50 mL M2	50 mL M16
A	5.0 mL	5.0
B	0.8 mL	5.0
C	0.5 mL	0.5
D	0.5 mL	0.5
E	4.2 mL	–
H_2O	39.0 mL	39.0
BSA	200 mg	200 mg

Filter through 0.2 μm filter

2.3.3 Alternative 1: Blastocyst Injection

1. Microinjection equipment.

   Olympus IX7 with Hoffman optical system.

   Eppendorf Transferman NK *x*2.

   Eppendorf Cell tram air.

   Eppendorf Cell tram Vario.

   Eppendorf ES cell transfer pipettes 15 μm inside diameter and 20 μm outside diameter, 20° tip angle, 1 mm long end.

   Eppendorf Holding pipettes 20–25 μm inside diameter, 120–130 μm outside diameter, 35° angle, 1 mm long end.
2. M16 medium.
3. M2 medium.
4. 60 × 15 mm sterile tissue culture dish.
5. Glass embryo handling pipettes.
6. Superovulated donor mice.
7. 0.3 M Mannitol Solution.
8. Mineral oil, embryo culture tested (Sigma).

2.3.4 Alternative 2: Aggregation of Morulae

1. M16 medium.
2. M2 medium.
3. Tissue culture dishes: 60 × 15 mm tissue culture dish, 100 × 20 mm tissue culture dish.
4. Superovulated donor mice.
5. 0.3 M Mannitol Solution.
6. Acid Tyrodes solution (Sigma. Catalogue Number T1788).
7. Mineral oil, embryo culture tested (Sigma. Catalogue Number M-8410).
8. Sterile plastic transfer pipette.
9. Darning needles DN -09 (BLS Biochemical Laboratory Services Ltd. Catalogue Number 11-9014-1).
10. Aspirator tube and mouthpiece (Sigma. Catalogue Number A5177).

2.3.5 Embryo Transfer to Pseudopregnant Mice

1. Surgical equipment:

   2 pairs fine forceps.

   2 pairs coarse forceps.

   Small scissors.

   Alligator clip (serrafine clip).

   Surgical stapler and staples.

   27G needle and 1 mL syringe.

Fine curved suture needle.

Suture holder.

Suture (Ethilon 5.0, 45 cm Reverse cutting C2 Nylon).

2. 125–170 μm glass pipette to transfer the embryos made by pulling a glass capillary or pasteur pipette over a flame.
3. 70 % ethanol.
4. 60 × 15 mm sterile tissue culture dish.
5. Mice: 8-week-old female mice mated with vasectomized males 2 days prior to embryo transfer.
6. 30G needles and 1 mL syringes.
7. Anesthetic (0.5 mL Ketamine, 0.25 mL Illium Xylazil-20 in total volume of 5 mL). Made fresh on day of use. Inject 10 μL per gram.
8. Heating pad and lamp.

3 Methods

3.1 Culture and Electroporation of ES Cells

3.1.1 Thawing ES Cells

1. To thaw cells, collect vial(s) from storage unit (−80 °C or liquid nitrogen).
2. Thaw in a 37 °C waterbath. Add contents of cryovial to 5 mL ES media (*see* **Note 1**).
3. Spin for 5 min at 250 × *g*.
4. Aspirate supernatant, flick to resuspend pellet plate on fibroblast feeder layer (*see* **Note 2**).

3.1.2 Passaging ES Cells

Optimally ES cells should be fed every day and split every second to third day at which time they should be 70–80 % confluent. It is important not to let the cells overgrow as this affects their pluripotency.

If the cells are to be passaged on 0.1 % gelatine, prepare the required number of dishes beforehand.

1. Assess cell morphology and density to determine the dilution required (*see* **Note 2**).
2. Aspirate media and rinse dish with PBS−. For a 60 mm plate use 1 mL or for a 100 mm dish use 3–5 mL. When aspirating, hold the dish at a slight angle and use a sterile Pasteur pipette held above the cell layer.
3. Aspirate PBS− and add trypsin. For a 60 mm plate use 0.5 mL, for a 100 mm dish use 1.5–2 mL.
4. Incubate at 37 °C for 2–3 min or until cells have become loosened from the bottom of the dish and assess under a microscope.

5. Inactivate trypsin using ES media, pipetting up and down (without generating bubbles) to ensure a single cell suspension (*see* **Note 3**).
6. Add cell suspension to a 15 mL tube and spin at 250 × *g* for 5 min.
7. Aspirate media without dislodging pellet. Flick to loosen pellet.
8. Prepare appropriate dilution and plate cells accordingly. For each 60 mm plate use 2–3 mL and for each 100 mm dish use 8–12 mL.

3.1.3 Electroporation and Drug Selection of ES Cells

Constructs encoding the shRNA are introduced into ES cells by electroporation and clones containing the construct are selected based on resistance to drug treatment. The choice of drug depends on the vector. Commonly used selective agents include G418 and puromycin. The concentration of selective agent used and the time course of selection should be determined empirically. Puromycin selection usually runs for 3–6 days and G418 selection for 5–10 days.

This protocol applies to electroporations that are carried out using a BioRad Gene Pulsar.

1. Thaw cells 3–5 days in advance to obtain a 100 mm dish of 70–80 % confluence.
2. Rinse cells with PBS-CMF and trypsinize as described in Subheading 3.2, **step 2**. Pipet ES cells thoroughly to ensure single cell suspension.
3. Remove 10–15 μL aliquot of cell suspension for counting on hemocytometer. Between 5×10^6 and 1×10^7 cells are required per electroporation.
4. Spin at 250 × *g* in a benchtop centrifuge for 5 min. Aspirate media.
5. Add 600–650 μL MT-PBS per electroporation. Flick pellet to resuspend.
6. Add cell suspension to 1.5 mL Eppendorf containing 25–50 μg linearized shRNA construct in maximum 100 μL volume MT-PBS or water.
7. Mix thoroughly. Add total contents to prechilled 0.4 cm cuvette.
8. Rest cells on ice for 8 min.
9. Prepare 100 mm gelatine dishes (*see* **Note 4**).
10. Set parameters on BioRad GenePulsar: 250 V, 500 μF for 25–50 μg DNA.
11. Dry outside of cuvette and place in GenePulsar.
12. Pressing both buttons simultaneously, hold in until machine beeps. Return cuvette to ice and rest for 10 min.

13. Dispense cuvette contents equally among the dishes and place in incubator.
14. Begin selection process in 24–48 h depending on recovery (*see* **Note 5**).
15. Most cells will die, leaving floating cell debris in the media. After several days drug resistant cells containing the plasmid will have proliferated to form colonies.
16. When this stage is reached, individual colonies are ready to be picked.
17. Aliquot 50 μL trypsin to a round bottom 96-well plate. Only aliquot the number of wells you plan pick.
18. Rinse the selection plate with 3–5 mL PBS then aspirate off. Add another 3–5 mL of fresh PBS. Do not aspirate.
19. Using a 20 μL pipette (set at 10–20 μL), and sterile 20–200 μL pipette tips, carefully pick an individual colony by scraping the colony from the bottom of the dish with the pipette tip, then sucking it into the pipette once dislodged. Transfer it to one of the trypsin wells in the 96-well plate. Repeat this for as many clones as required using a fresh pipette tip and fresh well for each colony.
20. When enough clones have been picked, incubate for 2–5 min at 37 °C. Pick 20–40 clones at one time.
21. Inactivate the trypsin with 150 μL media and pipette up and down to break up clumps of cells.
22. Transfer the entire volume of the well to a 48- or 24-well plate containing feeder cells.
23. Colonies should be fed each day until the majority of plates have reached 70–80 % confluence. Entire plates can then be frozen for future use, or trypsinized and moved to a larger surface area in order to make cell stocks.

3.1.4 Freezing Cells

A. From a 100-mm dish
   1. Harvest the cells by trypsin digestion and centrifugation.
   2. For each cryovial, add 0.5 mL ES media and 0.5 mL freeze media.
   3. Tightly seal the lids of the pre-labelled cryovials and place them in polystyrene holders or Nalgene holders for storage in −80 °C freezer. Vials must remain at least 24 h at −80 °C before being transferred to permanent storage at liquid nitrogen.

B. From a 48-well plate
   1. Aspirate media. Rinse each well with PBS-CMF and aspirate.

2. Add trypsin (100 μL), incubate for 2–3 min at 37 °C or until detached.
3. Inactivate trypsin with 200 μL ES media, pipette up and down to disaggregate cell clumps.
4. Add 300 μL freeze media to each well and mix gently.
5. Add mineral oil to just cover media in wells.
6. Keeping plate flat, wrap edges in parafilm to seal in liquids. Place plates on ice.
7. Freeze plates at −80 °C in Styrofoam containers, kept completely level and protected by several layers of paper towel.

3.2 Assessment of Knockdown in ES Cells or Embryoid Bodies

If the gene of interest is expressed in undifferentiated ES cells it is possible to determine the degree of knockdown by harvesting ES cells and extracting RNA or protein to measure transcript or protein levels compared to an appropriate control. If not, differentiation of ES cells into embryoid bodies (EB) enables knockdown efficiency to be tested. Firstly, determine the time of differentiation when expression of your gene is optimal by conducting a time course experiment of embryoid body differentiation.

1. Split ES cells from feeder layer plates onto gelatine (*see* **Note 6**).
2. Passage cells 1–3 times on to fresh gelatine plates until there are no fibroblasts in the culture plate.
3. Trypsinize cells until colonies begin to lift off plate. Gently dislodge cells without disrupting clumps and plate at 1:3–1:5 dilution onto bacterial petri dishes in LIF negative ES media. This is day 1 (*see* **Note 7**).
4. Feed cells on days 3, 5, 7, 9, 11, 13, and 15 or when the embryoid bodies are sticking to the dish surface (*see* **Notes 8** and **9**).
5. Harvest EB at desired time point of in vitro differentiation and collect in a 15 mL polypropylene tube.
6. Rest without disrupting for 2–5 min until all embryoid bodies settle at base of tube.
7. Carefully aspirate media and add 5 mL PBS.
8. Rest again.
9. Carefully aspirate all PBS and snap freeze cell pellet in liquid nitrogen. Store pellets at −80 °C or immediately extract RNA.
10. RNA can be extracted using methods optimized for small starting samples, including the RNeasy Micro kit (Qiagen). Expression of the gene of interest can be measured either by semiquantitative RT-PCR or quantitative real-time RT-PCR.
11. When the optimal time to assay for knockdown efficiency has been determined, perform the EB differentiation protocol on

ES cells transfected with shRNA plasmids. Harvest and extract RNA at the optimal time and measure the transcript or protein level of knockdown cell lines compared to controls and parental cell line.

3.3 Production of ES Cell-Derived Embryos by Tetraploid Complementation

3.3.1 Preparation of ES Cells

1. Aspirate media, rinse with PBS-CMF and aspirate.
2. Add minimal volume of trypsin (*see* **Note 10**).
3. Incubate at 37 °C for 2–3 min until cells become detached from surface.
4. Inactivate trypsin using ES media.
5. *Alternative A*

   If the cells are to be used for blastocyst injections, pipette up and down to ensure cells are in single cell suspension.

   Alternative B

   If the cells are to be used for morula aggregations, very gently pipette up and down (without generating bubbles) to ensure cells stay in grape-like clusters of 10 to 15 cells.
6. Collect cell suspension for each procedure into a 15 mL tube and place on ice.

3.3.2 Collection, Electrofusion, and Culture of Embryos

To obtain sufficient numbers of embryos for electrofusion prior to injection or aggregation, female mice are usually superovulated by injection with Folligon (serum gonadotrophin) and Chorulon (chorionic gonadotrophin). The volumes to be injected need to be optimized for each strain. Mice are plug checked and embryos collected at the 2-cell stage (day 1.5, *see* **Note 11**).

1. Prepare the surgical area, work area, and instruments so that the area is clean and sterile. Wipe down work area and instruments with 70 % alcohol then air dry.
2. Prepare culture dish. Place eight small separate drops of M16 into a sterile 60 × 15 mm culture dish. Carefully overlay with sterile mineral oil and incubate at 37 °C and 5 % CO_2.
3. Prepare the collection dish. Make two rows of large drops of M2 media and two rows of smaller drops of M2 in a sterile a 100 × 20 mm tissue culture dish lid (Fig. 1a; *see* **Note 12**).
4. Sacrifice the pregnant mouse by cervical dislocation. Note that euthanasia procedures must be approved by your Institutional Animal Care and Use Committee.
5. Lay the mouse on its back on an underpad and wipe down the abdomen with alcohol.
6. Using coarse forceps pinch the skin of the mouse around the midline and make a lateral incision with the small scissors. Using gloved clean fingers pinch the skin above and below the incision site and pull the skin toward the head and tail.

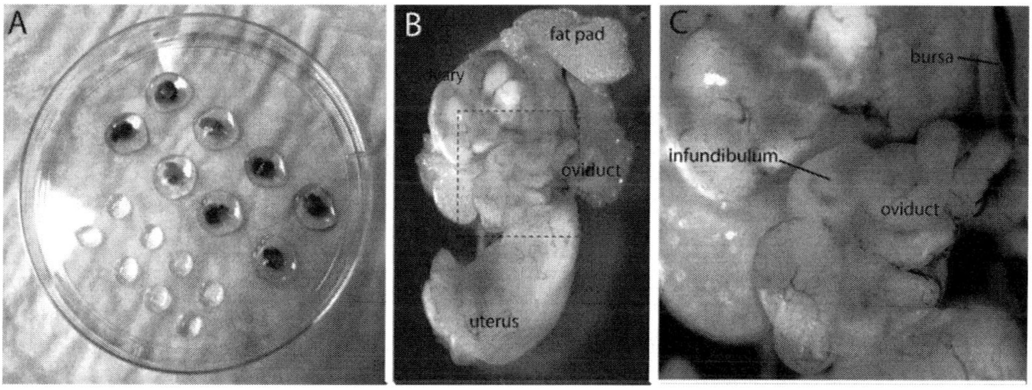

Fig. 1 Embryo collection. (**a**). Embryo collection dish with large and small drops of M2 media for flushing oviducts. The large drops contain dissected ovaries and oviducts. (**b**) Dissected ovary and oviduct. (**c**) Higher magnification view showing the opening at the infundibulum, where the needle is inserted for flushing the oviduct

The skin should easily come away, exposing the body wall underneath. Take a pair of fine forceps and pinch and lift the muscular body wall and once again make a lateral incision.

7. Continue to cut upwards along the midline being careful not to cut any internal organs. Make two incisions on either side of the midline towards the head at approximately 45° and push up the flaps of the body wall to expose the internal organs.
8. Using clean coarse forceps push the intestines up and out of the way to expose the reproductive organs: the uterus, oviduct, and ovary.
9. With fine forceps, grasp the uterus just below the oviducts and cut with scissors.
10. Cut away the fat pad to release the ovary and oviduct and place into one of the large M2 drops in the collection dish. Repeat on the other side (*see* **Note 13**).
11. To flush the oviduct, take two fine forceps and locate the infundibulum (the opening of the oviduct). Tear away the bursa to expose the entrance, and then insert the needle in the lumen and flush with a small amount of M2 (Fig. 1).
12. Lift the ovary/oviduct out of the drop and put aside.
13. Repeat with the remaining drops. The embryos will settle on the floor of the dish.
14. After each flushing collect the two cell embryos with a mouth pipette attached to a finely drawn Pasteur pipette, and place in a clean drop of M2 on the collection dish.
15. Take the embryos through several washes until, they are clean and free of debris.

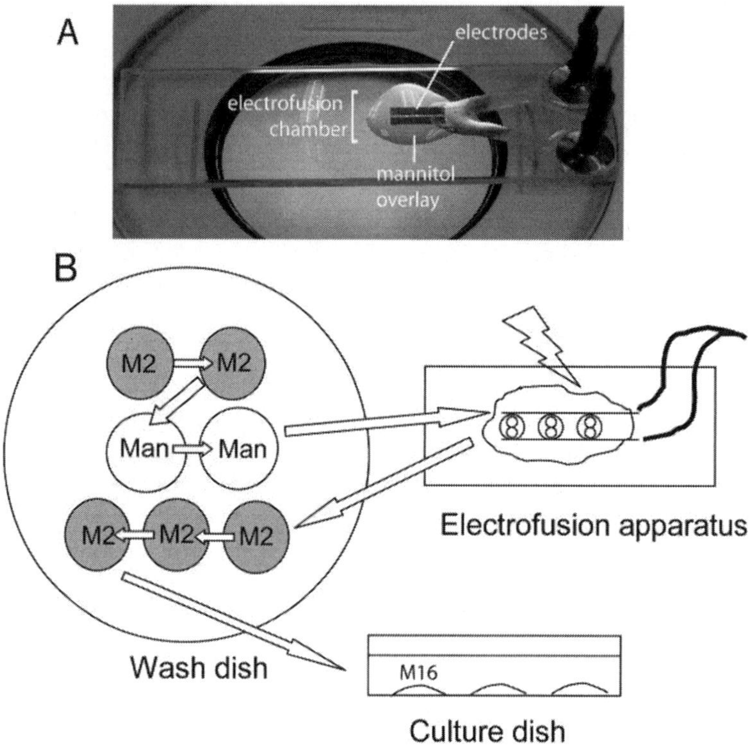

Fig. 2 Electrofusion. (**a**) Electrofusion slide showing the electrodes overlayed with mannitol solution. (**b**) Diagram of electrofusion process, showing the sequence of M2 and mannitol washes and the position of the 2-cell embryos between the electrodes on the electrofusion slide

16. Set the CF-150 Fusion Machine (BLS Hungary) to the following parameters: 40 V, 30 μs, two repeats, 1.0 attenuator. Set the non-electrolyte button to ON.
17. Prepare the fusion slide: Using a 250 μm Laszlo slide overlay the Mannitol solution onto the slide to cover the electrodes and attach the wires to the fusion machine (Fig. 2).
18. Prepare the wash dish. On the lid of a sterile 16 × 15 mm polystyrene tissue culture dish prepare two drops of M2 media, two drops of Mannitol, and three drops of M2 media.
19. Take the 2-cell embryos from the culture dish and pass through two washes of M2 media. Take no more than 20 embryos at a time (Fig. 2).
20. Place the embryos into the first drop of Mannitol solution and wait until they settle to the bottom.
21. Transfer to the second drop of Mannitol.
22. Pick up the embryos and then carefully place them between the electrodes of the fusion machine overlaid with the Mannitol solution.

23. Align the embryos by passing an AC current (1–2 V) so that the plane of contact of the two blastomeres is parallel to the electrodes.
24. Press the pulsing trigger button on the fusion machine.
25. Remove the embryos from the electrode and pass through two or three M2 washes.
26. Wash in M16 and place the electrofused embryos into the culture dish and return to the incubator.
27. Repeat with the remaining embryos.
28. Change all the wash drops and mannitol solutions every three batches.
29. Blastomeres should start to fuse (becoming one cell) in 30–60 min.
30. Discard any that did not fuse after 2 h.
31. Culture until blastocyst stage for micro-injection or 4-cell stage for aggregation in an incubator at 37 °C and 5 % CO_2.

3.3.3 Alternative 1: Blastocyst Injection

1. Prepare holding dish. Place eight small separate drops of M16 culture medium on the bottom of a sterile 60 × 15 mm tissue culture dish and overlay with mineral oil. Label dish as injected embryos. Leave in an incubator at 37° to warm and equilibrate.
2. Prepare injection dish. Place three medium sized drops of M2 media down the center of the lid of a 60 × 15 mm Sterile Tissue culture dish (*see* **Note 14**). Overlay this with enough mineral oil to just cover the drops.
3. Prepare micromanipulator. Attach the holding pipette and injection needle in the holders of the microscope (Fig. 3).
4. Turn the dial of the Cell Tram Vario until some of the oil is near the tip of the needle. Ensure there are no bubbles in the line.
5. Place a group of 10–20 blastocysts in the center of the first drop of M2 in the injection dish.
6. Place a small sample of ES cells in the same drop, above the blastocysts.
7. Place the dish on the microscope. Using a low power objective, align the holding pipette and bring it down through the oil into the drop of media into the same plane of focus as the blastocysts.
8. Gently attach one blastocyst to the holding pipette, using the Cell Tram Air.
9. Bring the injection needle down into the drop into the same plane of focus as the blastocysts.
10. Using the manipulators and the Cell Tram Vario, carefully pick up 10–15 ES cells into the injection needle.

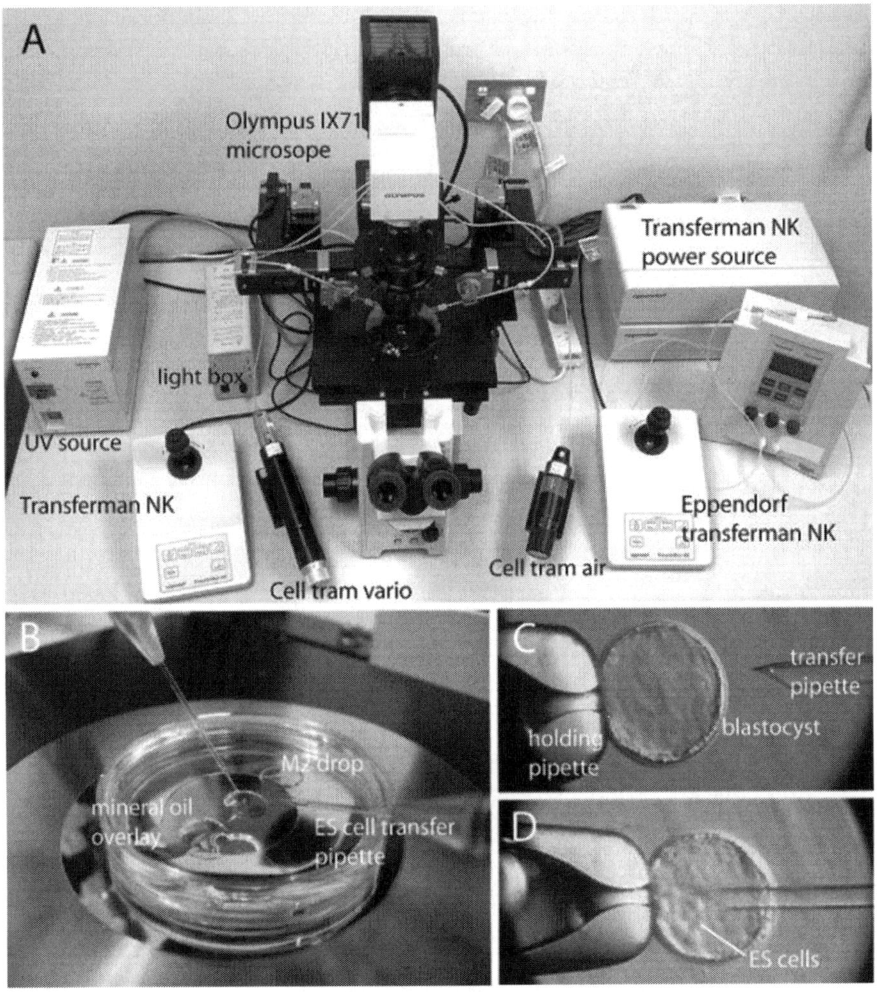

Fig. 3 Microinjection. (**a**) Layout of microinjection apparatus. (**b**) Close view of injection dish, holding pipette and ES cell transfer pipette. (**c**, **d**) Injection of ES cells into the cavity of a blastocyst. Images B-D Ruth Hardman

11. Bring the needle to the position of the holding pipette with the blastocyst attached.
12. Orientate the blastocyst so that the injection site faces the injection needle and make sure it is firmly but gently attached to the holding pipette (Fig. 3; *see* **Note 15**).
13. Align the injection site in the same plane as the needle using the ×20 objective.
14. With a smooth, quick motion, advance the needle towards and through the blastocyst into the blastocoel cavity. Gently release the ES cells into the cavity of the blastocyst with the Cell Tram Vario.

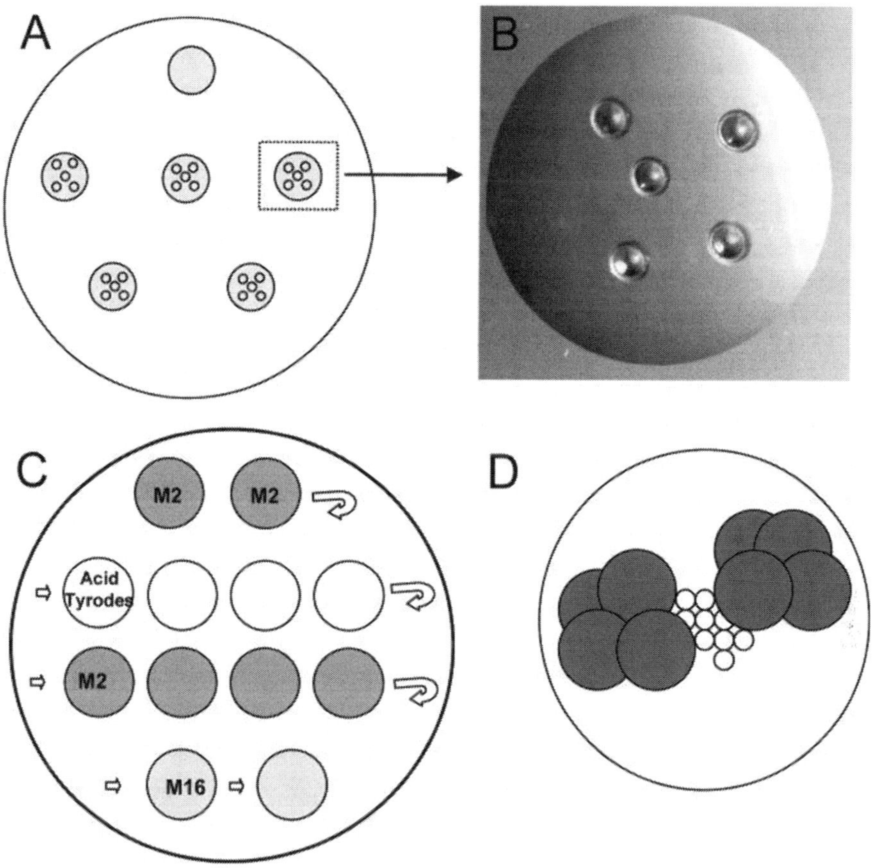

Fig. 4 Aggregation. (**a**) Arrangement of M16 drops on aggregation plate showing drops containing divots. (**b**) Individual drop of media showing divots made in plastic surface with a darning needle. (**c**) Diagram of wash plate showing order of media washes. (**d**) Aggregation of ES cells by sandwiching between two 4 cell tetraploid embryos

15. Withdraw the needle and move the injected blastocyst to the bottom of the drop, away from the non-injected blastocysts using the holding pipette and manipulators.
16. Repeat the procedure with the other blastocysts (*see* **Note 16**).
17. Transfer the successfully injected blastocysts to drops of M16 in a culture dish at 37 °C, 5 % CO_2 to recover.
18. Repeat using the next drop of the injection dish.
19. Transfer the injected blastocysts to pseudopregnant female mice.

3.3.4 Alternative 2: Aggregation with Morulae

1. Isolate the embryos that have reached the 4-cell stage using a 200 μm pipette and mouth piece, place in a new drop of M16 and return them to the incubator. Put between 10–25 embryos in each drop.
2. Set up aggregation dishes. Using a plastic transfer pipette, place six drops of M16 in a sterile 60 × 15 mm tissue culture

dish and overlay with mineral oil. Make five small divots using a clean, sterile darning needle in each of five drops, leaving one drop for ES cells (Fig. 4; *see* **Note 17**).

3. Return the dish to the incubator to allow equilibration of temperature and CO_2 levels (*see* **Note 18**).
4. Place a row of two drops of M2 and four drops of acid Tyrodes solution in a 100 × 20 mm tissue culture dish lid. Add another row of four drops of M2, followed by two drops of M16. Add a further four drops of M16 to a new 100 × 20 mm tissue culture dish lid. This will be used to rinse the embryos (*see* **Note 19**) (Fig. 4).
5. Remove ten 4-cell embryos from the incubator and place in the first drop of M2 using a 200 μm pipette and mouth piece.
6. Transfer the embryos to a new drop of M2 using the same pipette.
7. Transfer the embryos into the first drop of acid Tyrodes solution (*see* **Note 20**).
8. Allow the embryos to settle to the bottom of the drop before transferring to the next drop (*see* **Note 21**).
9. As soon as the zona pellucida has dissolved, transfer the zona-free 4-cell embryos to a new drop of M2.
10. When all of the embryos have been transferred to M2, rinse them through the remaining drops of M2 by allowing them to settle to the bottom of the drop.
11. After four M2 rinses, rinse in six drops of M16. Change the media and acid Tyrodes solution every 3–4 batches.
12. Remove the aggregation dish from the incubator and transfer all of the embryos to one of the drops containing five divots using a 200 μm pipette and mouthpiece.
13. Move one embryo into each of the five divots, and leave another at the edge of each divot before returning the dish to the incubator (*see* **Note 22**).
14. Using a sterile plastic transfer pipette, gently agitate the ES cell media to distribute cells evenly. Take a small aliquot of ES cells, and place in a 60 × 15 mm tissue culture dish using a plastic transfer pipette.
15. Using a 200 μm pipette and mouthpiece, select 5–10 ES cell aggregates containing 15–25 cells.
16. Rinse the ES cells in a drop of M16 media, before transferring to M16 drop with no divots in the aggregation dish.
17. Using a 200 μm pipette and mouth piece, transfer five ES cell aggregates to the drop containing ten embryos.
18. Gently place one cell aggregate on top of each of the embryos already in the divots.

19. Once the ES cell clumps are in position, transfer the remaining four cell embryos into the divot to form a sandwich between the embryos and ES cells (*see* **Note 23**).
20. Without moving the cells, carefully place the aggregation dish into the incubator. Cells should reach blastocyst stage within the next 24–36 h, ready to transfer into a pseudopregnant female.

3.3.5 Transfer to Pseudopregnant Mothers

1. Weigh each pseudopregnant mouse and inject the anesthetic. Return mouse to its cage until it is anesthetized.
2. Make five separate drops of M2 in a 60 × 15 mm tissue culture dish. Place 10–12 injected blastocysts into one of the drops and set aside.
3. Lay the mouse on its side (right side down) on the surgical table and lay a swab or tissue under the mouse. Check that it is properly anesthetized (*see* **Note 24**).
4. The ovary is located under a depression formed between the rib and the hind leg. Swab the incision site with alcohol. Part the hair slightly.
5. Using small scissors make a small incision in the skin and stretch the skin slightly (*see* **Note 25**).
6. Make a small cut in the fat layer with scissors and stretch slightly.
7. Holding the peritoneum with fine forceps, cut and stretch, avoiding visible blood vessels.
8. Pick up the suture with needle holders and pull the suture through one side of the peritoneum (*see* **Note 26**).
9. Using coarse blunt rounded forceps, enter the cavity and pick up the fat pad (which should be lying just underneath) and gently pull out the attached ovary/uterus to the surface (*see* **Note 27**).
10. Attach a small alligator clip to the fat pad to prevent it sliding back into the abdominal cavity.
11. Load the embryos into the transfer pipette. To do this, first take up a small amount of M2 media, then a small bubble of air. Pick up the blastocysts, another air bubble and then another small amount of media (*see* **Note 28**). This should be contained in no more than 0.25–0.5 cm area at the tip of the pipette.
12. Carefully press the pipette onto a piece of plasticine on the microscope stage.
13. Place the mouse under the dissecting microscope. Make a small hole in the uterus with a 27G needle and syringe by gently grasping the uterus with coarse blunt forceps, and piercing the uterus between the prongs of the forceps (*see* **Note 29**).

14. Gently place the pipette containing the embryos into the hole. Gently expel the medium containing embryos into the uterus. Stop before the medium is emptied from the pipette to avoid blowing air into the uterus.
15. Withdraw the pipette and return the mouse to the surgical table.
16. Remove the clip and gently push the ovary, uterus, and fat pad back into the cavity by only pushing the fat pad (*see* **Note 30**).
17. Using the suture holders and needle, suture the other side of the peritoneum and firmly tie two surgical knots. Cut the thread.
18. Wipe the area surrounding the wound with alcohol, wiping away from the wound at all times.
19. Using coarse forceps gently bring the two sides of the wound together. Apply one surgical staple to the outside skin.
20. Wrap the mouse in a tissue and place in a box on its side, on a heating pad with an overhead lamp. Place a tissue on the box lid to shield the eyes from the lamp.
21. Keep the mouse warm and under observation until awake.
22. Sacrifice pregnant mice and collect embryos at the appropriate developmental stage for phenotypic analysis (*see* **Notes 31** and **32**).

4 Notes

1. Cells need to be thawed quickly. Immerse vials in a small amount of water at 37 °C in a plastic beaker. To reduce pressure inside the tube briefly open and reseal the lid of the vial prior to placing them in the water.
2. ES cells are generally passaged every 2–3 days, with one media change in between. The colonies should not be touching and should be round or oval shaped, sitting on top of the fibroblast layer rather than looking flat and should have clear, diffractive edges. Dilutions and passage rates should be determined empirically to yield cultures with a reasonable density of uniformly sized colonies spread across a plate.
3. ES cell colonies need to be disaggregated at every passage. This requires extensive pipetting to ensure cells are not in clumps. Inattention to this results in uneven morphology and size in future plates and can possibly lead to propagation of a subpopulation of faster growing cells that can yield a mixed culture of cells and problems in future experiments.
4. The number of dishes required is based on the number of cells per electroporation. If the cell count is 1×10^7 or higher, use 3–4 dishes for recovery, for less than 1×10^7 use 2–3 dishes.

5. This step varies with each experiment and depends on the recovery of cells the next day and on the selective agent used. Cells should go into selection no later than 48 h post-electroporation.
6. ES cell morphology changes on gelatine plates. They become flatter and can lose the diffractive properties of the colony edges.
7. A high density 60 mm plate can be transferred onto a 100 mm petri dish or dilutions to a maximum of 1:5 can be made on the same surface areas.
8. Small tipped plastic transfer pipettes work very well for the removal and addition of media for embryoid bodies and can assist in dislodging adherent cells.
9. If embryoid bodies are sticking to dish surface, floating EBs can be transferred to fresh petri dishes. Lightly stuck EBs can be gently dislodged by forcefully adding fresh media.
10. Use just enough trypsin to cover the dish. If the cells are not spun down and resuspended, the trypsin will remain in the suspension even though it is inactivated by media. If cell density is low, it is crucial to keep total volume as low as possible.
11. Effective superovulation requires the optimization of times and volumes of injection. A good starting point is 5IU of Folligon (0.1 mL injection) at 1 pm Day 1; 5 IU of Chorulon (0.1 mL injection) at 1 pm Day 3. Mice are mated overnight with males on Day 3 and checked for plugs at Day 4. 10–15 super-ovulated mice should yield at least 150 morulae.
12. The large drops will house the dissected ovaries and oviducts to keep them moist, and the small drops will be used for the washes.
13. Do no more than three or four mice at a time. Time is crucial for viability of the embryos. If left too long, the ovaries will start to form a coagulated mass in the media, which makes collection of embryos difficult.
14. M2 at room temperature is best. If it is too warm the blastocysts become sticky and difficult to handle.
15. If the blastocyst is held too firmly, it will burst and collapse. If it is not secured well enough it will move and rotate during the injection. To find an appropriate injection site, find the thinnest part of the trophectoderm, preferably the junction between two cells and away from the inner cell mass.
16. After injection the blastocyst will collapse but should reexpand after about 1 h. This is a good indicator of the viability of the blastocyst and the success of your injections. Move any blastocysts that were not injected successfully aside and discard.

17. To avoid contamination, rinse the darning needle in 70 % ethanol and air dry. To make a divot, place the darning needle in the center of the drop and press down with just enough force to feel the divot maker move into the plastic dish. Hold the divot maker firmly and move with a circular motion. This allows not only a smooth well but also a wall leading into the depression. Do not twist. This will result in a well that is not smooth and will compromise aggregation efficiency. To reduce the risk cracking the dish while making divots, place a small piece of cardboard under the dish.
18. If M16 media has not been aerated with CO_2, dishes should be set up at least 3 h before you begin aggregating cells.
19. When the zona pellucida is removed, the embryos become sticky and are often difficult to remove from the acid Tyrodes solution without them sticking to the bottom of the dish. To prevent sticking, overlay the acid Tyrodes solution on a film of M2. First add four drops of M2 to the lid of the tissue culture dish, and then remove the M2 using a sterile plastic transfer pipette and replace with acid tyrodes.
20. Although there are four drops, not all of them need to be used if the zona dissolves rapidly.
21. When the zona pellucida starts to dissolve, the embryo has a compacted appearance and the zona pellucida may be difficult to see for a short time. It will then become visible as a thin membrane. At this stage, the embryo can be released by agitating with a mouth pipette.
22. Placing one of the host embryos in each of the divots before starting the aggregation process will allow the embryo to adhere slightly to the base of the divot, which makes it easier to "sandwich" ES cells and the remaining host embryo together.
23. The easiest way to "sandwich" cells is to place the tip of the pipette above the divot and gently blow out the cells and let them sink to the bottom of the well. Most of the time, the cells will settle onto the embryo if the ES cells sink into the divot.
24. Check by gently squeezing the back foot or gently tapping below the eye with your finger.
25. Cutting and stretching allows the tissue to tear around the blood vessels and prevents bleeding.
26. This aids in locating the peritoneum for closing the wound.
27. Do not pick up by the ovary or the uterus as this will damage or bruise the area.
28. The air bubbles helps visualization of the movement of the embryos when they are transferred to the uterus and protects them against leakage and loss of medium when inserting the pipette into the uterus.

29. The small hole should create a tiny blood mark, making it easier to locate during embryo transfer.
30. After pulling at the thread, the ovary, uterus, and fat pad should all slide back into the body easily. If the mouse is very large or fat or the incision is to small this can be difficult, but by gently encouraging the fat pad back with the forceps it will slide back in. Do not squeeze or squash the uterus as the embryos may come out.
31. The age of the embryos is calculated based on the time of mating of the pseudopregnant mother so that embryos are considered to be at embryonic day (E) 2.5 when transferred. Therefore, if embryos at E8.5 are required, they should be collected 6 days after transfer.
32. When examining the embryos for abnormal phenotypes it is important to carefully compare knockdown embryos to controls to ensure that any abnormalities are specifically due to expression of the shRNA and are not artifacts caused by the method of embryo production. Since ES cell-derived embryos can vary in age (up to 1 day in a litter in our experience) and in phenotype it may be necessary to collect a large number of embryos for analysis.

References

1. van de Wetering M, Oving I, Muncan V, Pon Fong MT, Brantjes H, van Leenen D, Holstege FC, Brummelkamp TR, Agami R, Clevers H (2003) Specific inhibition of gene expression using a stably integrated, inducible small-interfering-RNA vector. EMBO Rep 4: 609–615
2. Yu JY, DeRuiter SL, Turner DL (2002) RNA interference by expression of short-interfering RNAs and hairpin RNAs in mammalian cells. Proc Natl Acad Sci U S A 99:6047–6052
3. Kunath T, Gish G, Lickert H, Jones N, Pawson T, Rossant J (2003) Transgenic RNA interference in ES cell-derived embryos recapitulates a genetic null phenotype. Nat Biotechnol 21: 559–561
4. Soares ML, Haraguchi S, Torres-Padilla ME, Kalmar T, Carpenter L, Bell G, Morrison A, Ring CJ, Clarke NJ, Glover DM, Zernicka-Goetz M (2005) Functional studies of signaling pathways in peri-implantation development of the mouse embryo by RNAi. BMC Dev Biol 5:28
5. Wianny F, Zernicka-Goetz M (2000) Specific interference with gene function by double-stranded RNA in early mouse development. Nat Cell Biol 2:70–75
6. Cheng TL, Chang WT (2007) Construction of simple and efficient DNA vector-based short hairpin RNA expression systems for specific gene silencing in mammalian cells. Methods Mol Biol 408:223–241
7. Moffat J, Grueneberg DA, Yang X, Kim SY, Kloepfer AM, Hinkle G, Piqani B, Eisenhaure TM, Luo B, Grenier JK, Carpenter AE, Foo SY, Stewart SA, Stockwell BR, Hacohen N, Hahn WC, Lander ES, Sabatini DM, Root DE (2006) A lentiviral RNAi library for human and mouse genes applied to an arrayed viral high-content screen. Cell 124:1283–1298
8. Eakin GS, Hadjantonakis AK, Papaioannou VE, Behringer RR (2005) Developmental potential and behavior of tetraploid cells in the mouse embryo. Dev Biol 288:150–159
9. Eggan K, Jaenisch R (2003) Differentiation of F1 embryonic stem cells into viable male and female mice by tetraploid embryo complementation. Methods Enzymol 365:25–39
10. Coffinier C, Thépot D, Babinet C, Yaniv M, Barra J (1999) Essential role for the homeoprotein vHNF1/HNF1beta in visceral endoderm differentiation. Development 126:4785–4794
11. Duncan SA, Nagy A, Chan W (1997) Murine gastrulation requires HNF-4 regulated gene expression in the visceral endoderm: tetraploid rescue of Hnf-4(-/-) embryos. Development 124:279–287

12. Seibler J, Küter-Luks B, Kern H, Streu S, Plum L, Maue RJ, Kühn R, Brüning JC, Schwenk F (2005) Single copy shRNA configuration for ubiquitous gene knockdown in mice. Nucleic Acids Res 33:e67
13. Lickert H, Cox B, Wehrle C, Taketo MM, Kemler R, Rossant J (2005) Dissecting Wnt/beta-catenin signaling during gastrulation using RNA interference in mouse embryos. Development 132:2599–2609
14. Nagy A, Rossant J, Nagy R, Abramow-Newerly W, Roder JC (1993) Derivation of completely cell culture-derived mice from early-passage embryonic stem cells. Proc Natl Acad Sci U S A 90:8424–8428
15. George SH, Gertsenstein M, Vintersten K, Korets-Smith E, Murphy J, Stevens ME, Haigh JJ, Nagy A (2007) Developmental and adult phenotyping directly from mutant embryonic stem cells. Proc Natl Acad Sci U S A 104:4455–4460

Chapter 9

Generation of Tissue Organoids by Compaction Reaggregation

Julie M. Sheridan and C. Clare Blackburn

Abstract

Cellular reaggregation methods are commonly used to generate tissue organoids for use in biological studies. Using a modified method termed "compaction reaggregation," it is possible to establish reaggregates of reproducible size from defined input cell numbers with ease and without specialist equipment. Importantly, this method is suitable for the study of tissues that have proved refractory to reaggregation by other methods. With the option of juxtaposing cell populations, this method is useful for studies of tissue organization and structure.

Key words Aggregate, Reaggregate, CoROC, Organoid, Thymus

1 Introduction

Organ reaggregation has been used to study the cellular and molecular requirements for diverse processes in many tissue systems including limb [1], skin [2, 3], thymus [4], and lymph node [5]. With most solid tissues, it is possible to promote some cellular aggregation in suspension culture; however, to promote the degree of aggregation necessary to generate organotypic structures, it is often necessary to encourage cell–cell interaction by bringing the cells into close proximity and for this several strategies have been developed. Using gravity, the hanging drop method encourages aggregation of cells that gather at the base of a suspended droplet of medium. As the process is inefficient, many cells fail to aggregate and the resulting aggregates are generally small and of variable sizes. To encourage larger aggregate formation, centrifugation can be used to generate a cell pellet at the base of a well in a U- or V-bottomed multi-well plate, which aggregates upon subsequent culture [6]. This method incorporates pelleted cell types into the aggregate even if some of those cells are not themselves capable of aggregation such as those of some hematopoietic lineages [5, 6]. One further advantage of a centrifugation method is that using

Mark Lewandoski (ed.), *Mouse Molecular Embryology*, Methods in Molecular Biology, vol. 1092,
DOI 10.1007/978-1-60327-292-6_9,

© Springer Science+Business Media New York 2014

iterative cell addition and centrifugation steps it is possible to organize cell types in layers, a useful tool when studying cell–cell interactions. Although highly reproducible when using permissive cells, several tissue types reaggregate poorly or not at all under submerged culture conditions and require aggregation at the gas–liquid interface [7]. This can be achieved by generation of a viscous cell slurry, which is then deposited as a standing drop on a filter paper disc floating at the gas–liquid interface where an aggregate can form [4]. Although broadly applicable, aggregation in a standing drop is less controlled than in the above-described centrifugation method, and neither cell input numbers nor cell type position can be controlled [7].

A modified aggregation method, termed compaction reaggregation (CoROC), relies upon the generation of a cell pellet and subsequent extrusion as a single mass onto a membrane at the gas–liquid interface. CoROC combines the reproducibility and layering capabilities of the centrifugation method with the broad tissue applicability of aggregation at the gas–liquid interface. With this one method it is possible to investigate developmental processes in tissues that are known to be resistant to submersion reaggregation and for which composition and placement are key.

These instructions are for the generation of CoROCs of fetal thymus utilizing primary fetal thymus tissue as an example of a submersion-sensitive organ containing both aggregating (epithelial) and non-aggregating (hematopoietic) cell types. Though not necessary when using whole dissociated fetal thymus, when reaggregating purified epithelial cells fibroblasts are included to provide factors required for proper epithelial development as well as either immature thymocytes or lymphocyte precursors. These can be primary fibroblasts such as murine embryonic fibroblasts (MEF), or a fibroblast cell line, such as NIH/3T3. In the example given, NIH/3T3 cells are included as a separate cell population with which to illustrate the layering technique. A similar strategy to that outlined below has been successfully used for other tissue types including the aorta–gonad–mesonephros region [7, 8] and mammary gland (unpublished data) using tissue-appropriate dissociation methods and growth media. In cases where this has not been attempted before, it is recommended that these are determined for each tissue type using this method as a starting point.

2 Materials

2.1 Preparation of Cell Suspensions for Reaggregation of Mouse Fetal Thymus

1. Dulbecco's phosphate buffered saline solution (DPBS).
2. DPBS, modified, without calcium and magnesium (DPBS-CMF).
3. Solution of 0.025 % trypsin with 1 mM ethylenediamine tetraacetic acid (EDTA) in DPBS-CMF. Store aliquots at −20 °C.

4. Enzymatic dissociation mix. Solution of DPBS containing 2 mg/mL hyaluronidase, 0.7 mg/mL collagenase, and 0.05 mg/mL DNase I. Use immediately or store single use aliquots at −20 °C.
5. 0.4 % trypan blue solution prepared in DPBS-CMF, filtered through 0.45 μm filter unit. Store prepared working solution at room temperature.
6. Hemocytometer and coverslip.
7. Wash solution of DPBS-CMF containing 10 % fetal calf serum.
8. Tissues or cells for reaggregation.

2.2 Compaction Reaggregation at the Gas–Liquid Interface

1. 200 μL non-beveled tips (Axygen, Union City, CA).
2. 0.8 μm Isopore membrane filters (Millipore, Billerica, MA).
3. Parafilm M (Pechiney Plastic Packaging Company, Chicago, IL; *see* **Note 1**).
4. Sterile 6-well plate.
5. Cell growth medium (CGM): DMEM/F12 supplemented with 10 % FCS, 1× nonessential amino acids, 10 mM HEPES Buffer, 0.05 mM β-mercaptoethanol, and 100 IU/mL penicillin/streptomycin (all Invitrogen). Store prepared medium at 4 °C and use within 1 week.
6. Flask of NIH/3T3 fibroblasts at 80 % confluence. NIH/3T3 are maintained in DMEM containing 10 % FCS and 100 IU/mL penicillin/streptomycin and passaged when 80 % confluent with trypsin–EDTA.
7. Stereomicroscope (*see* **Note 2**).

3 Methods

3.1 Preparation of Cell Suspensions for Reaggregation

1. Materials for the preparation of a cell suspension are made ready (*see* **Note 3**).
2. A flask of NIH/3T3 should be harvested by incubation with trypsin–EDTA at 37 °C, washed in wash solution then centrifuged at 200 ×*g* for 5 min before being resuspended in CGM and placed on ice.
3. Primary fetal thymus tissue is dissected from embryonic day (E) 11.5–15.5 embryos under clean conditions and placed in an eppendorf tube containing wash solution at room temperature (*see* **Notes 4** and **5**).
4. Pre-warm enzymatic dissociation mix to 37 °C in heating block for 3 min.
5. The eppendorf tube containing the lobes is centrifuged at 200 ×*g* for 3 min to collect the lobes before the supernatant is

removed and replaced with pre-warmed enzymatic dissociation mix. As a guide, for E11.5–E13.5 tissue from fewer than 5 litters of mice, 0.5 mL enzymatic dissociation mix should suffice. For between 5 and 10 litters or when dissociating lobes from older embryos, increase this amount to 1 mL. Flick the tube to resuspend the lobe pellet and incubate at 37 °C for 10 min in a heating block.

6. Gently triturate the suspension five times, avoiding bubble formation, using a P1000 pipette to aid tissue break-up and release cells (*see* **Note 6**).
7. Incubate for a further 5 min at 37 °C.
8. Triturate the sample ten more times and visually inspect the resultant suspension to determine whether tissue fragments remain. At this point, E11.5–E13.5 lobes should have dissociated fully and therefore will require no further treatment (proceed to **step 12**). Should a further digestion be necessary, as would be the case with E14.5–E15.5 lobes, pre-warm 0.5 mL trypsin–EDTA (*see* **Note 7**).
9. Centrifuge the suspension containing the dissociated cells and remaining tissue fragments at 300 × *g* for 5 min to collect the cells.
10. Remove the supernatant and replace with pre-warmed trypsin/EDTA.
11. Resuspend the cell pellet by briefly flicking the tube and incubate for 3 min, then gently triturate the sample to ensure that all cells are released from the fragments.
12. Once satisfactorily dissociated, fill the remaining volume of the eppendorf tube with wash solution and spin the single cell suspension for 5 min at 300 × *g* to collect the cells (*see* **Note 8**).
13. Discard the supernatant containing the enzyme and resuspend the cell pellet in CGM for counting. Due to the small number of cells likely to be isolated from embryo dissections, as a guide, 200 μL volume would be a suitable resuspension volume for counting thymic cells from each litter of E13.5 embryos.
14. Cell types that will be added to the CoROC should be counted using a 1:1 mix of cells to 0.4 % trypan blue solution to enable trypan blue-excluding, viable cells to be enumerated.
15. A determination of input cell number and type should be made at this point taking into consideration the relative sizes of cells present to generate reaggregate is neither too small to study nor too large to maintain viability. A typical fetal thymus CoROC contains between 150,000 cells (50,000 thymus cells plus 100,000 NIH/3T3) and 300,000 cells (200,000 fetal thymus cells plus 100,000 NIH/3T3) with the majority of the aggregate size coming from the much larger fibroblast cells.

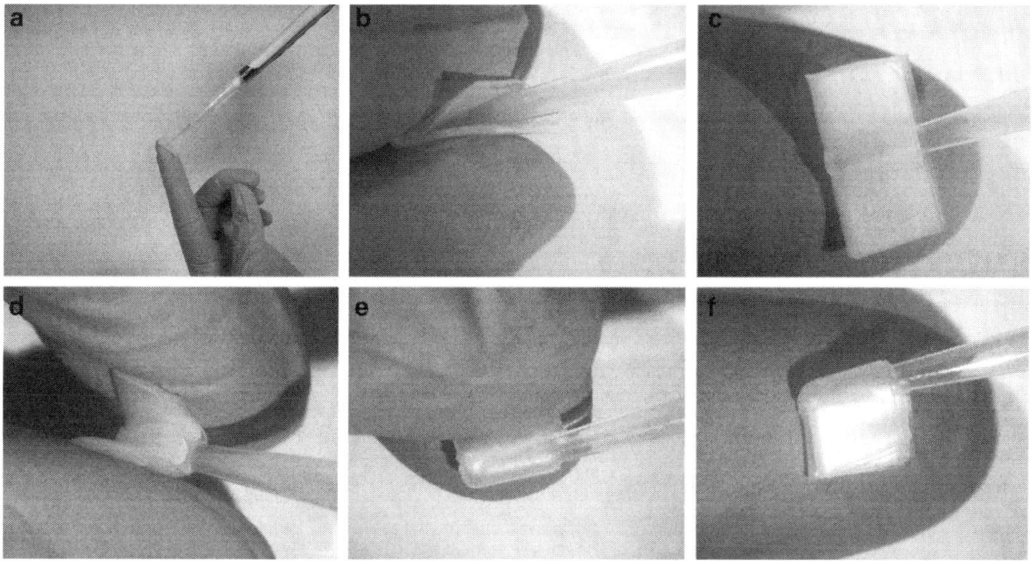

Fig. 1 Assembly of the sealed pipette tip in preparation for cell pellet generation. A cell suspension containing 100,000 thymic epithelial cells and 100,000 NIH/3T3 was drawn up into a 200 μL pipette tip that was subsequently sealed with Parafilm as shown. This assembly efficiently prevents leakage of the cell suspension prior to and during centrifugation that will result in a cell pellet at the aperture of the tip adjacent to the Parafilm seal

Such a mix would typically form a reaggregate of approximately 400–600 μM in diameter, an ideal size for reaggregation in a 200 μL tip (*see* **Notes 9** and **10**).

3.2 CoROC at the Gas–Liquid Interface

1. Aliquot cells to be included in each CoROC into a fresh tube so that all necessary cells are represented in one pool of cells of not greater than 100 μL (*see* **Note 11**).
2. Draw the suspension containing cells to be reaggregated into a non-beveled 200 μL pipette tip. Set down the pipette with the tip still attached to free up hands for the next stage.
3. Seal the narrow tip aperture with Parafilm as outlined below and shown in Fig. 1 (*see* **Notes 1** and **12**). Parafilm of approximate dimensions 1 cm × 5 cm is cut and folded twice to make a rectangle measuring 1 cm × 1.25 cm. Gently press the Parafilm over the narrow tip aperture so that the tip makes an indentation in the Parafilm that can be seen and felt on the reverse side (*see* Fig. 1a). It is important that the tip does not protrude so much as to weaken the Parafilm cover as this would result in leakage under the pressure of centrifugation. While maintaining even pressure on the Parafilm where it covers the aperture, fold the Parafilm using the tip as a pivot so that two sides of the Parafilm now lie against the tip wall on opposite sides with the Parafilm now being 8 ply (*see* Fig. 1b). Press these ends lightly together so that the Parafilm is now firmly attached to the tip

(*see* Fig. 1c). Fold the Parafilm once again using the tip as a pivot and then press together the two meeting ends (*see* Fig. 1d, e). This should leave 16-ply Parafilm square that securely encases the tip aperture and that overhangs the tip on one side (*see* Fig. 1f).

4. Transfer the assembly to a 50 mL centrifuge tube taking care not to dislodge the Parafilm seal. Centrifuge at $300\times g$ for 5 min so that a cell pellet is formed at the occluded aperture of the tip adjacent to the Parafilm seal.
5. Prepare the culture plate. Add 3 mL CGM to a 6-well plate and overlay a membrane disc taking care not to submerge any part of the disc or allow any media to flow over the edges of the disc, which can result in sinking of the assembly and dispersal of the cells to be aggregated. Position so that the floating membrane is in focus under a stereomicroscope (*see* **Note 2**).
6. Reattach the tip to a pipette that has been set to a pipette volume of 30 μL. It is necessary to attach the tip cautiously and incompletely as the action of reattachment can generate enough pressure within the tip to expel the pellet.
7. Unfold the Parafilm in preparation for removing it while maintaining slight pressure on the Parafilm at the tip end so that the aperture remains occluded until you are ready to proceed with the extrusion.
8. Position the pipette over the membrane and remove the Parafilm.
9. With the tip aperture almost touching the center of the floating disc, carefully and slowly, push the tip more firmly into position on the pipette so that a firm seal is made (*see* Fig. 2a). This action provides enough pressure within the tip to push the cell pellet outwards through the previously occluded narrow end onto the filter paper disc. Should the pressure be insufficient, extrusion can be forced by incrementally winding the volume setting of the pipette down as far as is required to expel the pellet. In either case, a pellet can be extruded onto the filter with minimal CGM (*see* Fig. 2b).
10. Culture the cell pellet at 37 °C for 24 h allowing time for the formation of a solid reaggregate. As the floating disc is susceptible to submersion, and the pellet itself is subject to disturbance, extreme care must be taken when moving the plate containing the floating assembly to the incubator. After successful reaggregation, the CoROC can be cultured or transplanted for further study.

3.3 Generating Cell Layers in a CoROC at the Gas–Liquid Interface

1. Cells to be reaggregated in separate layers should be resuspended in CGM so that the cumulative volume of cells to be included is no greater than 100 μL. For example, 40 μL thymus cells could be reaggregated with equal to or less than 60 μL NIH/3T3 cells.

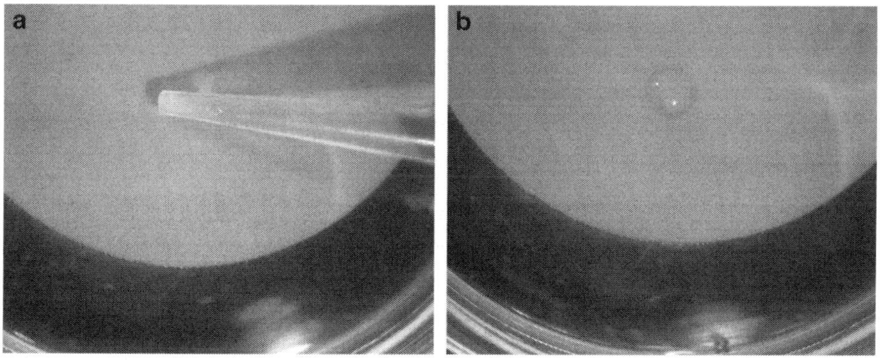

Fig. 2 Extrusion of the cell pellet onto an Isopore membrane floating at the gas–medium interface in a 6-well plate. Here, the cells will spontaneously aggregate within 24 h of culture

2. Draw the suspension containing the cells required for the first cell layer into a 200 μL pipette tip.
3. With the suspension inside, seal the narrow tip aperture with Parafilm as outlined in Subheading 3.2 and Fig. 1.
4. Transfer the assembly to a centrifuge tube taking care not to dislodge the Parafilm seal and centrifuge at 300 × *g* for 5 min so that a cell pellet is formed at the occluded aperture of the tip adjacent to the Parafilm seal.
5. Without removing the tip from the 50 mL tube, the suspension containing cells to be included in the second layer of the CoROC are carefully added to the open end of the tip without disturbing the first pellet that has already been established.
6. The assembled tube should be centrifuged once again to produce a cell pellet that lies on top of the first. Further cell additions and centrifugation steps will produce a multilayered pellet.
7. Continue as Subheading 3.2, **step 6**.

4 Notes

1. Although it is possible that other similar products may be used, the properties of Parafilm have proved ideal.
2. Where possible, the area used for reaggregate assembly should be clean to avoid microbial contamination. While this is best achieved using a stereomicroscope-equipped laminar flow hood, the authors typically use a clean area and aseptic technique without resulting in culture contamination
3. All solutions used in the preparation of samples and subsequent culture should be sterile.
4. Noon of the day of the vaginal plug was taken as day 0.5.

5. During dissection ensure that hair and dander contamination is eliminated, as this will be deleterious during enzymatic digestion.
6. The shear forces associated with bubbles can damage cells.
7. As with any dissociation procedure, care should be take not to "over digest" tissue, as this will reduce cell viability. It is important to note that damaged cells tend to adhere to released DNA forming clumps, which may be mistaken for remaining tissue fragments. In the event that this happens, clumps may be incubated with 0.05 mg/mL DNase I in wash solution to digest the DNA that holds these clumps together and limit the formation of further clumps.
8. The addition of FCS will inhibit the trypsin activity and thus halt further digestion.
9. As cell size directly influences reaggregate size, it is necessary to empirically determine optimal cell numbers for each tissue or cell type used.
10. The aperture of the tip in which the cells are compacted is critical. Although a wide bore 200 μL tip is capable of generating a cell pellet of larger dimensions, the authors have found that this generally results in cell pellets that when extruded are flatter and less reproducible in aggregation efficiency and therefore size. A 10 μL tip can be used to generate smaller CoROC; however, the force applied during extrusion commonly leads to an excess of extruded CGM along with the cells, which makes reaggregation less efficient. Therefore, in order to utilize different tip sizes for the purpose of generating CoROC of alternate dimensions, it is advisable to optimize this with a particular cell type and number in the first instance.
11. We have found that the greater the cell suspension volume used, the more chance that the pellet will extrude upon Parafilm removal prior to application of pressure. In such cases, the pellet is often expelled too quickly and results in CGM being also expelled, which disturbs the pellet and lowers aggregation efficiency. Therefore, it is advisable to use minimal volumes with less than 50 μL being ideal.
12. The folded Parafilm is critical to the success of the compaction process with most generation errors resulting from this stage of the process. Although various methods could be used, the outlined fold method has proved both simple and reliable.

Acknowledgements

We thank Biomed Unit staff for animal care and Samir Taoudi for discussion. This work was supported by the Leukaemia and Lymphoma Research (JMS and CCB) and the Medical Research Council (JMS and CCB).

References

1. Moscona A, Moscona H (1952) The dissociation and aggregation of cells from organ rudiments of the early chick embryo. J Anat 86:287–301
2. Moscona MH, Moscona AA (1965) Control of differentiation in aggregates of embryonic skin cells: suppression of feather morphogenesis by cells from other tissues. Dev Biol 11:402–423
3. Garber BB (1976) Control of epithelial development. Curr Probl Dermatol 6:154–190
4. Anderson G, Jenkinson EJ, Moore NC, Owen JJ (1993) MHC class II-positive epithelium and mesenchyme cells are both required for T-cell development in the thymus. Nature 362:70–73
5. White A, Carragher D, Parnell S, Msaki A, Perkins N, Lane P, Jenkinson E, Anderson G, Caamano JH (2007) Lymphotoxin a-dependent and-independent signals regulate stromal organizer cell homeostasis during lymph node organogenesis. Blood 110:1950–1959
6. Ng ES, Davis RP, Azzola L, Stanley EG, Elefanty AG (2005) Forced aggregation of defined numbers of human embryonic stem cells into embryoid bodies fosters robust, reproducible hematopoietic differentiation. Blood 106:1601–1603
7. Sheridan JM, Taoudi S, Medvinsky A, Blackburn CC (2009) A novel method for the generation of reaggregated organotypic cultures that permits juxtaposition of defined cell populations. Genesis 47:346–351
8. Taoudi S, Gonneau C, Moore K, Sheridan JM, Blackburn CC, Taylor E, Medvinsky A (2008) Extensive hematopoietic stem cell generation in the AGM region via maturation of VE-cadherin+CD45+ pre-definitive HSCs. Cell Stem Cell 3:99–108

Chapter 10

Ultrarapid Vitrification of Mouse Oocytes and Embryos

Mark G. Larman and David K. Gardner

Abstract

Cryopreservation facilitates long-term storage of gametes and embryos for numerous purposes. For example, cryobanking of unique mouse strains, particularly transgenic mice, offers important protection of valuable genetics. It also provides a practical solution for facilities trying to house large numbers of research animals or those looking to relocate without the risk of introducing an animal-derived pathogen. Furthermore, cryopreservation is currently being used for fertility preservation both in humans and as a safeguard for endangered animals. Ultrarapid vitrification offers an elegant, quick, and very reliable method for cryopreservation of mouse oocytes and embryos. Furthermore, research into the effects on mouse oocyte and embryo physiology has indicated that ultrarapid vitrification is superior to conventional slow freezing. High survival rates, embryo development, and viability are routinely achieved with the ultrarapid vitrification method described in this chapter.

Key words Blastocyst, Cryoloop, Cryopreservation, Embryo, Mouse, Oocyte, Vitrification

1 Introduction

The term vitrification is derived from the Latin *vitrum*, meaning glass. Thus, vitrification of cells requires the transition of the cytosol into an amorphous, glass-like solid. The absence of crystalline structures means that there is no formation of potentially lethal ice crystals, which damage the plasma membrane and intracellular organelles [1]. Vitrification of water is achieved if the molecules are cooled at a rate that is too rapid for them to organize themselves into ice crystals. In theory, vitrification of pure water can be achieved if the cooling rate is at least 10^8 °C/min. Direct application to liquid nitrogen (−196 °C) results in a cooling rate of approximately 10^4 °C/min. Therefore, the intracellular water concentration must be sufficiently reduced, so that a cooling rate of 10^4 °C/min is rapid enough to ensure that the remaining water molecules vitrify, rather than freeze. A reduction in water molecules is achieved through osmotic potential (using extracellular sugars), to draw water out of the cell. Permeable cryoprotectants (e.g., glycerol,

Mark Lewandoski (ed.), *Mouse Molecular Embryology*, Methods in Molecular Biology, vol. 1092,
DOI 10.1007/978-1-60327-292-6_10, © Springer Science+Business Media New York 2014

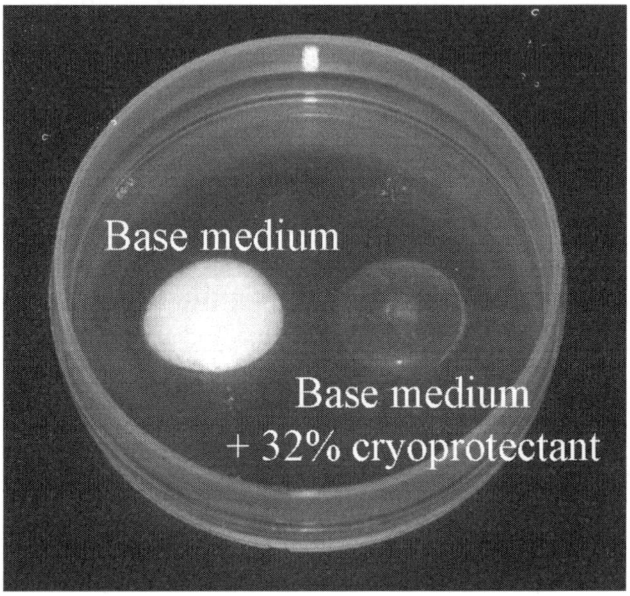

Fig. 1 The difference between freezing and vitrification. To indicate the difference between freezing and vitrification two different droplets of media were placed on a culture dish and submerged into liquid nitrogen. The droplet on the *left* is composed of the G-MOPS base medium. The droplet on the *right* is composed of the base medium containing 16 % (v/v) ethylene glycol and 16 % (v/v) DMSO. Upon removal from liquid nitrogen a photograph was immediately taken. The droplet on the *left* appears white from the ice crystals that have formed during the freezing process. The droplet on the *right* contains sufficient cryoprotectant, which prevents the water molecules from organizing themselves into a crystal lattice. Thus, the droplet containing cryoprotectant permits light to pass through and appear translucent

dimethyl sulfoxide, ethylene glycol, and/or propylene glycol) are introduced to disrupt interactions between the remaining water molecules.

Warming rates are equally important. If the cell is warmed too slowly the water will undergo devitrification, forming ice crystals. The cooling and warming rates required can be significantly reduced by increasing the percentage of cryoprotectant, but unfortunately these chemicals are, depending on their concentration and exposure time, cytotoxic. Therefore, it is necessary to use appropriate conditions whereby the cryoprotectant concentration and exposure time are adequate to introduce sufficient cryoprotectant into the cell, but without affecting embryo viability. Equally, the vitrification solution must also remove sufficient intracellular water, so that cooling and warming rates do not permit intracellular ice formation. As a visual accompaniment Fig. 1 shows two droplets of cryopreserved media in a culture dish. The droplet on the left is composed of the base medium used in the protocol described in this chapter. The droplet on the right is the Vitrification solution used to cryopreserve oocytes and embryos (32 % v/v cryoprotectant). The culture dish was then submerged into liquid nitrogen. Upon removal from liquid nitrogen a photograph was immediately taken to demonstrate that without appropriate levels of cryoprotectants, the water in the base medium will freeze. The frozen droplet has an opaque appearance due to the ice crystals, whereas the vitrified droplet is amorphous and glass-like and, therefore, translucent.

Vitrification of mammalian embryos was first reported by Rall and Fahy [2]. Since then the technique has been further developed to minimize the concentration of cryoprotectants. This has been achieved principally through the introduction of miniature devices, with high levels of temperature conduction, which hold submicroliter volumes. Studies to date have employed: electron microscope grid [3], open pulled straw [4], solid surface vitrification [5], cryotop [6], nylon mesh [7], and the cryotip [8]. The cryoloop [9] is another such device, which has been used extensively and optimized in our laboratory. Decreasing the volume of medium that is to be vitrified and direct application to liquid nitrogen significantly increases the cooling rates (approximately 20,000 °C/min). This has been termed ultrarapid vitrification and has been successfully applied to the cryopreservation of domestic and laboratory animal gametes and embryos [3, 9–12] and more recently to the clinical field of in vitro fertilization [13–15].

Recent publications in laboratory and domestic animals, as well as the human, have demonstrated that ultrarapid vitrification is superior to conventional slow freezing, with greater survival rates and more viable embryos [16–18]. Oocytes cryopreserved using slow freezing produce blastocysts with significantly fewer cells and reduced viability, compared to those undergoing ultrarapid vitrification [19]. Furthermore, oocyte and embryo metabolism, plasma membrane integrity, and protein expression are significantly altered by slow freezing compared to ultrarapid vitrification [19–21]. Additional evidence that ultrarapid vitrification imparts less overall cellular stress than slow freezing has recently been gleaned through the repeated cryopreservation of mouse embryos at successive stages of development (1-cell, 2-cell, 8-cell, and blastocyst) [22]. Using ultrarapid vitrification it was possible to re-cryopreserve mouse embryos at 4 success stages without loss of development in culture, or implantation potential. In contrast, mouse embryos could not survive three rounds of slow freezing. These data confirm that the cumulative stress of slow freezing significantly compromises oocyte and embryo physiology and ultimately viability [17]. Vitrification also offers several practical advantages over slow freezing [23, 24]. For example, vitrification requires no expensive equipment and can be performed with little preparation time and within a matter of minutes.

This chapter discusses the use of the cryoloop (*see* **Note 1**) for the ultrarapid vitrification of mouse oocytes and embryos.

2 Materials

2.1 Construction of the Cryoloop

The cryoloop is constructed from two components supplied by Hampton Research (www.hamptonresearch.com). The cryoloop vitrification device is composed of a cryovial (CrystalCap™) and a 20-µm nylon thread mounted on a stainless steel rod to give a loop diameter of 0.5–0.7 mm (Mounted Cryoloop™) (Fig. 2a). The end of the stainless steel rod is placed into the cryovial lid, which is then secured with an epoxy resin (*see* **Note 2**). Vitrolife no longer sells the cryoloop. The cryovial is secured and stored in a cryocane, which is submerged into liquid nitrogen held in a Dewar. The lid of the cryovial contains a metal base plate, which allows it to be screwed into the cryovial under liquid nitrogen with the use of the magnetic CrystalWand™ (Fig. 2b, c).

2.2 Preparation of Vitrification and Warming Solutions

The vitrification and warming solutions are composed of the same base medium; G-MOPS containing 5 mg/ml human serum albumin [25]. The composition is given in Chapter 11 of this volume entitled Mammalian Preimplantation Embryo Culture, by Gardner and Lane. This solution can be stored at 4 °C for 1 month as STOCK A. For ease of use, two further stock solutions are prepared, which can be stored at 4 °C for 1 month. STOCK B is the same as STOCK A, except that it includes 1 M sucrose.

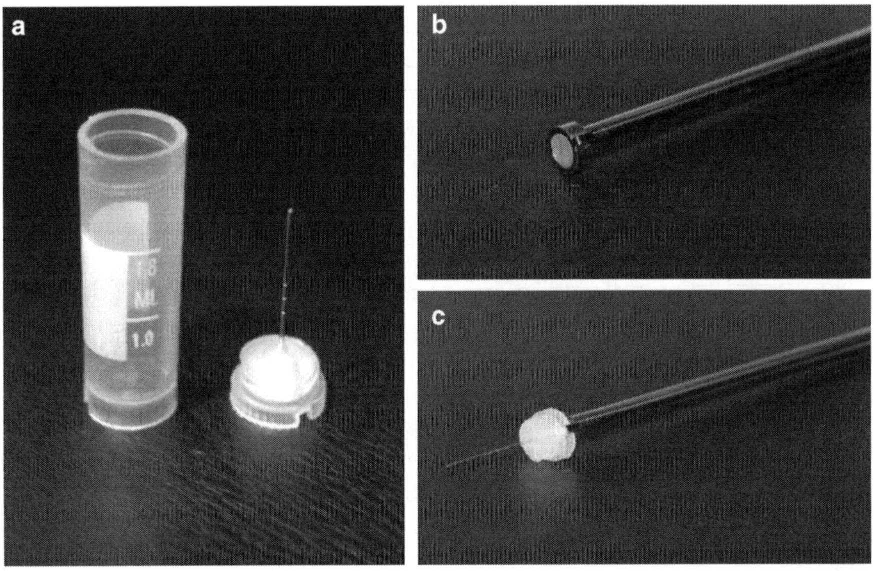

Fig. 2 The cryoloop vitrification device. (**a**) The cryoloop is composed of a metal rod with a nylon loop at one end. The other end of the metal rod is inserted into the lid of a cryovial. Once the cryoloop has been loaded with the vitrification solution the oocytes/embryos can be pipetted directly onto the thin film held across the nylon loop (*see* Fig. 3). (**b** and **c**) The cryoloop can then be stored in a cryovial by using the magnetic wand to screw the lid into the vial

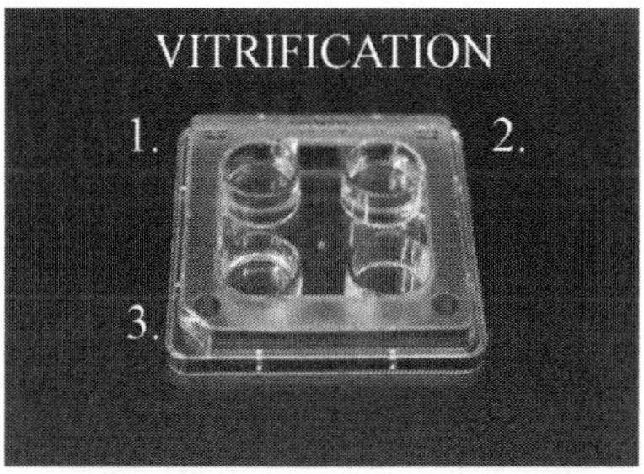

1. Holding Solution
1000μl STOCK A
2. Exposure Solution
840μl STOCK A, 80μl DMSO, 80μl EG
3. Vitrification Solution
680μl STOCK C, 160μl DMSO, 160μl EG

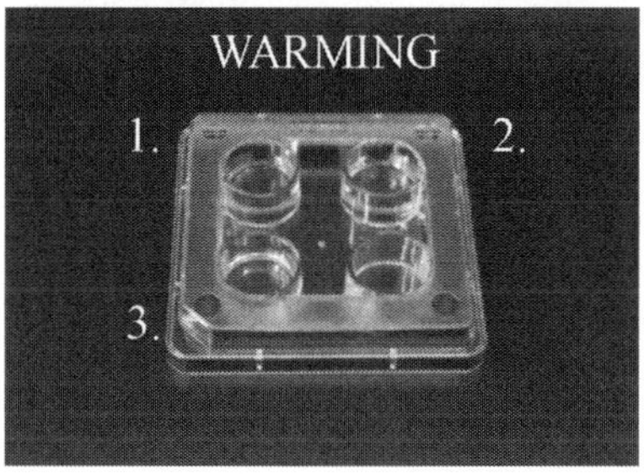

1. Warming Solution 1
750μl STOCK A, 250μl STOCK B
2. Warming Solution 2
875μl STOCK A, 125μl STOCK B
3. Warming Solution 3
1000μl STOCK A

Fig. 3 Setup of 4-well plate for vitrification and warming. The solutions for vitrification and warming are made in a 4-well plate. Vitrification: 1 ml of Holding Solution, Exposure Solution, and Vitrification Solution. Warming: 1 ml of Warming Solution 1, 2, and 3

STOCK C is the same as STOCK B, except that it includes 10 mg/ml Ficoll (400).

Thirty minutes before vitrification or warming is to be performed, the stock solutions are mixed (in a sterile environment) in the appropriate combinations; with cryoprotectants added to make the Exposure and Vitrification solutions. In a 4-well plate, 1 ml of each solution is made as follows (Fig. 3):

Vitrification

Holding solution: 1,000 μl of STOCK A.

Exposure solution: 840 μl of STOCK A plus 80 μl DMSO and 80 μl Ethylene glycol.

Vitrification solution: 680 μl of STOCK C plus 160 μl DMSO and 160 μl Ethylene glycol.

Note: both ethylene glycol and DMSO are considered toxic, so it is advisable to wear embryo-safe powder-free nitrile gloves during handling.

Warming

Warming solution 1: 750 μl of STOCK A plus 250 μl of STOCK B.

Warming solution 2: 875 μl of STOCK A plus 125 μl of STOCK B.

Warming solution 3: 1,000 μl of STOCK A.

Ensure that each solution is mixed well by pipetting up and down.

3 Methods

3.1 Oocyte and Embryo Collection

Production and collection of mouse oocytes and embryos is discussed in Chapter 11 entitled Mammalian Preimplantation Embryo Culture, by Gardner and Lane.

3.2 Vitrification of Oocytes and Early Embryos

1. Place 1 ml of each of the solutions for vitrification in separate wells of a labeled 4-well plate, e.g., Holding solution in well number 1; Exposure solution in well 2; Vitrification solution in well number 3.
2. Place the lid on the 4-well plate and allow the solutions to warm to 37 °C (*see* **Note 3**).
3. While the vitrification solutions are warming collect the following items:
   - (a) Cryoloops with cryovials.
   - (b) Magnetic wand.
   - (c) Cryocanes.
   - (d) Dewar containing liquid nitrogen.
   - (e) Pipettor with 20 μl tips.
   - (f) Embryo handling tool (*see* **Note 4**).
   - (g) Timer.
4. Label the cryovials with appropriate cryo-labeling system to allow subsequent identification.
5. The Dewar should contain enough liquid nitrogen to cover the cryocane up to the position that will be filled. A maximum of four cryovials can be attached to each cryocane.
6. Once the vitrification solutions have warmed to 37 °C move them onto a heated stereomicroscope stage that will maintain this temperature.
7. Oocytes/embryos are then moved from their culture media to the Holding solution using a pipette (*see* **Note 5**).
8. Move 1–10 mouse oocytes/embryos into the Exposure solution (*see* **Note 6**). Start the Timer. The embryos are incubated in the Exposure solution for a total time of 1 min.
9. Once 30 s has elapsed, pipette 20 μl of the Vitrification solution onto the underside of the 4-well plate lid (*see* **Note 7**).
10. Attach the cryoloop (i.e., cryovial lid) to the magnetic wand by lining up the tab and notch and then dip the nylon loop into the Vitrification solution. The "primed" cryoloop and wand can be left to one side on the microscope stage. When loading the cryoloop with oocytes/embryos the magnetic wand can be left flat on the microscope stage and rotated until the cryoloop is in the correct plane for loading.

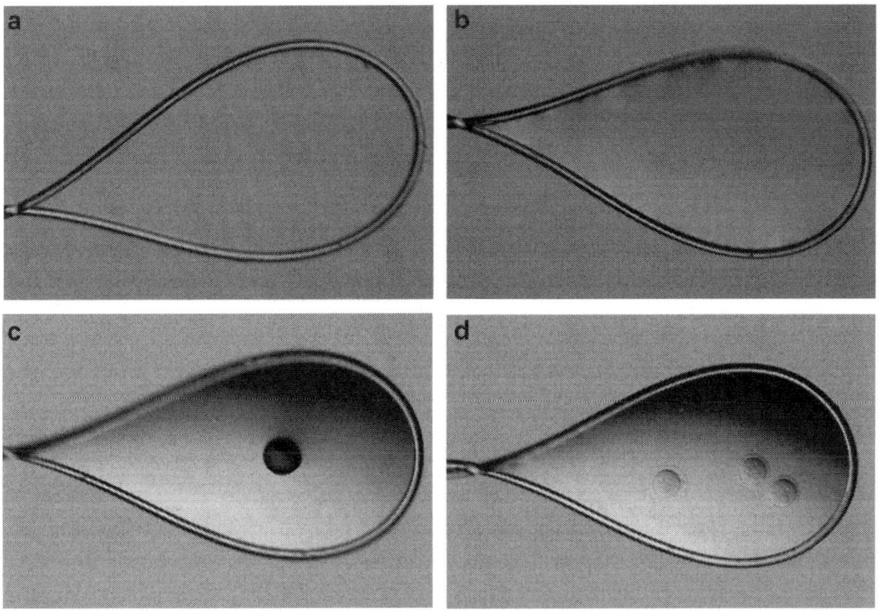

Fig. 4 Loading of the cryoloop. (**a**) The empty cryoloop is composed of a 20 μm diameter nylon thread looped to a metal rod. The diameter of the loop is between 0.5 and 0.7 mm. (**b**) When the cryoloop is dipped into the viscous vitrification solution there is sufficient surface tension to maintain a thin film across the loop. (**c**, **d**) A 100-μm diameter black sphere and mouse pronuclear oocytes, which have been pipetted onto the cryoloop from the Vitrification solution, give an indication of the size of the loop

11. Once 50 s has elapsed begin to collect the oocytes/embryos from the Exposure solution using the pipette (*see* **Note 8**).
12. The oocytes/embryos should leave the Exposure solution and enter the 20 μl droplet of the Vitrification solution as the 1 min expires.
13. Use the tip of the pipette to mix the solution once the oocytes and embryos have been released (*see* **Note 9**).
14. The total time, oocytes and embryos are exposed to the vitrification solution before being plunged into liquid nitrogen is 30–40 s. Therefore, once the oocytes and embryos have been exposed to the Vitrification solution for 20 s collect the embryos and pipette them onto the film of Vitrification solution held across the cryoloop (Fig. 4; *see* **Note 10**).
15. Using the magnetic wand, plunge the cryoloop into the cryovial containing liquid nitrogen and then screw the cryoloop into the cryovial that is attached to the cryocane and full of liquid nitrogen (*see* **Note 11**).
16. **Steps 8–15** can then be repeated to load the next cryoloop.
17. A total of four cryovials can be loaded onto each cryocane, which is then subsequently stored under liquid nitrogen in a suitable cryostorage unit.

3.3 Warming of Oocytes and Early Embryos

1. Place 1 ml of each of the solutions for warming in separate wells of a labeled 4-well plate e.g., Warming solution 1 in well number 1; Warming solution 2 in well 2; Warming solution 3 in well number 3.
2. Place the lid on the 4-well plate and allow the solutions to warm to 37 °C (*see* **Note 3**).
3. While the solutions are warming collect the following items:
   (a) Magnetic wand.
   (b) Dewar containing liquid nitrogen.
   (c) Embryo handling tool.
   (d) Timer.
4. Remove the cryocanes from the long term liquid nitrogen storage vessel. Place the cryocane into the Dewar containing liquid nitrogen so that all vials are submerged (*see* **Note 12**).
5. Use the magnetic wand to unscrew the lid from the cryovial. Rapidly, move the cryoloop into Warming solution 1 (*see* **Note 13**).
6. By touching the cryoloop to the surface of Warming solution 1 the oocytes/embryos will be removed from the cryoloop. As you observe the oocytes/embryos exit the loop start the timer. Oocytes/embryos remain in this solution for 1 min.
7. Once the 1 min has elapsed move the embryos into Warming solution 2. The oocytes/embryos will remain in this second solution for 2 min.
8. Once the 2 min has elapsed move the embryos into Warming solution 3. The oocytes/embryos will remain in this third solution for 5 min.
9. Move the oocytes or embryos into appropriate culture media if in vitro development is required.

3.4 Vitrification and Warming of Blastocysts

1. The protocol for vitrification of blastocysts is the same, except that blastocysts are incubated in the Exposure solution for 2 min. Therefore, make the 20 μl drop of Vitrification solution at 90 s (*see* **Note 14**).
2. The protocol for warming of blastocysts is the same except that blastocysts are incubated in Warming solution 1 for 2 min and Warming solution 2 for 3 min.

4 Notes

1. The cryoloop has been used extensively in our laboratory and is our preferred choice since it is an elegant and easy-to-use system. It allows one to monitor the oocytes and embryos as

they are pipetted onto the nylon loop and upon warming. However, the solutions described in this chapter can work with equivalent ultrarapid vitrification devices (as discussed in ref. 26). The majority of vitrification devices, including the cryoloop, are open systems, i.e., use direct contact with liquid nitrogen during vitrification and are not hermetically sealed for subsequent storage. Concerns have been raised with regard to viral and microbial contamination during storage in liquid nitrogen [27]. The Rapid-i™ (Vitrolife) has been developed with many of the advantages of the cryoloop, such as visualization of the embryos. However, instead of using direct contact with liquid nitrogen it uses supercooled air for vitrification with subsequent storage in a sealed straw (Fig. 5 [28]).

2. Some epoxy resins are embryo toxic, so it is recommended that a suitable bioassay (e.g., 1 cell mouse embryo assay 29) is performed to ensure that oocyte/embryo viability is not compromised.
3. Directly measure the temperature of the medium in the 4-well plate and adjust the temperature of the stage warmer to achieve a media temperature 37 °C. The solutions are warmed to 37 °C to maintain the oocytes and embryos at a physiological temperature. This is particularly important for the oocyte since exposure to room temperature will facilitate microtubule depolymerization, leading to meiotic spindle disassembly [15]. Having a protocol that uses 37 °C also means that the incubation times are shorter, meaning the oocytes and embryos spend less time out of the incubator.
4. A finely pulled glass pipette that has a diameter slightly bigger than the oocyte/embryo allows them to be moved between the solutions and onto the cryoloop in a minimal volume. The capillary action of the pipette can be broken by introducing air bubbles in the column of medium. This is simply achieved by rapidly moving the tip of the pipette in and out of the medium while applying light suction (Fig. 6).
5. Oocytes and embryos can remain in the Holding solution for 5–15 min. It is advised to culture embryos in media that has been supplemented with 0.125–0.5 mg/ml hyaluronan [30, 31]. The base medium contains hyaluronan, which has been shown to increase cryotolerance [31–35].
6. Due to the density of the Exposure solution, the oocytes and embryos will float to the top of the solution. It is advisable to collect the oocytes and embryos and replace them to the bottom of the drop to facilitate quick collection for transfer to the Vitrification solution.
7. As with **Note 5**, due to the density of the Vitrification solution the oocytes and embryos will float. Since the time the oocytes

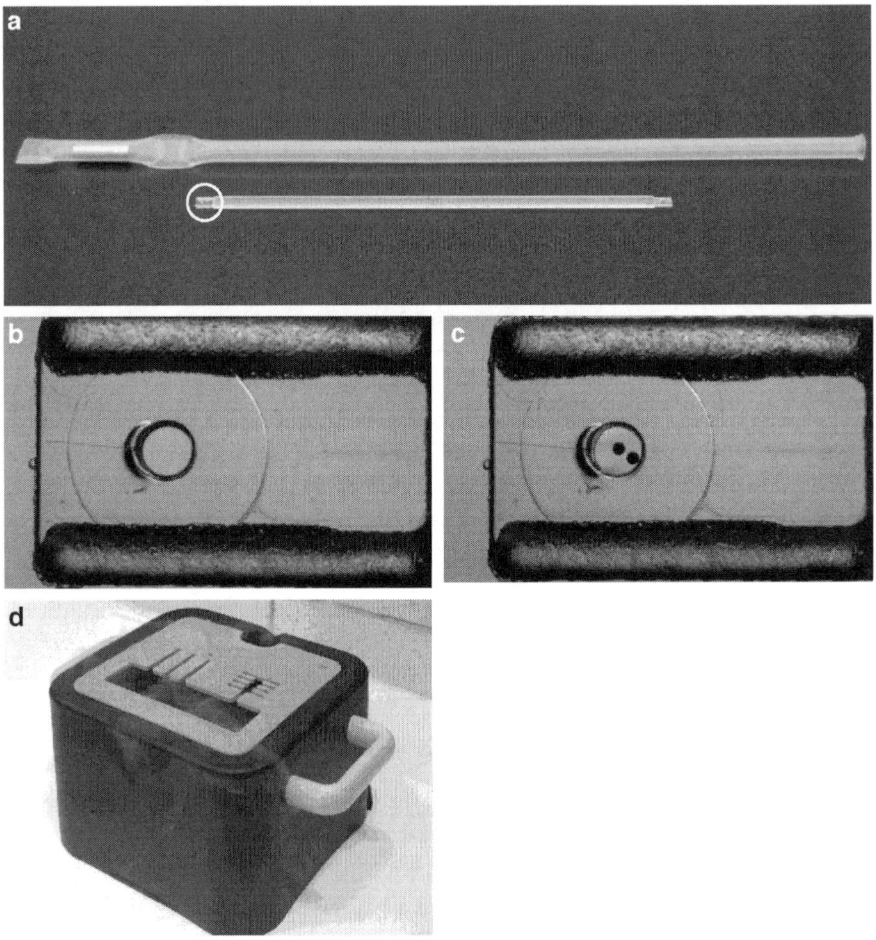

Fig. 5 The Rapid-i™: A closed vitrification device. The Rapid-i™ is a closed vitrification device that requires no direct contact with liquid nitrogen. (**a**) It is composed of a weighted storage straw that is sealed at one end and a plastic rod, which holds the embryos (*circled*). (**b**, **c**) The device is loaded with the embryos and vitrification medium by pipetting the embryos (or in this case 100 μm beads) directly into the 50 nl hole. The small volume and high viscosity of the vitrification solution means that the embryos remain securely in place. The plastic rod is then inserted into the straw that is held in a box containing liquid nitrogen (**d**). The air inside the straw is supercooled to around −190 °C causing instantaneous vitrification. The top of the straw is then sealed

and embryos spend in the Vitrification solution is critical, a 20 μl drop of the Vitrification is created. This restricts the movement of the oocytes and embryos and allows them to be quickly mixed with the Vitrification solution and gathered.

8. Since the oocytes and embryos are to be pipetted into a 20 μl drop, avoid carrying over a large volume of the Exposure solution, which will dilute the Vitrification solution. By maintaining the oocytes and embryos at the tip of the pipette will ensure that less than 100 nl is transferred. This is essential to ensure the most rapid cooling and warming rates possible.

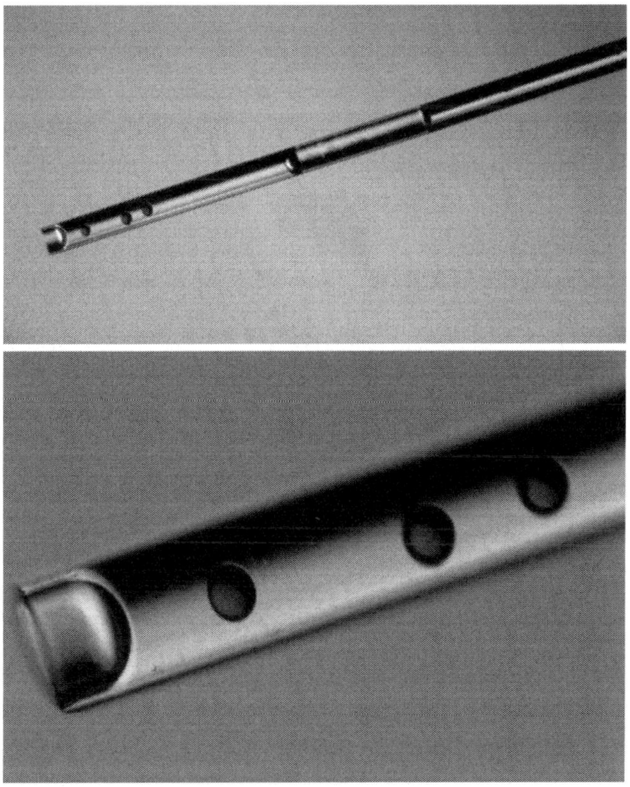

Fig. 6 Moving oocytes/embryos in a minimal volume. To minimize the transfer of medium when moving the oocyte/embryos between solutions and onto the cryoloop, it is important to be able to control the capillary action of the glass pipette. This is achieved by introducing air bubbles into the column of medium within the glass pipette. By moving the tip of the glass pipette in and out of a medium droplet while applying light suction breaks the capillary action with air bubbles. This allows the oocytes/embryos to be held as a group in a submicroliter volume

9. Due to the potential cytotoxicity, it is imperative to ensure that the time the oocytes and embryos are exposed to the Vitrification solution is 30–40 s. If the oocytes or embryos are removed to quickly insufficient water will be withdrawn. Longer than 40 s can cause a decrease in survival and viability.
10. The cryoloop holds a volume of 5–10 nl, which ensures rapid cooling/warming rates. To maintain the cooling/warming rates it is important to pipette the oocytes and embryos onto the cryoloop in a minimal volume. Maintaining a thin film will also help with visualization of the oocytes embryos.
11. To allow successive addition of each cryovial, the cryocane must be loaded from the bottom up. This allows the magnetic wand to have access to screw in the cryolid.
12. One of the advantages the cryoloop has is that it is stored in a 1.8 ml cryovial. This means that the cryocane can be

transiently removed from the long term storage vessel for identification without worrying about the drop devitrifying. To prevent the cryovial from exploding upon warming, there are two holes in the lid. Users should be made aware that liquid nitrogen can vent from these holes if the cryovial is warmed.

13. Since the metal stem of the cryoloop is very cold after storage in liquid nitrogen it is important to dip just the nylon loop into Warming solution 1. Preventing the cold metal touching the solution will ensure that there is no excessive bubbling of the medium.

14. Longer incubation times are required for the blastocyst to permit permeation of all cells and the blastocoel.

References

1. Mazur P (1963) Kinetics of water loss from cells at subzero temperatures and the likelihood of intracellular freezing. J Gen Physiol 47:347–369
2. Rall WF, Fahy GM (1985) Ice-free cryopreservation of mouse embryos at −196 °C by vitrification. Nature 313:573–575
3. Martino A, Songsasen N, Leibo SP (1996) Development into blastocysts of bovine oocytes cryopreserved by ultra-rapid cooling. Biol Reprod 54:1059–1069
4. Vajta G, Holm P, Kuwayama M, Booth PJ, Jacobsen H, Greve T, Callesen H (1998) Open pulled straw (OPS) vitrification: a new way to reduce cryoinjuries of bovine ova and embryos. Mol Reprod Dev 51:53–58
5. Dinnyes A, Dai Y, Jiang S, Yang X (2000) High developmental rates of vitrified bovine oocytes following parthenogenetic activation, in vitro fertilization, and somatic cell nuclear transfer. Biol Reprod 63:513–518
6. Hochi S, Terao T, Kamei M, Kato M, Hirabayashi M, Hirao M (2004) Successful vitrification of pronuclear-stage rabbit zygotes by minimum volume cooling procedure. Theriogenology 61:267–275
7. Matsumoto H, Jiang JY, Tanaka T, Sasada H, Sato E (2001) Vitrification of large quantities of immature bovine oocytes using nylon mesh. Cryobiology 42:139–144
8. Kuwayama M, Vajta G, Kato O, Leibo SP (2005) Highly efficient vitrification method for cryopreservation of human oocytes. Reprod Biomed Online 11:300–308
9. Lane M, Schoolcraft WB, Gardner DK (1999) Vitrification of mouse and human blastocysts using a novel cryoloop container-less technique. Fertil Steril 72:1073–1078
10. Yeoman RR, Gerami-Naini B, Mitalipov S, Nusser KD, Widmann-Browning AA, Wolf DP (2001) Cryoloop vitrification yields superior survival of Rhesus monkey blastocysts. Hum Reprod 16:1965–1969
11. Begin I, Bhatia B, Baldassarre H, Dinnyes A, Keefer CL (2003) Cryopreservation of goat oocytes and in vivo derived 2- to 4-cell embryos using the cryoloop (CLV) and solid-surface vitrification (SSV) methods. Theriogenology 59:1839–1850
12. Cai XY, Chen GA, Lian Y, Zheng XY, Peng HM (2005) Cryoloop vitrification of rabbit oocytes. Hum Reprod 20:1969–1974
13. Takahashi K, Mukaida T, Goto T, Oka C (2005) Perinatal outcome of blastocyst transfer with vitrification using cryoloop: a 4-year follow-up study. Fertil Steril 84:88–92
14. Desai N, Blackmon H, Szeptycki J, Goldfarb J (2007) Cryoloop vitrification of human day 3 cleavage-stage embryos: post-vitrification development, pregnancy outcomes and live births. Reprod Biomed Online 14: 208–213
15. Larman MG, Minasi MG, Rienzi L, Gardner DK (2007) Maintenance of the meiotic spindle during vitrification in human and mouse oocytes. Reprod Biomed Online 15:692–700
16. Vajta G (2000) Vitrification of the oocytes and embryos of domestic animals. Anim Reprod Sci 60–61:357–364
17. Gardner DK, Sheehan CB, Rienzi L, Katz-Jaffe M, Larman MG (2007) Analysis of oocyte physiology to improve cryopreservation procedures. Theriogenology 67:64–72
18. Balaban B, Urman B, Ata B, Isiklar A, Larman MG, Hamilton R, Gardner DK (2008) A randomized controlled study of human Day 3 embryo cryopreservation by slow freezing or vitrification: vitrification is associated with higher survival, metabolism and blastocyst formation. Hum Reprod 23:1976–1982

19. Lane M, Gardner DK (2001) Vitrification of mouse oocytes using a nylon loop. Mol Reprod Dev 58:342–347
20. Lane M, Maybach JM, Gardner DK (2002) Addition of ascorbate during cryopreservation stimulates subsequent embryo development. Hum Reprod 17:2686–2693
21. Larman MG, Katz-Jaffe MG, Sheehan CB, Gardner DK (2007) 1,2-propanediol and the type of cryopreservation procedure adversely affect mouse oocyte physiology. Hum Reprod 22:250–259
22. Sheehan CB, Lane M, Gardner DK (2006) The CryLoop facilitates re-vitrification of embryos at four successive stages of development without impairing embryo growth. Human Reprod 21(11):2978–2984
23. Kuleshova LL, Lopata A (2002) Vitrification can be more favorable than slow cooling. Fertil Steril 78:449–454
24. Vajta G, Nagy ZP (2006) Are programmable freezers still needed in the embryo laboratory? Review on vitrification. Reprod Biomed Online 12:779–796
25. Lane M, Gardner DK (2004) Preparation of gametes, in vitro maturation, in vitro fertilization, embryo recovery and transfer. In: Gardner DK, Lane M, Watson AJ (eds) A laboratory guide to the mammalian embryo. Oxford Press, New York, pp 24–40
26. Chen SU, Yang YS (2007) Vitrification of oocytes; various procedures. Tucker MJ, Liebermann J (eds), Vitrification in assisted reproduction: A User's Manual and Troubleshooting Guide. Chapter 7B. Informa Healthcare, pp 129-144
27. Bielanski A, Vajta G (2009) Risk of contamination of germplasm during cryopreservation and cryobanking in IVF units. Hum Reprod 24:2457–2467
28. Larman MG, Gardner DK (2011) Vitrification of mouse embryos with super-cooled air. Fertil Steril 95:1462–1466
29. Gardner DK, Reed L, Linck D, Sheehan C, Lane M (2005) Quality control in human in vitro fertilization. Semin Reprod Med 23:319–324
30. Gardner DK, Rodriegez-Martinez H, Lane M (1999) Fetal development after transfer is increased by replacing protein with the glycosaminoglycan hyaluronan for mouse embryo culture and transfer. Hum Reprod 14: 2575–2580
31. Lane M, Maybach JM, Hooper K, Hasler JF, Gardner DK (2003) Cryo-survival and development of bovine blastocysts are enhanced by culture with recombinant albumin and hyaluronan. Mol Reprod Dev 64:70–78
32. Palasz A, Alkemade S, Mapletoft RJ (1993) The use of sodium hyaluronate in freezing media for bovine and murine embryos. Cryobiology 30:172–178
33. Stojkovic M, Kolle S, Peinl S, Stojkovic P, Zakhartchenko V, Thompson JG, Wenigerkind H, Reichenbach HD, Sinowatz F, Wolf E (2002) Effects of high concentrations of hyaluronan in culture medium on development and survival rates of fresh and frozen-thawed bovine embryos produced in vitro. Reproduction 124:141–153
34. Balaban B, Urman B (2005) Comparison of two sequential media for culturing cleavage-stage embryos and blastocysts: embryo characteristics and clinical outcome. Reprod Biomed Online 10:485–491
35. Dattena M, Mara L, Bin TA, Cappai P (2007) Lambing rate using vitrified blastocysts is improved by culture with BSA and hyaluronan. Mol Reprod Dev 74:42–47

Chapter 11

Mammalian Preimplantation Embryo Culture

David K. Gardner and Michelle Lane

Abstract

Since the inception of modern embryo culture media over 50 years ago there have been significant developments in culture systems for the mammalian preimplantation embryo. Carbohydrate gradients have been shown to impact embryo physiology and viability, while amino acids have been determined to have specific temporal effects during the preimplantation period. Furthermore, due to the lability of amino acids at 37 °C and the subsequent release of embryo-toxic ammonium into the medium, the medium is renewed every 48 h not only to provide stage-specific nutrients but also to prevent toxicity. Subsequently stage-specific media are commonly employed. To facilitate the preparation of small volumes of media, whose formulations can be readily altered according to experimental design, the use of stock solutions is described, together with systems to facilitate the development of viable embryos.

Key words Blastocyst, Culture, Embryo, Media, Metabolism, Mouse, Physiology, Preimplantation

1 Introduction

Modern embryo culture techniques have their origins in Canberra, Australia, where some 50 years ago in the laboratory of Wes Whitten the first defined systems were devised and tested [1]. In the intervening five decades significant improvements in embryo culture media and systems have been made. Such improvements were made possible by an increase in knowledge regarding the physiology and metabolism of the preimplantation embryo [2, 3], together with a better understanding of the environment within the female reproductive tract [4, 5]. When considering culture conditions it is important to consider that the preimplantation period of embryo development is highly dynamic, and events during this time have consequences for subsequent fetal development [6], and, therefore, care must be taken to maintain a suitable stable environment in vitro. Commencing with the oocyte, and subsequently the pronucleate oocyte and cleavage stage embryo, there is limited transcription, protein synthesis, and energy production. All this is reflected in the embryo's idiosyncratic metabolism.

Mark Lewandoski (ed.), *Mouse Molecular Embryology*, Methods in Molecular Biology, vol. 1092,
DOI 10.1007/978-1-60327-292-6_11,

© Springer Science+Business Media New York 2014

Rather than utilizing glucose like most somatic cell types (and indeed the blastocyst), the embryo prior to compaction utilizes carboxylic and amino acids. Indeed it was held for some 40 years that the mammalian zygote had an absolute requirement for pyruvate as an energy source to support the first cleavage division [7]. Significantly, recent studies on the presence and activity of the malate–aspartate shuttle in the mouse embryo have revealed that lactate and aspartate can actually support the first cleavage division and complete embryo development in the absence of pyruvate [8]. However, it is not advocated that pyruvate be absent from an embryo culture medium. As embryo development proceeds, there is a gradual activation of the embryonic genome. In parallel, protein synthesis increases, and following compaction there is a significant increase in energy demand with the creation of the blastocoel, formed through the activity of Na^+/K^+ ATPases on the basolateral membrane of the trophectoderm. Subsequently, there is a significant increase in demand for glucose, not only as an energy source but also for the synthesis of triacylglycerols and phospholipids, and as a precursor for complex sugars of mucopolysaccharides and glycoproteins. Therefore, glucose metabolized through the pentose phosphate pathway (PPP) is required to generate the ribose moieties required for nucleic acid synthesis and the NADPH required for the biosynthesis of lipids and other complex molecules.

Consequently, in a period of just a few days, the embryo undergoes some remarkable metabolic transformations. In parallel with this, the environment of the female reproductive tract changes along the oviduct and uterus, creating gradients of nutrients and consequently different environments for the pre- and post-compacted embryo.

From the 1950s to the 1970s, the media were simple, containing carbohydrates, but lacking amino acids, and one formulation was used to support embryo development in static culture from the zygote to the blastocyst. In the intervening years the significance of amino acids in regulating embryo physiology and metabolism was established, and subsequently media formulations began to include amino acids [9–11]. One problem associated with conventional static culture in vitro is the unintentional introduction of artifacts. For example, amino acids are labile at 37 °C and spontaneously release ammonium into the medium. In tissue culture of somatic cells it is documented that this phenomenon results in compromised cell development [12]. Embryos, like somatic cells, have difficulty in dealing with an ammonium build up in the medium, and exhibit altered physiology, metabolism, gene expression, and compromised fetal development as a result of a chronic exposure to ammonium [13, 14]. To add insult to injury, embryos metabolize amino acids and produce more ammonium. Subsequently, all media containing amino acids need to be renewed every 48 h as even low levels of ammonium can compromise embryo function.

Consistent with the different environment created by the oviduct and uterus, carbohydrate and amino acid gradients affect embryonic development in vitro [15]. This, together with a need to alleviate ammonium toxicity, led to the introduction of a biphasic or sequential media system based on the use of two media, in which embryos are moved from one medium to another after 48 h. Such media were formulated not only to take into account the changing physiology of the embryo, but were designed to reduce intracellular stress. This chapter therefore outlines the use of such a system. The media presented can be used effectively for the embryos of the mouse, domestic animals and the human [16]. Indeed, in the mouse in vivo rates of development can be attained in vitro using such media. Perhaps most importantly, implantation rates of 65 % and pregnancy rates >80 % have been obtained in the human in an oocyte donor model [15], highlighting the efficacy of such a culture method.

For the sake of brevity, the specific protocols listed herein are those developed for the mouse. Other treatise should be referred to for specific details of other species [17]. The major difference for domestic animals is the temperature at which to perform the culture 38.5 °C, rather than 37 °C for the mouse and human.

It is essential to note at this point that success in the embryo laboratory is dependent upon more than the formulations of the media per se. Rather, it is paramount that quality control and assurance systems are in place to ensure that all contact supplies, chemicals, and oils are all prescreened before use [18]. Furthermore, it is imperative that mammalian embryos are never be exposed to any type of serum and their exposure to atmospheric oxygen (~20 %) be kept to a minimum during collection and manipulation. All cultures should be performed at a reduced oxygen concentration (typically 5–7 %) [15]. Preferably lights should be dimmed and no fluorescent lighting employed in the embryology laboratory.

2 Materials

1. Stereomicroscope with heated stage and magnification capacity of up to 40×.
2. Test tube warmers and heated stages (*see* **Note 1**).
3. Thermocouple to determine temperature in microdrop where embryo resides.
4. Inverted microscope with magnification capacity of up to 200×.
5. Multigas incubator (to control CO_2 and O_2) with infra-red (IR) sensor for CO_2, or some modular incubation system such as a desiccator or modular chamber (in conjunction with a pre-mixed gas cylinder, e.g., 6 % CO_2, 5 % O_2, 89 % N_2).
6. Fyrite/Infrared CO_2 analyzer to confirm the gas environment within the incubator.

7. Media for embryo manipulation and culture. When small amounts of media are required each week (for example 10–50 ml), or when one needs to alter the formulation for experimentation, it is practical to make media from stock solutions as described in Table 1. All chemicals should be of the highest and/or tissue culture grade and prescreened using an appropriate bioassay [18]. Prepare example embryo culture

Table 1
Stock solutions

	Mol weight	Weight (g)	Final conc. (mM)
Stock A			
10× stock use for 3 months			
Sodium chloride	58.44	5.2643	90.08
Potassium chloride	74.56	0.4101	5.50
Sodium phosphate	119.98	0.0300	0.25
Magnesium sulfate heptahydrate	246.47	0.2465	1.0
Penicillin		0.0600	
Stock B			
10× stock use for 1 month			
Sodium bicarbonate	84.01	2.1003	25.0
Phenol red (*see* **Note 2**)		0.0050	
Stock C			
100× stock use for 1 month			
Calcium chloride	147.02	0.2646	1.8
Stock D			
10× stock use for 1 month			
Glucose	180.16	0.0901	0.50
Sodium lactate (L-isomer only)	112.06	1.1766	10.50
Sodium pyruvate	110.04	0.0352	0.32
Stock E			
10× stock use for 1 month			
Glucose	180.16	0.5675	3.15
Lactate (L-isomer only)	112.06	0.6578	5.87
Sodium pyruvate	110.04	0.0110	0.1
Stock F			
100× stock use for 3 months			
EDTA disodium salt (Titriplex)	372.24	0.0037	0.01
Stock G			
100× stock (cleavage stage and ICM amino acids) use for 3 months			
Alanine	89.09	0.0089	0.1
Aspartate	133.1	0.0133	0.1
Asparagine–H_2O	150.14	0.0150	0.1
Glutamate	169.1	0.0169	0.1
Glycine	75.07	0.0075	0.1
Proline	115.13	0.0115	0.1
Serine	105.09	0.0105	0.1

(continued)

Table 1
(continued)

	Mol weight	Weight (g)	Final conc. (mM)
Stock H			
100× stock use for 3 months			
Taurine	125.14	0.0125	0.1
Stock I			
100× stock use for 3 months			
Alanyl-glutamine	217.2	0.1086	0.5
Stock J			
100× (ICM amino acids) use for 3 months			
Arginine–HCl	210.66	0.0632	0.3
Cystine	240.3	0.0120	0.05
Histidine–HCl–H_2O	191.62	0.0192	0.1
Isoleucine	131.17	0.0262	0.2
Leucine	131.17	0.0262	0.2
Lysine	146.19	0.0292	0.2
Methionine	149.21	0.0075	0.05
Phenylalanine	165.19	0.0165	0.1
Threonine	119.12	0.0238	0.2
Tryptophan	204.23	0.0051	0.025
Tyrosine	181.19	0.0181	0.1
Valine	117.15	0.0234	0.2
Stock K			
100× (vitamins) use for 3 months			
Choline chloride	139.62	0.0010	0.0072
Folic acid	441.4	0.0010	0.0023
Myo-Inositol	180.16	0.0018	0.01
Nicotinamide	122.12	0.0010	0.0082
Pantothenate	238.3	0.0010	0.0042
Pyridoxal	203.62	0.0010	0.0049
Riboflavin	367.37	0.0001	0.00027
Thiamine	337.27	0.0010	0.00296
Stock L			
10× stock use for 3 months			
MOPS (*see* **Note 3**)	209.3	4.814	23.0

Stock solutions that are 10× are prepared by weighing the grams for 1 l and dissolving in 100 ml in high-quality H_2O.
Stock solutions that are 100× are prepared by weighing the grams for 1 l and dissolving in 10 ml in high-quality H_2O
Stock G can be purchased in the form of Eagle's Nonessential amino acids (MEM). Alternatively, the amino acids can be weighed out individually and made as a stock solution. Due to the small weights, making the 100× stock in a total volume of 100 ml can be useful (i.e., 10x the weights).
Stock J can be purchased in the form of Eagle's Essential amino acids (MEM). Alternatively, they can be weighed out individually and made as a stock solution. Due to the small weights, making the 100× stock in a total volume of 100 ml can be useful (i.e., 10× the weights). In order to get the amino acids into solution HCl (1 M) is required, typically 0.5–1.0 ml of HCl is required for 100 ml of amino acid stock.
Stock L is adjusted to pH of 7.3 ± 0.5 by using a concentrated stock solution of NaOH (2–5 M)
All stock solutions should be filtered through a 0.2 μm filter immediately after preparation and stored at 4 °C. Always discard the first 2 ml as this contains potential harmful wash off from the filter

Table 2
Preparation of example embryo culture media from stock solutions

Stock	Cleavage medium (10 ml)	Blastocyst medium (10 ml)	Handling medium (10 ml)
H_2O	6.5	6.3	6.42
A	1.0	1.0	1.0
B	1.0	1.0	0.08
C	0.1	0.1	0.1
D	1.0	–	1.0
E	–	1.0	–
F	0.1	–	0.1
G	0.1	0.1	0.1
H	0.1	0.1	0.1
I	0.1	0.2	0.1
J	–	0.1	–
K	–	0.1	–
L	–	–	1.0

Stocks can be modified and added according to the requirements of the culture or experiment. Using stocks, media formulations can be changed readily to accommodate experimental treatments

media from stock solutions according to Table 2. When doing so, water should first be added to the culture flask/tube using a sterile pipette. Each component is then added using a displacement pipette.

If more or fewer stocks are used than in Table 2 then simply decrease or increase the amount of water added accordingly. Immediately after they are prepared media should be filtered through a 0.2 μm filter and then be stored at 4 °C for up to 4 weeks. Discard the first 2 ml of medium that passes through the filter.

Upon preparation of the Blastocyst medium, it is important to bring the ungassed pH of the medium to 8.0 by adding just a few μl of NaOH (1 M). The osmolality of Cleavage and Handling media is 273 ± 5, while that of the Blastocyst medium is 268 ± 5. After equilibration with 6 % CO_2, the pH of both Cleavage and Blastocyst media should be 7.3 ± 0.05.

Media should be supplemented with serum albumin and hyaluronan. Albumin can be added as either bovine serum albumin or human serum albumin or recombinant human albumin. If the albumin preparation to be added is a solution, then the amount of water should be shorted to compensate for the amount of albumin solution to be added. Albumin is typically added between 4 and 10 mg/ml. Hyauronan is typically added as a stock solution and can be acquired in two forms; extracts from rooster comb, or as a fermented product from bacteria. The latter is endotoxin-free and should be considered the preferred source and is typically added between 0.125 and 0.5 mg/ml (*see* **Note 5**).

8. Oil, paraffin (not mineral oil or blends).
9. Dishes; Falcon (Becton Dickinson 35 and 60 mm) or embryo tested dishes.
10. Hyaluronidase (500 IU/ml), preferred source recombinant e.g., cumulase.
11. Pipettes or glass capillaries, or gel loaders.
12. Pipetting system; mouth pipette, bulbs, syringes etc.
13. Blunt 32-gauge needle, cut down to around 1 cm and 1 ml syringe.

3 Methods

3.1 Superovulation of Mice

To increase the number of embryos available for an experiment and to reduce the number of animals that are required it is common to induce multiple ovulations in mice by the administration of hormones. Superovulation protocols also assist in the synchronization of embryo timing between females. Best results from superovulation protocols are achieved by using prepubertal females (3–5 weeks old). Two hormones are administered via the intraperitoneum. The first injection of pregnant mare's serum gonadotropin (PMSG) provides FSH that recruits follicles for ovulation. This is followed 48 h later by an injection of human chorionic gonadotropin (hCG), which mimics the LH surge and results in ovulation. The injection interval should be kept between 46 and 48 h. It is essential that the hCG injection is given prior to the release of the endogenous LH surge (especially important for animals >24 days old) which occurs around 15–20 h after the middle of the second dark cycle. Following hCG administration females are placed with males and allowed to remain overnight. Mating can be detected by the presence of a vaginal plug the following morning. Fertilized embryos can then be collected at discrete stages of development (*see* Table 3).

Response to superovulation procedures will vary significantly between strains of mice. A dose of 2.5–5 IU per mouse at 3–4 weeks old is typical. As response to hormones vary, it is important to start at a lower dose and then gradually increase to 5 IU, while assessing the number of embryos collected and any fragmentation.

Table 3
Timetable for embryo collection

Day of PMSG injection	Day of hCG injection	Zygote 9-10 a.m. (day of plug)	2 Cell 9-10 a.m. day 2	8 Cell 6-8 a.m. day 3	Morula 3-4 p.m. day 3	Blastocyst 8–9 a.m. day 4
Monday	Wednesday	Thursday	Friday	Saturday	Saturday	Sunday
Tuesday	Thursday	Friday	Saturday	Sunday	Sunday	Monday
Wednesday	Friday	Saturday	Sunday	Monday	Monday	Tuesday
Thursday	Saturday	Sunday	Monday	Tuesday	Tuesday	Wednesday
Friday	Sunday	Monday	Tuesday	Wednesday	Wednesday	Thursday
Saturday	Monday	Tuesday	Wednesday	Thursday	Thursday	Friday
Sunday	Tuesday	Wednesday	Thursday	Friday	Friday	Saturday

The light cycle in the mouse room is 12 h of daylight and 12 h of dark. Daylight is from 6:00 to 18:00. The room temperature is 21 ± 2 °C. Hormone injections are administered between 12 and 1 p.m.

The optimum dose will give the best-quality embryos, not necessarily the most.

Hormone solutions can be prepared by reconstituting lyophilized powder in sterile saline at a concentration of 50 IU/ml. This results in an injection of 0.1 ml per mouse to deliver 5 IU. Hormones can be stored for 2–3 weeks at −20 °C or at −80 °C for 2–3 months. There are significant differences in the activity of different batches of hormones therefore each new lot should be prescreened before general use by examining their effect on oocyte and embryo quality.

3.2 Dissection of Reproductive Tracts

1. Check for the presence of mating by visualization of a vaginal plug. Mated female mice can then be brought into the dissection area.
2. Instruments are washed, dried, and then sterilized by autoclaving.
3. A 35 mm petri dish containing 2–3 ml of pre-warmed handling medium with the lid on is placed on a warm stage at ≥37 °C adjacent to the dissection area. The temperature of the warming stage should be calibrated so the handling media within the petri dish is at 37 °C. Do not leave the dish on the plate for extended periods of time to avoid evaporation.
4. The mouse is sacrificed by cervical dislocation and placed ventral side-up on absorbent paper. Methods of euthanasia vary by institution but should be in compliance with local Institutional Animal Care and Use Committee guidelines.
5. Disinfect the abdomen with 70 % ethanol.
6. Cut the peritoneum to expose the body cavity.

7. Move the coils of gut to one side to expose the reproductive tract.
8. Hold the utero-tubal junction with watchmaker forceps and strip the uterus of surrounding connective tissue.
9. For collection of embryos from the oviduct, make a cut between the oviduct and the ovary (avoid cutting the ovary). For collection of zygotes and 2-cells, make a second cut below the utero-tubal junction to isolate the oviduct from the uterus. For collection of 8-cell embryos make the second cut around 1/3 of the way down the uterine horn. Place the dissected tissue into the collection dish of handling medium.
10. For collection of blastocysts from the uterus, make a cut below the utero-tubal junction of both uterine horns to isolate the oviduct from the uterus. Make a second cut through the cervix and place the connected uterine horns into the collection dish of handling media.

3.3 Collection of Zygotes

Before collection of the oviducts, warm 10 ml of handling medium and 1 ml aliquot of hyaluronidase to 37 °C (*see* **Note 4**). Prepare pulled Pasteur pipettes with an internal diameter of between 100 and 300 μm for manipulation of the embryos. A pipette that is just larger than the embryos is essential to ensure adequate washing and minimal transfer of medium between drops.

1. Place approximately 1 ml of handling medium into a new collection dish either a 35 mm petri dish or organ well dish.
2. Transfer an oviduct to the new collection dish, one at a time.
3. Stabilize the oviduct with fine forceps and locate the swollen ampulla.
4. Tear the ampulla of the oviduct to release the cumulus mass. Discard the oviduct and transfer a new oviduct to the collection dish. Repeat procedures until the cumulus mass from all oviducts are released. It is strongly recommended that a maximum of three mice are dissected at a time in order to ensure rapid collection and processing of the embryos.
5. Add the 500 μl of warmed hyaluronidase solution to the dish to remove cumulus cells. Expose for 20–60 s.
6. During the exposure of the cumulus complex to hyaluronidase, set up three washing drops of handling medium approximately 50 μl in a second collection dish (alternatively the lid of a collection dish can be used).
7. Collect the embryos using the pipette as soon as they begin to become denuded (*see* **Note 4**).
8. Wash denuded embryos through the three drops of handling medium.
9. Embryos are then ready to be placed into culture.

3.4 Collection of 2-Cell Embryos

Collect oviducts as above. Attach a blunt 32-gauge needle to a 1 ml syringe loaded with pre-warmed Handling medium.

1. Place approximately 1 ml of handling medium into a new collection dish either a 35 mm petri dish or organ well dish.
2. Transfer an oviduct to the new collection dish, one at a time.
3. Stabilize the oviduct with fine forceps and locate the infundibulum (opening of the oviduct by the ovary).
4. Insert the blunt 32-gauge needle and hold in position with fine forceps.
5. Gently flush the oviduct with ~0.25 ml of Handling medium. It is strongly recommended that a maximum of three mice are dissected at a time in order to ensure rapid collection and processing of the embryos.
6. Set up three washing drops of Handling medium approximately 50 μl in a second collection dish (alternatively the lid of a collection dish can be used).
7. Wash 2-cell embryos through the three drops of handling medium.
8. Embryos are then ready to be placed into culture.

3.5 Collection of 8-Cell Embryos

Collect oviducts with the top third of the uterus attached as above. Attach a blunt 32-gauge needle to a 1 ml syringe loaded with pre-warmed Handling medium.

1. Place approximately 1 ml of handling medium into a new collection dish either a 35 mm petri dish or organ well dish.
2. Transfer an oviduct connected to the top third of the uterus to the new collection dish, one at a time.
3. Stabilize the oviduct and uterus with fine forceps and locate the infundibulum.
4. Insert the blunt 32-gauge needle and hold in position with fine forceps.
5. Gently flush the oviduct with ~0.5 ml of Handling medium. It is strongly recommended that a maximum of three mice are dissected at a time in order to ensure rapid collection and processing of the embryos.
6. Set up three washing drops of Handling medium approximately 50 μl in a second collection dish (alternatively the lid of a collection dish can be used).
7. Wash 8-cell embryos through the three drops of handling medium.
8. Embryos are then ready to be placed into culture.

3.6 Collection of Blastocysts

Collect uteri as above. Attach a blunt 32-gauge needle to a 1 ml syringe loaded with pre-warmed Handling medium.

1. Place approximately 1 ml of handling medium into a new collection dish either a 35 mm petri dish or organ well dish.
2. Transfer the uteri from one mouse to the new collection dish.
3. Stabilize the cervix with fine forceps.
4. Insert the blunt 32-gauge needle into one uterine horn through the cervix and hold in position with fine forceps.
5. Gently flush the uterus with ~0.5–1.0 ml of Handling medium. It is strongly recommended that a maximum of three mice are dissected at a time in order to ensure rapid collection and processing of the embryos.
6. Repeat **steps 4** and **5** for the contralateral uterine horn.
7. Set up three washing drops of Handling medium approximately 50 μl in a second collection dish (alternatively the lid of a collection dish can be used).
8. Wash blastocysts through the three drops of handling medium.
9. Embryos are then ready to be placed into culture.

Embryos are very sensitive to their environment. Therefore, it is important to minimize the time from extraction of the cumulus masses, cleavage stage embryos or blastocysts and their placement in culture. It is advisable to keep this time period to a maximum of 10 min. Therefore, it is preferable to do multiple smaller embryo collections rather than one large collection.

3.7 Embryo Culture

Afternoon before culture

1. After 4 p.m. on the day before embryo culture, culture dishes for pronucleate oocytes should be prepared. Rinse the pipette tip by taking up the required volume and then expelling into a discard dish. Using this pre-rinsed sterile tip place the required number of 10 μl drops of culture medium into the petri dish. Rinsing the tip is performed by taking up and expelling media. This is a crucial step and should not be overlooked. Immediately cover drops with 3.5 ml paraffin oil to avoid evaporation. Always set up several wash drops in each dish. Optimally there should be one initial wash drop followed by a wash drop for every culture drop. Prepare no more than 2–3 dishes at one time. Once the initial 10 μl drops have been laid down and covered with oil, each drop is then built up to 20 μl. First rinse the tip and then make each drop up to the final volume of 20 μl. Use a new tip for each drop to avoid contaminating the culture media with oil.
2. Immediately place the dish in the incubator at 6 % CO_2 and in reduced oxygen (5–7 %). Gently remove the lid of the dish and set at an angle on the side of the plate. Dishes must equilibrate in the incubator with a semi-opened lid for a minimum of 4 h

(this is the minimal measured time for the media to reach correct pH under oil) and for a maximum of 18 h to prevent excessive ammonium production.

Day 1

1. Pronucleate oocytes are washed in handling medium and then further washed in the wash drops of cleavage medium. Washing entails picking up the embryos two to three times in a minimal volume and moving them around within the drop in order to minimize medium carryover to the next drop. Embryos should be washed successively through the two wash drops in the culture dish before being transferred to the culture drops. The time that the culture dish is removed from the incubator needs to be kept to a minimum and should be less than 5 min.
2. Typically ten embryos are cultured per 20 μl drop. If embryos are to be cultured individually, the droplet volume can be reduced to 10 μl. The minimum volume recommended is 2 μl and only then if the covering paraffin oil has been screened using a microdrop assay.
3. Dishes should be returned to the incubator and cultured for 48 h.

Day 2

1. After 4 p.m. on the day before embryo changeover into blastocyst medium, culture dishes should be prepared. Again rinse the pipette tip by taking up the required volume and then expelling into a discard dish. Using this pre-rinsed sterile tip place 10 μl drops of culture medium into the petri dish. Always set up several wash drops in each dish. Optimally there should be one initial wash drop followed by a wash drop for every culture drop. Immediately cover drops with paraffin oil to avoid evaporation. Prepare no more than 2–3 dishes at one time. Using a new tip for each drop, first rinse the tip and then round each drop up to the final volume (final culture volume is 20 μl).
2. Immediately place the dish in the incubator at 6 % CO_2 and in reduced oxygen (5–7 %). Gently remove the lid of the dish and set at an angle on the side of the plate. Dishes must equilibrate in the incubator with a semi-opened lid for a minimum of 4 h (this is the minimal measured time for the media to reach correct pH under oil) and for a maximum of 18 h to prevent excessive ammonium production.

Day 3

1. When embryos are changed over from cleavage medium to blastocyst medium, it is essential that embryos be thoroughly washed from one medium into the other.

2. Embryos are removed from the drops of cleavage medium and washed twice in wash drops of blastocyst medium. Embryos should be washed successively in the two wash drops in the culture dish and then transferred to the new culture drop. Remember, the time that the culture dish is removed from the incubator needs to be kept to a minimum, less than 5 min.
3. Dishes should be returned to the incubator and cultured for a further 48 h.

4 Embryo Scoring

Development of embryos should be monitored at a specific time on each successive day of development. It is important to stick to such times in order to increase the accuracy of assessing the embryonic stage reached under different culture conditions. For example, a blastocyst formed on the morning of day 4 is different to one that develops by 4 p.m. on the same day. Photomicrographs of cultured mouse embryos are illustrated in Fig. 1.

5 Maintenance of Equipment

The maintenance of equipment is a key aspect to providing an environment that ensures consistent high rates of embryo development. As discussed, all heated stages should be calibrated to ensure that the temperature within the culture drop is 37 °C. This will often mean (dependent on dish type) that the heated stages will be set 2–3 °C higher than the desired final temperature in the drops.

It is critical to maintain the incubator because the embryo will spend the majority of its time in this environment. Incubator carbon dioxide and oxygen levels, as well as temperature, should be checked at least once a week. Levels should be checked using calibrated equipment, e.g., a Fyrite device is commonly used to check carbon dioxide levels. Although popular, Fyrite measurement has a variance of up to ±1.0 %, and thus electronic meters with infrared (IR) sensors with a variance of only ±0.2 % should preferably be used. Whichever technology is favored, based on an acceptable tolerance level, it is essential that the first step to any equipment reading is to calibrate (or check) against a known carbon dioxide concentration, e.g. a certified gas cylinder, before taking the incubator reading. Similarly, temperature should be checked against a certified thermometer. All readings are best performed before an incubator is opened for the first time in a day, i.e., taken from an undisturbed environment.

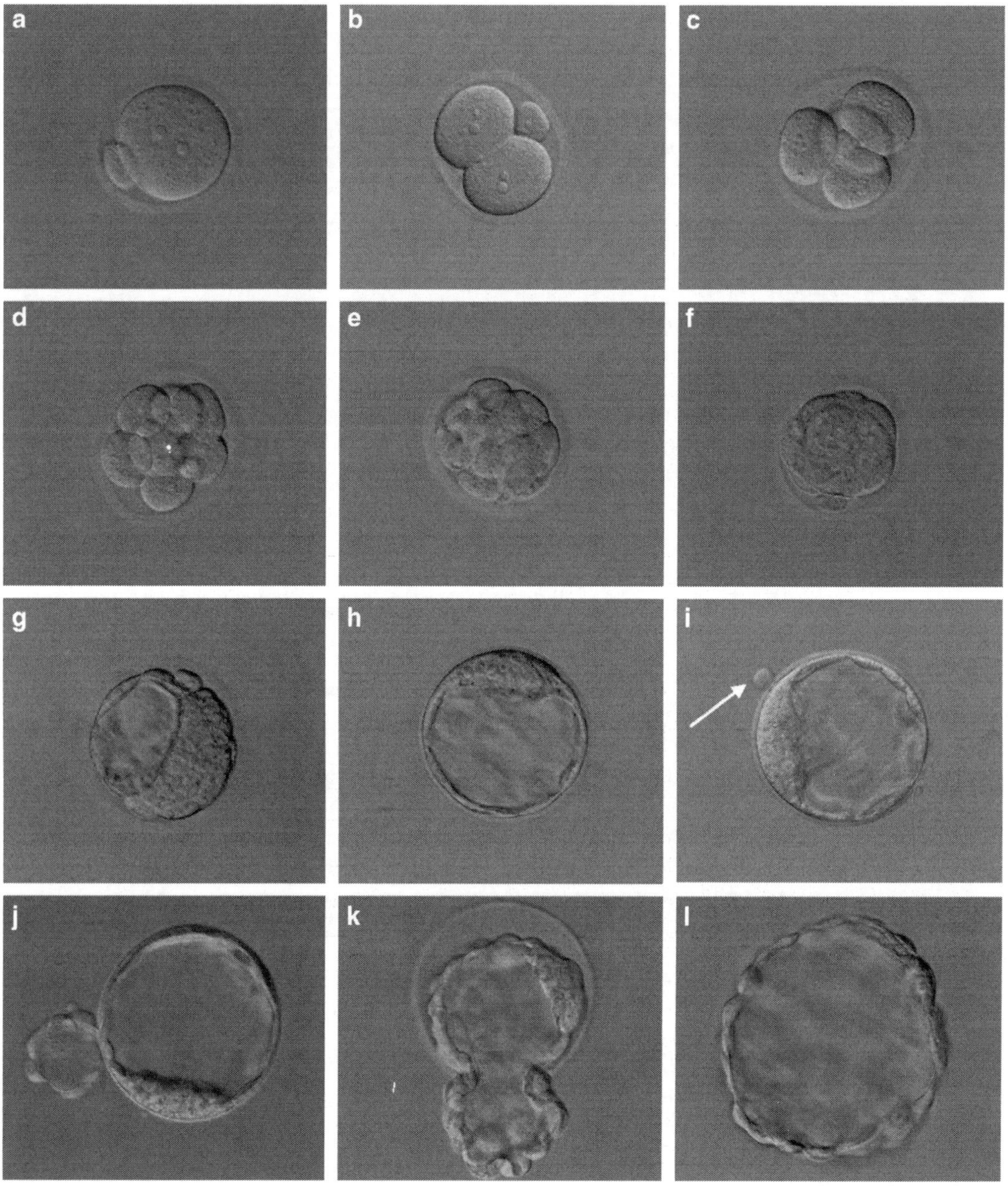

Fig. 1 Preimplantation development of the mouse embryo in culture. (**a**) Pronucleate oocyte at 10 a.m. Day 1. (**b**) Two-cell embryo at 8 a.m. Day 2. (**c**) Four-cell embryo at 4 p.m. Day 2. (**d**) Eight-cell embryo at 8 a.m. Day 3. (**e**) Beginnings of compaction at 8 a.m. Day 3. (**f**) Morula at 4 p.m. Day 3. (**g**) Early blastocyst at 8 a.m. Day 4. (**h**) Expanded blastocyst at 4 p.m. Day 4. (**i**) Initiation of hatching at 4 p.m. Day 4. A piece of trophectoderm can be seen herniating (marked with *arrow*) through the zona pellucida. (**j**) Hatching blastocyst at 8 a.m. Day 5. A group of trophectoderm cells are clearly visible at the 9 o'clock position. (**k**) Hatching blastocyst at 10 a.m. Day 5. Half of the blastocyst has emerged through the zona pellucida. (**l**) Hatched blastocyst at 4 p.m. Day 5. PMSG and hCG were both administered at noon

6 Notes

1. The temperature of all microscope and heating stages need to be greater than 37 °C to ensure that the temperature within the culture fluid at the site of the embryos is actually 37 °C.
2. Phenol red is a useful pH indicator, although some people prefer to exclude it due to concerns regarding its purity. However, in general it is considered a practical component of media. In order to calibrate the color of the medium to its pH, Sorenson's phosphate buffer standards are prepared as follows:

   Preparation of Color Standards for pH of Media.

   Stock A: 9.08 g KH_2PO_4 (0.067 M), 5 mg phenol red in 1 l of water.

   Stock B: 9.46 g Na_2HPO_4 (0.067 M), 5 mg phenol red in 1 l of water.

pH at 18 °C	Solution A (ml)	Solution B (ml)
6.6	62.7	37.3
6.8	50.8	49.2
7.0	39.2	60.8
7.2	28.5	71.5
7.4	19.6	80.4
7.6	13.2	86.8

   Measure the pH with meter, and adjust pH as required, i.e., add solution A to lower the pH (make more acidic), Add solution B to increase the pH (make more alkaline).

   The pH standards should be filter-sterilized and can then be kept for up to 6 months.

   Please note that less phenol red can be added to the culture media if desired. In this case simply reduce the levels of phenol red in the Sorenson's buffers accordingly.

3. Although the protocol described uses MOPS, an alternative buffer is HEPES. If HEPES is chosen over MOPS, then simply substitute based on molarity.
4. Exposure of embryos to hyaluronidase for too long, or at too high a concentration, is highly detrimental to the embryo. Therefore, ensure that the embryos are removed and washed as soon as the cumulus cells detach from the zona pellucida.

This can be facilitated with gentle pipetting. Do not sit and watch waiting for all the cumulus to simply fall off!

5. Hyaluronan is not currently considered a routine component of mammalian culture media. However, there is growing evidence that this macromolecule helps to maintain cellular ultrastructure and improve cryotolerance and viability of embryos (reviewed in ref. 15).

References

1. Whitten WK (1956) Culture of tubal mouse ova. Nature 177:96–97
2. Leese HJ (1991) Metabolism of the preimplantation mammalian embryo. Oxf Rev Reprod Biol 13:35–72
3. Gardner DK (1998) Changes in requirements and utilization of nutrients during mammalian preimplantation embryo development and their significance in embryo culture. Theriogenology 49:83–102
4. Gardner DK, Lane M, Calderon I, Leeton J (1996) Environment of the preimplantation human embryo in vivo: metabolite analysis of oviduct and uterine fluids and metabolism of cumulus cells. Fertil Steril 65:349–353
5. Harris SE, Gopichandran N, Picton HM, Leese HJ, Orsi NM (2005) Nutrient concentrations in murine follicular fluid and the female reproductive tract. Theriogenology 64: 992–1006
6. Lane M, Gardner DK (1994) Increase in postimplantation development of cultured mouse embryos by amino acids and induction of fetal retardation and exencephaly by ammonium ions. J Reprod Fertil 102:305–312
7. Biggers JD, Whittingham DG, Donahue RP (1967) The pattern of energy metabolism in the mouse oocyte and zygote. Proc Natl Acad Sci U S A 58:560–567
8. Lane M, Gardner DK (2005) Mitochondrial malate-aspartate shuttle regulates mouse embryo nutrient consumption. J Biol Chem 280:18361–18367
9. Spindle AI, Pedersen RA (1973) Hatching, attachment, and outgrowth of mouse blastocysts in vitro: fixed nitrogen requirements. J Exp Zool 186:305–318
10. Bavister BD, Arlotto T (1990) Influence of single amino acids on the development of hamster one-cell embryos in vitro. Mol Reprod Dev 25:45–51
11. Gardner DK, Lane M (1993) Amino acids and ammonium regulate mouse embryo development in culture. Biol Reprod 48:377–385
12. Schneider M, Marison IW, von Stockar U (1996) The importance of ammonia in mammalian cell culture. J Biotechnol 46: 161–185
13. Gardner DK, Lane M (2005) Ex vivo early embryo development and effects on gene expression and imprinting. Reprod Fertil Dev 17:361–370
14. Lane M, Gardner DK (2005) Understanding cellular disruptions during early embryo development that perturb viability and fetal development. Reprod Fertil Dev 17: 371–378
15. Gardner DK (2008) Dissection of culture media for embryos: the most important and less important components and characteristics. Reprod Fertil Dev 20:9–18
16. Gardner DK, Lane M (2003) Towards a single embryo transfer. Reprod Biomed Online 6: 470–481
17. Gardner DK, Lane M (2012) Culture systems for the human embryo. In: Gardner DK, Weissman A, Howles C, Shoham Z (eds) Textbook of assisted reproductive techniques laboratory and clinical perspectives, Fourth Edition. Informa Healthcare, London, pp 218–239
18. Gardner DK, Reed L, Linck D, Sheehan C, Lane M (2005) Quality control in human in vitro fertilization. Semin Reprod Med 23:319–324

Chapter 12

Serum-Free Culture of Mid-gestation Mouse Embryos: A Tool for the Study of Endoderm-Derived Organs

Julie Gordon, Billie A. Moore, C. Clare Blackburn, and Nancy R. Manley

Abstract

The experimental manipulation of mid-gestation mouse embryos is an important tool for the study of developmental biology. However, such techniques can be challenging due to difficulties accessing the embryos in utero, and therefore the ability to maintain mid-gestation mouse embryos in vitro has proved invaluable. Described here is an example of a whole embryo culture system, where a serum-free medium is used to support the development of mouse embryos in vitro from embryonic day 10.5 (E10.5) to E11.5. During this time the embryos increase in size and undergo developmental progression, as determined by morphological and molecular criteria. This makes it an ideal environment in which to support and maintain mid-gestation mouse embryos following experimental manipulations.

Two applications of this whole embryo culture system are described here. In the first, protein-soaked beads are carefully positioned in the pharyngeal region of an E10.5 embryo, allowing the concentration of specific proteins to be altered within the tissue. In the second technique, morpholino oligonucleotides are electroporated into the pharyngeal region of the embryo at E10.5, creating an efficient system for the knockdown of gene function in the target cells. These techniques demonstrate the use of in vitro techniques to study organogenesis within the pharyngeal region of the mouse embryo, but with some modification they could be adapted to target any region of the endodermal gut tube.

Key words Serum-free whole embryo culture, Mouse embryogenesis, Bead implants, Electroporation, Morpholinos, Endoderm

1 Introduction

The manipulation of mouse embryos has traditionally been a challenge because it is difficult to access specific structures, particularly internal ones, of embryos in utero. However, advances in embryo culture have made such experiments possible, allowing embryos to develop in vitro, and therefore making them amenable to many experimental procedures [1–6]. Here, we describe a serum-free whole embryo culture for mid-gestation mouse embryos that we have developed and used in our laboratory [7, 8] in an attempt to overcome the inconsistencies involved in the use of the more traditional 100 % rodent serum used in whole embryo culture [9–13].

Mark Lewandoski (ed.), *Mouse Molecular Embryology*, Methods in Molecular Biology, vol. 1092,
DOI 10.1007/978-1-60327-292-6_12, © Springer Science+Business Media New York 2014

Then, two applications of the whole embryo culture system are described. In the first, protein-soaked beads are placed in the region of the third pharyngeal pouch, an endodermal structure that will form the common thymus–parathyroid primordium. Bead implant experiments are a traditional method for administering a localized dose of a protein of choice to a specific region within the embryo for over-expression and ectopic expression studies. The experiment is fast and easy to set up and is limited only by the proteins available and the ability to access the target tissue within the embryo. Second, we describe proof-of-concept experiments that allow a rapid and efficient assay of gene function during mouse embryogenesis. Morpholino oligonucleotides are a well-established method for knocking down gene function [14], and while they are commonly used in zebrafish [15], their use in mouse embryos has remained limited [16, 17]. This is due largely to the inaccessibility of developing mouse embryos and to difficulties in delivering the morpholinos to the desired target structures. For more accessible structures such as the eye and central nervous system it is possible to perform DNA injection and electroporation in utero [18, 19], but this would not be as straightforward for internal structures such as the endoderm-derived organs. There are reports where DNA injection and electroporation have been combined with the culture of mid-gestation mouse embryos, but again this has been done primarily in studies of the central nervous system, where the neural tube is easily accessible [12, 20]. Here, these techniques are applied to the study of endoderm-derived organs in the mouse embryo. Injection and electroporation of morpholino oligonucleotides into the pharyngeal endoderm, combined with whole embryo culture, allows the knockdown of gene function in the target cells of choice, after which the effects on organogenesis can be analyzed.

2 Materials

Whole embryo culture

1. Rolling bottle whole embryo culture apparatus (BTC Engineering, Cambridge, UK).
2. 95 % O_2/5 % CO_2 cylinder with low flow regulator.
3. Serum-free whole embryo culture medium: KnockOut DMEM (Gibco) containing 10 % KnockOut Serum Replacement (Gibco), 1× N2 supplement (Gibco), 2 % cell culture grade albumin (Sigma), 25 units/ml penicillin, and 25 μg/ml streptomycin. This medium should be made fresh as needed.
4. DMEM (Gibco) or M2 (Sigma) medium for dissection.
5. Fine dissecting forceps (no. 5, Roboz Surgical).

6. Dissecting stereomicroscope.
7. 50 or 35 mm petri dishes for dissection.
8. Plastic transfer pipets.

Bead implants

9. Affi-Gel Blue Gel 100-200 mesh beads (Bio-Rad Laboratories).
10. Protein(s) of choice.
11. 1.5 ml plastic tubes.
12. PBS (Phosphate-Buffered Saline).

Morpholino injection and electroporation

13. Picopump microinjector (World Precision Instruments).
14. Fine glass needles: use a micropipet puller (Sutter Instruments) to pull a thin-walled glass capillary tube (Clark Electromedical Instruments), giving a needle length of about 1 cm.
15. EMC830 Square Wave Electroporation System (BTX).
16. Pair of 0.5 mm diameter (L-shaped (Gold tip)) "Genetrodes" (BTX Model 516).
17. Genetrode holder with shaft (BTX).
18. Clamp stand.
19. Morpholinos (GeneTools LLC, fluorescein-tagged, optional but advised, *see* **Note 10**).
20. 0.1 % Fast Green solution.
21. Fluorescent microscope for analysis (if using fluorescein-tagged morpholinos).

3 Methods

3.1 Serum-Free Whole Embryo Culture

Described below is a procedure for culturing whole E10.5 mouse embryos in a serum-free medium for up to 30 h. A thorough analysis of embryonic development during the culture period has been described elsewhere [7], and this section is focused on providing a detailed protocol for embryo, media, and equipment preparation.

3.1.1 Embryo Isolation

1. Sacrifice the mother by cervical dislocation (*see* **Note 1**), and dissect E10.5 embryos from the uterus in DMEM or M2 culture medium. The age of the embryos should be confirmed by somite count (E10.5 = 34–38 somites) and can also be assessed by morphological criteria such as overall size, limb bud and pharyngeal arch morphology [21]. *See* **Note 2**.
2. Remove the decidua completely, but leave the placenta and yolk sac intact (Fig. 1a).

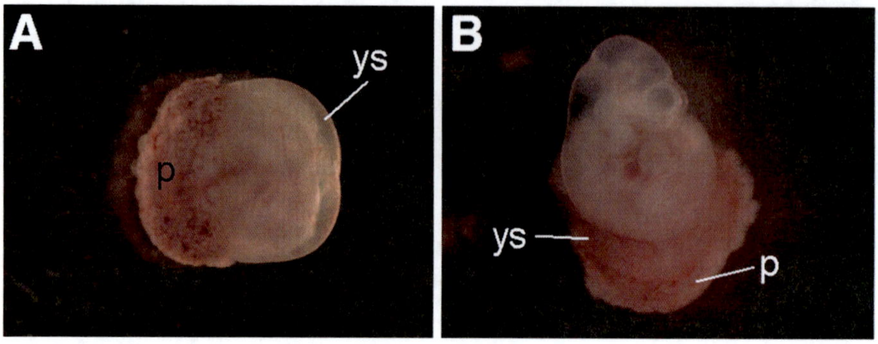

Fig. 1 Whole embryo culture. (**a**) E10.5 embryo inside the yolk sac, with the placenta attached. (**b**) E10.5 embryo after removing from the yolk sac. The yolk sac is still attached and intact. *p* placenta, *ys* yolk sac

3. To free each embryo from the yolk sac and amnion, make a small slit close to the placenta, taking care to maintain the integrity of the main blood supply of the yolk sac (Fig. 1b). This ensures that the embryos are exposed to all nutrients in the medium while maintaining good blood circulation.
4. The embryos are now ready for culture and/or manipulation as required. It is very important to avoid tearing the placenta or damaging major blood vessels in the yolk sac, as this can compromise oxygen and nutrient flow to the embryo. At this time it is important to remove tissue debris from the culture media as this can also affect embryo survival.

3.1.2 Culture

1. Prepare the serum-free culture medium fresh—allow 3 ml per embryo.
2. Equilibrate the medium for 20 min at 37 °C in 95 % O_2/5 % CO_2. The medium should be placed into the culture vials and allowed to rotate in the incubator until the embryos are ready to be added.
3. At this time the embryos can be manipulated as necessary for the specific experiment (*see* Subheadings 3.2 and 3.3).
4. Use a plastic transfer pipet to transfer the embryos to the rolling bottle culture apparatus (BTC Engineering, Cambridge, UK). Optimal conditions are one embryo in 3 ml serum-free medium, per 5 ml vial.

3.1.3 Recovery and Analysis

1. After the desired culture period (*see* **Note 3**), remove the embryos from the culture bottles.
2. Assess embryo growth and development according to overall increase in size, presence of eye pigmentation, and appearance of hindlimb buds [7, 21]. Discard any embryos that have not developed successfully.
3. Continue to process the remaining embryos as appropriate.

3.2 Protein-Soaked Bead Implantation Experiments

In this section, we describe the implantation of protein-soaked beads to the region of the developing third pharyngeal pouch at E10.5, which will form the common thymus–parathyroid primordium. This is just one application of the technique, and the protein-soaked beads may be positioned anywhere in the embryo according to the specific experimental requirements.

3.2.1 Prepare Embryos for Culture

1. As described in Subheading 3.1, carefully dissect embryos from the uterus, leaving the yolk sac, placenta, and their blood vessels intact.
2. Prior to manipulation, embryos may be stored in M2 or DMEM medium at 37 °C.

3.2.2 Prepare Affi-Gel Mesh Beads

1. Remove a small number (around 20 μl) of beads to the required number of 1.5 ml tubes (one for each protein, plus a negative control). Centrifuge gently and remove most of the residual storage solution.
2. Wash the beads several times with sterile PBS to remove any preservatives that may have been used to prevent microbial contamination (e.g., sodium azide).
3. Prepare the protein(s). For many experiments, titration of the protein concentration is necessary.
4. Add the protein solution to each 1.5 ml tube. Include a buffer-only negative control (use the buffer used to prepare protein dilutions, usually PBS).
5. Allow the beads to soak in the protein solution for 30 min at room temperature.
6. Unused soaked beads will be stable at 4 °C for up to 1 week, and therefore may be stored for use in future experiments.

3.2.3 Placement of Beads in the Region of the Third Pharyngeal Pouch

1. Position the embryo on its right side, with the left-side third pharyngeal pouch visible (*see* **Notes 4** and **5**).
2. Pipette a small volume (around 5 μl) of the protein-soaked beads into the media, close to the embryo. Allow them to sink to the bottom of the dish and select one of the larger-sized beads (*see* **Note 6**). Pick the bead up with fine forceps and place on the surface of embryo close to third pharyngeal cleft.
3. Use one arm of the forceps to push the bead under the cleft and then push into the body (Fig. 2a), as close to the third pouch as possible without damaging blood vessels (Fig. 2b, and *see* **Notes 7–9**).
4. Transfer the embryo immediately to rolling culture (*see* Subheading 3.1.2).

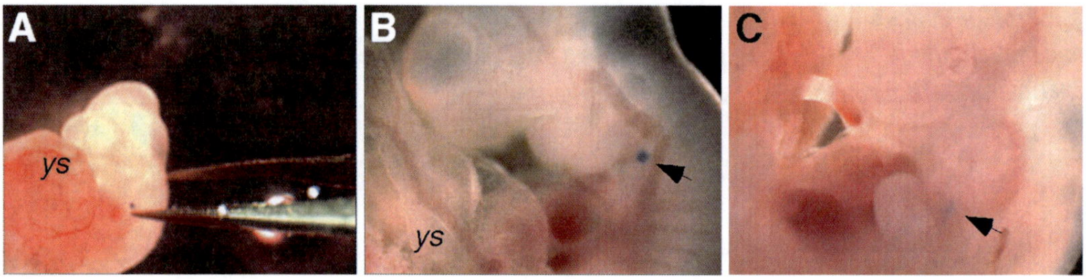

Fig. 2 Bead implantation to the region of the third pharyngeal pouch. (**a**) forceps are used to position a single protein-soaked bead to the area of the third pharyngeal pouch, on one side of the embryo only. (**b**) following placement, the bead is clearly visible within the embryo (*arrow*). (**c**) after the desired culture period, the embryo has developed and the bead can still be seen within the embryo (*arrow*). *ys* yolk sac

3.2.4 Recovery and Analysis

1. Following culture for the desired amount of time, remove the embryos and rinse in PBS. The location of the bead can be confirmed at this time (Fig. 2c).
2. Assess the growth and survival of the embryos, and discard any that did not develop well during the culture period.
3. The embryos can now be analyzed as required. For example, the embryos may be fixed overnight in 4 % paraformaldehyde and processed for in situ hybridization to determine the molecular effects of the protein(s). Or they may be processed for histological analysis.

3.3 Electroporation of Morpholino Oligonucleotides into the Third Pharyngeal Pouch Epithelium

This section describes the establishment of an efficient system for manipulation of gene expression in developing mid-gestation mouse embryos, via delivery of nucleic acids, including morpholino antisense oligonucleotides. Specifically, we show here the combination of the use of antisense morpholino oligonucleotides with electroporation in whole embryo culture. The location of the endodermal gut tube on the interior of the embryo makes it less accessible than the central nervous system and therefore not suited to the in utero manipulations reported by others [18, 19]. After testing many conditions, and altering all parameters sequentially, these conditions were found to be optimal for successful morpholino uptake by target cells without adverse effects on embryo survival. At E10.5 the third pharyngeal pouch is essentially a single layer of epithelial cells surrounding a lumen that is continuous with that of the pharynx. The morpholino solution is injected directly into the lumen of the pouch, which functions as a "reservoir" so that all cells are exposed to the injected material, helping promote successful electroporation into the target cells. The same principle may be used to target other cells or tissues with any charged molecule within the embryo, but specific optimization will be necessary (*see* **Notes 10** and **11**).

3.3.1 Prepare Embryos for Whole Embryo Culture

1. Dissect the E10.5 embryos as described in Subheading 3.1.
2. Store in M2 or DMEM medium until ready to inject/electroporate. If a large number of embryos are being used, they may be transferred to a 37 °C incubator and removed one at a time, or in small batches, for injection.

3.3.2 Injection of DNA into the Lumen of the Third Pharyngeal Pouch

1. Load the glass needle with the morpholino solution at the desired concentration (*see* **Note 12**). Addition of a small amount of a 0.1 % Fast Green solution permits visualization during and after the injection.
2. Store the needle in a safe place between injections, for example by taping it to the base of the microscope.
3. Prepare the electrodes: position them so that they are visible down the microscope. The negative electrode should be at the top and the positive electrode at the bottom, so that they will contact the dorsal and ventral surfaces of the embryo respectively. They should be gripped in a Genetrode Holder (BTX) so that they are maintained at the same distance apart (*see* **Note 13**).
4. Bring each embryo in turn, along with a small drop of culture medium, to the electrodes.
5. Carefully position the embryo between the electrodes (Fig. 3a). The embryo should be placed on one side, with the head to the right and the heart pointing down. The electrodes should be in contact with its ventral and dorsal surfaces, and aligned with the third pharyngeal pouch. Orient the embryo the same way every time, so that the pouch on the right side of the embryo is always the experimental one.
6. Inject a small volume of morpholino solution into the lumen of the third pouch. Around 15 nl is sufficient, although the exact amount is not important. It is, however, important that the same amount is injected each time (*see* **Notes 14** and **15**).
7. The embryos should be electroporated immediately following injection (*see* **Note 14**).

3.3.3 Electroporation of Morpholinos into Epithelial Cells of the Third Pharyngeal Pouch

1. Set the ECM830 square wave electroporator (BTX) to deliver three 50 ms square wave pulses of 30 V.
2. Some embryos may be analyzed immediately following electroporation to determine effective cell uptake (*see* **Notes 16–18**). Embryos should be rinsed in PBS to remove residual injected material. They can be analyzed and/or photographed using a fluorescent microscope and then processed for paraffin sectioning if necessary. Figure 3b, c shows a representative embryo with a restricted domain of intense fluorescence around the injected pouch, but not around the pouch on the opposite non-injected side. Sagittal sections show that this fluorescence is in the desired cells (Fig. 3d, e).

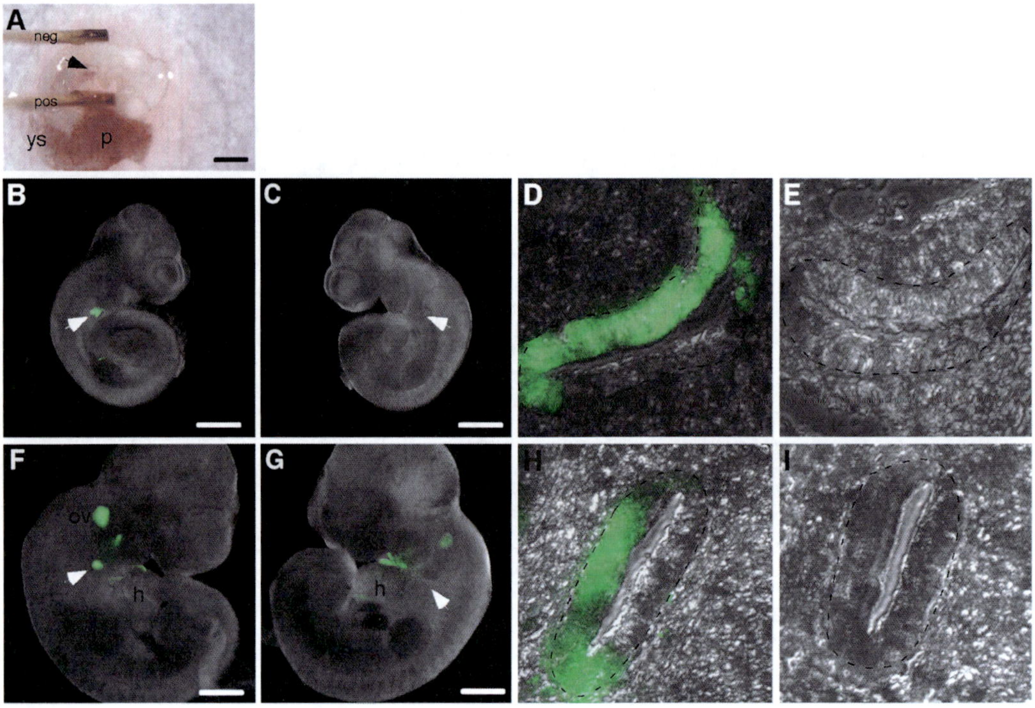

Fig. 3 Electroporation of fluorescein-tagged morpholinos into cells of the third pharyngeal pouch. (**a**) an E10.5 embryo positioned between two parallel electrodes, so that the pouch (*arrow*) and electrodes are aligned. (**b–e**) Immediately following injection and electroporation, an area of fluorescence is visible around the experimental pouch (*arrow* in **b**), but not on the opposite side (*arrow* in **c**). Sagittal sections demonstrate that this fluorescence is intracellular on the injected side (**d**, *outline*) and absent from cells on the opposite side (**e**, *outline*). (**f–i**) after 30 h in whole embryo culture fluorescence is still apparent in the region of the experimental primordium (*arrow* in **f**) but not on the opposite side (*arrow* in **g**). In this example, some fluorescence is also visible in cells on the surface of the embryo, around the heart (h) and trapped within the otic vesicle (ov). Again, sagittal sections show fluorescence in cells of the injected primordium (**h**, *outline*) but not in the primordium on the opposite side (**i**, *outline*). Note the increase in size and developmental progression after the culture period. (**b–i**) Merged bright-field/fluorescence images. Scale bars = 1 mm. *neg* negative electrode. *pos* positive electrode. *h* heart. *ov* otic vesicle

3. Experimental embryos should be transferred to rolling culture bottles immediately following electroporation (*see* Subheading 3.1.2).
4. Repeat the injection and electroporation process for each embryo.

3.3.4 Recovery and Analysis

1. After 30 h, remove the embryos from the culture bottles. Use a transfer pipet to transfer them to a petri dish containing PBS.
2. Screen embryos for good growth and development. At this time, discard any embryos that did not survive the culture.

3. Using a fluorescent microscope, select embryos for further analysis. Embryos should have retained a localized area of intense fluorescence at the injection site (Fig. 3f, g). Sagittal sections show that this fluorescence is intracellular (*see* **Note 18** and Fig. 3h, i).
4. Process experimental embryos as desired (*see* **Note 19**).

4 Notes

Serum-free whole embryo culture

1. It is preferable to sacrifice the pregnant female by cervical dislocation. CO_2 asphyxiation will adversely affect the embryos. Note that euthanasia procedures must be approved by your Institutional Animal Care and Use Committee.
2. Somites are blocks of mesodermal tissue visible on either side of the neural tube from around E8.0 of mouse development. As each somite pair takes about 1.5–2 h to form, counting them is an extremely accurate method for staging embryos, much more so than number of days post coitum, or morphological features alone.
3. The maximum time that an E10.5 mouse embryo will survive, grow, and develop normally under these culture conditions is 30 h. This may be due in part to poor placental development, or to the absence of an as-yet-unidentified serum component, both of which will compromise the nutrient supply to the embryo. Changing the media after 24 h will allow survival and some development up to 40 h; however, different regions and structures within the embryo develop at different rates and so it is no longer an accurate representation of embryonic development.

Bead implants

4. It is important to position the embryo the same way in every experiment. This ensures that the same pouch is used as the experimental each time, making all experiments consistent, and allowing the other pouch to be used as an internal control. The natural curvature of the embryo and the tail mean that resting the embryo on the right side, and therefore placing the bead on the left side, is easier.
5. The blood flow in the pharyngeal region is from left to right. Dependent on the permeability of the protein or reagent used with the bead implants, further controls may be necessary to evaluate whether or not the opposite side is also being affected by the treatment. For example, moving the bead to the right side where blood is flowing away may be better, although as discussed in **Note 4**, this is more difficult due to the shape of the embryo.

6. Use beads of similar size each time—the larger ones are easier to handle and ensure that a maximum dose of protein reaches the target site.
7. In some cases, during placement of the bead, it will enter the lumen of pouch and move through into pharynx. When this happens, the bead will usually be expelled out of the embryo's mouth, and another attempt should be made at placing the bead in the correct position.
8. During placement of the beads, sometimes they are accidentally placed inside a pharyngeal artery. In such instances, the bead will be visibly moving, and can potentially be carried around the body. This may affect the experimental results, and so it is necessary to bear this in mind during analysis, and in fact it is often better to discard these embryos.
9. Placement of the bead in this way will almost always result in the bead being at the dorsal end of the pouch. However, with practice, more precise positioning of the bead around the pouch is possible.

Morpholino electroporation

10. For each new experiment, it is necessary to establish the optimal DNA concentration: a simple titration experiment comparing cellular uptake (fluorescence) with embryo survival is sufficient. In the experiments described here, morpholino concentrations of 5–500 mM were tested. 50 mM was optimal and was therefore used in all subsequent experiments. The number of embryos with fluorescence in the target cells versus viability after 30 h in culture was measured.
11. The key requirement for accurate targeting of the desired cells is specific positioning of the electrodes. The size of the electrodes is very important, as is their position relative to the target site. The injection of DNA does not have to be as precise, but still should be as close to the target site as possible. For this reason, injection of a sufficient volume anywhere into the endodermal gut tube could be used to target numerous sites, simply by altering the position of the electrodes. In the studies performed here, injection not followed by electroporation resulted in no fluorescence in cells of the third pouch, indicating that morpholino entry was facilitated by electroporation only.
12. To load the morpholino solution into the needle, place a small drop into the bottom of a petri dish. Attach the needle to a 1-ml syringe and slowly draw the solution in. It is better to fill the needle once, with enough solution to inject all embryos in the experiment.
13. If the electrodes are too close together, the embryo will have to squeeze between them and will be damaged. If they are too far apart, they will not be in contact with the embryo, and electroporation will be inefficient.

14. Microinjector settings and needle diameter must be consistent to ensure that a single pulse always delivers the same volume. A pulse time of 60 ms was used in the experiments described here.
15. Several measures can be taken to minimize leakage of the injected material from the lumen of the pouch. Leakage back out of the pouch results in loss of material and poor uptake by target cells, whereas leakage into the pharynx and into the opposite pouch means those cells may also take up the injected material, and will not be able to be used as an internal control.
    (a) Position the electrodes prior to injection. This means that the embryo does not have to be moved between injection and electroporation.
    (b) Do not place the needle too far into the pouch. This will ensure a shallow injection and prevent leakage through the pharynx to the pouch on the opposite side.
    (c) Inject a small volume.
    (d) Electroporate immediately following injection.
16. Fluorescein-tagged morpholino oligonucleotides (Gene Tools LLC) allow assessment of their distribution after electroporation. This is useful for all experiments, but is particularly valuable for optimization of the technique.
17. Analysis prior to culture is not necessary for every experiment, but is essential in the beginning to establish the efficiency of the technique.
18. Sectioning of the embryos following injection and electroporation allows for a more detailed assessment of morpholino distribution.
19. Depending on the requirements of the analysis, some embryos might be processed for fluorescence to confirm morpholino uptake, and others for histological analysis, x-gal staining, in situ hybridization, etc. It is important that controls are done for each experiment so that the appropriate embryos are used in the analysis: a control morpholino should have no effect on development, and in cases where an equivalent structure on the opposite side of the embryo is being used as an internal control, this should also have developed normally.

Acknowledgements

This work was supported by LRF and MRC (for J.G. and C.C.B.) and by NIH, NICHD and NIAID (for N.R.M. and B.M.).

References

1. New DAT (1990) Whole embryo culture, teratogenesis and the estimation of teratologic risk. Teratology 42:635–642
2. New DAT, Coppola PT, Cockroft DL (1976) J Embryol Exp Morphol 36:133–144
3. Quinn P, Horstman FC (1998) Is the mouse a good model for the human with respect to the development of the preimplantation embryo in vitro? Hum Reprod 13:173–183
4. Kinder SJ, Tan S-S, Tam PPL (2000) Cell grafting and cell fate mapping of the early-somite stage mouse embryo. Methods Mol Biol Dev Biol Protocols 135:425–437
5. Tam PPL (1998) Postimplantation mouse development: whole embryo culture and micro-manipulation. Int J Dev Biol 42:895–902
6. Takahashi M, Osumi N (2010) The method of rodent whole embryo culture using the rotator-type bottle culture system. J Vis Exp 42. http://www.jove.com/index/details.stp?id=2170
7. Moore-Scott BA, Gordon J, Blackburn CC, Condie BG, Manley NR (2003) New serum-free in vitro culture technique for midgestation mouse embryos. Genesis 35:164–168
8. Gordon J, Wilson VA, Blair NF, Sharp L, Manley NR, Blackburn CC (2002) Functional evidence for a single endodermal origin for the thymic epithelium. Nat Immunol 5:546–553
9. Cockroft DL (1991) Culture media for postimplantation embryos. Reprod Toxicol 5:223–228
10. Van Maele-Fabry G, Picard JJ, Attenon P, Berthet P, Delhaise F, Govers MJAP, Peters PWJ, Piersma AH, Schmid BP, Stadler J, Verhoef A, Verseil C (1991) Interlaboratory evaluation of three culture media for postimplantation rodent embryos. Reprod Toxicol 5:417–426
11. Akamatsu W, Okano HJ, Osumi N, Inoue T, Nakamura S, Sakakibara S-I, Miura M, Matsuo N, Darnell RB, Okano H (1999) Mammalian ELAV-like neuronal RNA-binding proteins HuB and HuC promote neuronal development in both the central and the peripheral nervous system. Proc Natl Acad Sci U S A 96:9885–9890
12. Osumi N, Inoue T (2001) Gene transfer into cultured mammalian embryos by electroporation. Methods 24:35–42
13. Sadler TW, New DAT (1981) Culture of mouse embryos during neurulation. J Embryol Exp Morphol 66:109–116
14. Heasman J (2002) Morpholino oligonucleotides: making sense of antisense? Dev Biol 243:209–214
15. Nasevicius A, Ekker SC (2000) Effective targeted gene "knockdown" in zebrafish. Nat Genet 26:216–220
16. Coonrod SA, Bolling LC, Wright PW, Visconti PE, Herr JC (2001) A morpholino phenocopy of the mouse mos mutation. Genesis 30:198–200
17. Mellitzer G, Hallonet M, Chen L, Ang SL (2002) Spatial and temporal "knock down" of gene expression by electroporation of double-stranded RNA and morpholinos into early postimplantation mouse embryos. Mech Dev 118:57–63
18. Garcia-Frigola C, Carreres MI, Vegar C, Herrera E (2007) Gene delivery into mouse retinal ganglion cells by in utero electroporation. BMC Dev Biol 7:103–112
19. Saito T (2006) In vivo electroporation in the embryonic mouse central nervous system. Nat Protoc 1:1552–1558
20. Akamatsu W, Okano HJ, Osumi N, Inoue T, Nakamura S, Sakakibara S, Miura M, Matsuo N, Darnell RB, Okano H (1999) Mammalian ELAV-like neuronal RNA-binding proteins HuB and HuC promote neuronal development in both the central and the peripheral nervous systems. PNAS 96:9885–9890
21. Kaufman MH (1992) The atlas of mouse development. Academic, London

Chapter 13

Genetically Encoded Probes Provide a Window on Embryonic Arrhythmia

Yvonne Norine Tallini, Kai Su Greene, Bo Shui, Calum William Russell, Jane Constance Lee, Robert Michael Doran, Junichi Nakai, and Michael I. Kotlikoff

Abstract

Supraventricular tachycardias are the most prevalent group of arrhythmias observed in the fetus and infant and their incidence increases through early childhood. The molecular pathogenesis of embryonic cardiac dysfunction is poorly understood, due in part to the absence of imaging techniques that provide functional information at the cellular and molecular levels in the developing mammalian heart, particularly during early heart formation. The combination of protein engineering, genetic specification, and high-resolution optical imaging enables new insights into cardiac function and dysfunction during cardiac development. Here we describe the use of GCaMP2, a genetically encoded Ca^{2+} indicator (GECI), to determine the processes of cardiac electrical activation during cardiac organogenesis. Transgenic specification of GCaMP2 in mice allows sufficient expression for Ca^{2+} imaging as early as embryonic day (e.d.) 9.5, just after the heart begins to function at e.d. 8.5. Crosses with knockout lines in which lethality occurs due to cardiac dysfunction will enable precise determination of the conduction or excitation–contraction coupling phenotypes and thereby improve the understanding of the genetic basis of heart development and the consequence of gene mutations. Moreover, lineage-specific targeting of these sensors of cell signaling provides a new window on the molecular specification of the heart conduction system. We describe mouse lines and imaging methods used to examine conduction in the pre-septated heart (e.d. 10.5), which occurs through dramatically slowed atrioventricular (AV) canal conduction, producing a delay between atrial and ventricular activation prior to the development of the AV node. Genetic constructs including single and bi-allelic minimal promoter systems, and single allele BAC transgenes, enable general or lineage-specific targeting of GCaMP2. High-resolution imaging of embryonic heart conduction provides a new window on one of the most complex events in the mammalian body plan.

Key words GCaMP2, Genetically encoded Ca^{2+} sensors, In vivo imaging, Fluorescence imaging, Embryo, Cardiac, Heart

1 Introduction

In the heart, changes in intracellular Ca^{2+} are at the center of pacemaker rate regulation, excitation–contraction coupling and gene transcription [1, 2]. The sequential electrical activation,

Mark Lewandoski (ed.), *Mouse Molecular Embryology*, Methods in Molecular Biology, vol. 1092,
DOI 10.1007/978-1-60327-292-6_13, © Springer Science+Business Media New York 2014

pacemaking, conducting, and myocardial cells result in stereotypical Ca^{2+} transients in each cell type, allowing determination of the coordinated activation of the heart by monitoring cellular increases in Ca^{2+}. The genetically encoded Ca^{2+} sensor, GCaMP2, is a bright, high signal sensor that functions effectively for in vivo physiological recordings in the mouse [3–5]. Here we describe embryonic recording methods and genetic strategies that enable the recording of Ca^{2+} transients in the live embryo or isolated embryonic heart.

In the mammalian body plan the heart is the first organ to function, and circulation of blood through the developing embryo must occur at a stage prior to complete cardiac organogenesis. Peristaltic-like contractions are initiated within the primitive heart tube at approximately e.d. 8.5 in the mouse and over the next 13 days, the heart continues to pump while major structural and functional changes such as chamber septation and node development occur. Methods to study cardiac function at the earliest stages of organogenesis have been limited by an absence of techniques that enable the resolution of cellular activation in real time. Synthetic fluorescent sensors, for example, do not load well, cannot be effectively targeted to individual cell types, and do not persist in living organs. This review describes genetic and optical strategies that enable in vivo and ex vivo imaging of heart cell function during development.

2 Materials

2.1 BAC Clones

1. BAC clones are purchased from Children's Hospital Oakland Research Institute Web site http://bacpac.chori.org/.
2. Vector NTI Advance Software (Invitrogen, Carlsbad, CA) (*see* **Note 1**).

2.2 PCR, Electrophoresis, and DNA Purification

1. Oligonucleotides are ordered from Integrated DNA Technologies (www.IDTDNA.com).
2. Taq DNA Polymerase (Invitrogen) (*see* **Note 2**). Store all contents at −20 °C.
3. Purchase all four 2′-deoxynucleoside 5′-triphosphate (dNTP; Invitrogen) at 100 mM concentrations. A working stock of 2.5 mM for each nucleic acid is prepared by adding 25 μl of each 100 mM nucleic acid to 900 μl of polymerase chain reaction (PCR) grade water. Store at −20 °C.
4. Electrophoresis buffer: Tris/Acetate/ethylenediaminetetraacetic acid (EDTA) (TAE) buffer: A 50× TAE stock concentration buffer is made by adding 242 g tris(hydroxymetyl) aminomethane base, 57.1 ml glacial acetic acid, 100 ml of 0.5 M EDTA, pH 8.0, bring to final volume of 1 l using water

(*see* **Note 3**). Store at room temperature (RT). Prepare 0.5× TAE by adding 20 ml of 50× TAE to 1.98 l of water. Store at RT.

5. Ultrapure agarose and 1 kb DNA PLUS ladder (Invitrogen).
6. Ethidium bromide (Etbr; Sigma-Aldrich) make a stock concentration of 10 mg/ml. Purchase in tablet form and prepare according to manufacturer's instructions in a screw top eppendorf tube. Protect from light by covering eppendorf tube with aluminum foil. Store at RT (*see* **Note 4**).
7. 6× loading dye: Combine 0.25 % bromophenol blue (Sigma-Aldrich), 0.25 % xylene cyanol FF (Sigma-Aldrich), and 30 % Glycerol (Sigma-Aldrich) (*see* **Note 5**).
8. QIAquick Gel Extraction Kit (Qiagen, Valencia, CA).
9. BAC DNA microinjection buffer: 10 mM Tris-HCL pH 7.5, 0.1 mM EDTA (Sigma-Aldrich), 30 μM spermine (Sigma-Aldrich), 70 μM spermidine (Sigma-Aldrich), 100 mM NaCl. The buffer was made fresh from the following stocks: 1 M Tris–HCl, pH 7.5; 0.5 M EDTA, pH 8.0; 5 M NaCl;1,000× polyamine stock, 30 mM spermine and 70 mM spermidine.

2.3 Recombineering Bacterial Strain and Media

1. The bacterial strain SW105 is modified from DH10B and can be obtained from NCI-Frederick Web site http://ncifrederick.cancer.gov/research/brb/recombineeringInformation.aspx (*see* **Note 6**). Store at −80 °C in 20 % glycerol (*see* **Note 7**).
2. EZNA Plasmid Mini Prep Kit (Omega Bio-Tek, Inc, Norcross, GA) solution 1, solution 2, and solution 3.
3. Chloramphenical (CM; Amersham Life Science, Piscataway, NJ): Add 250 mg to 10 ml of 100 % ethanol (stock concentration 25 mg/ml). Aliquot 1 ml of dissolved CM into 1.5 ml eppendorf tubes (*see* **Note 8**). Store at −20 °C.
4. Ampicillin, sodium salt (Amp; EMD Chemicals, Inc., Gibbstown, NJ), make a working dilution of 5 g ampicillin in 50 ml water (100 mg/ml), dissolve, filter-sterilize, and aliquot into 1 ml aliquots in 1.5 ml eppendorf tubes. Store at −20 °C.
5. Kanamycin (Kan; Gibco/BRL, Bethesda, MD) is purchased as a 10 mg/ml stock. Pipette 1 ml aliquots into 1.5 ml eppendorf tubes and store at −20 °C.
6. LB (Difco Luria-Bertani Miller powder broth; BD, Sparks, MD) liquid media: Add 12.5 g of LB powder to 500 ml of water in a 1 l flask. Cover top with aluminum foil, label, date, autoclave on liquid setting for 30 min, and cool at RT (*see* **Note 9**).
7. Isopropanol, 70 % ethanol, and glycerol (Sigma).

2.4 Targeting Construct

1. PBS-GCaMP2-IRES-GCaMP2-FRT-Neo/Kan-FRT can be obtained from the Kotlikoff laboratory (http://www.vet.cornell.edu/BioSci/Faculty/Kotlikoff).

2. T4 DNA ligase (Fermentas Life Sciences, Glen Burnie, MD). Store at −20 °C.
3. Restriction enzymes: EcoRI, SmaI (for arm1), SpeI and NotI (for arm2) (*see* **Note 10**). All restriction enzymes are obtained from Fermentas Life Sciences and are stored at −20 °C.
4. DH5α competent cells (Invitrogen). Aliquot 50 μl into 1 ml cryopreservation tubes with screw caps. Store at −80 °C.

2.5 BAC Recombineering

1. CM/LB media: Prepare 500 ml LB media (Subheading 2.3, **item 6**), when reaches RT add 250 μl CM (final concentration 12.5 μg/ml). Store at RT.
2. To make antibiotic resistant LB plates add 12.5 g LB powder broth, 7.5 g agar powder (Alfa Aesar, Johnson Matthey Co., Heysham, Lancaster) to 500 ml water in a 1,000 ml flask. Cover top with aluminum foil, label, date, and autoclave for 35 min on liquid cycle. Cool to 55 °C then add antibiotics in the following dilution:

   CM/LB plates add 250 μl CM, final concentration 12.5 μg/ ml.

   Kan/LB plates add 1.5 ml Kan, final concentration 30 μg/ ml.

   Amp/LB plates add 500 μl Amp, final concentration 100 μg/ ml.

   CM/Kan/LB plates add 250 μl CM and 1.25 ml Kan, final concentration 12.5 μg/ml (CM) and 25 μg/ml (Kan).

   Pour approximately 20 ml of media into 100 mm petri dishes. Cool overnight at RT. Store at 4 °C.
3. Low salt bacterial (LSB) culture media: Add 2.5 g yeast extract (FisherBiotech, Fair Lawn, NJ), 5 g BactoTryptone (BD), and 2.5 g NaCl (Sigma-Aldrich) to 500 ml of water in a 1 l flask. Autoclave as described in Subheading 2.3, **item 6**.

2.6 Removal of Neomycin/Kanamycin Cassette

1. Make a 10 % stock concentration of L-(+)-Arabinose (Sigma-Aldrich) by adding 1 g of L-(+)-Arabinose to 10 ml of water. Make new each time (*see* **Note 11**).

2.7 DNA Isolation

1. QIAGEN Large-Construct Kit (Qiagen).
2. Puregene (Gentra Systems, Valencia, CA).

2.8 Probe and Scale

1. Sturdy gross anatomy blunt tip probe, angle 45°, diameter point 1.2 mm (Fine Science Tools, Foster CA).
2. Counting Pocket Balance (ISC BioExpress, Kaysville, UT).

2.9 Perfusion System, Dissecting Microscope, and Imaging System

1. *See* Fig. 1 for perfusion system setup.
   (a) Heating circulating water bath working temperature range of 25–100 °C (World Precision Instruments, Sarasota, FL).
   (b) 1 l glass reservoir (Harvard Apparatus, Holliston, MA).

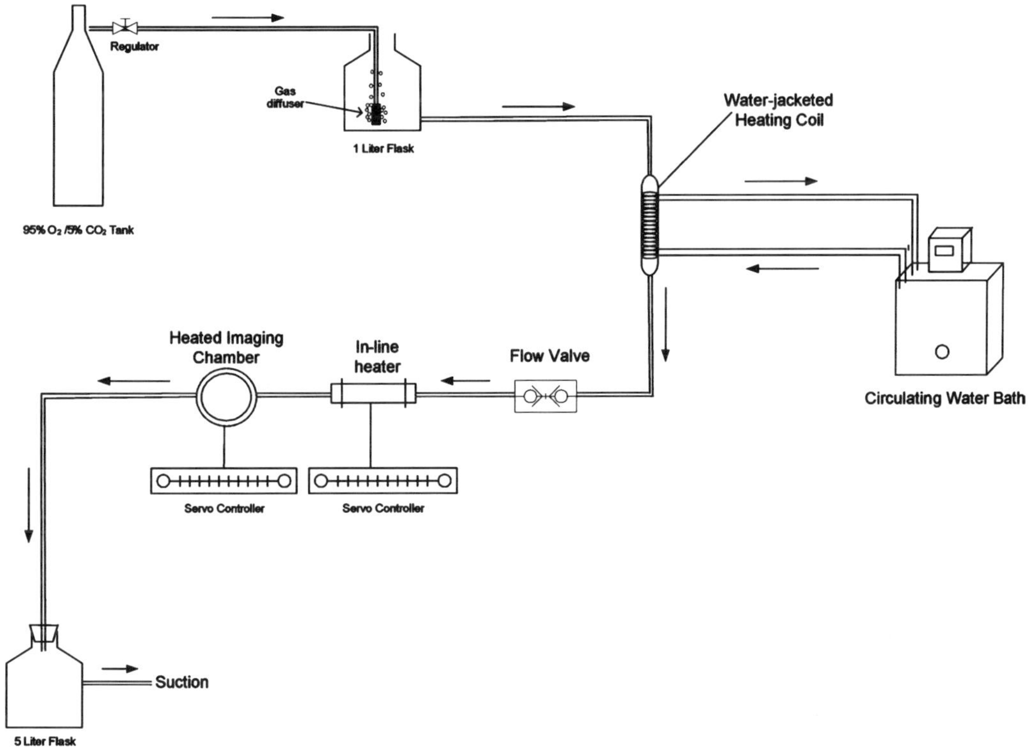

Fig. 1 Schematic of perfusion system setup. Solution is oxygenated at the main reservoir. Preheating of solution is accomplished by warming solution through two stages: first through the heat exchange coil and then through the inline heater so that the temperature entering the imaging chamber is 37 °C. Keep tubing from the inline heater to the chamber as short as possible to minimize heat loss. To add different solutions or drugs to the chamber a manifold can be inserted prior to the flow valve

(c) One 10 ml heating coil (Radnoti, Monrovia, CA).

(d) Vacuum solution flow valve with on-off switch (Harvard Apparatus).

(e) Solution in-line heater that will accommodate flow rates up to 10 ml/min (Harvard Apparatus)

(f) 5 l vacuum Flask (Radnoti).

(g) Microprobe (Harvard Apparatus).

(h) Various size tubings, disconnect parts, clamps, and lab stands (VWR, Bridgeport, NJ).

(i) Customized heated chamber with a Sylgard bottom (see below).

2. The setup for the dissection is shown in Fig. 2.

(a) Binocular dissection scope working distance of at least 11 cm with boom stand. Several microscope companies manufacture dissecting scopes.

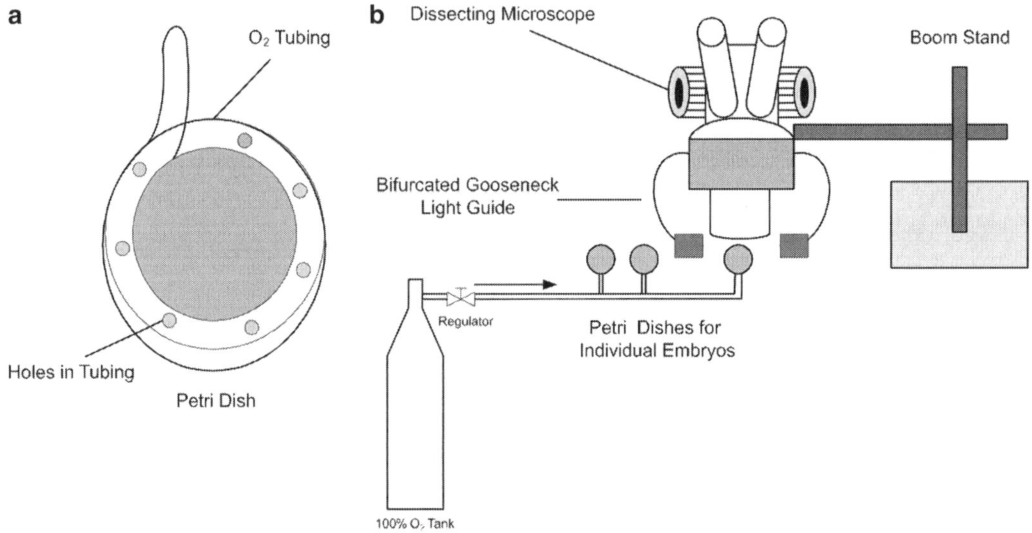

Fig. 2 Diagram of dissecting setup. (**a**) Epoxy tubing around the outer edge of a 60 × 15 mm petri dish and poke holes. (**b**) Embryos are oxygenated in individual petri dishes and maintained on ice until imaging

(b) Fiber optic cold light sources with bifurcated gooseneck light guide (4.7 mm fiber diameter) and standard focusing lens (Warner Instruments Inc., Hamden, CT) (*see* **Note 12**).

(c) Polyurethane rectangular ice bucket (Fisher Scientific, Pittsburgh, PA)

(d) Glass petri dishes (Fisher Scientific) 60 mm × 15 mm with tubing glued in a circle around the outer edge with holes poked in the tubing.

3. Figure 3 shows a schematic of the macro fluorescent imaging system.

   (a) We have a custom built chamber to include a port to hold the inflow and outflow tubing. Chambers may be purchased from Radnoti, Warner's Instruments (Hamden, CT), or Harvard Apparatus. It is important the chamber has enough depth to add a black Sylgard bottom (Sylgard, 184 silicone elastomer curing agent, Dow Corning Co., Midland, MI), to hold 3–5 ml of solution and allow room for inflow and outflow tubing (*see* **Note 13**).

   (b) These instructions assume the use of an MVX10 (Olympus) macro zoom fluorescent microscope coupled to an electron multiplied cooled charged coupled device camera (EMCCD) (Ixon; Andor Technology, South Windsor, CT). Other macro zoom fluorescent microscopes and imaging cameras are available through alternative manufacturers.

   (c) Dichroic filter, excitation 472/30 nm and emission 520/35 nm (*see* **Note 14**).

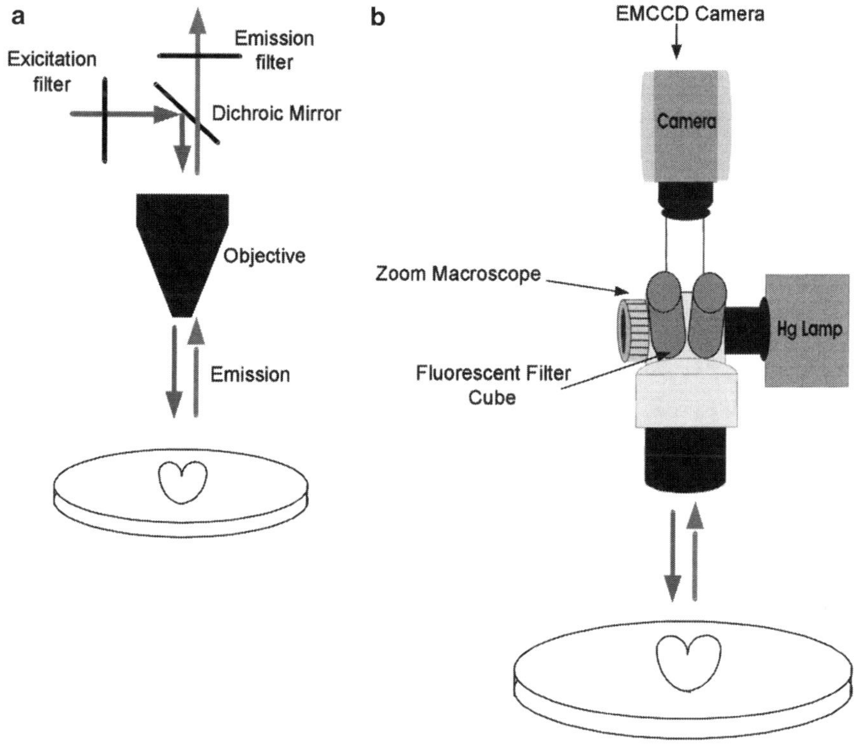

Fig. 3 Graphic of macro zoom fluorescent imaging system. (**a**) Dichroic mirror (**b**) Embryonic hearts containing GCaMP2 are excited at 488 nm and image sequences are obtained from the emission at 510/15 nm onto an EMCCD camera that is coupled to the macro zoom fluorescent microscope

2.10 Solutions and Dissection Instruments

1. Dulbecco's phosphate buffered saline 1× (Gibco/Invitrogen).
2. Imaging solution in mM (all chemicals from Sigma-Aldrich): NaCl 112; KCl 5; $MgSO_4{\cdot}7H_2O$) 1.2; $KH_2PO_4$1; $NaHCO_3$ 25; $CaCl_2{\cdot}2H_2O$ 1.3; glucose 50. Add all components to 600 ml of water and adjust final volume to 1,000 ml with water in a graduated cylinder. Equilibrate warmed imaging solution pH for 30 min by placing a pyrex gas dispersion tube (12 mm outer diameter, coarse porosity fritted cylinder, Corning Glass Works, Corning, NY) that is connected to 95 % O_2/5 % CO_2 into the 1 l pyrex beaker. Make fresh daily.
3. Embryo holding solution in mM (all chemicals from Sigma-Aldrich): NaCl 136; KCl 5.4; $MgCl_2{\cdot}6H_2O$ 1; NaH_2PO_4 0.33; $CaCl_2{\cdot}2H_2O$ 2.5; glucose 10; and 4-(2-hydroxyethyl)-1-piperazineethanesulfonic acid 10. Add all components to 600 ml of water and bring to 1,000 ml final volume in a 1-l graduated cylinder. Make fresh weekly.
4. Mouse induction chamber (Summit Anesthesia Solutions, Bend, OR).
5. Squeeze bottle with 70 % ethanol (VWR).

6. The following dissection tools can be obtained from Fine Science or other surgical instrument providers: serrated tip straight forceps, tip diameter 2 mm, length of forcep 12 cm; straight fine tip forceps (Dumont No. 5 or No. 5 Biologie forceps); curved serrated forceps, 0.17 mm × 0.1 mm; finely serrated forceps, serrated 1.5 mm wide; spring loaded straight scissors, 8 mm cutting edge, 0.2 mm tip diameter; straight fine iris scissors, 9 cm; perforated round spoon, diameter 15 mm, deep 3 mm; and curved large scissors, blade length 42 mm.
7. Glass petri dishes; 100 mm × 20 mm and 60 mm × 15 mm (*see* **Note 15**).
8. Insect pins; stainless steel pins 0.10 mm diameter (Fine Science Tools) (*see* **Note 16**).

3 Methods

Functional imaging of embryonic heart development in vivo requires a high level of transgene expression. We have found that minimal promoter strategies often do not produce sufficient sensor expression or highly restricted lineage specification. BAC transgenesis allows the efficient use of transcriptional elements of a gene without requiring full knowledge of those elements. Proper choice of BAC DNA can result in very high levels of sensor expression with exquisite lineage specificity [5, 6]. We first describe the steps involved in BAC recombineering to generate genetically encoded Ca^{2+} indicator mice in which expression is driven by Connexin40 promoter elements (Cx40^{BAC}-GCaMP2 transgenic mice) [5]. We will then outline the techniques used to obtain embryos for imaging. Finally, we will describe the wide-field imaging system used to acquire Ca^{2+} signals in the embryonic heart.

3.1 BAC Clone

1. Figure 4 outlines the general steps involved in BAC recombineering.
2. BAC clones are found at the National Center for Biotechnology Information (NCBI) mouse genomic resources MapViewer and Clone Finder. Web link: http://www.ncbi.nlm.nih.gov/genome/guide/mouse/.
3. Choose a BAC clone in which the entire gene of interest is flanked by as much 5′ DNA sequence as possible.
4. Construct a BAC map and include the BAC vector and the BAC sequence from Map Viewer using Vector NTI software.

3.2 Confirmation of BAC Clone

1. Streak the BAC clone onto a CM/LB plate using a sterile loop and incubate overnight at 32 °C. Only clones that are resistant to CM will grow.

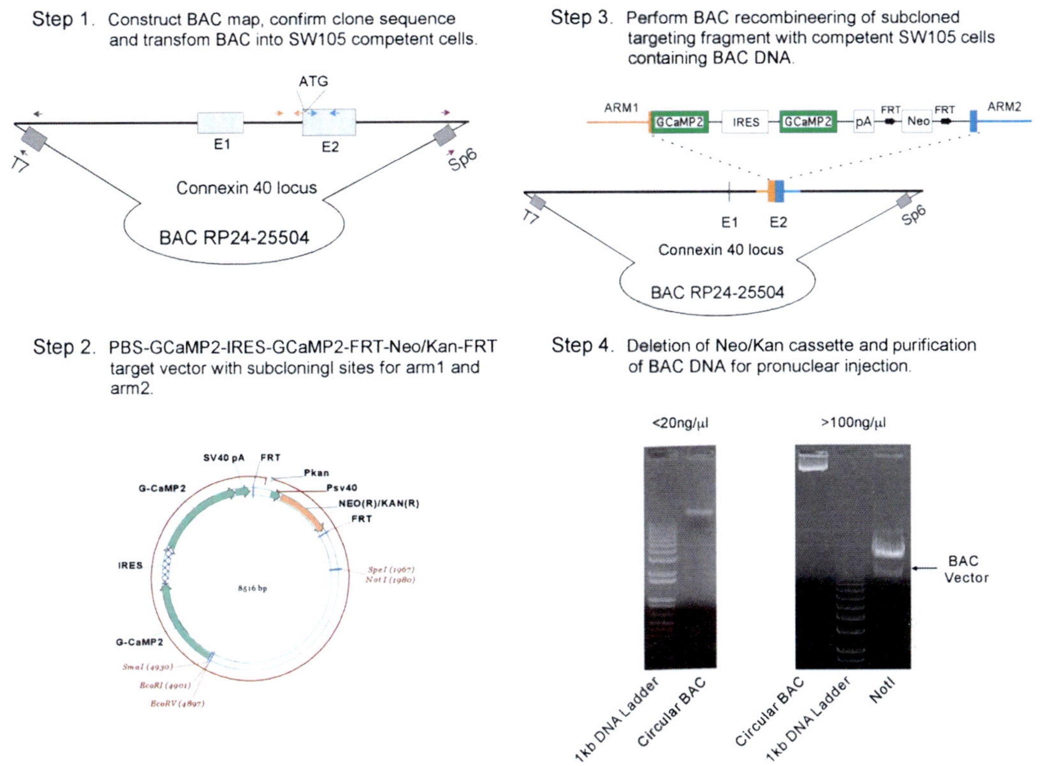

Fig. 4 Steps involved in BAC recombineering. *Step 1.* Construct BAC map using VectorNTI. Confirm BAC sequence by PCR and enzyme digestion. Primer pair sets are color coded (i.e., arm1 is *orange*, arm2 is *blue*, T7 into BAC is *grey* while Sp6 into BAC is *purple*), *arrows* denote orientation and indicate the regions within the BAC for sequence confirmation. Following BAC clone confirmation, transform SW105 competent cells with BAC DNA. *Step 2.* Subclone arm1 using EcoRI/SmaI enzyme sites and arm2 with SpeI/NotI into the Kotlikoff Laboratory PBS-GCaMP2-IRES-GCaMP2-FRT-Neo/Kan-FRT plasmid. Release targeting vector by EcoRI/NotI enzyme digestion. *Step 3.* Homologous recombination occurs after electroporation of the targeting cassette into competent SW105 cells containing the BAC DNA, inserting the GCaMP2 directly after the ATG start codon. Note the color coded arm1 and arm2 regions correlate to the sequence confirmed in *Step 4.* Picture of our purified circular Cx40^{BAC}-GCaMP2 DNA used for pronuclear injection. Integrity of the intact circular BAC DNA was verified by one band on a 0.8 % agarose gel. Mobility of the circular BAC DNA is dependent on the concentration, low concentration (<20 ng/µl) BAC DNA after dissolving overnight migrates differently than high concentration (>100 ng/µl) BAC DNA. Digestion with NotI results in two bands (linearized BAC and plasmid vector)

2. Prepare a BAC DNA miniprep by picking a single colony using a sterile toothpick and drop the entire toothpick into a labeled 5 ml CM/LB liquid culture (*see* **Note 17**). Place in 32 °C shaking (250 rpm) water bath overnight (*see* **Note 18**).
3. The next morning isolate BAC DNA by using solutions from EZNA Plasmid Mini Prep Kit. Spin 5 ml cell culture at 10,000×*g*, 5 min, at RT. Pour off supernatant and add 250 µl of solution 1 to pellet, mix, and transfer to 1.5 ml eppendorf tube. Add 250 µl solution 2, mix well and in less than 5 min

add 250 μl solution 3, mix well, and put on ice for 10 min. Spin down at 20,000 × *g*, 10 min, at RT. Transfer supernatant to a new tube, if necessary spin one more time, and add 750 μl isopropanol, mix. Put on ice for 10 min then spin at 20,000 × *g*, 10 min, at RT. Pour off supernatant and DNA pellet should be seen on bottom. Wash with 1 ml 70 % ethanol, spin at 20,000 × *g*, 5 min, at RT. Pour off ethanol, dry pellet at RT, and resuspend in 50 μl water (*see* **Note 19**).

4. Enzyme-digest BAC DNA with one restriction enzyme (as determined using VectroNTI) by adding 1 unit of enzyme / 1 μg of DNA (*see* **Note 20**). Incubate at 37 °C for 1–2 h.
5. Prepare a 0.8 % agarose gel by adding 0.8 g agarose to 100 ml of 0.5× TAE buffer. Microwave for 1 min, mix, microwave for 30 s, mix. Be sure that agarose is completely dissolved. Cool to 65 °C, add 5 μl Etbr (stock 10 mg/ml; final concentration, 0.5 μg/ml), mix, and pour into gel tray. Gel is ready to be used when agarose has solidified and turned opaque, approximately 30 min.
6. Run enzyme-digested BAC DNA on 0.8 % TAE agarose gel using MiniGel Electrophoresis System $GelX_L$ Plus (Labnet; or similar gel electrophoresis system). To each 20 μl DNA sample add 4 μl of 6× loading dye, mix, and load 15 μl of mixture into gel wells. Include one well for 1 kb PLUS DNA molecular weight marker. Run in 0.5× TAE buffer at 100 V for 30 min (*see* **Note 21**). Compare the gel pattern determined by VectorNTI with the gel pattern obtained by enzyme digestion (*see* **Note 22**).
7. Confirm the BAC sequence using PCR. Design four sets of 16–25 mer oligionucleotide primer pairs using Vector NTI software (*see* **Note 23**).
   (a) Primer pair set 1: a forward primer located within the T7 sequence of the vector and a reverse primer in the BAC sequence.
   (b) Primer pair set 2: a forward primer located within the BAC sequence and a reverse primer in the SP6 site of vector.
   (c) Primer pair set 3: forward primer within the BAC region that will be used for homologous recombination in arm1 and a reverse primer within the BAC region that will be used for arm1.
   (d) Primer pair set 4: a forward primer within the BAC region that will be used for homologous recombination in arm2 and a reverse primer within the BAC region that will be used for arm2.
8. Perform PCR using Taq DNA Polymerase, the parameters have been customized for the Mastercycler Personal (Eppendorf, Westbury, NY) but can be easily adapted to any

thermal cycler. Prepare a PCR master reaction mixture, 20 μl/ total volume for each reaction.

13.0 μl PCR grade water

0.6 μl $MgCl_2$ (stock 50 mm), final concentration 1.5 mM

2.0 μl 10× PCR buffer, final concentration 1×

2.0 μl dNTPs (stock 2.5 mM), final concentration 250 nM

0.6 μl antisense primer (stock 10 μM), final concentration 300 nM

0.6 μl sense primer (stock 10 μM), final concentration 300 nM

0.2 μl Taq polymerase (stock 5 U/μl), final concentration 1 U.

19 μl Total Volume

To each PCR tube add 19 μl of master mix, 1 μl of DNA sample and mix (*see* **Note 24**). Keep Taq cold at all times. Make master mix immediately before each use.

9. Load PCR samples into thermal cycler and program PCR machine with the following cycling parameters:
   Step 1: 94 °C 2 min.
   Step 2: 94 °C 30 s.
   Step 3: 55–60 °C 30 s.
   Step 4: 72 °C 30–60 s (repeat **steps 2–4**, 30–40 times).
   72 °C 10 min and 4 °C hold (*see* **Note 25**).
   Save in memory as BAC clone (*see* **Note 26**). The PCR cycling parameters has been customized for the Mastercycler Personal (Eppendorf) but should be comparable in other thermal cyclers.
10. Run PCR products on 1.2 % TAE agrose gel. A right BAC clone will show positive PCR bands. Purify PCR product per QIAquick Gel Extraction Kit protocol. Sequence the PCR product and compare sequence results to the BAC sequence (*see* **Note 27**).

3.3 BAC Transformation into Recombineering Bacterial Strains

1. These instructions assume the use of the SW105 bacterial strain obtained from NCI-Frederick (*see* **Note 28**).
2. Make recombineering strain competent cells by streaking SW105 cells onto a LB plate with a sterile loop. Incubate at 32 °C overnight (*see* **Note 29**).
3. Pick a colony using a sterile toothpick and place the toothpick into a 5 ml LSB bacterial culture tube and grow at 32 °C overnight.
4. The next day transfer 1 ml of the overnight culture to 50 ml LSB in a 250 ml sterile flask and grow until optical density reads at 600 reads 0.5–0.7 on a spectrophotometer (*see* **Note 30**).
5. Transfer the culture to 50 ml Falcon tube and spin down the cells at 6,000 × *g*, 5 min at 0–4 °C.

6. Discard the supernatant and add 1 ml ice-cold 10 % glycerol to the pellet. Swirl tube in ice-water to dissolve the cells, add 30 ml ice-cold 10 % glycerol, mix well (*see* **Note 31**). Spin down the cells at 6,000 × *g*, 5 min at 0–4 °C. Repeat washing two more times.
7. After the last washing, discard the supernatant and resuspend the cells in 200–500 μl of ice-cold 10 % glycerol and mix. Aliquot 50 μl of cell into screw top tubes (*see* **Note 32**). Cells can be used the same day or frozen at −80 °C for future use.
8. Take one tube of 50 μl SW105 competent cells on ice, mix with 5 μl of BAC miniprep DNA (Subheading 3.2, **steps 2–5**), place in ice-cold 0.1 cm electrode cuvette. Set the Bio-Rad Gene Pulser II-Pulse Controller Plus at 1.75 kV, 25 μF, and pulse controller 200 Ω. The time constant will be displayed and typically ranges from 4.0 to 5.
9. After electroporation, immediately add 1 ml of LB into the cuvette, transfer the mixture to 15 ml bacterial culture tube and shake at 32 °C, 250 rpm for 1 h.
10. Transfer the mixture to a 1.5 ml eppendorf tube and pellet the cells (6,000 × *g*, 30 s, RT). Discard supernatant and resuspend cells in 150 μl of LB.
11. Spread the cells onto CM/LB plate and incubate at 32 °C for 36–48 h.
12. Pick colonies with a sterile toothpick and confirm the BAC DNA is within the SW105 competent cells by performing miniprep DNA protocol (Subheading 3.2, **steps 2–5**) followed by PCR strategy (Subheading 3.2, **steps 7–10**).

3.4 Target Vector and Subcloning

1. Determine a location for the two arms, 500–1,000 bp in length, using VectorNTI. These arms will be inserted into the target vector and will allow for recombination to occur in the BAC. Arm1 and arm2 flank the start codon ATG sequence (*see* **Note 33**).
2. Design primers to amplify both arm1 and arm2 using VectorNTI. Using the BAC DNA as a template, amplify arm1 and arm2 by PCR, gel-purify both products, (Subheading 3.2, **step 10**) obtain sequence, and compare to VectorNTI sequence.
3. Our laboratory has made the targeting vector PBS-GCaMP2-IRES-GCaMP2-FRT-Neo/Kan-FRT with predefined restriction enzyme sites, and two flipase recognition target (FRT) sequences that flank neomycin/kanamycin resistant nucleotide sequences [5]. Restrictions sites to insert arm1 are EcoRI and SmaI and for arm2 are SpeI and NotI.
4. Design four primers using VectorNTI that include the enzyme sites.

(a) arm1-forward primer: 6 bp random nucleotides+ EcoRI sequence + 18–22 bp BAC sequence.

(b) arm1-reverse primer: 6 bp random nucleotides+ SmaI sequence + 18–22 bp BAC sequence (reverse primer sequence should be just before ATG).

(c) arm2-forward primer: 6 bp random nucleotides + SpeI sequence + 18–22 bp BAC sequence (forward primer sequence should start after ATG).

(d) arm2-reverse primer: 6 bp random nucleotides + NotI sequence + 18–22 bp BAC sequence (*see* **Note 34**).

Determine PCR product size of primer pair "1" + "2" and "3" + "4" using VectorNTI.

5. Prepare 100 μl PCR reactions using BAC DNA as the template pairing arm1-forward primer with arm1-reverse primer, and arm2-forward primer with arm2-reverse primer (Subheading 3.2, **step 8**) (*see* **Note 35**).
6. Enter PCR cycling parameters into thermal cycler as follows (*see* **Note 36**):

   Step 1: 94 °C 2 min.
   Step 2: 94 °C 30 s.
   Step 3: 56–58 °C 30 s.
   Step 4: 72 °C 1 min (repeat **steps 2–4**, 5 times).
   Step 5: 94 °C 30 s.
   Step 6: 72 °C 1 min 30 s (repeat **steps 5–6**, 35 times).
   Step 7: 72 °C 10 min.
   Step 8: 4 °C.
   Save in memory as SubClone.
7. Run PCR product on a 1.2 % TAE agrose gel and purify with QIAquick Gel Extraction Kit. Send a PCR sample from each arm product for sequence confirmation.
8. Digest gel-purified arm1 PCR product and the targeting vector with EcoRI and SmaI using the enzyme digestion technique described in Subheading 3.2, **steps 4–6** (*see* **Note 37**).
9. Purify digested arm1 and targeting vector with QIAquick Gel Extraction Kit.
10. Ligate arm1 and the targeting vector using the instructions provided by the T4 DNA ligase manufacturer, however, incubate at 16 °C overnight.
11. Thaw one 50 μl aliquot of DH5α competent cells on ice. Add 5 μl of ligation products of targeting vector and arm1, incubate on ice for 30 min. Heat shock at 42 °C for 1 min. Place back on ice for 5 min. Add 1 ml of LB media and shake for 1 h at 37 °C. Spin down 30 s at 6,000 × *g*. Remove supernatant and add 150 μl of LB media to cells. Spread the cells on Amp/LB plate and incubate plates overnight at 37 °C.

12. To subclone arm2, enzyme-digest the gel-purified arm2 PCR product and the target vector that contains arm1 using SpeI and NotI (Subheading 3.2, **steps 4–6**). Purify digested arm2 and targeting vector that contains arm1 with QIAquick Gel Extraction Kit, followed by ligation and transformation (Subheading 3.4, **steps 10–11**).
13. To confirm the subcloned targeting vector construct that now contains both subcloned arm1 and arm2, design oligonucleotide primers that are located before arm1 and after arm2. Perform PCR and sequence product according to Subheading 3.2, **steps 8–10**.
14. Once sequence information is confirmed, release the targeting fragment by EcoRI and NotI digestion, run samples on 0.8 % agrose gel (Subheading 3.2, **steps 4–6**) and purify with QIAquick Gel Extraction Kit.

3.5 BAC Recombineering

1. Streak SW105 containing BAC on CM/LB plate and incubate at 32 °C overnight.
2. Pick one colony using a sterile toothpick and place in 5 ml of CM/LB liquid media, shake (250 rpm) in 32 °C water bath overnight.
3. The next morning transfer 1 ml overnight culture to 50 ml CM/LB liquid in 250 ml flask and continue shaking at 32 °C until optical density at 600 reads 0.5 on spectrophotometer.
4. Shake at 42 °C for 15 min.
5. Immediately remove flask and swirl in ice-water until culture is cold, leave in ice-water for another 5–10 min (*see* **Note 38**). Continue making the competent cells following the directions at Subheading 3.3, **steps 5–7**.
6. Add 5 μl targeting cassette from Subheading 3.4, **step 14** into 50 μl of SW105 containing BAC competant cells, mix well and place in ice-cold 0.1 cm electrode cuvette. Set the Bio-Rad Gene Pulser II-Pulse Controller Plus at 1.75 kV, 25 μF, and pulse controller 200 Ω. Immediately, add 1 ml LB to the cuvette and transfer the mixture to a 15 ml bacterial culture tube, shake at 32 °C, 250 rpm for 1 h.
7. Spread cells on CM/Kan/LB plate and incubate at 32 °C for 36–48 h.
8. To confirm the BAC targeting clone, design three different primer pairs as indicated:
   (a) Primer pair set 1: a forward primer before arm1 that is located within the BAC sequence (name: Before-arm1-F) and a reverse primer in the IRES (name: IRES-R).
   (b) Primer pair set 2: forward primer from the Neo/Kan (name: Neo/Kan-F) and a reverse primer after arm2 located within the BAC sequence (name: After-arm2-R).

(c) Primer pair set 3: T7 (name: T7-F) primer paired with primer within the BAC.

9. Perform PCR and check PCR product size (Subheading 3.2, **steps 6–9**) using primer pairs designed in Subheading 3.5, **step 8** with the BAC targeting clone as the DNA template. Isolate DNA fragment with QIAquick Gel Extraction Kit, obtain product sequence and compare to sequence in VectorNTI.

3.6 Deletion of FRT-Neo/Kan-FRT (see Note 39)

1. Begin SW105 competent cells procedure as described in Subheading 3.3, **steps 2–3**. When optical density at 600 reaches 0.4–0.5 on the spectrophotometer add 0.5 ml of 10 % L-(+)-Arabinose to 50 ml LB culture (final concentration is 0.1 %), shake 1 h at 32 °C.
2. Continue the SW105 competent cells procedure detailed in Subheading 3.3, **steps 5–7**. Electroporate BAC targeted clone DNA (1–5 μl) to SW105 competent cells (1.75 kV, 25 μF, and pulse controller 200 Ω; Subheading 3.3, **step 8**). Add 1 ml LB to the cuvette and transfer the mixture to a 15 ml bacterial culture tube, shake 250 rpm at 32 °C, for 1 h.
3. Plate cells onto CM/LB plate and Kan/LB resistant plate, grow at 32 °C, 36–48 h. Using a sterile toothpick, touch each colony (10–20 colonies) and draw two small parallel x's on the CM/LB plate and Kan/LB resistant plate. Colonies with removed Neo/Kan cassette will be visible on the CM/LB plate but absent on the Kan/LB plate.
4. Confirm the removal of Neo/Kan nucleotides from the BAC targeted clone using primers designed in Subheading 3.5, **step 8**. Perform PCR (Subheading 3.2, **steps 8** and **9**) using the BAC DNA lacking the Neo/Kan cassette as the template. Check the PCR product size on 1.2 % agrose gel, isolate DNA with the QIAquick Gel Extraction Kit, obtain PCR product sequence and compare to VectorNTI sequence.

3.7 Preparing BAC Targeting DNA for Injection

1. Pick one colony of targeted BAC clone into 5 ml CM/LB media and grow 3–5 h at 32 °C, shaking 250 rpm. Transfer the 1 ml into a 500 ml CM/LB media incubate at 32 °C, shaking 250 rpm, overnight.
2. Purify the BAC DNA using the QIAGEN Large-Construct Kit per manufacture instructions for pronuclear injection (*see* **Note 40**).
3. Run the purified circular BAC DNA and NotI digested BAC DNA on a 0.8 % agarose gel to evaluate the integrity, quality and quantity of BAC DNA.
4. Perform microinjection using circular BAC DNA diluted into 0.5 ng/μl in microinjection buffers.

5. When potential founders are obtained, isolate genomic DNA from tails or toes with the instructions provided by Puregene. To identify founders perform PCR with the primer pair sets in Subheading 3.5, **step 8** on the isolated genomic DNA and run the PCR product on 1.2 % agrose gel. A founder will have all three PCR product bands (*see* **Note 41**).

3.8 Timed Matings

1. In the afternoon, after 2:00 p.m., add 1–2 females, 8–20 weeks of age to a singly housed male >7 weeks old.
2. The following morning at 8:00 a.m. check for a vaginal plug by looking for a hard white waxy substance at the vaginal opening or inside the vaginal canal. Insert a sturdy gross anatomy blunt tip probe into the vaginal opening to check for a plug deep within the vaginal canal. The morning a vaginal plug is observed is considered e.d. 0.5.
3. For e.d. 9.5–11.5 stage embryos, weigh dam on morning plug is observed and record weight. Place in a new cage and note the date of the plug. Reweigh the pregnant dam the morning of the experiment. The dam should gain a minimum of 2–3 g by e.d. 10.5 if pregnant. Palpation of the abdomen can also be performed and the embryos will feel like small hard beads.
4. Leave females with the same male until a plug is observed (*see* **Note 42**).

3.9 Experimental SetUp

1. Prepare 2 l of imaging solution and 1 l of embryo holding solution.
2. Set the circulating water bath so the outflow from the tip of needle is 37 °C (*see* **Note 43**).
3. Place imaging chamber under the macro zoom fluorescent microscope and hook up inflow and outflow tubing, turn on both heating chamber and suction.
4. Fill 1 l flask with 100 ml water and run through perfusion system.
5. Fill the 1 l flask reservoir with 1 l of imaging solution, turn on 95% O_2/5 % CO_2 and adjust the regulator to gently bubble the solution. Run imaging solution through the perfusion setup for 5 min to prime lines and eliminate air bubbles in the system (*see* **Note 44**).
6. Allow perfusion system to pre-warm until temperature at the outflow tip reaches 37 °C (approximately 15 min). Use a flexible Teflon probe attached to a digital thermometer to test temperature of solution in the center of the bath.
7. Fill the polyurethane bucket with ice, place under the dissecting scope, illuminate with gooseneck lamp, and focus.
8. Fill two 100 mm × 20 mm petri dishes with cold PBS.

9. Fill 6–8 60 mm × 15 mm glass petri dishes with embryo holding solution and set in ice rectangular bucket. Sequentially number each glass petri dish with a piece of tape. Attach each glass petri dish to the oxygen line and turn on oxygen.
10. Sequentially number 6–8 1.5 ml eppendorf tubes and place in rectangular ice bucket.
11. Illuminate the imaging chamber with the gooseneck lamp and focus.
12. Turn on fluorescent lamp and camera.

3.10 Dissection of Embryos

1. Euthanasia procedures must be approved by your Institutional Animal Care and Use Committee. The recommended procedure to euthanize pregnant dams that will result in viable embryos is light asphyxiation by CO_2 followed by cervical dislocation.
2. Place dam with ventral surface facing up on a piece of absorbent paper and spray ethanol (70 %) on the abdomen of the mouse. Grab a fold of skin with a pair of serrated tip straight forceps in the lower midline of the abdomen and pull up. With curved large scissors cut the skin starting at the lower abdomen towards the chest exposing the abdominal muscle layer. Perform the same cutting action on the abdominal muscle as with the skin. Once the muscle has been cut, the embryos will be visible on each side of the abdomen within the left and right uterine horns. The embryos look like beads on a string within the uterine horns.
3. Remove the uterus, intact, by grabbing with finely serrated forceps at the base of the uterus (cervix), pulling up and cutting underneath each uterine horn with straight fine iris scissors.
4. Place uterus containing embryos in petri dish #1 (100 mm × 20 mm) containing cold PBS. Add additional ice-cold PBS to cover uterus. Gently swirl the uterus in PBS to wash away blood or fecal contaminants.
5. Remove the uterus from petri dish #1 and place in petri dish #2 (100 mm × 20 mm) containing cold PBS. Cut between each embryo using straight fine iris scissors.
6. Move the separated embryos that are still contained within the uterus with a perforated round spoon into individual glass petri dishes (60 mm × 15 mm) containing ice-cold embryo holding solution that is oxygenated, making sure that they are covered with solution and place on ice.
7. Under the dissecting microscope expose the embryo and umbilical cord by sliding curved serrated forceps between the muscle layer and decidua and hold firmly. Using spring loaded straight scissors cut the tissue then the sac and finally the umbilical cord. Leave embryos on ice and only dissect an embryo just prior to imaging.

8. Transfer the embryo to the preheated imaging chamber (37 °C) using a perforated round spoon while the perfusion system is running (*see* **Note 45**).
9. Slightly different methods are used for imaging embryos e.d. 9.5–11.5 compared to >e.d.12.5.
   (a) 9.5–11.5 e.d.: It is easiest to image this stage while they are placed on their side and imaging occurs on the lateral surface. Adjust the solution in the chamber to just cover the embryo. Carefully, remove the outer membrane around the heart using spring loaded straight scissor and place an insect pin between the atrioventricular canal and the out flow tract. For a ventral view stabilize the embryo placing pins on both sides of the head and tail (*see* **Note 46**). Alternatively, the heart can be removed by cutting straight across the outflow tract as superior as possible and then superior to the atria with spring loaded straight scissors. An example of an in vivo lateral and in vitro view taken at e.d. 10.5 is shown in Fig. 5.
   (b) >e.d. 12.5: Using insect pins, pin the embryo through the head into the sylgard to obtain a ventral view using non-serrated straight fine tip forceps. If at all possible, tilt the head away from you to obtain a better angle for the dissection. Once the head has been satisfactorily pinned, with a pair of forceps, unfurl the embryo by gently pushing the tail down. Once the embryo has uncurled, put another pin through the base of the tail region to stabilize the embryo (*see* **Note 47**). Once the embryo is satisfactorily laid out (the pins can be moved to give a better view), with the spring loaded straight scissors make a transverse incision just above the liver (the big red blotch in the mid portion of the embryo) cutting from one side across to the other. Grab tissue to gently pull up on the thorax and cut toward the head, vertically, along both sides of the thorax. This vertical incision should end approximately at the top of the forelimbs and another horizontal incision made to remove the flap. Often times the tissue does not come away in once nice, clean flap, but may be cut as small snippets. Figure 6 shows grey scale images obtained at e.d. 15.5 demonstrating the placement of pins and open view of the heart.
10. At this point the atria and ventricles should be clearly visible and contracting at a rate of >100 beats/min.

3.11 Imaging

1. Prescreen embryos that are positive for the expression of GCaMP2 by using the GFP fluorescent filter cube in the stereo fluorescent scope (illuminating them in blue wavelength light and viewing the green wavelength protons). If the embryos are

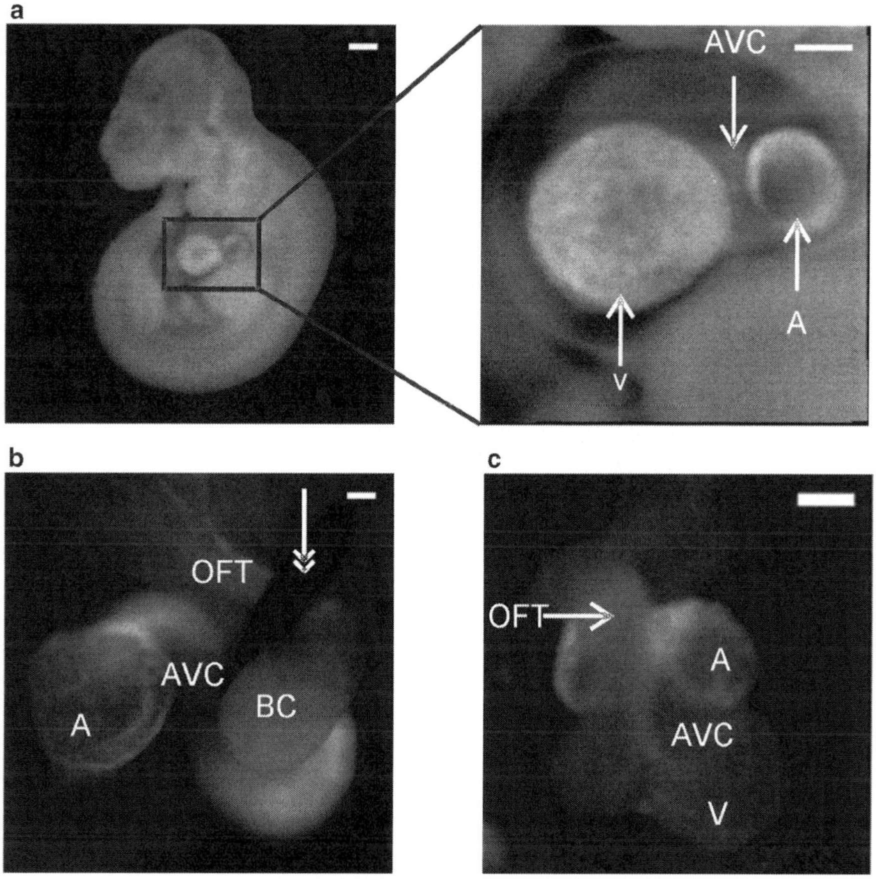

Fig. 5 Greyscale images of e.d. 10.5. (**a**) Lateral view with *left* side facing up. To the *right* is a blow-up of the heart. (**b**) The tissue enclosing the heart is removed and the heart is pulled slightly away from the embryo and a pin placed between the outflow tract (OFT) and the atrioventricular canal (AVC) thus leaving the heart in vivo. White double headed arrow denotes placement of pin. (**c**) In vitro view of the heart. *A* atria, *AVC* atrioventricular canal, *V* ventricle, *BC* bulbous cordis, *OFT* outflow tract. Scale bar: (**a**) 500 μm blow up 100 μm, (**b**) 50 μm, and (**c**) 100 μm

positive there will be a change in the Ca^{2+} dependent fluorescence as indicated by going from dark green to very bright green (*see* **Note 48**).

2. Once a positive embryo has been identified record image sequences using a frame rate of at least 67 Hz and a sequence duration of at least 6 s (*see* **Note 49**). Examples of sequential colorized Ca^{2+}-dependent fluorescent images taken at e.d. 10.5 and 15.5 are shown in Fig. 7.
3. After the last image sequence, cut off a small piece of the tail and place in an eppendorf tube. Place at –20 °C until ready to extract DNA for genotyping.
4. Discard embryo per institution biohazard protocol.

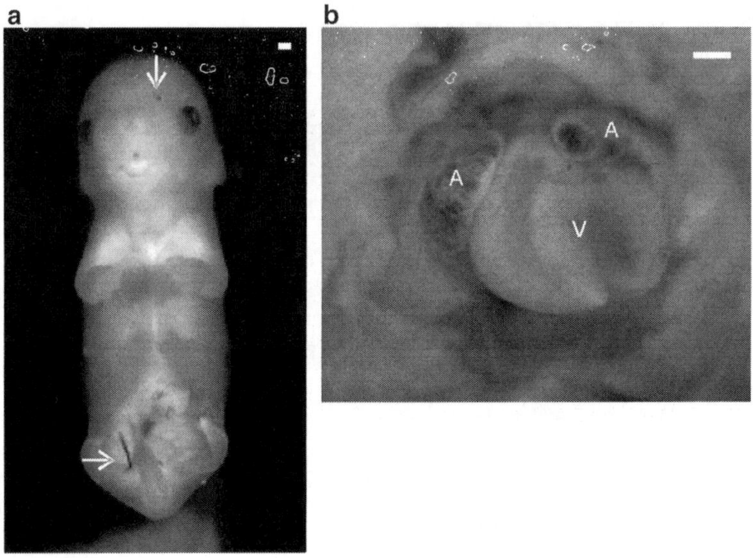

Fig. 6 Brightfield embryo image at stage e.d. 15.5. (**a**) Embryos are pinned with ventral surface facing up. *White arrows* denote location of pins. (**b**) Ventral view of heart after removal of chest wall. *A* atria and *V* ventricle. Scale bar: (**a**) and (**b**) 500 μm

5. Clean instruments and imaging chamber with ethanol (70 %) then rinse well with PBS. Refill chamber with imaging solution and repeat **step 7**, Subheading 3.10 to **step 5**, Subheading 3.11 for each individual embryo.

3.12 Clean-Up

1. Fill reservoir with 200 ml and tissue organ chamber with Millipore water, perfuse through system.
2. Fill reservoir with 100 ml and tissue organ chamber with 70 % ethanol. Soak for 20 min then run through the system.
3. Repeat Subheading 3.12, **step 1** twice (*see* **Note 50**).

4 Notes

1. Vector NTI software is available for both IBM and Mac platforms.
2. We purchase the largest size, aliquot 200 μl into 1.5 ml eppendorf tubes and store at –20 °C.
3. Water used to make all solutions, unless otherwise stated, has a resistance of 18.2 MΩ cm and <5 parts per billion of total organic content.
4. EtBr is a mutagen. Handle with caution.

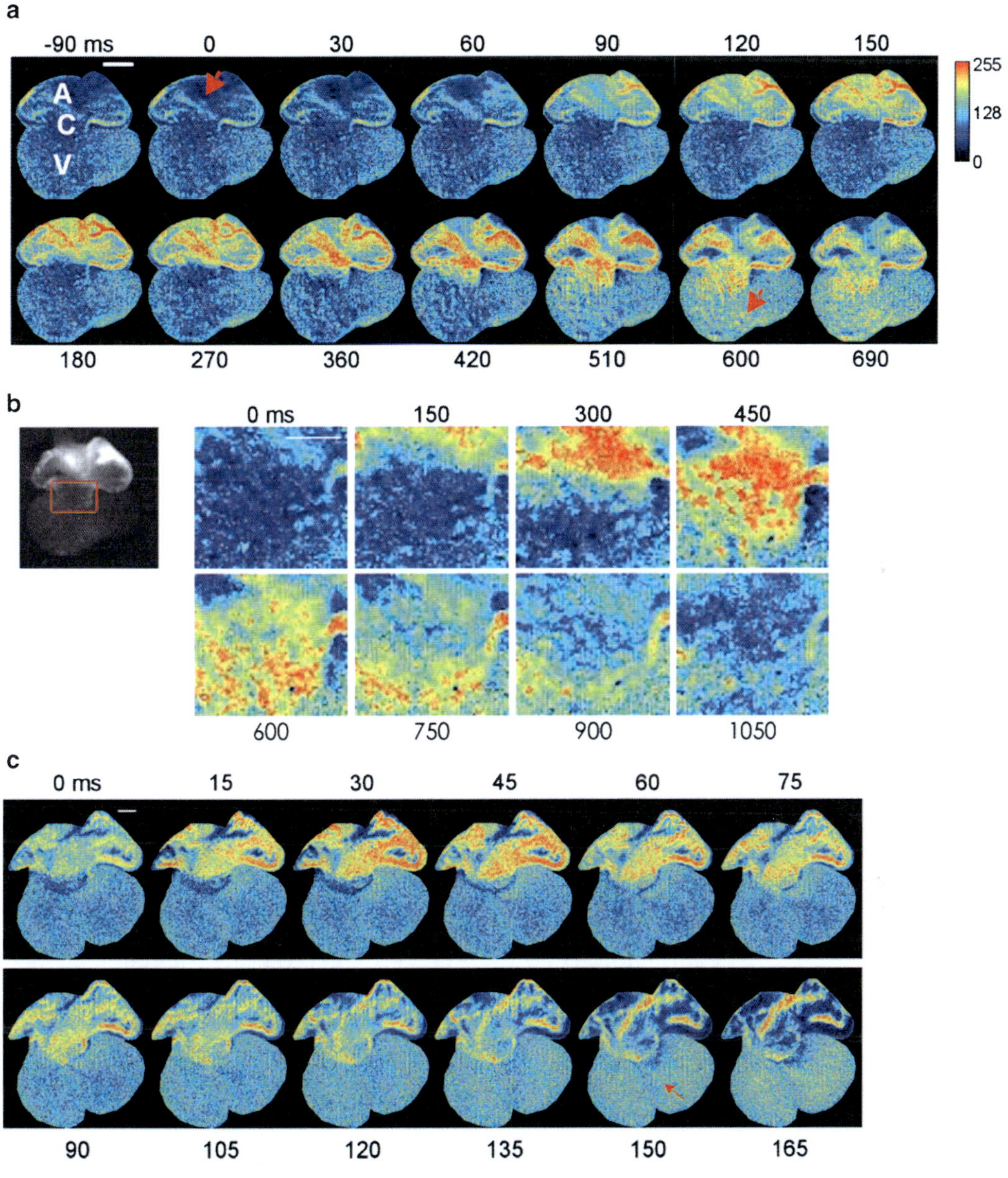

Fig. 7 Ca^{2+} signaling in the developing heart. (**a**) Sequential images of activation of e.d. 10.5 heart during spontaneous beating (dorsal surface). *Arrow* in second image shows initial activation of atrium. Note slow passage of Ca^{2+} wave through the AV canal (AVC) and progression into ventricle. Ventricular activation at apex (*arrow* at 510 ms) occurs independently after passage through AVC. (**b**) Passage of cytosolic Ca^{2+} wave through the AVC region shown in box on image at *left*. Shown is the e.d. 10.5 dorsal surface. (**c**) Images of Ca^{2+} fluorescence during spontaneous activation of an e.d. 13.5 heart (dorsal surface, 67 Hz). Note the Ca^{2+} wave ends in a ring at base of compacted AVC and does not extend into the ventricle. *Arrow* shows breakthrough of Ca^{2+} wave at ventricular apex. *A* atria, *C* atrioventricular canal, and *V* ventricle. Scale bars: 250 μm. Color scale in (**a**) applies throughout. (Reprint with modification from ref. 3)

5. Glycerol is very thick. Pipette slowly to withdraw an accurate amount.
6. EL250 and EL350 bacterial strains are no longer available. SW105 covers the same function as EL250 while SW106 covers EL350 function. We have recently used both bacterial strains successfully.
7. When bacterial cells arrive we remove a small amount with a sterile loop and follow the procedures to grow up bacterial cells. We use this as our stock of SW105 for the remainder of the BAC protocols. Freeze more tubes of cells in 20 % glycerol to have as additional stocks.
8. Chemical is carcinogenic. Handle with extreme caution.
9. All LB media must be autoclaved the day it is made. All liquid media should be mixed immediately after coming out of the autoclave but be careful of sudden boil ups.
10. EcoRV, EcoRI and SmaI can be used for arm1 subclone, we used EcoRI and SmaI for the connexinBAC-GCaMP2 construct.
11. Flipase recognition target (FRT) is a short sequence that in the presence of Flipase recombination enzyme (FLP) recombines and removes the nucleotide sequences between FRT sites.
12. Test out dissecting microscopes with ring lights prior to purchase. We have found there is too much reflected light making the dissections more difficult.
13. If black sylgard is not available put black paper on the bottom of the dish then pour the clear sylgard on top.
14. A red fluorescent filter cube (RFP) or dual band filter, FITC-TRITC, (green and red) can be useful to determine autofluorescence. If green is observed using the red filter then it is autofluorescence.
15. Use a deep sided petri dish for embryos older than e.d. 15.5.
16. The anodized pins will rust in salt solutions. Use sparingly if purchased.
17. Pick 4–6 colonies.
18. The BAC is in *E.coli* strain DH10B and can be cultured at 37 °C. It is not necessary in this step to grow the BAC at 32 °C.
19. This protocol is based on the alkaline lysis with SDS principle; therefore, solutions from other mini/maxi prep kits may be used.
20. A rule of thumb is an enzyme should be less than 10 % of total volume.
21. Bromophenol blue migrates more than half the way down the gel and is approximately equivalent to the migration of 400–500 bp double-stranded DNA.

22. Occasionally not all of the bands are clear due to genomic sequence error but the majority of the bands will be correct.
23. The primer we use for T7 is 5′-TAA TAC GAC TCA CTA TAG GG-3′ and for SP6 is 5 ′-ATT TAG GTG ACA CTA TAG-3′.
24. Always use clean gloves, aerosol tips, PCR pipettes, and PCR quality reagents. To reduce possible contamination purchase and aliquot ultra pure water from Invitrogen and have a designated area for PCR preparation. Design a PCR calculator in Microsoft Excel. This allows for easy manipulation of each individual concentration.
25. Adjust extension time using 1 min extension time per 1 kb of PCR product. Times also may need to be adjusted based on GC content of PCR product.
26. For GC-rich sequences we use Phusion High-Fidelity DNA polymerase (Biolabs).
27. Exposure to ultraviolet light for as short a time as possible to reduce the damage of PCR product.
28. EL250 cells were used for the Cx40BAC-GCaMP2 transgenic mouse but are no longer available. Therefore, throughout this document we will refer to SW105 cells instead.
29. If the colonies are too small incubate one more day.
30. Follow spectrophotometer manufacturer instructions to dilute your sample.
31. Make sure that all steps take place on ice and glycerol is ice-cold.
32. Do not use a flip top eppendorf tube.
33. You can also flank the exon that contains the ATG start codon.
34. We use the random sequence AGTCAC. The nucleotide sequence for each of the restriction sites are as follows EcoRI—GAATTC; SpeI—ACTAGT; NotI—GCGGCCGC and SmaI CCCGGG.
35. We have adjusted the PCR reaction volume to increase the amount of PCR product.
36. Adjust temperature to optimized annealing and adjust extension time using 1 min per 1 kb of PCR product as a rule of thumb.
37. Either arm can be subcloned first unless there is an enzyme restriction site conflict.
38. Cool mixture as quickly as possible by swirling.
39. In our hands, no deleterious effects have been observed if the Neo/Kan sites are left within the recombined BAC.
40. Several other methods are available to isolate high quality BAC DNA. We have successfully injected circular BAC DNA via pronuclear injection.

41. Once we have verified the transgene has been transmitted to offspring then genotyping is performed using one primer pair.
42. To increase the likelihood of a plug, house a male singly for 1 week, without changing bedding. Group house five females for 2 weeks then place two females with a singly housed male, a plug should be observed by 3 days. This takes advantage of what is termed the Whitman effect. We do not superovulate due to stress on the dam that would most likely occur while carrying at least 20 pups.
43. Routinely check the temperature of the perfusate.
44. Attach a syringe onto the end of the tubing and withdraw solution to help eliminate bubbles.
45. Use a non-perforated spoon or pipette tip to transfer e.d. 9.5 embryos.
46. If the embryo does not have to be moved, put a drop of Cell-Tak on the bottom of the imaging chamber then place embryo on top of Cell-Tak (Collaborative Research, Boston, MA).
47. Immobilization can be obtained by placing the head between two stationary pins.
48. It is easy to mistake the change in blood volume and contraction movement as a positive embryo when first learning how to genotype using fluorescence. If the embryo is positive the change in fluorescence ($\Delta F/F0$) will be greater than 25 % compared to a negative heart which is typically <5 %.
49. Bleaching of GCaMP2 can occur if images are acquired for several minutes.
50. If precipitate forms in the solution wash glassware and tubing in 1 M HCl and then rinse thoroughly. Once or twice a year replace all tubing.

Acknowledgments

We would like to thank Dr. Junichi Nakai for the development of GCaMP2 and Dr. Guy Salama for aid in developing optical recording methods. This work was supported by NIH grants HL45239 and DK65992.

References

1. Berridge MJ (2003) Cardiac calcium signalling. Biochem Soc Trans 31:930–933
2. Berridge MJ, Bootman MD, Roderick HL (2003) Calcium signalling: dynamics, homeostasis and remodelling. Nat Rev Mol Cell Biol 4:517–529
3. Tallini YN, Ohkura M, Choi BR, Ji G, Imoto K, Doran R, Lee J, Plan P, Wilson J, Xin HB, Sanbe A, Gulick J, Mathai J, Robbins J, Salama G, Nakai J, Kotlikoff MI (2006) Imaging cellular signals in the heart in vivo: cardiac expression of the high-signal Ca2+ indicator

GCaMP2. Proc Natl Acad Sci U S A 103: 4753–4758

4. Roell W, Lewalter T, Sasse P, Tallini YN, Choi BR, Breitbach M, Doran R, Becher UM, Hwang SM, Bostani T, von Maltzahn J, Hofmann A, Reining S, Eiberger B, Gabris B, Pfeifer A, Welz A, Willecke K, Salama G, Schrickel JW, Kotlikoff MI, Fleischmann BK (2007) Engraftment of connexin 43-expressing cells prevents post-infarct arrhythmia. Nature 450:819–824

5. Tallini YN, Brekke JF, Shui B, Doran R, Hwang SM, Nakai J, Salama G, Segal SS, Kotlikoff MI (2007) Propagated endothelial Ca2+ waves and arteriolar dilation in vivo: measurements in Cx40BAC GCaMP2 transgenic mice. Circ Res 101:1300–1309

6. Tallini YN, Shui B, Greene KS, Deng KY, Doran R, Fisher PJ, Zipfel W, Kotlikoff MI (2006) BAC transgenic mice express enhanced green fluorescent protein in central and peripheral cholinergic neurons. Physiol Genomics 27:391–397

Chapter 14

Microscopic Computed Tomography-Based Skeletal Phenotyping for Genetic Model Organisms

Suresh I. Prajapati, Lisa Nevell, and Charles Keller

Abstract

Forward and reverse genetics now enable researchers to understand embryonic and postnatal gene functioning in a wide range of species. Some genetic mutations cause obvious morphological change, whereas other mutations can lead to more subtle phenotypes and might be overlooked without adequate observations and quantifications. Due to the increase in number of genetic model organisms examined by the growing field of *phenomics*, standardized but sensitive methods for quantitative analysis are increasingly necessary in the everyday practice of analyzing ever-increasing quantities of phenotypic data. In this chapter, we have presented platform-independent parameters for the use of microscopic X-ray computed tomography (microCT) for phenotyping species-specific skeletal morphology of a variety of different genetic model organisms.

Key words Skeletal phenotyping, microCT, Genetic model organisms, Embryogenesis

1 Introduction

The introduction of rapid genome sequencing empowers researchers to understand the sequence and structure of genes from multiple species, yet we are only at the entry point of an exciting new era of understanding gene function. Phenotypic variations can be achieved by intentional genomic modification of model organisms; however, detailed phenotypic analysis is often extremely slow, and standardization is only beginning [1].

At the macroscopic and microscopic scale, small animal micro-computed tomography (microCT) is ideally suited to anthropomorphic studies [2–4]. microCT also allows more rapid analysis of certain phenotypes such as skeletal defects and defects of the vasculature system [5]. Even the fine structure of soft tissues can also be observed and characterized with the use of metal-containing contrast agents [6]. Hence, microCT can be considered a good general performance phenotyping tool.

Mark Lewandoski (ed.), *Mouse Molecular Embryology*, Methods in Molecular Biology, vol. 1092,
DOI 10.1007/978-1-60327-292-6_14, © Springer Science+Business Media New York 2014

In this chapter, we demonstrate that microCT is a specifically powerful and practical tool for the skeletal analysis of diverse model genetic organisms [7]. Guidelines for reproducible, species-specific qualitative and quantitative phenotypic analysis of several genetic model organisms at different stages of development are presented here. Species studied include *Mus musculus* (mouse), *Xenopus laevis* (frog), *Danio rerio* (zebrafish), *Myotis lucifugus* (bat), *Monodelphis domestica* (opossum), *Gallus domesticus* (chicken), *Anas domesticus* (duck), and *Microcebus murinus* (mouse lemur). The guidelines presented here are optimized with respect to the resolution, signal–noise, and artifact reduction—all important to detect features of frequent interest for each organism. Furthermore, investigators will benefit greatly from making use of freely available image analysis software for image processing which can significantly enhance the information content of scans.

2 Materials

2.1 Animals (See Note 1)

- In general, detailed high-resolution whole-body skeleton scans on most microCT scanners are slow. To avoid motion artifacts from breathing, euthanized specimens are preferred unless your scanner can perform respiratory gating. Live animal microCT scans require special attention to maintaining body heat and oxygenation. For the safety of the animal, most live animal scans should be limited to 10–40 min.
- Methods of euthanasia vary by institution but should be in compliance with local IACUC guidelines. Whole undissected fresh or fixed specimen can be scanned so long as fixation does not decalcify tissue (i.e., Do not use Bouin's solution or EDTA-containing fixative).

1. *Mus musculus* (Mouse)—Newborn or Adult.
2. *Xenopus laevis* (Frog)—Adult.
3. *Danio rerio* (zebrafish)—Adult.
4. *Myotis lucifugus* (Bat)—Adult.
5. *Monodelphis domestica* (Opossum)—Adult.
6. *Gallus domesticus* (Chicken)—Fetus (Day 14).
7. *Anas domesticus* (Duck)—*Fetus* (Day 19).
8. *Microcebus murinus* (Lemur)—Adult.

2.2 Scanners

- microCT instruments from any manufacturers can be used to scan specimens. The key is to use the smallest field of view for which your specimen will fit, at the best resolution for that field of view. Maximize the number of different views (view angles) and the frames per view that are within your time and budget constraints.

2.3 Freely Available Image Analysis Software

1. Basic visualization and quantification:
   MicroView© version 2.1.2, GE Healthcare (http://microview.sourceforge.net).
2. Advanced visualization:
   BioImage, University of Utah Scientific Computing Institute (http://www.sci.utah.edu/cibc/software/#bioimage).
3. Advanced visualization and quantification:
   Seg3D, University of Utah Scientific Computing Institute (http://www.sci.utah.edu/cibc/software/seg3d.html).

3 Methods

3.1 Preparation of Mouse Specimens

Newborn and adult *Mus musculus* specimen can be scanned using the parameters detailed in Table 1.

3.2 Preparation of Adult Frog

Adult *Xenopus laevis* specimens can be scanned using the parameters detailed in Table 2.

3.3 Preparation of Adult Zebrafish

Adult *Danio rerio* specimens can be fixed in 10 % buffered formalin and scanned using the parameters detailed in Table 3.

3.4 Preparation of Adult Bat

Adult *Myotis lucifugus* specimens can be scanned using the parameters detailed in Table 4.

Table 1
Parameters for *Mus musculus*

Specimen (common name)		Scanner	Current (μA)	Voltage (kVp)	Exposure time (ms)	Number of views	Frames per view	Scan time	File size	Scanner resolution setting (focal spot)
Mus musculus (mouse)	Newborn	SCANCO μCT40	144	55	300	2,000	5	25 h	10 GB	10 μm
	Adult	GE eXplore Locus	450	80	100	720	5	2 h	1.9 GB	93 μm

Table 2
Parameters for *Xenopus leavis*

Specimen (common name)		Scanner	Current (μA)	Voltage (kVp)	Exposure time (ms)	Number of views	Frames per view	Scan time	File size	Scanner resolution setting (focal spot)
Xenopus laevis (frog)	Adult	GE eXplore Locus	450	80	100	400	5	20 min	84 MB	93 μm

Table 3
Parameters for *Danio rerio*

Specimen (common name)		Scanner	Current (μA)	Voltage (kVp)	Exposure time (ms)	Number of views	Frames per view	Scan time	File size	Scanner resolution setting (focal spot)
Danio rerio (zebrafish)	Adult	SCANCO μCT40	144	55	300	2,000	10	60 h	15 GB	6 μm

Table 4
Parameters for *Myotis lucifugus*

Specimen (common name)		Scanner	Current (μA)	Voltage (kVp)	Exposure time (ms)	Number of views	Frames per view	Scan time	File size	Scanner resolution setting (focal spot)
Myotis lucifugus (bat)	Adult	GE eXplore Locus	450	80	500	900	10	5 h	1.5 GB	46 μm
	Bone and joint	SCANCO μCT40	144	55	300	1,000	5	12 h	500 MB	12 μm

Table 5
Parameters for *Monodelphis domestica*

Specimen (common name)		Scanner	Current (μA)	Voltage (kVp)	Exposure time (ms)	Number of views	Frames per view	Scan time	File size	Scanner resolution setting (focal spot)
Monodelphis domestica (opossum)	Adult	GE eXplore Locus	450	80	100	720	5	5 h	487 MB	93 μm

3.5 Preparation of Adult Opossum

Adult *Monodelphis domestica* specimens can be scanned using the parameters detailed in Table 5.

3.6 Preparation of Fetal Chick

White leghorn chicken eggs (*Gallus domesticus*) can be incubated in a Lyon R-COM-20 incubator (Lyon Technologies, Inc., Chula Vista, CA) at 38 °C and 60 % humidity from day 0 to day 14. On day 14, the fetus is harvested and fixed in 10 % PBS buffered formalin overnight at room temperature. The fetus can be scanned using the parameters detailed in Table 6.

3.7 Preparation of Fetal Duck

White Pekin duck eggs (*Anas domesticus*) can be incubated in a Lyon R-COM-20 incubator at 38 °C and 60 % humidity from day 0 to day 19. On day 19, the fetus is harvested and fixed in 10 % PBS buffered formalin overnight at room temperature. The fetus can be scanned using the parameters detailed in Table 7.

Table 6
Parameters for *Gallus domesticus*

Specimen (common name)		Scanner	Current (μA)	Voltage (kVp)	Exposure time (ms)	Number of views	Frames per view	Scan time	File size	scanner resolution setting (focal spot)
Gallus domesticus (chicken)	Fetus	SCANCO μCT40	144	55	300	2,000	5	22 h	1.5 GB	15 μm

Table 7
Parameters for *Anas domesticus*

Specimen (common name)		Scanner	Current (μA)	Voltage (kVp)	Exposure time (ms)	Number of views	Frames per view	Scan time	File size	Scanner resolution setting (focal spot)
Anas domesticus (duck)	Fetus	SCANCO μCT40	144	55	300	1,000	5	18 h	2 GB	36 μm

Table 8
Parameters for *Microcebus murinus*

Specimen (common name)		Scanner	Current (μA)	Voltage (kVp)	Exposure time (ms)	Number of views	Frames per view	Scan time	File size	scanner resolution setting (focal spot)
Microcebus murinus (mouse lemur)	Adult	GE eXplore Locus	450	80	100	720	5	2 h	946 MB	93 μm

3.8 Preparation of Adult Lemur

Adult lemur (*Microcebus murinus*) can be scanned using the parameters detailed in Table 8.

3.9 Quantitative Analysis

2-Dimensional transfer function (2DTF) visualizations can be performed with open source software BioImage, from the University of Utah Scientific Computing Institute (http://www.sci.utah.edu/cibc/software/#bioimage). Quantification and segmentation of anatomical features is performed using open source software such as MicroView© version 2.1.2, GE Healthcare (http://microview.sourceforge.net) or Seg3D from the University of Utah Scientific Computing Institute (http://www.sci.utah.edu/cibc/software/seg3d.html).

4 Note

1. Universal precautions are necessary for all biological specimens, and all animal studies should only be conducted with proper institutional animal care and use committee approval.

References

1. Brown SD, Hancock JM, Gates H (2006) Understanding mammalian genetic systems: the challenge of phenotyping in the mouse. PLoS Genet 2:e118
2. Turner CH, Sun Q, Schriefer J, Pitner N, Price R, Bouxsein ML, Rosen CJ, Donahue LR, Shultz KL, Beamer WG (2003) Congenic mice reveal sex-specific genetic regulation of femoral structure and strength. Calcif Tissue Int 73: 297–303
3. Silha JV, Mishra S, Rosen CJ, Beamer WG, Turner RT, Powell DR, Murphy LJ (2003) Perturbations in bone formation and resorption in insulin-like growth factor binding protein-3 transgenic mice. J Bone Miner Res 18: 1834–1841
4. Moverare S, Venken K, Eriksson AL, Andersson N, Skrtic S, Wergedal J, Mohan S, Salmon P, Bouillon R, Gustafsson JA, Vanderschueren D, Ohlsson C (2003) Differential effects on bone of estrogen receptor alpha and androgen receptor activation in orchidectomized adult male mice. Proc Natl Acad Sci U S A 100: 13573–13578
5. Paulus MJ, Gleason SS, Easterly ME, Foltz CJ (2001) A review of high-resolution X-ray computed tomography and other imaging modalities for small animal research. Lab Anim (NY) 30:36–45
6. Johnson JT, Hansen MS, Wu I, Healy LJ, Johnson CR, Jones GM, Capecchi MR, Keller C (2006) Virtual histology of transgenic mouse embryos for high-throughput phenotyping. PLoS Genet 2:e61
7. Vasquez SX, Hansen MS, Bahadur AN, Hockin MF, Kindlmann GL, Nevell L, Wu IQ, Grunwald DJ, Weinstein DM, Jones GM, Johnson CR, Vandeberg JL, Capecchi MR, Keller C (2008) Optimization of volumetric computed tomography for skeletal analysis of model genetic organisms. Anat Rec (Hoboken) 291:475–487

Chapter 15

Gene Transfer Techniques in Whole Embryo Cultured Post-implantation Mouse Embryos

Daisuke Sakai and Paul A. Trainor

Abstract

Gene transfer techniques such as electroporation and lipofection are powerful systems for investigating gene function. In this chapter we focus on the methods and applications of gene transfer into specific cells and tissues of post-implantation mouse embryos.

Key words Gene transfer, Electroporation, Lipofection, Whole embryo culture

1 Introduction

Understanding the genetic regulation of embryogenesis involves analyzing gene function through gain-of-function and loss-of-function experiments. Traditionally this has been achieved through the generation of transgenic and knockout mouse models. However, the efficient and successful generation of genetically engineered mice requires the routine supply of large numbers of preimplantation embryos and pseudopregnant recipients. More recently, new methods such as electroporation and lipofection have been devised for the in vivo alteration of gene function in diverse species. Electroporation is a mechanical method used to introduce polar molecules into a host cell through the cell membrane. In this procedure, an electric pulse temporarily disturbs the phospholipid bilayer, allowing molecules like DNA to pass into the cell [1]. In contrast lipofection is a technique that uses liposomes to introduce genetic material into a cells and tissues [2]. Liposomes are phospholipid vesicles that easily integrate with the membranes of cells. The introduction of DNA, RNA, and morpholinos via electroporation and lipofection provides convenient and efficient alternatives to transgenic and knockout mouse models for the rapid analysis of gene function during embryonic development. Here we describe protocols for electroporation and lipofection in post-implantation

Mark Lewandoski (ed.), *Mouse Molecular Embryology*, Methods in Molecular Biology, vol. 1092,
DOI 10.1007/978-1-60327-292-6_15, © Springer Science+Business Media New York 2014

mouse embryos and their potential applications which make the mouse an attractive model for embryological studies.

2 Materials

2.1 Whole Embryo Culture of Post-implantation Mouse Embryos

1. Tyrode's salt solution.
2. Dissection tools.
3. DMEM/F12/Glutamax (Invitrogen).
4. Rat serum (Harlan).
5. Antibiotics (penicillin and streptomycin).
6. Incubator with rotating drum (BTC Engineering).
7. Specialty gas mixtures (Praxair).

2.2 Electroporation of Post-implantation Mouse Embryos

1. Tyrode's salt solution.
2. Chamber-type or forceps-type electrodes.
3. Electroporator (e.g., Electro-Square Porator CUY21, BTX, NEPA gene).
4. Fine-tip injection needles pulled from wide-bore glass capillaries (outer diameter = 1.0 mm, inner diameter = 0.75 mm, length = 100 mm, Sutter Instrument).
5. Genetic Material (DNA, RNA, RNAi, morpholinos, etc.).

2.3 Lipofection of Post-implantation Mouse Embryos

1. Tyrode's salt solution.
2. LipofectAMINE2000 (Invitrogen) or FuGENE (promega) or Turbofect (Thermo Scientific).
3. 5 % Sucrose in Opti-MEM (Invitrogen) is filtrated and stored at 4 °C.
4. Fine-tip injection needles are pulled from glass capillaries (outer diameter = 1.0 mm, inner diameter = 0.75 mm, length = 100 mm, Sutter Instrument).
5. Genetic Material (DNA, RNA, RNAi, morpholinos, etc.).

3 Methods

3.1 Whole Embryo Culture of Post-implantation Mouse Embryos

1. Humanely terminate pregnant female mice by cervical dislocation or other Institutional Animal Care and Use Committee (IACUC) or regulatory body approved method.
2. Lay each mouse supine and wipe abdomen with 70 % ethanol.
3. Using sharp scissors make a small ventral incision through the skin and body wall at the level of the hindlimbs. Next, make a v-shaped incision by cutting diagonally towards the left and right edges of the ribs and reflect the body wall tissue.

Table 1
Parameters required for each developmental stage

Age (E)	MeMedia	Gas/oxygenation	Yolk sac	Electroporation	Distance
7.5 [3]	DR50	5 % O_2, 5 % CO_2, 90 % N_2	Intact	26 V, 30 ms, 3 P	0.5 cm
7.5–8.5 [4]	DR50	5 % O_2, 5 % CO_2, 90 % N_2	Intact	10–15 V,50 ms, 3–5 P	0.5 cm
8.5–9.5 [5]	DR50	5 % O_2, 5 % CO_2, 90 % N_2 20 % O_2, 5 % CO_2, 75 % N_2	Intact	15–20 V,50 ms, 3–5 P	0.5–1.0 cm
9.5–10.5 [5]	DR75	20 % O_2, 5 % CO_2, 75 % N_2 65 % O_2, 5 % CO_2, 35 % N_2	Intact or open	15–20 V, 50 ms, 5 P	1.0–1.5 cm
10.5–11.5 [5]	DR100 (+2 mg/ml glucose)	65 % O_2, 5 % CO_2, 35 % N_2 100 % O_2	Intact or open	15–20 V,50 ms, 5 P	1.5–2.0 cm

D DMEM/F12/Glutamax supplemented with penicillin/streptomycin, *R50*, *R75*, *R100* rat serum percentage composition of culture media, *P* pulses, *V* voltage, *ms* millisecond

4. Using blunt forceps displace the intestine and other internal organs to one side to reveal the uterus. Pinch the uterus but not the individual deciduas contained within. Cut away the mesometrium (blood supply to the uterus and embryos) and sever the uterine horns from the cervix and oviducts.
5. Place the uterine horns into warm or room temperature Tyrode's solution and cut the uterus into sections containing a single deciduum.
6. With finely sharpened #5 forceps, extract each deciduum by gently tearing away the muscle tissue of the uterine wall.
7. Transfer the deciduas to a fresh dish of Tyrode's solution and remove the maternal tissue. Carefully remove Reichert's membranes to expose the yolk sac. Do not damage the placenta or blood vessels in the yolk sac. Leave the yolk sac and amnion intact or simply attached depending on the culture conditions required for the desired embryonic stage to be manipulated (refer to Table 1).
8. If opening the membranes, try to avoid severing the large vessels in the yolk sac. Usually a small opening can be made in the yolk sac and amnion immediately above the head though which the embryo can be pushed out. Also it is recommended to leave the placental tissue attached after allantois–chorion fusion has taken place to facilitate better nutrient exchange and circulation during in vitro culture.
9. Transfer freshly dissected embryos into bottles of warm equilibrated media (30 min minimum) containing the stage specific

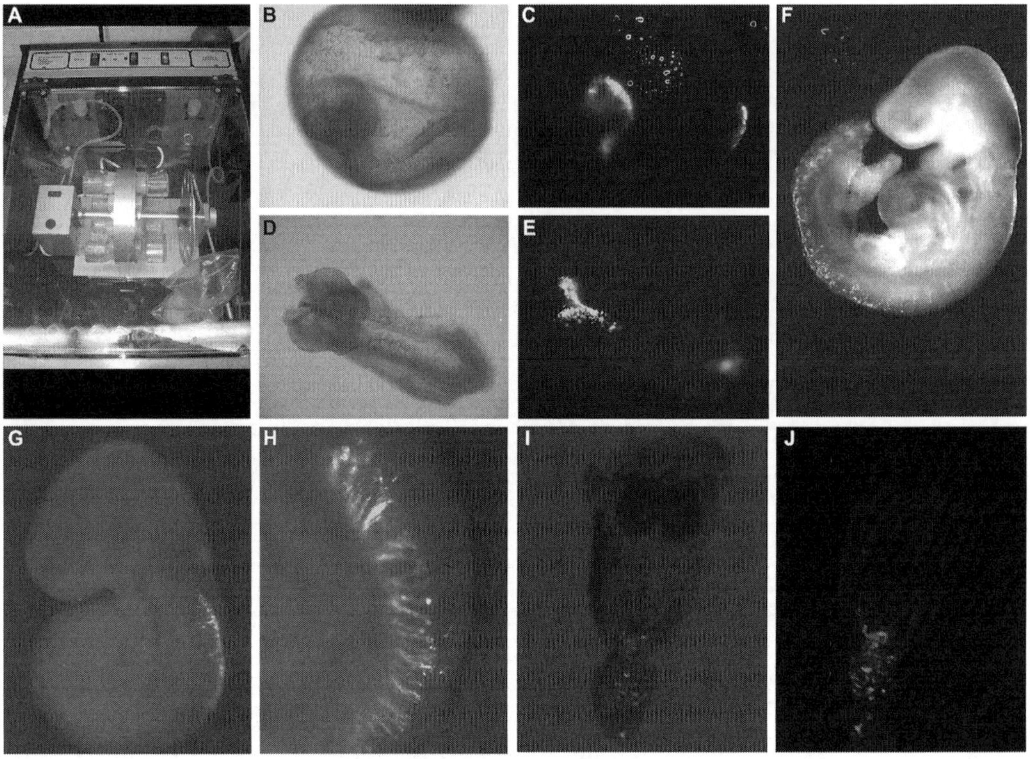

Fig. 1 Gene transfer into whole embryo cultured post-implantation mouse embryos. (**a**) BTC Engineering rotating drum and culture incubator. (**b**) Bright-field and (**c**) fluorescence lateral images of E8.5 embryo with extraembryonic membranes intact that was electroporated at E7.5 with *pMES* into the open neural plate and cultured for 24 h. (**d**) Bright-field and (**e**) fluorescence dorsal images of same embryo as (**b**, **c**) with the yolk sac and amnion removed. (**f**) Combined bright-field/fluorescence image of E9.5 embryo with extraembryonic membranes removed post 24 h culture that was electroporated at E8.5 with pMES into the primitive streak. (**g**) Combined bright-field/fluorescence and (**h**) higher magnification image of E11.0 embryo cultured for 36 h post-electroporation with *pMath1-tau-IRES-GFP* into the closed trunk neural tube of an E10.0 embryo showing labeling of commissural neurons and their axons. (**i**) Combined bright-field/fluorescence and (**j**) fluorescence dorsal images of E8.5 embryo cultured for 24 h post lipofection with *pMES* into the primitive streak and allantois of a gastrulating E7.5 embryo

concentration of rat serum under the appropriate atmospheric conditions (refer to Table 1).

10. Attach bottles containing embryos to the rotating drum inside a BTC incubator (Fig. 1a).
11. Replace culture media each 24 h and adjust for rat serum concentration and oxygenation as required for each developmental stage (refer to Table 1) (*see* **Notes 1–3**).

3.2 Electroporation of Post-implantation Mouse Embryos

Detailed below is the method for electroporation of the neural plate and neural tube in mouse embryos. However this approach is generally applicable to any cells, tissues or organs in the embryo provided a suitable luminal space is available to be injected with the DNA, RNA, or morpholino molecules of choice (Fig. 1b–h).

3.2.1 Preparation of Genetic Material

1. Clone the gene or sequence of interest into an expression plasmid which has a constitutively active promoter such as the cytomegalovirus (CMV) or β-actin (BA) promoter. pcDNA series (Invitrogen), pCAX [5], and pEF-BOS [6] plasmids are frequently used for mouse embryo electroporation. It is preferable to use a plasmid that also contains a lineage reporter (eg IRES-GFP) such as pMES [7] to visualize the efficiency of electroporation as well as the specificity and autonomy of gain-of-function and loss-of-function analyses. If such a plasmid is not available, co-transfection with another expression vector containing a fluorescent or other lineage reporter, such as pEGFP-N1 (Clonetech) can be used to the same effect.
2. Electroporation is not limited to DNA plasmids but can also be used to transfer various forms of interfering RNAs such as shRNA (short hairpin RNA) and miRNA (micro RNAs) as well as morpholinos. The only criterion is that the material to be electroporated is charged (positively or negatively).
3. Store genetic material to be electroporated as frozen aliquots for one time use. Avoid freeze–thaw cycles which will diminish the efficiency of gene transfer.
4. Prior to injection, add Fast Green to the genetic material at a final concentration of 0.025–0.05 % (*see* **Notes 6**, 7).

3.2.2 Injection and Electroporation of Post-implantation Mouse Embryos

1. Remove embryos from roller culture one bottle at a time and transfer into Tyrode's solution in a 6 cm or larger petri dish. Orient the embryo as desired to visualize the neural plate or tube or tissue of interest for injection.
2. Break the tip off a finely pulled glass needle to the desired width or use a microforge to fashion specific widths for injection.
3. Front end load or back fill the needle with genetic material depending on the injection method to be used. The injection can be performed via a mouth aspirator, hydraulic syringe, Picospritzer, or FemtoJet as desired.
4. Inject genetic material adjacent to the open neural plate (E8.5 embryos) or into the lumen of neural tube (e9.5 or older) or tissue of interest. For embryos with intact membranes the needle needs to pass through the yolk sac and amnion and typically the amniotic cavity is filled with genetic material. The volume injected will depend on the stage of the embryo being manipulated but is typically in the 0.5–2.0 μl range.
5. Place the electrodes on either side of the embryo. For neural tube electroporation, the electrodes should be aligned perpendicularly to the anterior–posterior axis of the neural tube.

6. Electroporate embryos individually using the conditions described in Table 1 which are stage specific as well as electrode type specific (forceps versus chamber). The standard time gap between each pulse is approximately 1 s.
7. Transfer electroporated embryos back into the same roller bottles and reattach to the rotating drum for in vitro culture (Fig. 1a) (*see* **Notes 4**, **5**).

3.3 Lipofection of Post-implantation Mouse Embryos

Electroporation as a technique appears to work best with epithelial cells in contact with a tissue lumen that functions as a reservoir for injected genetic material. However, with respect to mesenchymal cell populations, electroporation is much less efficient at achieving successful gene transfer. Lipofection however provides an alternative strategy for efficient gene transfer of mesenchymal cells. The following protocol describes a method for lipofection mediated gene transfer of cranial and trunk mesenchyme but is equally applicable to any cell, type, tissue or organ of interest in the embryo (Fig. 1i, j). The lipofection method is not limited to DNA transfer but similar to electroporation can also be used for various forms of interfering RNAs such as shRNA (short hairpin RNA) and miRNA (micro RNAs) [8] as well as morpholinos.

3.3.1 Preparation of Genetic Material–Lipofectamine Complex

1. Dilute 1 μl of Lipofectamine2000 in 50 μl of 5 % sucrose/Opti-MEM and incubate for 5 min for room temperature.
2. Dilute 1–2 μl of DNA (500 ng/μl) or other genetic material (concentration must be empirically optimized) in 50 μl of 5 % sucrose/Opti-MEM (*see* **Notes 6**, **7**).
3. Combine the diluted genetic material with diluted Liopfectamine2000 and incubate for 20 min at room temperature.
4. Add Fast Green to the complex at a final concentration of 0.025–0.05 %.
5. The working stock of genetic material–lipofectamine complex should be prepared fresh for each experiment.

3.3.2 Injection and Lipofection of Post-implantation Mouse Embryos

1. Remove embryos from roller culture one bottle at a time and transfer into Tyrode's solution in a 6 cm or larger petri dish. Orient the embryo as desired to visualize the head or tissue of interest for injection.
2. Break the tip off a finely pulled glass needle to the desired width or use a microforge to fashion specific widths for injection.
3. Front end load or back fill the needle with genetic material depending on the injection method to be used. The injection can be performed via a mouth aspirator, hydraulic syringe, Picospritzer, or FemtoJet as desired.

4. Inject genetic material into the cranial or trunk mesenchyme at the desired axial level and position. For embryos with intact membranes the needle needs to pass through the yolk sac and amnion. The volume injected will depend on the stage of the embryo being manipulated but is typically between 0.1 and 1.0 μl. Some expansion of the mesenchyme is to be expected but do not inject excessive genetic material–lipofectamine complex as this will mechanically perturb craniofacial development or if the injection is performed in the primitive streak it could disrupt the process of gastrulation.
5. Transfer injected embryos back into the same roller bottles and reattach to the rotating drum for in vitro culture.

4 Notes

1. For additional details about stage specific embryo dissections, see *Manipulating the Mouse Embryo* [9].
2. A major limitation of whole embryo culture technique is that mouse embryos will only grow in vitro for a maximum of 2–3 days. When electroporation or lipofection is performed in E7-10 embryos, in vitro culture does not permit subsequent analyses to be carried out during the later phases of fetal development or even postnatally, to assess long term cell fate and differentiation. However, this obstacle has been overcome by both exo utero [10] and ultrasound-guided in utero [11, 12] approaches which have successfully and efficiently transferred DNA fragments up to 11 kb in length. Alternatively, at the completion of whole embryo culture the cells, tissues, or organs of interest can be explanted in tissue culture or transplanted under the kidney capsule to assess long term cell fate and differentiation.
3. Chemically defined serum-free media can be used to culture E10.5 embryos successfully for 24 h, which negates the need for rat serum in some experiments [13]. Unfortunately, this media does not adequately support in vitro culture of embryos younger than E10.5.
4. Conditions for electroporation such as voltage, duration, number of pulses, and distance of electrodes should be optimized for each developmental stage and tissue to be manipulated as well as for the type of electrodes being used (forceps versus chamber).
5. Parameters such as cell death should be measured in any analyses involving electroporation to account for nonspecific effects of the technique.
6. For both electroporation and lipofection, it is very important to use highly purified nucleic acids. This can be achieved by

using commercially available plasmid extraction kits such as Qiagen Maxi prep kit.

7. Purified DNA for electroporation is dissolved in PBS to a concentration of between 1 and 5 mg/ml. Highly concentrated DNA solutions are viscous which have the advantage of limiting diffusion with the lumen of the neural tube or other tissues of interest.

Acknowledgements

Work in the Trainor laboratory is supported by the Stowers Institute for Medical Research, the March of Dimes (#6-FY08-265), and the National Institute for Dental and Craniofacial Research (DE016082).

References

1. Sugar IP, Neumann E (1984) Stochastic model for electric field-induced membrane pores. Electroporation. Biophys Chem 19:211–225
2. Felgner PL, Gadek TR, Holm M, Roman R, Chan HW, Wenz M, Northrop JP, Ringold GM, Danielsen M (1987) Lipofection: a highly efficient, lipid-mediated DNA-transfection procedure. Proc Natl Acad Sci U S A 84:7413–7417
3. Mellitzer G, Hallonet M, Chen L, Ang SL (2002) Spatial and temporal 'knock down' of gene expression by electroporation of double-stranded RNA and morpholinos into early postimplantation mouse embryos. Mech Dev 118:57–63
4. Davidson BP, Tsang TE, Khoo PL, Gad JM, Tam PP (2003) Introduction of cell markers into germ layer tissues of the mouse gastrula by whole embryo electroporation. Genesis 35:57–62
5. Osumi N, Inoue T (2001) Gene transfer into cultured mammalian embryos by electroporation. Methods 24:35–42
6. Mizushima S, Nagata S (1990) pEF-BOS, a powerful mammalian expression vector. Nucleic Acids Res 18:5322
7. Swartz M, Eberhart J, Mastick GS, Krull CE (2001) Sparking new frontiers: using in vivo electroporation for genetic manipulations. Dev Biol 233:13–21
8. Sheehy NT, Cordes KR, White MP, Ivey KN, Srivastava D (2010) The neural crest-enriched microRNA miR-452 regulates epithelial-mesenchymal signaling in the first pharyngeal arch. Development 137:4307–4316
9. Nagy A, Gertsenstein M, Vintersten K, Behringer RR (2003) Manipulating the mouse embryo. Cold Spring Harbor Laboratory, Cold Spring Harbor
10. Fukuchi-Shimogori T, Grove EA (2001) Neocortex patterning by the secreted signaling molecule FGF8. Science 294:1071–1074
11. Saito T, Nakatsuji N (2001) Efficient gene transfer into the embryonic mouse brain using in vivo electroporation. Dev Biol 240:237–246
12. Saito T (2006) In vivo electroporation in the embryonic mouse central nervous system. Nat Protoc 1:1552–1558
13. Moore-Scott BA, Gordon J, Blackburn CC, Condie BG, Manley NR (2003) New serum-free in vitro culture technique for midgestation mouse embryos. Genesis 35:164–168

Chapter 16

Segmentation and Quantitative Analysis of Individual Cells in Developmental Tissues

Kaustav Nandy, Jusub Kim, Dean P. McCullough, Matthew McAuliffe, Karen J. Meaburn, Terry P. Yamaguchi, Prabhakar R. Gudla, and Stephen J. Lockett

Abstract

Image analysis is vital for extracting quantitative information from biological images and is used extensively, including investigations in developmental biology. The technique commences with the segmentation (delineation) of objects of interest from 2D images or 3D image stacks and is usually followed by the measurement and classification of the segmented objects. This chapter focuses on the segmentation task and here we explain the use of ImageJ, MIPAV (Medical Image Processing, Analysis, and Visualization), and VisSeg, three freely available software packages for this purpose. ImageJ and MIPAV are extremely versatile and can be used in diverse applications. VisSeg is a specialized tool for performing highly accurate and reliable 2D and 3D segmentation of objects such as cells and cell nuclei in images and stacks.

Key words Image analysis, Image segmentation, Developmental biology, ImageJ, MIPAV, VisSeg, Thresholding, Snakes, Levelset, Dynamic programming

1 Introduction

Quantitative analysis of images is fast becoming an integral part of biological research, because such an approach extracts considerably more and accurate information from the sample. Usually the information serves as input for hypothesis testing complex biological processes by mathematical modeling, leading to improved understanding of the mechanisms and in some cases to subsequently improved diagnosis and treatment of major human diseases. However, most microscope images of biological samples are still analyzed manually, by subjective visual inspection.

Image Segmentation. In general, quantitative analysis of images commences with segmentation (delineation) of the objects of interest, such as cells or cell nuclei. Segmentation can be performed either on 2D images or 3D image stacks and during the past 20

Mark Lewandoski (ed.), *Mouse Molecular Embryology*, Methods in Molecular Biology, vol. 1092,
DOI 10.1007/978-1-60327-292-6_16, © Springer Science+Business Media New York 2014

years, many algorithms have been developed to perform this task. Introductory information about segmentation and other aspects of image analysis is readily available in many textbooks [1–3] and Web sites [4].

Sample Preparation and Image Acquisition. A prerequisite for obtaining quantitative 3D measurements from a sample is that the labeled sample is transparent. This is so that when imaging at a particular depth through the sample, the part of the sample in front of the depth does not obscure the image. Using flourescent labels ensure that this is the case, as well as bestowing two additional advantages: (1) Fluorescent labeling is quantitative because the intensity of the fluorescent signal is proportional to the concentration of the labeled molecules, (2) Multiple molecular species can be simultaneously imaged by employing fluorescent labels with different spectral properties. Thus, by using fluorescent labeling, one can visualize objects of interest in the biological samples, such as individual cells, subcellular organelles, and specific proteins of interest. Furthermore, fluorescent protein genetic constructs and certain fluorescent labels can be applied to living samples, thus enabling the extraction of structural, temporal, and spatial information about them.

Fluorescence microscopy advanced significantly in the 1980s with the advent of confocal microscopy, allowing direct acquisition of in focus 3D images from thick samples, and a decade later by the introduction of multi-photon microscopy enabling deeper penetration into living tissue.

For accurate quantification, particularly for 3D images, it is extremely important to prepare the samples with great care so as to avoid artifacts during imaging. In terms of image acquisition the objective lens is probably the most important component of the microscope since it determines the quality of images in terms of spatial resolution and sensitivity. It is best to use high numerical aperture (NA) oil immersion lenses for depths up to 50 μm and a water immersion lens for greater depths. Also, in order to capture as much spatial detail as possible from the sample it is necessary to use a voxel (Volume Picture Element: smallest distinguishable box in a 3D image) size less than half the Nyquist frequency ((x,y,z) size less than (100, 100, 250) nm for a high-NA oil lens.). However, in practice a larger voxel size may be needed to reduce sample exposure. Sample preparations for the samples presented in this chapter are in Appendix.

Examples of Image Analysis Applications in Developmental Biology. To illustrate the diverse applications of image analysis in developmental biology we present a few examples:

1. Merks et al. [5] investigated vasculogenesis during the development of the circulating system in vertebrates, by comparing a

computation (in silico) model to an in vitro system. The in vitro system was HUVEC (human umbilical vein endothelial) cells grown in Matrigel over time, which mimicked 2D vasculogenesis in flat regions such as the yolk sac. Images of this system were analyzed by segmenting the vascular cords (using thresholding and morphological operations), followed by removal of spurious branch points and enumerating the number of nodes (branch points) and lacunae (connected regions separated by vascular cords). The computational model showed that it was necessary to consider the elongated shape of endothelial cells in order to correctly predict the network growth of the vasculature. Quantitative agreement between the computational model and the in vitro system was observed, leading to a more in-depth understanding about the process of vasculogenesis in developing vertebrates.

2. Heid et al. [6] envisioned the need for 4D computational reconstruction of individual cells and nuclei in developing embryos in order to understand the processes of cell division, cell differentiation, definition of body axis, tissue reorganization, and the generation of organ systems during development of the organism. Consequently, they built a semi-interactive, computer-assisted 3D system to reconstruct and enable motion analysis of cells and nuclei in a developing embryo. Since all the components of an embryo are reconstructed individually, analysis and mathematical modeling of all or selected nuclei, any single cell lineage, and any single nuclear lineage becomes very easy. The tool also provides a new method for reconstructing and motion analyzing in 4D of every cell and nucleus in a live developing embryo. The novel features of their tool were demonstrated by reconstructing and analyzing cell motion in the *Caenorhabditis elegans* embryo through the 28-cell stage. They anticipate that their tool will have applications for analyzing the effect of drugs, environmental perturbations, and the effect of mutations on cellular and nuclear dynamics during embryogenesis.

3. Angiogenesis is the process by which new blood vessels develop and in vitro models of this process are essential for further understanding the biological mechanisms as well as for screening for angiogenic agents and inhibitors. To address this need, Blacher et al. [7] developed computer-assisted tools to quantify angiogenesis in the aortic ring assay; a common in vitro model. Their system is capable of determining the aortic ring area, its shape, number of microvessels, total number of branchings, the maximal microvessel length, microvessel distribution, the total number of isolated fibroblast-like cells and their distribution. The method can quantify spontaneous angiogenesis and can perform analysis of complex microvascular networks induced by

vascular endothelial growth factor. By analyzing the distribution of fibroblast like cells they concluded that during spontaneous angiogenic response, maximal fibroblast-like cell migration delimits microvascular outgrowth and Batimastat (an angiogenic inhibitor) prevents sprouting of endothelial cells, but does not block fibroblast-like cell migration.

4. The spatiotemporal pattern of gene expression at the individual cell level reveals important information about gene regulation during embryo development. The patterns can be obtained using in situ hybridization (ISH) to label specific mRNA sequences. In order to further understand these patterns, Peng et al. [8] developed image analysis algorithms to extract gene expression features from ISH images of the *Drosophila melanogaster* embryo. Their algorithm clusters genes sharing the same spatiotemporal pattern of expression, suggesting transcription factor binding site motifs for genes that appear to be co-regulated. Their image analysis procedure recapitulates known co-regulated genes and gives 99+% correct developmental stage classification accuracy despite variations in morphology, rotation, and position of the embryo relative to the 3D image.

Outline of this Chapter. In this chapter three freely available software packages are covered that can be used to do the initial segmentation of the cell nuclei in cell cultures or tissue sections. They are ImageJ [9], MIPAV (Medical Image Processing, Analysis, and Visualization) [10], and VisSeg [http://ncifrederick.cancer.gov/Atp/Omal/Flo/]. The methods include segmentation of both 2D and 3D images. ImageJ and MIPAV are general image processing software packages offering a wide range of functionalities and can be used in a number of different applications apart from segmentation. However here we cover only the small portion of each package related to segmentation. VisSeg is specialized software for very accurate and reliable 2D and 3D segmentation of cell nuclei. All the segmentation methods described here are semiautomatic since they require modest human intervention to perform the segmentation. This has the advantage that all objects will be correctly segmented relative to the gold standard of the human visual system.

2 Image Preprocessing

Before performing the segmentation it is advisable to reduce the image noise using a smoothing filter so that the segmentation algorithm is able to find the edges more reliably and accurately. The following segmentation procedures include noise removal as a first step.

3 ImageJ

ImageJ is developed in Java and can be used with any operating system such as Windows, Linux, Mac OS 9, or Mac OS X which has Java enabled. ImageJ and its Java code are available free in the public domain. The main ImageJ Web site is http://rsb.info.nih.gov/ij/ and it can be downloaded at http://rsb.info.nih.gov/ij/download.html. ImageJ provides a wide range of basic image processing operations for example image enhancement, image filtering, cropping, scaling and resizing, image measurement tools, color conversion and image editing tools to mention a few. Users can automate tasks and create custom tools using the macros in ImageJ. The software can also automatically generate macro code using the command recorder. Apart from the macros, users can contribute and develop extensions for ImageJ by implementing new plugins for the software. The plugins that have already been developed can be found at http://rsb.info.nih.gov/ij/plugins/ and detailed documentation can be accessed at http://rsb.info.nih.gov/ij/docs/index.html. The ImageJ graphical user interface is shown in Fig. 1. We demonstrate segmentation with ImageJ using the fluorescent nuclear channel 3D image of a MCF-10A cell culture sample.

3.1 Thresholding Based Segmentation Using ImageJ

1. Reading the image: Go to file and open the image file that has to be processed. ImageJ reads a variety of image formats including most of the microscopy image formats.
2. Image denoising: Go to Process > Filters > Gaussian Blur. Use an appropriate radius for the operation and perform the blurring. In this case a radius of 2.5 has been used. However the user has to adjust the radius depending on the amount of noise in the image. Figure 2 shows a single slice before and after denoising. Other denoising algorithms are available for this purpose.
3. 3D Object counter plugin: The plugin has to be downloaded and installed as per instructions on the ImageJ Web site. Once installed go to Plugins > 3D objects counter. Adjust the "Threshold" slider to select the threshold intensity value that segments the image objects by partitioning it into regions of

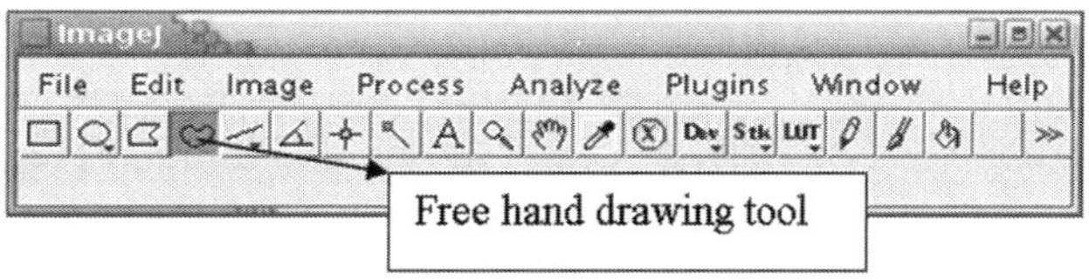

Fig. 1 ImageJ graphical user interface (GUI) and the free hand drawing tool

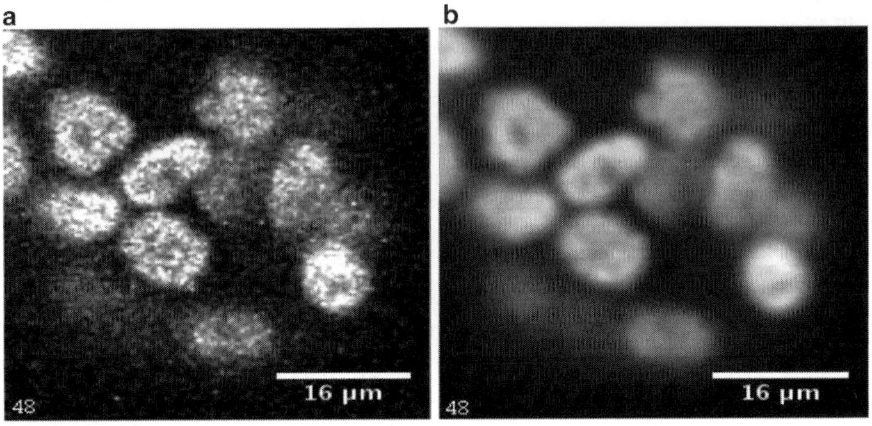

Fig. 2 A single slice before and after denoising. (**a**) Original slice from the MCF-10A sample. (**b**) Denoised version of A using a Gaussian filter radius of 2.5

bright objects and darker background. One can vary the slice number too using the proper slider so that one can check whether the threshold value is actually performing a good segmentation across the different slices.

4. Output selection: Next specify the size range of the segmented objects to be measured using the "Min number of voxels" and "Max number of voxels" parameters. One can then select the desired output parameters from the plugin by choosing "Particles" (actual segmented objects), "Edges" (object boundaries), "Geometric center," "Intensity based center," "Numbers" (object number), and "Log summary" (log of all the objects segmented). Press "OK" once all the parameters and options have been specified.
5. Output: The algorithm goes through the stack and numbers all the connected objects after thresholding (Fig. 3). The output log shows various parameters for the segmented objects such as volume, coordinates of object center etc. Although Fig. 3 shows a single slice from the sample, the algorithm performs a 3D segmentation.

Note: It is not always possible to segment all the objects successfully using this method since it uses a simple thresholding procedure for the purpose.

3.2 Active Contour Segmentation Using ImageJ

Active contours [11, 12] is an extremely popular modeling method that is used in segmentation and tracking of objects in images and video. It makes use of geometric and probabilistic modeling for analyzing objects. It is far more sophisticated than the previous method since it makes effective use of certain prior information such as local edge strengths and boundary smoothness. However, it

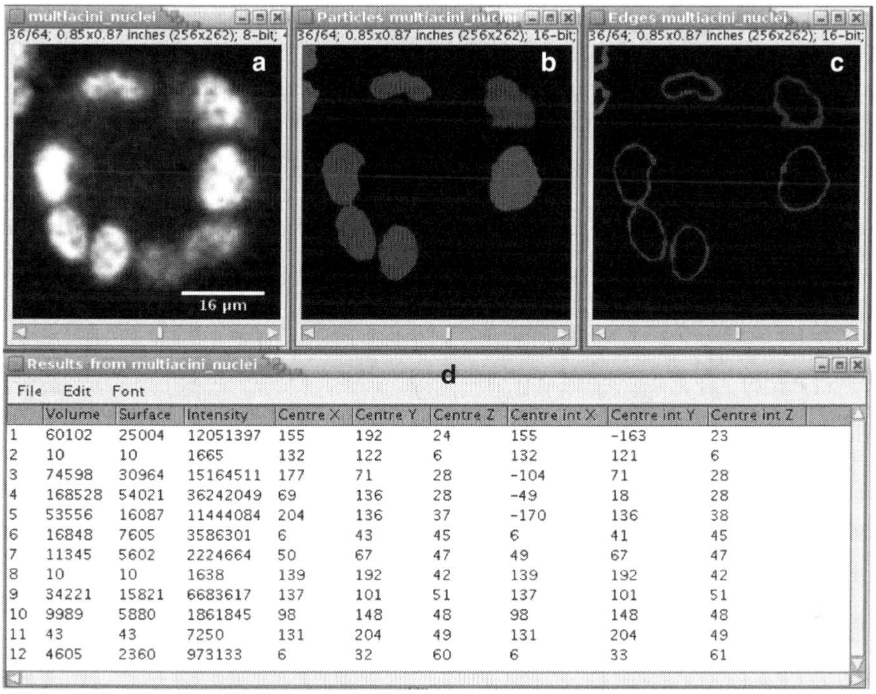

	Volume	Surface	Intensity	Centre X	Centre Y	Centre Z	Centre int X	Centre int Y	Centre int Z
1	60102	25004	12051397	155	192	24	155	-163	23
2	10	10	1665	132	122	6	132	121	6
3	74598	30964	15164511	177	71	28	-104	71	28
4	168528	54021	36242049	69	136	28	-49	18	28
5	53556	16087	11444084	204	136	37	-170	136	38
6	16848	7605	3586301	6	43	45	6	41	45
7	11345	5602	2224664	50	67	47	49	67	47
8	10	10	1638	139	192	42	139	192	42
9	34221	15821	6683617	137	101	51	137	101	51
10	9989	5880	1861845	98	148	48	98	148	48
11	43	43	7250	131	204	49	131	204	49
12	4605	2360	973133	6	32	60	6	33	61

Fig. 3 ImageJ output. (**a**) Original slice from the MCF-10A sample. (**b**) Detected objects shown in slice 36 using the 3D Object Counter plugin. (**c**) Edge delineation of the detected samples. (**d**) Object log showing various statistics obtained from the 3D Object Counter

processes the image locally instead of using the global information, requiring an initial approximation of the boundary.

The first two steps are the same as the previous procedure. Then we use the ABsnake plugin that uses active contours to calculate the segmentation of a single object across different slices. This plugin will segment one object at a time. The plugin has to be downloaded and installed first.

1. Initialize Snake algorithm: Choose the object to segment and go through the slices to identify the slice range over which it is clearly visible. Go to the first slice in which the object is visible and use the free hand drawing tool (Fig. 1) to draw an approximate boundary for that slice.
2. ABSnake plugin: Go to Plugins > ABSnake. Specify the slices over which the object is clearly visible in the "First slice" and "Last slice" boxes. The "Gradient threshold" value will differ depending upon the image you are going to segment. Choose the color of the boundary and the other options that are available in the plugin window. Once done press "OK."
3. Output: The output shows the slices with the computed boundaries (Fig. 4). The procedure is then repeated for other nuclei.

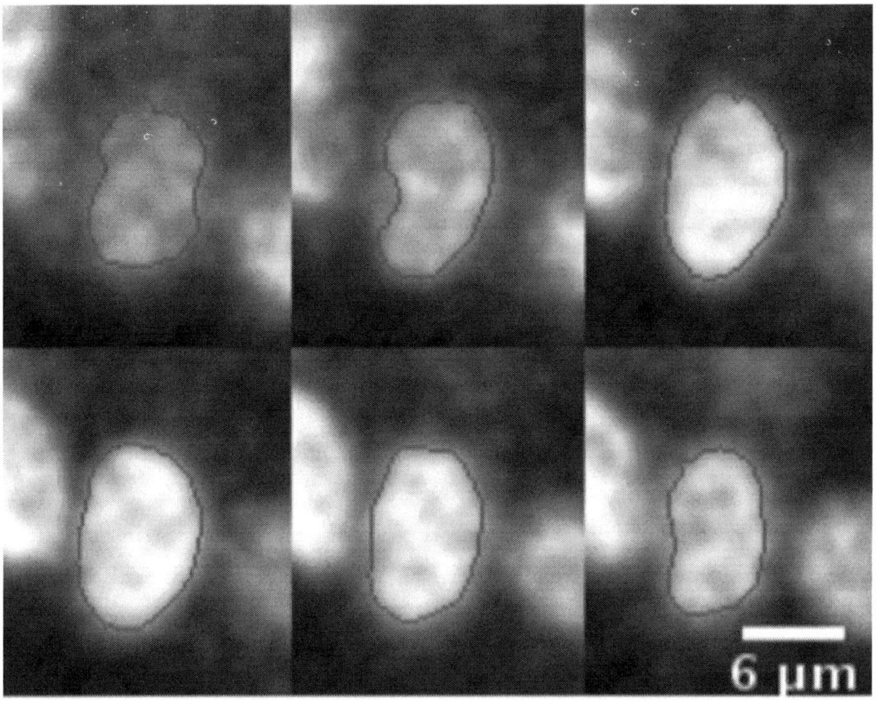

Fig. 4 ABSnake plugin. Figure shows six slices out of 12 slices from a sample segmentation of a single nucleus in the MCF-10A sample using the ABSnake plugin in ImageJ

Notes: we have mentioned two different methods for nuclei segmentation using ImageJ. Both of them are somewhat limited in performance, because they segment each 2D slice image independently in a 3D stack image. However, they show good results in most cases. The second method is often sensitive to the initial boundary selection.

4 MIPAV

MIPAV [13] is a general purpose application that is intended for medical and biological microscopic image analysis (Fig. 5). It is also a Java based application and can be used on any Java-enabled platform such as Windows, Unix, or Mac OS X. The main MIPAV site is http://mipav.cit.nih.gov/index.php and it can be downloaded from http://mipav.cit.nih.gov/clickwrap.php. MIPAV can also be extended using plugins and scripts for custom applications. The plugins are functions written in Java using the MIPAV application programming interface (API). The scripts can be recorded using MIPAV and later used on different datasets. MIPAV offers tools for a wide range of operations that can be used for processing images which include filtering, segmentation, measurement and registration. It also has integrated visualization capabilities where one can

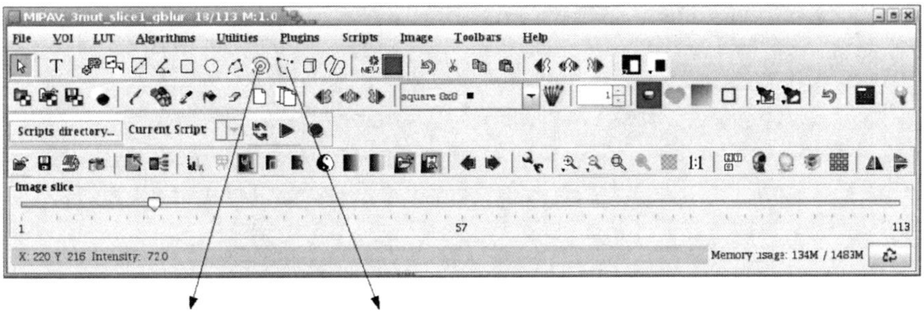

Fig. 5 MIPAV GUI indicating the Levelset and Livewire segmentation tools

visualize multidimensional data. The detailed documentation can be obtained at http://mipav.cit.nih.gov/documentation.php.

First we will show 2D segmentation techniques using the Levelset tool and Livewire tool, and then we extend to 3D images. However, for this software the algorithms also analyzed each 2D slice in a 3D stack independently. Tools for segmenting the images using thresholding are also available in this software, but are not described here.

4.1 2D Method Using the Levelset Tool

1. Reading the image: MIPAV like ImageJ reads a wide variety of image formats. For opening an image go to File > Open Image (A) from disk and select the image that will be analyzed.
2. Blurring to reduce noise: To perform blurring of the image go to Algorithms > Filters (spatial) > Gaussian blur. In the Gaussian blur window choose the degree of blurring necessary by specifying the "Scale of the Gaussian" in X, Y, and Z directions. In the examples presented here, we use 2 for all the dimensions. After choosing all the other options press "OK."
3. Segmentation using Levelset: Change the slider to change the image slice and choose the slice that is to be segmented. Click on the Levelset tool and position the mouse arrow near the boundary of one of the nuclei to be segmented. This shows a border around the nuclei in yellow. If the border is satisfactory click once and the border color changes to red giving the 2D segmentation for the nucleus (Fig. 6a). If the border shown in yellow is not satisfactory it can be modified by changing the position of the mouse arrow near the border of the nuclei. By repeatedly choosing the Levelset tool all the nuclei in the image slice can be segmented one by one.

4.2 2D Method Using the Livewire tool

1. Reading the image and performing blurring: Follow **steps 1** and **2** mentioned in the previous method (Subheading 4.1).
2. Segmentation using Livewire: Choose the Livewire tool from the MIPAV GUI. This will show a new window where the

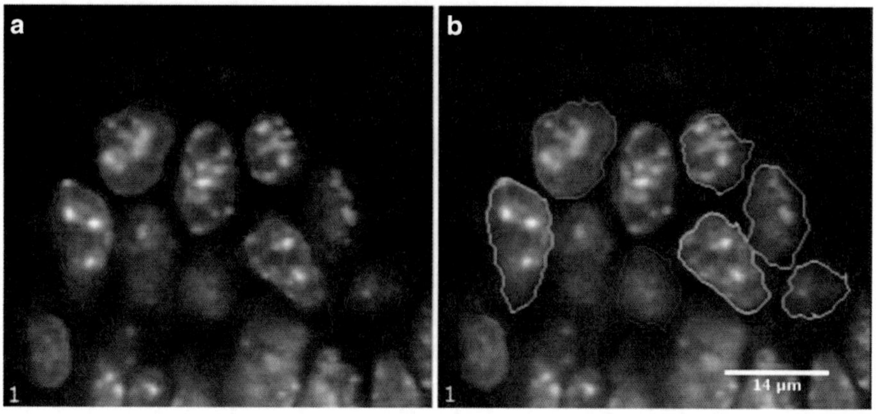

Fig. 6 Levelset tool (**a**) 2D segmentation using the Levelset tool in MIPAV of slice 1 of the somite stage E8.5 mouse embryo image (**b**) 2D segmentation using Livewire tool in MIPAV of the same sample

MIPAV cost function has to be chosen. In case of volume labeled objects such as fluorescent labeled cell nuclei "Gradient magnitude and direction" should be used. On the other hand, when the objects are surface labeled "Intensity" should be used. Since the nuclei in the example are volume labeled we use "Gradient Magnitude and direction". Press "OK." Go to the image window and click on a boundary point of the nucleus that is to be segmented first. Move the mouse arrow approximately along the nuclei boundary in a clockwise or anticlockwise direction. The software will keep on tracing the actual boundary of the nuclei using a dynamic programming (DP) based optimization algorithm [14]. The boundary trace is the path between the first anchor point and the current position of the mouse arrow that has the highest gradient magnitude per voxel compared to any other path between the anchor point and mouse position. To force the boundary to go through a certain point click once to add an anchor point at that location. While tracing around the nuclei keep on adding the anchor points to segment the nuclei satisfactorily. Position the final anchor point near the position of the first anchor point and the boundary will be completed in red. This procedure can be repeated to segment all the nuclei in the slice (Fig. 6b).

4.3 3D Segmentation Using MIPAV

1. Initialization: For doing 3D segmentation using MIPAV choose an object to be segmented and go through the slices to identify a slice that is approximately in the middle. Perform the 2D segmentation using either one of the tools mentioned namely, Levelset tool or the Livewire.
2. Propagating the boundary to other slices: After 2D segmentation in the previous step, select the contour by clicking on it. Then go to VOI > Evolve boundary 2D > Active contour.

In the new window one can specify all the parameters to be used for evolving the boundary. In the "Evolve Boundary" options select "Propagate to adjacent slices" and "Replace Original Contour." This will propagate the boundary on to the other slices. Press "OK." After the computation is complete the boundaries of the object in different slices can be viewed by changing the slice number. Usually the algorithm segments in a range of slices adjacent to the starting slice, but omits slices further away from the starting slice even though the object is still visible. In this case the boundary can be propagated further by selecting the last segmentation contours (at the top or bottom) one by one and running the algorithm once more with the same settings.

3. Go to next object: For segmenting the next object, press the new contour button and the GUI will show a new color for the new contour. After this initialization of the new contour follow the steps from 1. In this way all the objects can be segmented.

5 VisSeg

VisSeg is available freely at http://ncifrederick.cancer.gov/Atp/Omal/Flo/. It is a specialized software that is specifically used for semiautomatic 2D and 3D segmentation. It starts with a 2D segmentation of an object in a user selected plane, using DP to locate the border with an average intensity per unit length greater than any other path around the object in that plane. The next step extends the 2D segmentation to the adjacent slices using a combination of DP and combinatorial searching. Once the algorithm has completed the segmentation the user can interactively correct the segmentation by adding anchor points at any location in the 3D-image. The final surface is forced to go through anchor points. The computation algorithms are written in C language and the graphical user interface was developed using Qt (http://qt.digia.com/) and VTK (Visualization Toolkit) (http://www.vtk.org/). For installation of the software one has to install these supporting packages first. The details of the algorithm is available in [15].

General Notes: VisSeg reads images as 8-bit single channel ICS (Image Cytometry Standard) [16]. Some modifications have to be done to the ".ics" file in order to provide all the relevant information to the software. We will cover those in the next section. VisSeg can only accurately segment 3D objects that are point convex. Point convex means that all straight lines radiating in any direction from a user-marked internal point (see below) intersect the boundary of the object exactly once. Point convex objects can contain concavities, but cannot be too irregular in their shape.

Method

1. Conversion and header additions: If necessary use appropriate software, for example MIPAV to convert the image channel that has to be segmented into an 8-bit single channel ICS image. VisSeg requires two more pieces of information to be in the ICS file, the aspect ratio (*z*-axis-resolution/*x*[*y*]-axis-resolution) and the "dyed type" (volume or surface depending on whether the surface is stained or the entire object is stained). Open the ".ics" file using any word processor and add the following lines to the end

   ...

   history dyedType volume[or surface]

   history aspectRatio 3.5 [or another number depending on the resolution of the axes]

   ...

2. Load the image: Open VisSeg and go to File > Open and choose the ICS image file that has to be segmented. The loaded image will be displayed using volume rendering (Fig. 7a) and in a single slice viewer (Fig. 7b).
3. Data Exploration Controls:
   - In the 3D Rendering Window (Fig. 8a), rotate the image using the mouse left button, pan the image using mouse middle button and zoom using mouse wheel or right button.
   - In 2D Manipulation Window (initially it shows a z-slice) (Fig. 8b), move up and down the *z*-axis using the mouse middle button, zoom using the mouse wheel or right button outside the big white box and adjust the brightness/contrast using the right button inside the box. One can also rotate the 2D slice using the mouse middle button. To do that, grab the image edge (the red box) using the mouse middle button and rotate. After rotation one can again move up and down the volume using the middle button within the big box. As one rotates the 2D slice the 3D volume rendering window updates and shows the corresponding 2D slice through the volume.
   - The view can be reset to the original view using CTRL+R or go to Edit > Reset Camera.
4. 2D Segmentation: 2D segmentation only works on a "z slice", an image slice that is orthogonal to the depth axis of the image, so it is better to reset the camera before doing the segmentation. After resetting the camera go to the slice where the nuclei to be segmented is clearly visible. At this point check the "Segmentation Mode" box. Use the mouse left button to

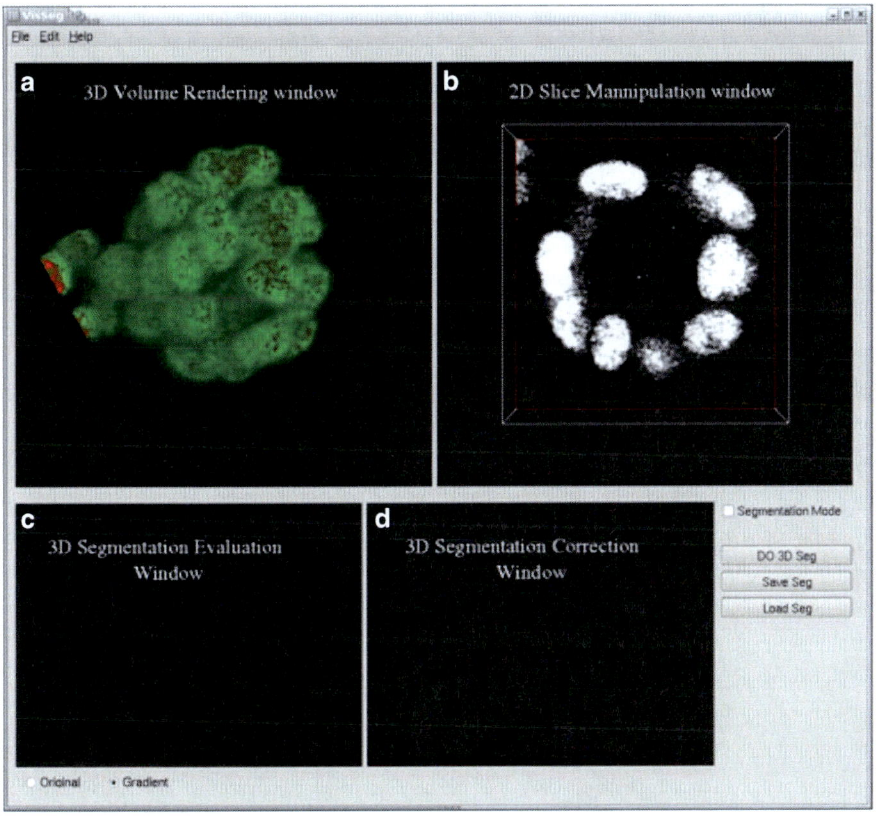

Fig. 7 Loading the image in VisSeg. (**a**) Initial VisSeg GUI 3D volume rendering window showing a MCF-10A sample with nuclear staining. (**b**) 2D slice manipulation window. (**c**) 3D segmentation evaluation window. (**d**) 3D segmentation correction window (refer to Fig. 10c, d for explanations)

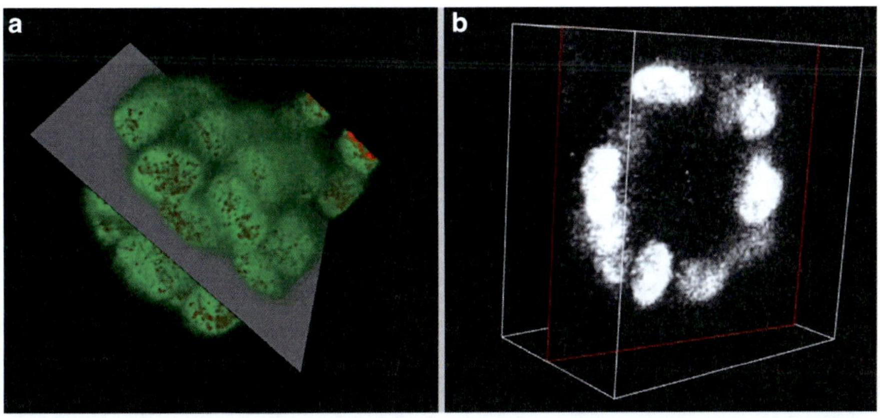

Fig. 8 The 3D rendering window and the 2D manipulation window. (**a**) View of the volume rendering of the nuclear stained MCF-10A sample after view manipulation and showing a 2D plane of view (**b**) 2D slice manipulation window showing the corresponding 2D plane

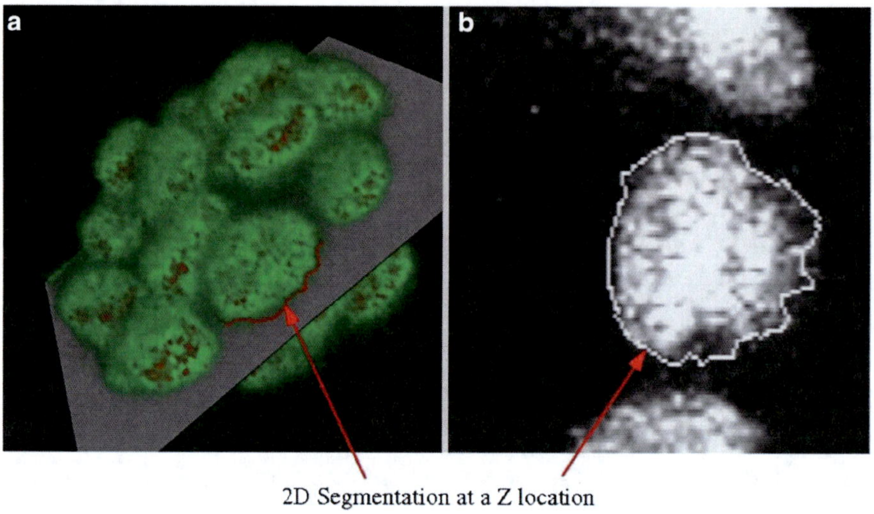

Fig. 9 2D segmentation. (**a**) 2D segmentation of one of the nuclei as viewed in the 3D manipulation window. (**b**) The boundary is shown in a single Z-slice in the 2D slice manipulation window for the MCF-10A sample

mark a point inside the nucleus. Then use the mouse middle button to mark an edge point. The initial 2D segmentation is done immediately after marking the edge point. Correction points can be added using the mouse right button and the updated 2D segmentation is done automatically. The 2D boundary is shown as a red border in the volume rendered image (Fig. 9a) and as a white border in the slice viewer (Fig. 9b).

5. 3D Segmentation: Following 2D segmentation press "Do 3D Seg." After the 3D segmentation is done the screen will show a surface rendering of the segmented object overlaid on the volume rendering (Fig. 10a) and a white border in the slice viewer (Fig. 10b).
6. 3D Segmentation evaluation windows (Fig. 10c, d): The window in Fig. 10c shows the intensities or gradient magnitude from the original image over the entire surface of the 3D segmented object in a single view known as the "global" view. The white regions suggest a good segmentation, while the dark regions show places where the segmentation is possibly wrong and might need manual correction. One can switch between the original intensity values or gradient magnitude values using the radio buttons in the bottom left corner. The gradient image is only used when the object is volume labeled, as is in this example. When the object is surface stained the original image itself is used instead of the gradient image.
7. Evaluation: Move the mouse pointer over the evaluation window (Fig. 10c) with the right button pressed. The corresponding 2D slice of the original or the gradient image is

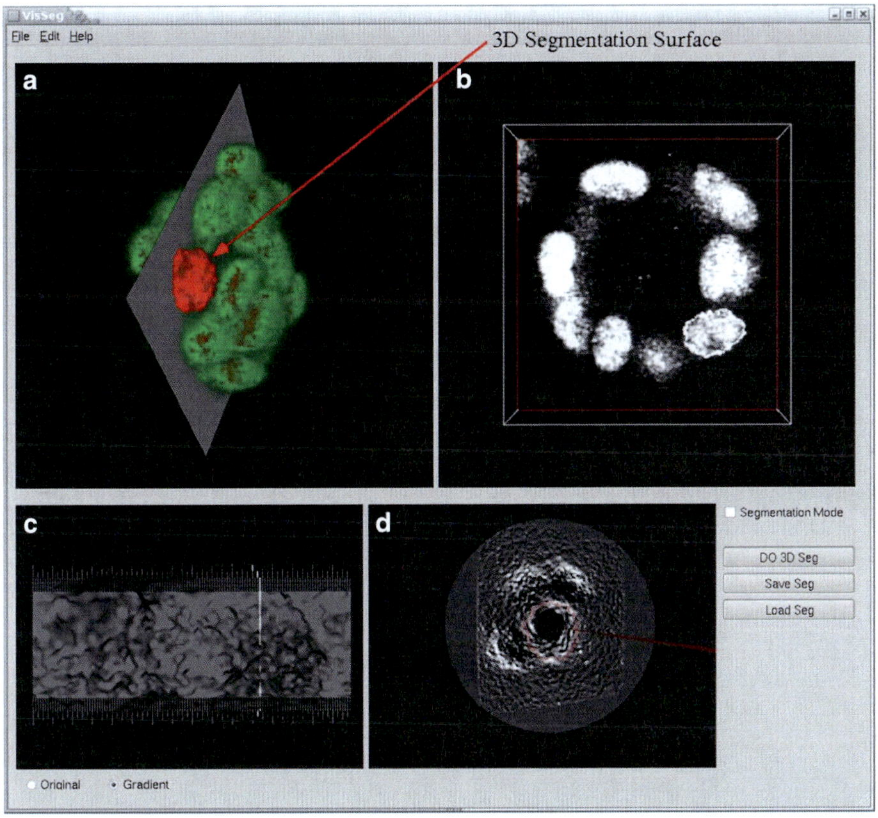

Fig. 10 3D Segmentation. (**a**) VisSeg 3D volume rendering window showing the segmentation surface after completion of 3D segmentation of a nuclei. (**b**) 2D slice showing the 2D segmentation in a specific slice. (**c**) 3D segmentation evaluation window showing the segmentation quality. (**d**) 3D segmentation correction window where additional correction points can be added

shown in the 3D segmentation correction window (Fig. 10d) overlaying the original image and the border of the segmented object. This view clearly shows any deviations between the visually observed object surface and the segmented surface.

8. Correction: The red closed path in the 3D segmentation correction window (Fig. 10d) shows the current segmentation result. If the result is not satisfactory, one can add correction points in this window using the mouse right button so that the final segmentation goes through those points. Any number of correction points can be added. After marking the correction points, press "Do 3D Seg" to correct the 3D segmentation. Figure 11 shows full or partial segmentation of the two samples.

9. Undo: During the segmentation process one can undo the current segmentation using CTRL+U or Edit > Undo. This will sequentially delete the result in a decreasing order of object id.

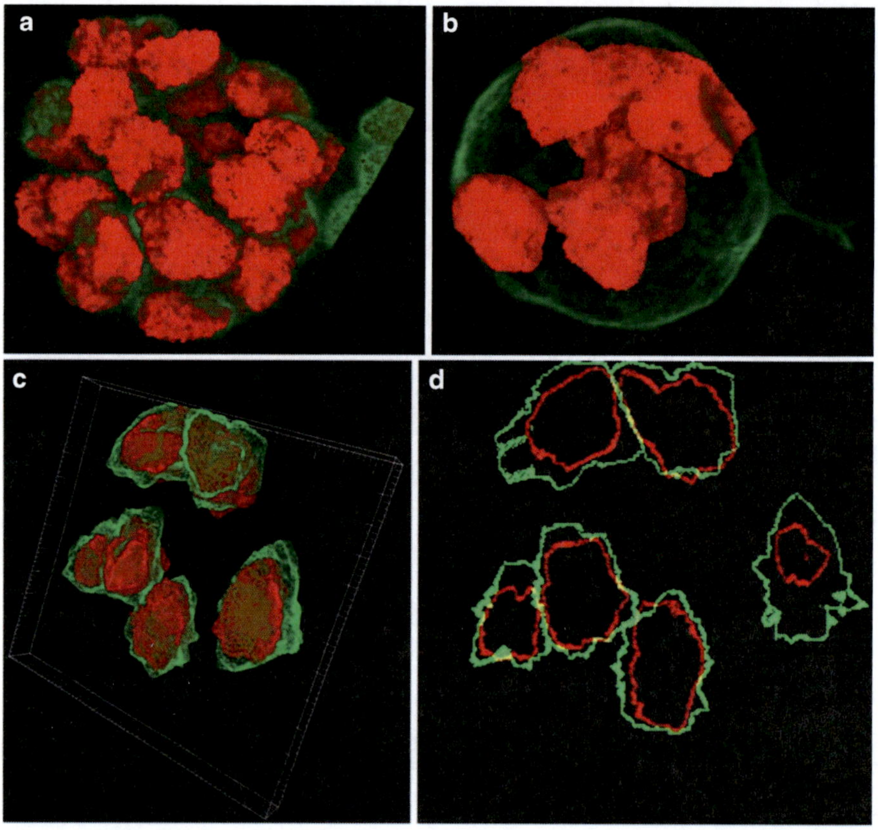

Fig. 11 Segmentation correction. (**a**) Segmentation of all nuclei in the MCF-10A sample using volume staining channel. (**b**) Partial segmentation of whole cells in the MCF-10A sample using surface staining channel. (**c**) Partial segmentation of the somite stage mouse embryo sample with surface staining visualized with Imaris (http://www.bitplane.com/go/products/imaris). (**d**) Slice 13 of the mouse embryo sample showing the nuclear (*red*) and surface (*green*) segmentation

10. Load and Save Segmentation: While performing the segmentation one can save the segmentation done until that point of time by pressing "Save Seg." The segmentation will be saved as BASE_FILE_NAME.seg file. At a later time the user can load this segmentation and continue segmenting it. To do this first load the original image and then press "Load Seg."

6 Concluding Remarks

This chapter illustrates methods and related software for segmentation of individual cells and cell nuclei in developmental tissues. The first two software packages, namely, ImageJ and MIPAV are extremely versatile and offer various image processing and analyzing tools. Although only the relevant portions have been covered here they can be used for a number of other applications. VisSeg on the other hand is a specialized software package which can perform highly accurate and reliable 2D and 3D segmentation of cells and cell nuclei in images with volume or surface fluorescence staining. Segmentation is the starting point for most of the quantitative analysis methods and the methods illustrated in this chapter is a good starting point for such an analysis. Although many other segmentation methods are available we have covered some of the most common and effective ways for segmenting objects in biological images. This kind of quantitative analysis will surely help in gathering more information about complex biological processes.

Acknowledgments

Fluorescence imaging of the MCF-10A sample was performed at the National Cancer Institute Fluorescent Imaging Facility. The cell line was a gift from Dr S. Muthuswamy (Cold Spring Harbor Laboratory). This project was funded in whole or in part with federal funds from the National Cancer Institute (NCI), National Institutes of Health under contract N01-CO-12400. The content of this publication does not necessarily reflect the views or policies of the Department of Health and Human Services and nor does mention of trade names, commercial products, or organizations imply endorsement by the US Government. NCI-Frederick is accredited by AAALAC International and follows the Public Health Service Policy for the Care and Use of Laboratory Animals. Animal care was provided in accordance with the procedures outlined in the "Guide for Care and Use of Laboratory Animals" (National Research Council; 1996; National Academy Press, Washington, D.C.). This research was supported [in part] by the Intramural Research Program of the NIH, National Cancer Institute, Center for Cancer Research. We would like to thank Dr. Tom Mistelli for his continuous support and feedback.

Appendix

Sample preparation and image acquisition. Two biological samples were used in this study. One was an early somite stage (E8.0–8.5) mouse embryo obtained from an NIH Swiss female mouse. The embryo was fixed with 2 % paraformaldehyde/phosphate buffered saline (PBS) for 30 min at room temperature and stained with Oregon Green 488 Phalloidin (O-7466, Molecular Probes, Eugene, OR, USA) after treatment with 0.1 M glycine/0.2 % Triton X-100/PBS for 10 min. Phalloidin predominantly labels filamentous actin found abundantly at the cell surface underlying the cell membrane. The embryo was counter stained with the DNA dye 4′,6-diamidino-2–phenylindole (DAPI). The posterior primitive streak region of the embryos was manually dissected and mounted with SlowFade Light Antifade kit (S-7461, Molecular Probes) on a glass slide. Images were acquired using a 40×, 1.3 numerical aperture oil objective lens, and pinhole of one airy unit on an LSM 510 confocal microscope (Carl Zeiss Inc., Thornwood, NY, USA). Excitation was with a 488-nm laser light and emitted light between 500 and 550 nm was acquired. Thus, the effective thickness of the optical sections was 425 nm at the coverslip but increased further away from the coverslip. Pixel size was 0.14 μm in the x and y dimensions. DAPI was excited with two photon excitation at 780 nm and emitted light between 420 and 480 nm was collected.

The other biological sample used in this chapter was a 3D cell culture model of early breast tumorigenesis. MCF-10A.B2 a human mammary epithelial cell line was grown on basement membrane extract (Trevigen Inc., Gaithersburg, MD) for 20 days. MCF-10A. B2 cells express a synthetic ligand-inducible active ErbB2 variant. To induce tumorigenesis the synthetic ligand (AP1510; ARIAD Pharmaceuticals Inc)) was added for the final 10 days of culture. The acini structures formed by growth in 3D culture were fixed in 2 % paraformaldehyde/PBS for 20 min at room temperature and permeabilized by a 10 min incubation in 0.5 % Triton X-100 at 4 °C. Following three 15 min rinses in 100 mM glycine/PBS, the acini were blocked in IF buffer (0.1 % bovine serum albumin/0.2 % Triton X-100/0.05 % Tween-20/PBS) containing 10 % fetal bovine serum (Invitrogen) for 1 h. The cells were then incubated overnight in anti-integrin alpha 6 antibody (Chemicon International). Integrin alpha 6 stains for basolateral polarity, it stains the cell surface membrane with strong basal and weaker lateral staining. After three 20 min washes in IF buffer the cells were incubated in Alexa Fluor 488 donkey anti-rat (Invitrogen). Samples were mounted in DAPI containing VECTASHIELD mounting media (Vector Laboratories Ltd.) after three further 20 min washes in IF buffer. Images were acquired using a 63× 1.4-NA oil objective lens on an LSM 510 confocal microscope.

References

1. Pawley J (2006) Handbook of biological confocal microscopy, 3rd edn. Springer, New York
2. Russ JC (2002) The image processing handbook, 4th edn. CRC, Boca Raton, FL
3. Castleman KR (1995) Digital image processing. Prentice Hall, Englewood Cliffs, NJ
4. Delft University of Technology, Quantitative Imaging Group interactive image processing course. http://www.tnw.tudelft.nl/fileadmin/Faculteit/TNW/Over_de_faculteit/Afdelingen/Imaging_Science_and_Technology/Research/Research_Groups/Quantitative_Imaging/Education/doc/FIP2_3.pdf
5. Merks RMH, Brodsky SV, Goligorksy MS, Newman SA, Glazier JA (2006) Cell elongation is key to in silico replication of in vitro vasculogenesis and subsequent remodeling. Dev Biol 289:44–54
6. Heid PJ, Voss E, Soll DR (2002) 3D-DIASemb: a computer-assisted system for reconstructing and motion analyzing in 4D every cell and nucleus in a developing embryo. Dev Biol 245:329–347
7. Blacher S, Devy L, Burbridge MF, Roland G, Tucker G, Noël A, Foidart J-M (2001) Improved quantification of angiogenesis in the rat aortic ring assay. Angiogenesis 4:133–142

8. Peng H, Long F, Zhou J, Leung G, Eisen MB, Myers EW (2007) Automatic image analysis for gene expression patterns of fly embryos. BMC Cell Biol 8(Suppl 1):S7
9. ImageJ website http://rsb.info.nih.gov/ij/
10. MIPAV website http://mipav.cit.nih.gov/index.php
11. Kass M, Witkin A, Terzopoulos D (1988) Snakes: active contour models. Int J Comput Vision 1:321–331
12. Blake A, Isard M (1998) Active contours. Cambridge University Press, New York, NY
13. McAuliffe MJ, Lalonde FM, McGarry D, Gandler W, Csaky K, Trus BL (2001) Medical image processing, analysis and visualization in clinical research. Proceedings 14th IEEE symposium on computer-based medical systems, pp 381–386
14. Baggett D, Nakaya M, McAuliffe M, Yamaguchi TP, Lockett S (2005) Whole cell segmentation in solid tissue sections. Cytometry A 67A: 137–143
15. McCullough DP, Gudla PR, Harris BS, Collins JA, Meaburn KJ, Nakaya M, Yamaguchi TP, Misteli T, Lockett SJ (2008) Segmentation of whole cells and cell nuclei from 3D optical microscope images using dynamic programming. IEEE Trans Med Imaging 27:723–734
16. Dean P, Mascio L, Ow D, Sudar D, Mullikin J (1990) Proposed standard for image cytometry data files. Cytometry 11:561–569

Chapter 17

Protein/Peptide Transduction in Metanephric Explant Culture

Sergey Plisov, Honghe Wang, Nadya Tarasova, Nirmala Sharma, and Alan O. Perantoni

Abstract

While gene targeting methods have largely supplanted cell/explant culture models for studying developmental processes, they have not eliminated the need for or value of such approaches in the investigator's technical arsenal. Explant culture models, such as those devised for the metanephric kidney and its progenitors, remain invaluable as tools for screening regulatory factors involved in tissue induction or in the inhibition of progenitor specification. Thus, some factors capable of inducing tissue condensations or nephronic tubule formation in explants of metanephric mesenchyme have been identified through direct treatment of cultures rather than lengthy genetic engineering in animals. Unfortunately, renal progenitors are largely refractory to most contemporary methods for gene manipulation, including transfection and viral transduction, so the applications of explant culture have been rather limited. However, methods for protein or peptide transduction offer greatly improved efficiencies for uptake and expression/regulation of proteins within cells and tissues. Biologically active TAT- or penetratin-fusion proteins/peptides are readily taken up by most cells in metanephric explants or monolayer cultured cells (Plisov et al., J Am Soc Nephrol 16:1632–1644, 2005; Osafune et al., Development 133:151–161, 2006; Wang et al., Cell Signal 22:1717–1726, 2010; Tanigawa, Dev Biol 352:58–69, 2011), allowing a direct functional evaluation of theoretically any protein, including biologically active enzymes and transcription factors, or any targeted interactive domain within a protein.

Key words Protein transduction, Cell-penetrating peptides, TAT, Penetratin, Kidney, Stat, Cited

1 Introduction

The successful ectopic expression of a gene dictates not only an efficient delivery system for a variety of cell types but also strong expression of the transgene once delivered. Furthermore, it is difficult to titrate expression levels accurately with genetic constructs, and there is a nagging problem of cell transfectability especially in primary tissues. To circumvent these complications, techniques have been developed that instead efficiently deliver entire biologically active fusion proteins or peptides directly using polycationic

Mark Lewandoski (ed.), *Mouse Molecular Embryology*, Methods in Molecular Biology, vol. 1092,
DOI 10.1007/978-1-60327-292-6_17,

© Springer Science+Business Media New York 2014

cell-penetrating peptides (CPPs)/protein transduction domains. The HIV-1 Transactivator of Transcription (TAT) protein or *Drosophila* homeoprotein Antennapaedia (penetratin) contain an arginine-rich domain, which, when fused with a biologically active protein/peptide, oligonucleotide, or siRNA, greatly enhances the translocation of the fused element into virtually any mammalian cell [1]. The wide uptake spectrum is thought to result from binding of the CPP to negatively charged cell-surface proteoglycans, such as heparan sulfate-modified proteins. In the case of TAT protein, the active 11-amino acid sequence (residues 47–57) has been used to transduce not only peptides but also biologically active full-length proteins (some as large as 120 kDa) into cultured tissues and into rodents systemically following intraperitoneal injection [2, 3]. On the other hand, the 16-amino acid domain in penetratin (residues 43–58) has been applied primarily to peptide or oligonucleotide [1]. These shorter peptides, however, can be generated by direct chemical synthesis using D-amino acids, which are more resistant to proteolytic degradation within the cell and therefore tend to be significantly more active.

Uptake by cells of TAT- or other CPP-fusion proteins is a multistep process involving TAT-domain binding to cell surface molecules, such as glycans, followed by macropinocytosis, a type of actin-mediated fluid phase endocytosis that is carried out by membrane protrusions that engulf fluid surrounding the cell [4]. This event appears to be independent of glycan interactions, since TAT-protein transduction can occur in glycan-deficient cells [5]. The process then brings the TAT cargo into endosomal vesicles, which must be perturbed in some manner to allow release of the fusion protein into the cytoplasm or the nucleus for biological activity. It has been demonstrated that chemical permeabilization of endosomal membranes can induce the release of cargo/content, markedly decreasing the effective dose required for biological activity following transduction. It is this release from endosomes that appears to be the rate-limiting step in protein transduction, suggesting that activity of CPP-fusion proteins can be enhanced by endosomolytic agents.

While efforts to use protein transduction for therapeutic purposes are ongoing, the application of this technology remains fairly limited despite its apparent efficacy [6–9]. In our own studies, we have observed highly efficient uptake of a FITC-labeled Cited1 TAT-fusion protein (Fig. 1) or a rhodamine-labeled penetratin-Stat1-fusion peptide (Fig. 2). In both cases, transduction is observed in the preponderance of cells within renal progenitor explant cultures, and both reagents regulate differentiation in the transduced tissues [6, 8]. Moreover, we were able to design a peptide to specifically target Stat1 but not Stat3 transcriptional activation, and a similarly targeted Stat3 peptide that inhibited Stat3 but not Stat1 signaling, [8, 10] so these molecules can be highly

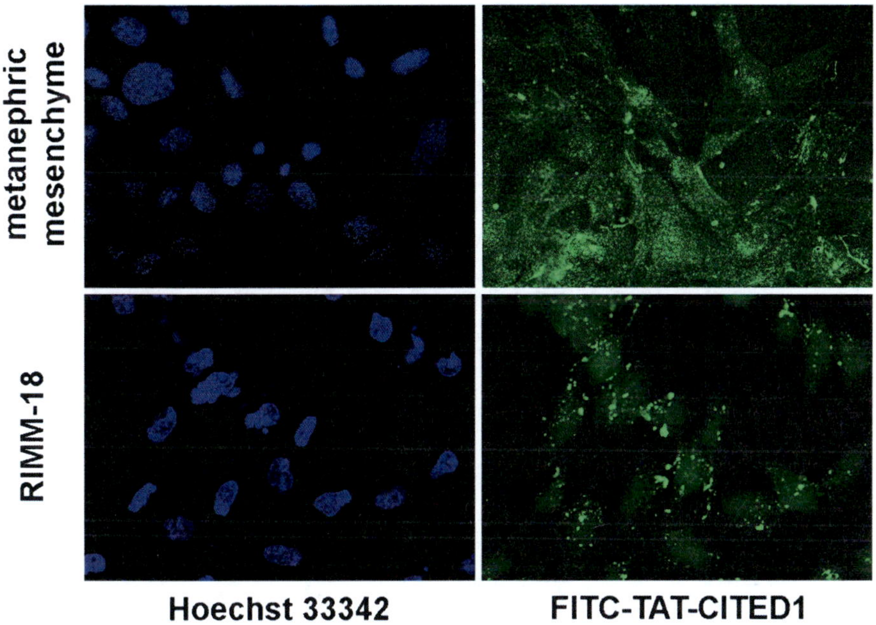

Fig. 1 Full-length TAT-Cited1 is taken up by most cells in explant or monolayer cultures. Cells were nuclear stained with Hoechst 33342 or treated overnight with FITC-labeled TAT-Cited1 protein and imaged by confocal microscopy. Cells in explants of metanephric mesenchyme (*upper panels*) or monolayer cultures of the immortalized line Rat Inducible Metanephric Mesenchyme-18 (RIMM-18, *lower panels*) show a fairly uniform intracellular distribution of labeled protein. *Large bright spots* of labeling are surface precipitates

specific. Finally, using this approach, it is also possible to target a particular domain within a protein and alter its ability to interact with other macromolecules at that region, thus providing a tool for biochemically dissecting a specific domain-related protein function in viable cells. For example, helix-C, a putative 13th Armadillo repeat in β-catenin, is believed to be principally responsible for its interaction with a TCF family member in activating canonical Wnt signaling. A peptide designed to target that domain completely blocks TCF-dependent transcription without apparently affecting β-catenin's critical role in cell adhesion, since Wnt-stimulated mesenchymal cells convert to epithelial tubules despite the presence of the peptide inhibitor [9]. This approach should allow us to independently examine each functional domain within any given protein.

Production of TAT proteins is generally accomplished by expression in and purification from bacterial cultures, and protocols are well described by the developer of this technology, Steven Dowdy [2]. The advantages of this approach have already been described, and its application seems virtually unlimited, as both nuclear and cytoplasmic proteins have been targeted. The complications of this approach are somewhat similar to those encountered in efforts to concentrate a protein by ultrafiltration, i.e., the tendency of concentrated proteins or peptides to self-associate

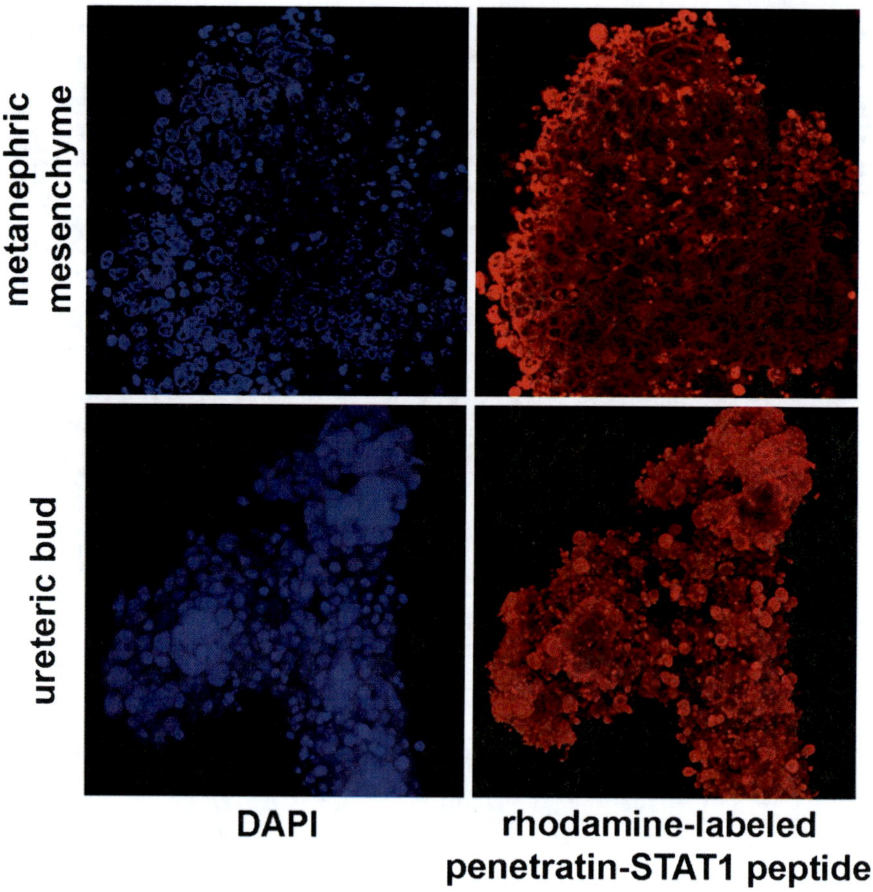

Fig. 2 A 26-amino acid penetratin-Stat1 fusion peptide readily permeates and is taken up by cells in cultured tissue explants. Cells were nuclear stained with DAPI or treated for 3 h with a rhodamine-labeled penetratin peptide and imaged by confocal microscopy. Cells in explants of intact metanephric mesenchymes (*upper panels*) or ureteric buds (*lower panels*) incorporated labeled peptide in virtually all cells, although labeling was most prominent in cells at the tissue periphery

and precipitate, rendering them inactive. This is combated with the use of strong denaturants such as urea, which must then be removed or greatly diluted prior to tissue treatment. As Dowdy indicates, there is no single protocol for every TAT-fusion protein. Conditions must be optimized for each protein produced to ensure uptake and biological activity. We will outline the procedures that we have found useful in producing fusion proteins or peptides for explant culture studies. These tissues are quite susceptible to toxicity, so removal of bacterial contamination is critical. Furthermore, they are cultured on small-pored polycarbonate filters floated on culture medium, so tissue penetration can be a problem. This seems a greater problem for intact kidney rudiments as opposed to separated progenitors, i.e., ureteric buds or metanephric mesenchymes, as FITC-labeled peptides are restricted only from the ureteric bud in intact metanephroi (not shown). However, it would seem that even this problem can be circumvented by microinjection directly into the lumen of the ureteric bud [11].

2 Materials

2.1 Sequences for TAT Cloning

1. Forward and reverse oligonucleotide primers for gene of interest. Primers should be designed to include restriction sites that allow for proper frame reading and orientation within the vector. We have used primers for the transcriptional co-activator Cited1 as the gene of interest. Rat Cited1-Forward primer: 5′-d(GAT**GGTACC**ATGCCAACTATGTC GAGGC)-3′; Rat Cited1-Reverse primer: 5′-d(GAT**CTCGA G**TCAGCAGCCAGAGGGAAAATC)-3′. These contain sequences that are recognized in duplex by restriction enzymes **KpnI** and **XhoI**.
2. Novex 10 % polyacrylamide/Tricine/SDS gels (Invitrogen).
3. Agarose, GenePure LE (BioExpress).
4. Ethidium bromide solution (10 mg/ml stock).
5. TE buffer: 10 mM Tris–HCl, 10 mM EDTA, pH 8.0; TAE buffer (50× stock from Quality Biological, Inc).
6. Na-acetate.
7. 5× DNA loading buffer: 80 % glycerol, 10 mM Tris–HCl, 10 mM EDTA, pH 8.0.
8. Ultrafree-DA centrifugal filter units (Millipore).

2.2 Cloning into TAT Vectors

1. pTAT and pTAT-HA vectors contain a 6× Histidine coding sequence upstream of the TAT or TAT-HA coding region followed by a polylinker region with unique cloning sites to insert genetic sequences of interest. These vectors were obtained from Stephen F. Dowdy (Howard Hughes Medical Institute and Department of Cellular and Molecular Medicine, University of California, San Diego School of Medicine, La Jolla, CA 92093-0686).
2. 10× Multicore buffer (Promega).
3. Various restriction enzymes for modifying PCR fragment and confirming insert identity.
4. DNA ligase.
5. 2× Rapid Ligation Buffer (Promega).
6. Calf intestinal alkaline phosphatase.
7. *Escherichia coli* strain TOP10 (Invitrogen).
8. Ampicillin is dissolved in distilled water at 100 mg/ml (1,000× stock) and stored frozen.

9. LB medium is prepared using 25 g powdered LB mix (Sigma) dissolved in 1 l distilled water and autoclaved. For selection, ampicillin is added to cooled medium to a final concentration of 100 μg/ml.
10. LB agar plates are prepared from 35 g powdered LB agar (Sigma) dissolved in 1 l distilled water and autoclaved. Add ampicillin as above for selection.
11. Super Optimal Catabolite (SOC) medium (Invitrogen).
12. T7 forward primer: 5′-d(TAATACGACTCACTATAGG)-3′.
13. Qiagen Qiaprep Spin MiniPrep kit.

2.3 Bacterial Production of Fusion Protein

1. L-arabinose.
2. *Escherichia coli* strains BL21(DE), BL21 Star™ (DE3)pLysS cells [F–*ompT* *hsd*SB(rB–, mB–) *gal dcm rne*131 (DE3) pLysS (CamR)], and BL21-AI™ F–*ompT* *hsd*SB(rB–, mB–) *gal dcm ara*B::*T7RNAP-tetA* (Invitrogen).
3. Denaturing Buffers: 8 M urea, 100 mM NaCl, 10 mM imidazole, and 20 mM HEPES, pH 7.4, 6.0, 5.3, or 4.0.
4. Nickel-chelating ProBond resin (Invitrogen).
5. FITC-labeling kit (Invitrogen).

3 Methods

The Methods describe the creation, expression, and purification of TAT-fusion proteins as well as the use of the TAT-fusion proteins to treat explanted metanephric progenitors. Full-length murine Cited1 cDNA is used as an example and was produced from E11.5–12.5 mouse or E13.5–14.5 rat embryos. The pTAT–*Cited1* bacterial expression vector was generated by cloning *Cited1* cDNA in frame into the vector pTAT-HA (hemagglutinin antigen tag), which includes 5′-ATG-(His)×6-TAT-HA-MCS cassette to enable purification and detection of the recombinant TAT-fusion protein.

3.1 Gene Cloning into TAT Vectors

1. Gene sequences are amplified using standard RT-PCR methods from total RNA derived from E11.5–12.5 mouse or 13.5–14.5 rat embryos (*see* **Note 1**). Primers (shown in Subheading 2) include sequences that encode sites for restriction enzyme digestion. Amplified PCR fragments are prepared for insertion into a preexisting multiple cloning site by restriction digestion with enzymes KpnI and XhoI. This provides for a proper reading frame and orientation in the vector. Incubation mixes included the *Cited1* PCR fragment (40 μL, ~2 μg), 10× Multicore buffer (5 μL), 10 U/μL KpnI (2 μL), and 10 U/μL XhoI (2 μL). Mixes are incubated for 3 h at 37 °C.

2. Digested fragments are then mixed with 5× DNA loading buffer and gel purified on 0.8 % agarose/TAE gel containing ethidium bromide (0.5 μg/mL).
3. Gels are run for 1 h at 100 V and visualized with UV light.
4. The appropriate band is cut from the gel with a clean scalpel or razor blade.
5. To purify the DNA fragment (~700 bp for *Cited1*) from the agarose, the block containing the DNA is placed in an Ultrafree®-DA filter and spun for 10 min at 2,700 × *g*.
6. The flow-through containing purified cDNA insert is collected and the DNA concentration determined.
7. To prepare the pTAT-HA vector for cDNA sequence insertion, it is first linearized with restriction enzymes KpnI and XhoI in the following reaction mix: pTAT-HA (10 μg), distilled water (26 μL), 10× Multicore buffer (10 μL), 10 U/μL KpnI (2 μL), and 10 U/μL XhoI (2 μL). This is incubated for 3 h at 37 °C.
8. The linearized vector is then isolated by precipitation with three volumes of ice-cold 100 % ethanol in the presence of 0.3 M Na-acetate, centrifuged at 21,000 × *g*/4 °C, washed in 70 % ethanol, and dissolved in 50 μL TE buffer.
9. The purified linearized vector is then dephosphorylated to prevent vector self-ligation. This is accomplished in the following reaction: vector (10 μg in 50 μL TE buffer), distilled water (38 μL), 10× calf intestine phosphatase buffer (10 μL), and 10 U/μL calf intestine phosphatase (2 μL). This is incubated for 1 h at 37 °C.
10. The enzyme is removed with phenol/chloroform extraction and ethanol precipitation, as described above and the dephosphorylated vector dissolved in 50 μL TE buffer. The vector is further gel purified in agarose to remove the short XhoI/KpnI-generated DNA fragments as described in **steps 2–6** of this section and diluted to 1 μg/10 μL in distilled water.
11. Ligation of the insert into the vector is accomplished in the following reaction mix: 1 μL (0.1 μg) of pTAT-HA from **step 10** of this section, 2 μL (~0.5 μg) of PCR fragments from **step 6** of this section, 5 μL of 2× Rapid Ligation Buffer, 1 μL of distilled water, and 1 μL of DNA ligase. The mix is incubated overnight at 10 °C.

3.2 Generating pTAT Constructs in *E. coli* Cells

1. For bacterial transformation, TOP10 chemically competent cells (50 μL) are incubated on ice for 1 h with 1 μL of the ligation mixture from **step 11** in the previous section.
2. This mixture is then heat shocked at 42 °C for 45 s, returned to the ice, diluted with 1 mL SOC medium, and incubated for 1 h at 37 °C with shaking.

3. Cells are then pelleted at 10,600 × *g* and resuspended in 50 μL of SOC medium.
4. The suspension is plated on LB agar containing 100 μg/mL ampicillin and plates incubated at 37 °C overnight. Plates with colonies may be stored for several days at 4 °C prior to analysis.
5. For colony analysis, 10–20 colonies are selected from the plates with sterile tips and re-plated on a fresh LB-agar Master plate and labeled for later identification.
6. The remainder of each colony is resuspended in a PCR tube with 20 μL PCR mix as follows for confirmation by PCR amplification: 2 μL of 10× PCR buffer, 0.5 μL of 25 mM dNTPs, 12 μL of distilled water, 1 μL of 50 mM MgCl2, 0.5 μL of 25 μM T7 Forward primer, 0.5 μL of 25 μM *Cited1* Reverse primer, 0.05 μL of 5 U/μL Taq polymerase, and ~0.2 μL of transformed *E. coli* cells.
7. PCR reactions are run as described in Subheading 3.1, **step 6** and analyzed according to Subheading 3.2, **steps 2** and **3**.
8. Colonies (2–4), which produce the appropriately sized PCR fragment (~700 bp for *Cited1*), are then selected from the Master plate and grown overnight in LB/Amp medium.
9. Plasmid DNA is then purified using a Qiagen Qiaprep Spin MiniPrep kit, according to the manufacturer's instructions, analyzed with restriction enzymes (KpnI, XhoI, and others as appropriate), and sequenced to verify the identity and integrity of the plasmids.

3.3 Production and Purification of TAT-Fusion Protein

1. Mix 1 μL of 10 ng/μL TAT-Cited1 plasmid DNA with 50 μL of competent *E. coli* BL21-AI cells and follow the transformation protocol described above in Subheading 3.3, **steps 1–4** with this new cell line (*see* **Note 2**).
2. Colonies (2–4) are selected and grown overnight in 5 mL LB/Amp at 37 °C with shaking.
3. The overnight cultures are then diluted 1/100 in LB/Amp and grown to an OD_{600} that approximates 0.6–0.8 optical units against a LB blank.
4. As mentioned, BL21-AI cells require L-arabinose to induce the expression of T7 RNA polymerase and therefore TAT-Cited1 protein. To accomplish this, arabinose is added to a final concentration of 0.2 % and cultures are shaken at 37 °C for an additional 4 h.
5. Cells are then pelleted by centrifugation at 5,000 rpm for 30 min at 4 °C, resuspended, and washed in cold PBS, and again pelleted.

6. The washed pellet is resuspended and lysed in 8 mL of Denaturing Lysis Buffer (8 M urea, 20 mM HEPES, pH 7.4, 100 mM NaCl, and 10 mM imidazole). To facilitate dissociation, the suspension is also sonicated on ice, 4× for 1 min at the maximum power setting followed by a 2-min interval to cool the lysates (*see* **Note 3**).
7. Bacterial debris is removed by centrifugation (10,600 × *g* for 15 min) and the supernatant collected and incubated overnight with 1 mL nickel-chelating ProBond resin, which was washed with Denaturing Lysis Buffer.
8. A 1-mL slurry of resin with bound TAT fusion-protein is poured into a 10-mL column such as provided with the Invitrogen ProBond kit and washed first with 6 mL of Denaturing Lysis Buffer. The effluent should be collected in the event that the TAT-fusion protein does not bind to the column matrix.
9. The column resin is then washed 2× with 6 mL of Denaturing Wash1 Buffer (8 M urea, 20 mM HEPES, pH 6.0, 100 mM NaCl, and 10 mM imidazole) and the effluent collected from each wash.
10. The resin is further washed 2× by mixing with 4 mL Denaturing Wash2 Buffer (8 M urea, 20 mM HEPES, pH 5.3, 100 mM NaCl, and 10 mM imidazole) and the effluent collected from each wash.
11. For elution, 5 mL of Denaturing Elution Buffer (8 M urea, 20 mM HEPES, pH 4.0, 100 mM NaCl, and 10 mM imidazole) are passed through the column and collected as 1 mL fractions. 10 μL of each wash and elution fractions are analyzed in 10 % SDS-PAGE gels for quantity and purity by staining with Coomassie blue. Dowdy reported that unfolding and rapid refolding of purified TAT-fusion proteins improves protein transduction efficiency probably by unmasking the TAT-PTD domain. He recommends either desalting on a PD-10 column or elution from an HPLC ion-exchange column. To rapidly remove denaturant (urea) and salts from the TAT-Cited1 preparation, we instead precipitated the protein with ethanol.
12. For this, combine fractions containing TAT-Cited1 and precipitate protein with nine volumes of ice-cold ethanol. Keep on dry ice for 30 min. Do not allow the solution to freeze.
13. The solution is then spun at 5,000 × *g* for 30 min/4 °C to collect the protein, and the precipitate is washed with ice-cold 90 % ethanol and dissolved in 0.2 M $NaHCO_3$, pH 5.5 for FITC labeling or phosphate-buffered saline (PBS) for use in culture.
14. The dissolved preparation in then spun for 10 min/4 °C at 21,000 × *g* to remove insoluble protein and quantified.

15. It is often useful to FITC-label proteins in order to assess uptake and intracellular localization. This was performed on the same preparations using a FITC-labeling reagent (Invitrogen/Molecular Probes). For this, 100 μL of TAT-fusion protein (200 μg) in 0.2 M $NaHCO_3$ buffer, pH 5.5, is mixed with 20 μL of active FITC-labeling reagent (10 mg/mL DMSO) for 30 min at room temperature. To remove unincorporated label, the protein is precipitated with nine volumes of ice-cold ethanol and washed in 90 % ice-cold ethanol. Protein is solubilized in sterile PBS and stored at 4 °C for use in culture. For long-term storage, proteins should be stabilized in 10 % v/v glycerol and frozen at −80 °C in small aliquots to avoid multiple freeze–thaw cycles (*see* **Note 4**).
16. TAT-fusion proteins are generally applied to cultures (monolayer or explant) at concentrations between 0.05 and 1 μM and are often toxic at higher concentrations. Levels for should optimized though for each case (*see* **Notes 5–7**).

3.4 Penetratin Fusion Peptide Synthesis

Another approach to molecular regulation involves the targeting of specific protein-binding domains using peptidomimetics or decoys that mimic the normal binding partner of a protein [12]. It is well beyond the scope of this chapter to describe the chemical modeling that is used to rationally design these reagents. Suffice it to say that by increasing our understanding of these interactions and relationships, we also gain insight into targeting them. In our own work, we have applied structural information from X-ray crystallography and NMR studies in the design of peptides that disrupt oligomerization of the Stat protein. Specifically, we have a generated a library of cell-permeable peptides that target the 12-amino acid second α-helix of Stat1 or Stat3 [8, 10]. The peptides produced consist of the 14–16 amino acids covering helix 2 of the Stat molecule fused at the C-terminus with 16 amino acids from penetratin (residues 43–58). Since the synthesis requires instrumentation that is not readily available to most molecular biology labs, we will provide primarily experimental details sufficient for a chemistry lab. Such peptides may be custom synthesized by businesses with this expertise, for example AnaSpec.

1. Peptides are generated by solid-phase peptide synthesis on a 433A Peptide Synthesizer (Applied Biosystems) equipped with a conductivity monitoring unit utilizing Fmoc amino acid derivatives (AnaSpec). The synthesis is performed with conditional blocking of unreacted amino groups with acetic anhydride for easier purification of the resulting peptides (*see* **Note 8**).
2. Peptides are cleaved from the resin with 87.5 % trifluoroacetic acid containing 5 % water, 5 % thioanisol, and 2.5 % triisopropyl-

silane (v/v), precipitated with cold diethyl ether, washed five times with ether, and lyophilized overnight.

3. Peptides are dissolved in dimethylformamide and purified by HPLC on a preparative (25 × 250 mm) Atlantis C18 reverse-phase column (Agilent Technologies) in a gradient of 0.05 % (v/v) trifluoroacetic acid in water and acetonitrile containing 0.05 % trifluoroacetic acid.
4. Fractions are analyzed by electrospray LC/MS in an Agilent 1100 series instrument (Agilent Technologies) with the use of a Zorbax 300SB-C18 Poroshell column and a gradient of 5 % acetic acid in water and acetonitrile.
5. Fractions containing more than 95 % pure product are combined and freeze-dried from 5 % acetic acid to ensure conversion into acetate salts.
6. The purity and structure are confirmed by LC/MS with separation in a Zorbax 300SB-C18 analytical column and Microsorb-MW 300A C8 (Varian) column.
7. For fluorescence labeling, purified peptides with C-terminal Cys are reacted with maleimide-tetramethylrhodamine (Invitrogen) added at 10 % excess in acetonitrile–PBS, 1:1, pH 7.2.
8. After overnight incubation at 25 °C, the reaction mixture is injected onto a preparative HPLC column and labeled peptides are purified as described above.
9. Peptides are solubilized in 100 % DMSO at 10 mM and stored at −20 °C (*see* **Note 9**).

3.5 Tissue Explantation

1. E11.5–12.5 mouse or E13.5–14.5 rat embryos from timed-mated animals are generally used for studies of the developing kidney. These periods represent the early stages of metanephric development for both species. Metanephroi can be removed from embryos during this period, and progenitor ureteric bud separated from metanephric mesenchyme if necessary.
2. Separations of metanephric progenitor populations and conditions for tissue cultivation are well described in ref. 13.
3. Tissues are placed on type IV collagen-coated 0.1 μM pore-size polycarbonate filters (*see* **Note 10**). Filters are collagen coated by soaking the surface with a concentrated solution of mouse collagen (BD Biosciences) and allowing them to dry.
4. Intact metanephroi are best cultured in Dulbecco's modified Eagle's medium–Ham's F12 medium supplemented with 10 % fetal bovine serum. Separated progenitors are better maintained in serum-free or low serum (1–2 %) conditions supplemented with factors described in our previous publication shown above. Tissues are grown in a humidified incubator with 7 % CO_2 at 37 °C.

4 Notes

1. The sequence of interest from *Cited1* was PCR amplified in a mix of 10× PCR buffer (2 μL), 25 mM dNTPs 0.5 μL, 50 mM MgCl2 (1 μL), distilled water (12 μL), cDNA (0.25 μg), 0.1 mM Forward primer (0.5 μL), 0.1 mM Reverse primer (0.5 μL), and 5 U/μL Taq polymerase (0.05 μL). For *Cited1*, incubation mixes were heated to 95 °C for 3 min, and amplified using 37 cycles of 92 °C/1 min, 50 °C/1 min, and 72 °C/2 min with a final elongation step of 72 °C for 5 min.
2. To produce recombinant TAT-Cited1 fusion protein, we initially used the recommended *E. coli* line BL21(DE). This cell line requires isopropyl β-D-1-thiogalactopyranoside (IPTG) in the growth medium to induce expression of T7 RNA polymerase, which, in turn, drives the transcription of TAT-mRNA. The BL21(DE) cells failed to express TAT-Cited1 protein. Apparently these cells are leaky for T7 RNA polymerase expression and therefore produce some TAT-Cited1 protein even in the absence of IPTG induction. The protein was toxic for the cells expressing it and selected against their growth in culture, resulting in the overgrowth of cells that did not express it. To reduce the expression of TAT-Cited1 in cells prior to induction with IPTG, we used another *E. coli* cell line BL21 Star™ (DE3)pLysS. These cells express pLysS to lower the basal levels of T7 polymerase that drives expression of the target protein (TAT-Cited1). The expression of TAT-Cited1 in these cells was also poor. The only cell line that allowed strong inducible expression of TAT-Cited1 was BL21-AI™. The strain carries a chromosomal insertion of a cassette containing the T7 RNA polymerase gene in the araB locus, allowing expression of T7 RNA polymerase to be tightly regulated by the araBAD promoter, which is active only if L-arabinose is present in the growth medium. This cell line is recommended for expression of toxic genes as it provides the tightest control of T7 expression.
3. A protease inhibitor cocktail should be added to preparations especially in the absence of denaturant.
4. TAT-fusion proteins are unstable and suffer the same technical issues faced in protein purification in general. They tend to precipitate at high concentrations or with repeated freeze–thaw cycles. For this reason, preparations should be centrifuged prior to addition to culture medium, and confocal microscopy should be applied to confirm cellular uptake. Furthermore, biological activity, when possible, should be assessed after one freeze–thaw cycle
5. Uptake of TAT-fusion proteins occurs quite rapidly, often detectable within minutes of application and peaks in a matter of hours.
6. As mentioned in the Introduction, transduced proteins/peptides collect in endosomes and their release may be rate

limiting in elicited a biological effect. For that reason, it may be useful to try endosomolytic agents such as Endo-Porter (Gene Tools, LLC) to facilitate their intracellular release and enhance their biological activity.

7. Because of potential toxicity, some effort should be directed to producing appropriate control reagents. These could include a TAT-fusion protein generated from a mutated biologically inactive form of the protein of interest or another protein not relevant to the biological system under investigation, e.g., β-galactosidase.
8. During the de novo synthesis of peptides with C-terminal Cys, deprotection is performed with 20 % piperidine containing 0.1 M HOBt to avoid the formation of dehydro-alanine. The synthesizer should be reprogrammed to deliver appropriate amounts of deprotection mixture composed of 0.25 M HOBt in 50 % piperidine in *N*-methyl-pyrrolidone.
9. Peptides are generally effective at concentrations below 1 μM. Higher levels may cause toxicity, so biological effects should be optimized by titration.
10. Larger pore-sized filters distort the visual field and prevent visualization by phase-contract microscopy.

References

1. Murriel CL, Dowdy SF (2006) Influence of protein transduction domains on intracellular delivery of macromolecules. Expert Opin Drug Deliv 3:739–746
2. Becker-Hapak M, Dowdy SF (2003) In vivo protein transduction: delivery of a biologically active protein into the mouse. Curr Protoc Cell Biol Chapter 20, Unit 20 2
3. Schwarze SR, Ho A, Vocero-Akbani A, Dowdy SF (1999) Science 285:1569–1572
4. Gump JM, Dowdy SF (2007) TAT transduction: the molecular mechanism and therapeutic prospects. Trends Mol Med 13:443–448
5. Gump JM, June RK, Dowdy SF (2010) Revised role of glycosaminoglycans in TAT protein transduction domain-mediated cellular transduction. J Biol Chem 285:1500–1507
6. Plisov S, Tsang M, Shi G, Boyle S, Yoshino K, Dunwoodie SL, Dawid IB, Shioda T, Perantoni AO, de Caestecker MP (2005) Cited1 is a bifunctional transcriptional cofactor that regulates early nephronic patterning. J Am Soc Nephrol 16:1632–1644
7. Osafune K, Takasato M, Kispert A, Asashima M, Nishinakamura R (2006) Identification of multipotent progenitors in the embryonic mouse kidney by a novel colony-forming assay. Development 133:151–161
8. Wang H, Yang Y, Sharma N, Tarasova NI, Timofeeva OA, Winkler-Pickett RT, Tanigawa S, Perantoni AO (2010) STAT1 activation regulates proliferation and differentiation of renal progenitors. Cell Signal 22:1717–1726
9. Tanigawa S, Wang H, Yang Y, Sharma N, Tarasova N, Ajima R, Yamaguchi TP, Rodriguez LG, Perantoni AO (2011) Wnt4 induces nephronic tubules in metanephric mesenchyme by a non-canonical mechanism. Dev Biol 352:58–69
10. Timofeeva OA, Gaponenko V, Lockett SJ, Tarasov SG, Jiang S, Michejda CJ, Perantoni AO, Tarasova NI (2007) Rationally designed inhibitors identify STAT3 N-domain as a promising anticancer drug target. ACS Chem Biol 2:799–809
11. Tufro A, Teichman J, Woda C, Villegas G (2008) Semaphorin3a inhibits ureteric bud branching morphogenesis. Mech Dev 125: 558–568
12. Fry D, Sun H (2006) Utilizing peptide structures as keys for unlocking challenging targets. Mini Rev Med Chem 6:979–987
13. Perantoni AO (2003) The ureteric bud. Tissue-culture approaches to branching morphogenesis and inductive signaling. Methods Mol Med 86:179–192

Chapter 18

Detection of Cells Programmed to Die in Mouse Embryos

Rocío Hernández-Martínez, Rodrigo Cuervo, and Luis Covarrubias

Abstract

Programmed Cell Death (PCD) is a broad term used to describe a series of events that culminate in the death of specific cells. In the embryo it occurs at predictable stages and tissues. During mouse development, PCD is a mechanism to preserve the homeostasis of the growing organism, and also is needed for the morphogenesis of a variety of structures. Apoptosis or PCD type I shows a sequence of morphological and biochemical changes such as plasma membrane blebbing, increase in mitochondrial membrane permeability, caspase activation, chromatin condensation, and phagocytosis. Many of these changes can be used to determine the occurrence of apoptosis in different type of samples. For example, apoptosis has been visualized in whole embryos and tissue sections using vital dyes, and by detection of degraded DNA or active caspases. In the present report, we compare these methods during the course of interdigital cell death in the mouse limbs. We discuss which method is the most suitable to detect a particular stage of apoptosis, which in some cases may be relevant for the interpretation of data. We detail combined protocols to observe mRNA expression or protein and cell death in the same tissue sample. Furthermore, we discuss some of the methodological problems to analyze autophagic cell death or PCD type II during embryo development.

Key words Apoptosis, Vital stain, LysoTracker, Phagocytosis, Cleaved caspase-3, TUNEL, Mouse limb

1 Introduction

Programmed Cell Death (PCD) is a term that defines the predictable, in time and space, death of cells during development. This process is genetically controlled, requires the expenditure of energy (i.e., is ATP-dependent), and occurs in a way that ensures that neighboring cells remain unaffected. Under a different concept, PCD is used by many researchers to refer to a genetically controlled process, not necessarily predictable, that occurs in living cells, including those in the healthy or diseased adult; a more proper denomination to this concept should be physiological cell death [1]. According to the morphology of the dying cell, PCD has been classically classified in three types: PCD type I (or apoptosis), PCD type II (or autophagy), and PCD type III (this is also known as lysosomal, and may represent a subtype of PCD type II) [2]. Of note it is that apoptosis is

Mark Lewandoski (ed.), *Mouse Molecular Embryology*, Methods in Molecular Biology, vol. 1092,
DOI 10.1007/978-1-60327-292-6_18, © Springer Science+Business Media New York 2014

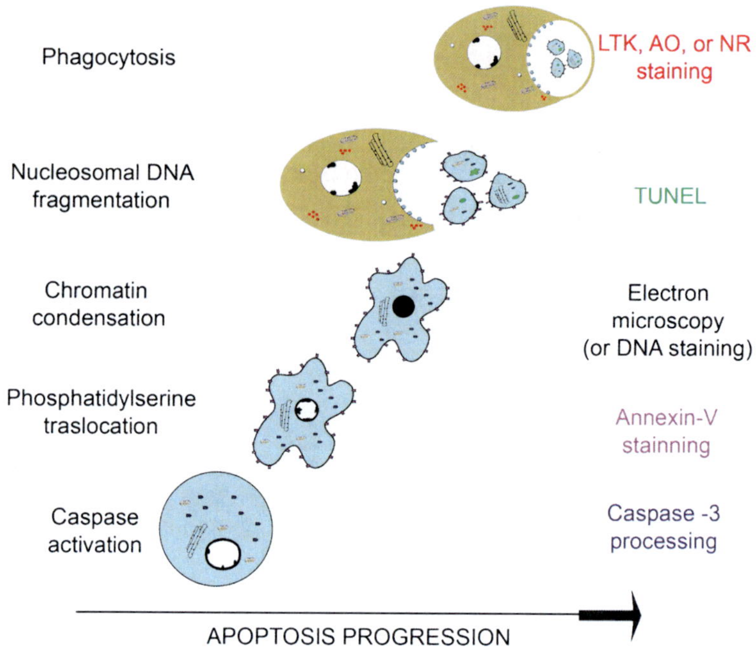

Fig. 1 Methods to detect apoptosis along the process of cell degeneration. See text for more details

frequently used as a synonymous of PCD; this should be avoided since now it is clear that different types of cell death can occur during development. PCD is essential in the control of growth and morphogenesis of developing organs, in the degeneration of transitory structures, and the establishment of the final cell number in adult tissues. Despite the known fact that cell death occurs at different stages of mouse development, only in few cases the actual cell death function has been directly determined [3–5].

1.1 Cell Death by Apoptosis

The process more commonly observed during development, and also the one that has been better characterized, is apoptosis. Apoptotic cell death describes a type of cell death with a series of morphological and biochemical changes conserved in different organisms [6, 7] (Fig. 1). Among the morphological aspects that are commonly observed in apoptotic cells are: cell shrinkage which is associated with the loss of cell attachment to the extracellular matrix and to their neighboring cells, as well as with changes in water balance, making organelles look more tightly packed; plasma membrane blebbing, which is probably the main cause of cell fragmentation into the apoptotic bodies; chromatin condensation; and phagocytosis by specialized cells. At the molecular level apoptotic cells expose phosphatidylserine on the outer leaflet of the plasma membrane, show active caspases and frequently their DNA is degraded with a nucleosomal pattern.

The process of apoptosis can be divided in three phases: activation, execution, and cell degradation [6]. Apoptosis may be activated in many ways through a variety of signaling pathways that target the apoptotic regulatory machinery. It is thought that there are no specific apoptotic pathways for activation, but rather, a signaling pathway activates apoptosis in a context-dependent fashion. The conditions that favor apoptosis are not well defined, but the influence of the environment on the activation of apoptosis can be exemplified with the ability of c-myc and other oncogenes to activate apoptosis when cells are under conditions not permissive for proliferation (e.g., low serum, absence of p53) [8]. Different type of stimuli activates apoptosis during mouse embryonic development [7]. Typical factors that promote cell death are Bmp family members and retinoic acid, whereas a large number of factors promote survival such as Ngf and Fgf family members. Life or death of cells depends on the balance of death and survival factors [9]. It is commonly thought that cells are prompt to die so they continuously require survival factors to live. This appears to be the case for the establishment of the number of sympathetic nervous system neurons which naturally are produced in excess; those neurons that innervate the target receive the pro-survival factor NGF, while the others die due to the lack of factors that promote their survival [10]. A different situation occurs for the interdigital cells of the developing limb whose fate (life or death) depends on the balance of a pro-death (RA) and pro-survival factors (Fgf8) [4].

The execution of apoptosis is tightly bound to the activation of caspases: autoprocessing cysteine-proteases that cut many substrates during apoptosis [11, 12]. Most of the morphological and biochemical features of apoptotic cells are consequences of caspase activation. Two major pathways regulate caspase activation, the extrinsic and the intrinsic, which activate the initiator caspases 8 and 9, respectively [13]. The extrinsic pathway starts with the activation of receptors at the plasma membrane (e.g., receptors for FasL or TRAIL), which then serve as a module for clustering and recruitment on the cytosolic membrane face of other proteins that allow caspase-8 activation. The intrinsic pathway, on the other hand, starts at the mitochondria where major changes in membrane permeability cause the release of cytochrome c, among other proteins (see below). Cytochrome c in this case works as a cofactor that promotes Apaf oligomerization forming a wheel-shaped signaling platform that activates caspase-9 [14]. Antiapoptotic and proapoptotic Bcl2 family members control the mitochondrial events. Bcl2 and Bax are prototypes of antiapoptotic and proapoptotic Bcl2 proteins, respectively. Bax contributes to form the pores that change mitochondria membrane permeability; whereas Bcl2 interacts with Bax preventing the formation of those mitochondrial pores [15]. The so-called BH3-only family members such as Bid are also relevant regulators of the intrinsic apoptotic pathway that appear to promote cell death by preventing the interaction of Bcl2 with Bax. Apoptosis is not initiated until caspase-3 (or other executioner caspases such as 6 and 7) is processed and activated by either caspase-8 or caspase-9. Activated executioner caspases cleave many substrates and produce the dismantling of the cell [11]. Most if not all survival factors act before caspase activation.

The loss of mitochondrial integrity during apoptosis produces, in addition to the release of cytochrome c, also, among others, of the Apoptosis Inducing Factor (AIF) and endonuclease G (endoG) [16]. These two latter proteins as well as the Caspase-Activated DNAse (CAD; also known as the nuclear DNA fragmentation factor, DFF) participate in the chromatin condensation and DNA fragmentation events that occur during the terminal stages of apoptosis; AIF and endoG both can function in a caspase-independent manner. Caspase-3 specifically cleaves the inhibitor of CAD (ICAD), consequently, CAD activates. CAD in combination with AIF catalyzes the initial DNA cleavage into 50 to 300 kb long genomic fragments, and next, endoG together with CAD continues cleaving the DNA generating the typical oligonucleosomal sized fragments [17].

The degradation of the apoptotic cells is always due to phagocytosis: professional macrophages or non-professional phagocytes (e.g., neighboring cells) engulf and digest apoptotic bodies without the generation of an inflammatory response [18]. The process of apoptotic-cell engulfment involves a series of steps including: (a) the production of signals by the apoptotic cells to recruit phagocytes; (b) the recognition of the apoptotic cells by the phagocyte; (c) the cytoskeletal reorganization and internalization of the target cell by phagocytes; (d) the processing and degradation of the ingested apoptotic cell. Signals that trigger phagocyte migration to sites of apoptosis ("find-me" signal) usually are diffusible extracellular factors such as lysophosphatidylcholine (LPC) and nucleotides (ATP and UTP) [18]. Apoptotic cell recognition by the phagocytic cell is achieved by a huge number of cell surface molecules that act alone or in combination as "eat-me" signals [18]. The best-studied "eat-me" signal in different organisms is the externalization of phosphatidylserine to the surface of the apoptotic cell. Although the mechanism of phosphatidylserine translocation to the outer leaflet of the cell during apoptosis is not completely understood, it has been associated with the loss of aminophospholipid translocase activity and nonspecific flip-flop of phospholipids of various classes [19]. The phagocytic cell carries out a series of metabolic reactions to digest the apoptotic bodies. After the engulfment, it has been observed an increase in endosome and lysosome activities [20, 21]. Lysosomes and endosomes are specialized organelles in which occurs the degradation of phagocytosed and autophagocytosed cell material. Lysosomal proteases mediate protein degradation and, through endocytic traffic, all this degraded material is recycled. It has been observed that high lysosomal activity correlates with known regions in the embryo where both phagocytosis and apoptosis occurs [21, 22].

1.2 The Basics of Apoptotic Cell Death Detection in Mouse Embryos

Detection of apoptosis during mouse development mostly restricts to the execution phase. Previous to caspase activation, level, post-translational modifications, or intracellular distribution of Bcl2 family members frequently occur [23]. However, up to now, these changes can only be used as a sign that an apoptotic process may occur; thus, if a particular modification in a Bcl2 family member associates with apoptosis, it needs to be supported by other markers. All the methods that are described below relate to caspase activation or downstream events (Fig. 1).

Caspase-3 is considered the most important executioner caspase. Caspase-3 cleavage can easily be detected by western-blot [24]; however, this requires dissecting the embryo region where cell death supposes to be occurring, with the additional inconvenience that the amount of protein needed for the assay is relatively large, thus, the amount of tissue becomes a limiting factor. Caspase activity can also be measured using cell extracts from the embryo region of interest and novel specific substrates [25]. For more detailed analysis of active caspase in the embryo, antibodies that recognize specifically the active fragment of caspase-3 have been developed allowing the detection of active caspase by immunohistochemistry [26]. For the spatiotemporal real time tracking of the apoptotic process, recently, a genetically encoded caspase sensor (apoliner) was developed [27]. Apoliner comprises two fluorophores, the monomeric red fluorescent protein (mRFP) and the enhanced green fluorescent protein (eGFP), linked by an efficient and specific caspase-3 sensitive site. Upon caspase activation, the sensor is cleaved and eGFP translocates to the nucleus, leaving mRFP at membranes. This novel tool detects apoptosis before the detection of other markers, including anti-cleaved caspase-3 immunoreactivity. It is important to mention that cells with apoptotic morphology have been observed without detection of caspase activation (called caspase-independent apoptotic cell death) [28].

The dismantling of internal nuclear structures produces chromatin condensation. Chromatin condensation can be observed with permeable or semipermeable dyes that stain nucleic acids, such DAPI or propidium iodide, though, it is not always easy to define apoptotic cells at low power microscopy. Chromatin condensation is easily observable by electron microscopy as black and dense nucleus (pyknotic) in apoptotic cells. Accompanying chromatin condensation, DNA is fragmented, as mentioned above, generating initially large fragment and then a nucleosomal degradation pattern. When this occurs, a DNA ladder is visualized under ultraviolet illumination after agarose gel electrophoresis and staining with ethidium bromide [17, 29]. In spite of this expected characteristic, evident DNA ladders have not always been observed in samples obtained from embryo regions undergoing apoptotic cell death [30, 31]. This may be due to technical difficulties during the extraction of intact fragmented DNA from small pieces of tissue. DNA fragmentation can be analyzed in cells, tissue sections and in the whole embryo using the technique called TUNEL (terminal deoxynucleotidyl transferase (TdT)-mediated dUTP nick end labeling) [32, 33]. The principle of TUNEL detection is the use of TdT to add fluorescent or biotinylated nucleotides to the 3′-hydroxyl groups in DNA generated by the apoptotic nucleases. It is important to note that the TUNEL technique is not specific for nucleosomal DNA degradation but essentially detects abundant DNA degradation. Therefore, the TUNEL technique does not definitively distinguish among different types of cell death. DNA degradation is a late event also observed in autophagic cell death [34, 35].

The phosphatidylserine exposed by the apoptotic cell can be detected by using the property of Annexin-V to interact strongly and specifically with phosphatidylserine residues after binding to Ca^{2+} ions [19, 36]. Presently there are a variety of recombinant forms of Annexin-V commercially available with different covalently bound labels for easy detection. It is important to remark that detection of apoptotic cells by labeled Annexin-V relies on intact cells; fixatives or any other treatment that allows the pass of the protein to the intracellular compartment eliminate the selective detection of apoptotic cells.

LysoTracker (LTK), Acridine Orange (AO), Nile Blue Sulfate (NBS), and Neutral Red (NR) are lysosomotropic dyes, which become protonated and are retained in acidic compartments (pH around 5) [21]. All these dyes are frequently used to analyze the lysosomal function in the embryo [20, 21]. Under fluorescence microscopy cells stained with these dyes but NBS show an increase in fluorescence intensity when the lysosomal activity increases [20]; lysosomes appear like bright dots. LTK Red, AO, and NR when activated by blue light (577 nm) give a distinct red fluorescence (peak at about 590 nm); however, some caution must be taken because at high dye concentration the nonspecific emission and background increase. Using the NBS, the blue lysosome staining needs to be observed under light microscopy (NR can also be observed under light microscopy). From the above dyes, only LTK Red (DND-99) version is suitable for aldehyde fixation [20]; AO, NBS, and NR are vital stains that require the immediate observation of the sample after incubation with the dye. It is important to mention that AO also stains nucleic acids which are visualized at 525 and 650 nm, DNA and RNA, respectively [37], property that have also been used to detect apoptotic cells [38, 39].

1.3 Autophagy-Dependent Cell Death

Autophagy has been implicated as part of the PCD type II. Cells dying through this mechanism engulf their own cytoplasmic material and organelles to destroy them in the lysosomes [40]. At the morphological level cells under autophagy show autophagosomes characterized by a double membrane surrounding the engulfed material. Recently, many molecules that control autophagy have been characterized in several organisms including mammals [41]. In contrast with apoptosis, autophagy is not a cell death mechanism per se; actually, a survival function has been determined in vivo [42, 43]. The occurrence of autophagy has been reported in degenerating tissues of vertebrate embryo since more than 25 years ago [44, 45]; however, because in several of those regions apoptosis is essential [46, 47], these observations are waiting for a confirmation with new methods. Still much needs to be learned about the conditions at which autophagy contributes to the death of a cell, especially in vivo.

The best method to detect autophagy is using electron microscopy to search for the double membranes of autophagosomes. However, this method is not suitable for assessing whole embryo or large number samples. Furthermore, it is sometimes difficult to distinguish autophagosomes and autolysosomes from phagosomes in phagocytes, which may misidentify autophagy in place of apoptosis [48]. Recently, based on the molecular events involved, a more specific method has been developed to monitor autophagy. When autophagy occurs, conversion of LC3 (also known as Atg8) from the cytosolic form (LC3-I) to the autophagosome-associated form (LC3-II) occurs, which can be detected by immunoblotting of LC3 proteins [48]. This change in intracellular localization of LC3 protein (visualized as punctate structures) can be detected by immunofluorescence or, indirectly, by detecting the localization of the recombinant protein LC3-GFP after being introduced into cells (e.g., via production of transgenic mice) [49]. Cautions should be taken when using LC3 to identify autophagic cells in situ, since endogenous and over-expressed LC3 can easily aggregate and, consequently, a misinterpretation of the punctuated signal in cells may take place [48]. Detection of autophagy using lysosomal activity markers alone is not recommended because, as mentioned above, phagocytic activity is also associated with high lysosomal activity [20, 48]. However the combination of LTK staining with TUNEL detection may allow to distinguish apoptotic from autophagic cells.

In some cell types and circumstances, apoptotic and autophagic cell death occur simultaneously, so a mixed cell death phenotype is observed; the extent of this overlap has not been determined yet [50].

1.4 Limb Morphogenesis

Mouse limbs provide an excellent model to investigate the role of cell death in morphogenesis. Particularly, the degeneration of the interdigital mesenchyme is one of the best-studied models of PCD [51]. A detailed analysis of the stepwise events of the apoptosis observed during interdigital cell death (ICD) could provide useful information relevant to other tissues undergoing PCD. Initiation of ICD is associated with the disappearance of ectodermal Fgf8, which function as a survival factor on distal mesenchymal cells [4]. Retinoic acid, on the other hand, is essential for the death of interdigital cells [4, 52]. Therefore, Fgf8 and retinoic acid antagonistically regulate ICD. Bmps are also relevant in this process negatively regulating the expression of Fgf8 in the ectoderm [53], and contributing to the death of proximal interdigital cells, evident in the chick limb [52]. During ICD Bax expression, active caspases, condensed chromatin, degraded DNA, and phagocytosis are among the apoptosis-associated events observed [4, 30, 31]. Below, apoptosis detection is exemplified using mouse limbs during ICD, though protocols described can be applied to other organs with few modifications.

2 Materials

2.1 Disposables

1. Centrifuge tubes 1.5 mL (Eppendorf, Germany).
2. Screw-cap microcentrifuge tubes (cat. # EF4264DC, Daigger, USA).
3. Four-well plates for cell culture (cat. # 176740, Nunc, USA).

2.2 Vital Stains (AO, RN, and NBS)

1. Phosphate buffered saline (PBS): Prepare a 10× PBS stock by dissolving the following in 800 mL deionized H_2O (dH_2O): 80 g NaCl, 2 g of KCl, 14.4 g Na_2HPO_4, and 2.4 g KH_2PO_4. Adjust pH to 7.4 the volume to 1 L with additional dH_2O. Sterilize by autoclaving. Prepare 1× PBS by dilution of one part with nine parts of dH_2O.
2. Acridine Orange (AO) stock solution: Prepare a stock solution of 1 mg/mL AO (cat. # A6014 Sigma-Aldrich, USA) in PBS. Store at 4 °C protected from light.
3. Neutral Red (NR) stock solution. Prepare a stock solution of 1 mg/mL NR (cat. # N4638 Sigma-Aldrich, USA) in PBS. Store at 4 °C protected from light.
4. LysoTracker Red DND-99 (LTK Red). Commercially available as 1 mM stock solution (cat. # L-7528 Molecular Probes-Invitrogen, USA). Store at 4 °C.
5. Paraformaldehyde (PFA): Prepare a 4 % PFA stock solution in PBS. Dissolve 4 g of PFA in 90 mL of dH_2O in an Erlenmeyer flask under heat and with constant rocking with a stirring bar; do this in a fume hood. When the solution reaches about 70 °C

remove from the heater and add three drops of NaOH 10 M and mix. Let the solution cool for a few minutes and add 10 mL of 10× PBS. Make 1 mL aliquots and store at −70 °C.

2.3 Whole-Mount Immunohistochemistry of Cleaved Caspase-3

1. PFA: Prepare a 8 % PFA stock solution in PBS in the same way as the 4 % PFA but weigh 8 g of PFA instead of 4 g. Store 1 mL aliquots at −70 °C.
2. PBN (0.1 % v/v NP40 in PBS).
3. Absolute methanol.
4. 25 % v/v Methanol in PBN; 50 % v/v methanol in PBN; 75 % v/v methanol in PBN.
5. Hydrogen peroxide.
6. PBSST (5 % v/v goat serum and 0.1 % v/v Triton in PBS).
7. Primary antibody: Anti-mouse Cleaved Caspase-3 (cat. # 9464 Cell signaling, USA).
8. Secondary antibody: Biotinylated Anti-rabbit (cat. # PK-6101 VECTASTAIN Elite ABC Kit (Rabbit IgG), Vector, USA).
9. Reagent A (cat. # PK-6101 VECTASTAIN Elite ABC Kit (Rabbit IgG), Vector, USA).
10. Reagent B (cat. # PK-6101 VECTASTAIN Elite ABC Kit (Rabbit IgG), Vector, USA).
11. 3,3-Diaminobenzidine tetrahydrochloride (DAB; cat. # 2042D, Research Inorganics, USA).
12. PBT (0.1 % v/v Tween and 0.2 % bovine serum albumin (BSA) in PBS).
13. 50 % v/v Glycerol in PBS; 75 % v/v glycerol in PBS.

2.4 Cleaved Caspase-3 Immunohistochemistry Combined with TUNEL on Tissue Sections (see Note 1)

1. 1× PBS.
2. Paraplast tissue embedding medium (cat. # 8889–503002 Oxford, USA).
3. 4 % PFA 1 mL aliquots.
4. Absolute ethanol, 90 % v/v ethanol in PBS, 70 % v/v ethanol in PBS, 50 % v/v ethanol in PBS.
5. Xylene (cat. # 9490–03, J.T. Baker, USA).
6. 1:1 v/v Absolute ethanol–xylene.
7. Primary antibody: Anti-mouse Cleaved Caspase-3 (cat. # 9464 Cell signaling, USA).
8. Secondary antibody: anti-rabbit Alexa-Fluor 594 (cat. # A-11012 Molecular Probes-Invitrogen, USA).
9. TUNEL (In Situ Cell Death Detection Kit, Fluorescein; cat. # 11835246001, Roche, USA). Reaction mixture: 50 μL of "Enzyme solution" to 450 μL of "Label solution".

10. Proteinase K stock solution (10 mg/mL) in dH_2O.
11. Proteinase K working solution (10 μg/mL in 10 mM Tris–HCl pH 8). Made on the day of use.
12. Blocking solution (10 % v/v goat serum, 0.1 % v/v Triton in PBS).
13. 50 % v/v glycerol in PBS.
14. Microscopy slides (cat. # 2948-75X25, Corning, USA), cover it with poly-L-lysine (cat. # P8920, Sigma-Aldrich, USA) by sinking the slide in a 100 μg/mL poly-L-lysine solution in dH_2O and then air-drying at room temperature (RT).
15. Microscope cover glasses (cat. # G15972K, Daigger, USA).
16. Humidified slide chamber. Place wetted paper towel or Kimwipes on the bottom of a square plastic box with hermetic lid and then, above, place two pipettes in parallel (slides will be on pipettes). Protect from light by covering all the camber with aluminum foil.
17. ImmEdge hydrophobic barrier pen (cat. # H-4000, Vector, USA).

2.5 In Situ Hybridization Combined with LysoTracker Staining

2.5.1 Preparation of Single-Stranded RNA Probes

1. DEPC-treated water (DEPC-H_2O): 0.1 % v/v DEPC in dH_2O. Prepare by stirring the solution with a clean magnetic bar overnight (ON), at RT in a fume hood. Next day inactivate the DEPC by autoclaving.
2. 10× Transcription buffer is provided with the RNA polymerase (Roche, USA).
3. Dithiothreitol.
4. DIG RNA labeling mix (cat. # 11277073910, Roche, USA).
5. RNAsin (cat. # N261A, Promega, USA).
6. SP6, T3, or T7 RNA polymerase (Roche, USA).
7. 4 M LiCl prepared with DEPC-H_2O.
8. Absolute ethanol.
9. Linearized plasmid containing the specific probe sequence (1 μg/μL). Remember to cut the plasmid with the right restriction enzyme according to the promoter to be used to produce the antisense RNA.
10. 70 % v/v Ethanol in DEPC-H_2O.

2.5.2 In Situ Hybridization

1. DEPC-H_2O (prepared as described above).
2. 10× PBS stock solution: Prepare as described above but use DEPC-H_2O. To obtain 1× PBS also use DEPC-H_2O.
3. PBT (0.1 % v/v Tween in PBS).
4. 25 % v/v Methanol in PBT; 50 % v/v methanol in PBT; 75 % v/v methanol in PBT.

5. Proteinase K stock solution (10 mg/mL) in DEPC-H_2O.
6. 20× SSC pH 5 stock solution. Prepare by dissolving 17.53 g NaCl, 8.82 g Na-Citrate in 100 mL of DEPC-H_2O. Adjust to pH 5 with citric acid.
7. 10 % SDS stock solution. Prepare in DEPC-H_2O.
8. 5 M NaCl stock solution prepared in DEPC-H_2O.
9. 1 M Tris–HCl pH 7.5 stock solution. Prepare in dH_2O and autoclave.
10. 1 M Tris–HCl pH 9.5 stock solution. Prepare in dH_2O and autoclave.
11. 1 M $MgCl_2$ stock solution. Prepare in DEPC-H_2O.
12. Glycine.
13. Glutaraldehyde.
14. Formamide.
15. Tetramisole hydrochloride.
16. tRNA yeast (cat. # R8508, Sigma-Aldrich, USA).
17. Heparin stock solution (10 mg/mL in DEPC-H_2O).
18. Pre-hybridization solution (prepared from stock solutions): 50 % Formamide, 50 μg/mL yeast tRNA, 1 % SDS, 5× SSC pH 5, 50 μg/mL heparin. Prepare on the day of use.
19. Washing Solution 1 (prepared from stock solutions): 50 % formamide, 1 % SDS, 5× SSC pH 5. Prepare on the day of use.
20. Washing Solution 2 (prepared from stock solutions): 0.5 M NaCl, 0.1 % Tween 20, 10 mM Tris–HCl pH 7.5. Prepare on the day of use.
21. Washing Solution 3 (prepared from stock solutions): 50 % formamide, 2× SSC pH 5. Prepare from the stocks the day of use.
22. Sheep serum (cat. # S2263, Sigma-Aldrich, USA).
23. Embryo powder. Dissect some embryos at the stages of the samples to be analyzed, and with a sterile ceramic mortar and a pestle make a fine powder in liquid N_2. With a sterile spoon take the powder and transfer it into a 50 mL Falcon-tube, and then add four volumes of cold acetone. Incubate for 30 min on ice. Centrifuge at 10,000 × *g* for 10 min. Remove the supernatant; add again four volumes of cold acetone. Centrifuge at 10,000 × *g* for 10 min. Remove the supernatant and leave the pellet dry at RT for 2 h. Keep the powder at −20 °C.
24. Anti-Digoxigenin-AP antibody (cat. # 11093274910, Roche).
25. 10× TBST (100 ml): 8 g NaCl, 0.2 g KCl, 25 mL 1 M Tris–HCl pH 7.5, 10 mL Tween-20. Autoclave. Dilute to 1× and add tetramisole hydrochloride to 2 mM (0.48 mg/mL) on the day of use.

26. NTMT: 100 mM NaCl, 100 mM Tris–HCl pH 9.5, 50 mM $MgCl_2$, 0.1 % Tween-20, 2 mM tetramisole hydrochloride (all are final concentrations). Prepare from stocks on day of use.
27. NBT (cat. # 11383213001, Roche, USA).
28. BCIP (cat. # 11383221001, Roche).

3 Methods

3.1 Vital Stains

1. Take the sample with fine tweezers and sink it in the acridine orange staining solution (AO stock solution 2 μL in 1 mL PBS), neutral red staining solution (NR stock solution 2 μL in 1 mL PBS), or LTK staining solution (LTK stock solution 1 μL in 1 mL PBS) keep for 15 min in a regular incubator at 37 °C. All the staining solutions are prepared on the day of use (*see* **Note 2**).
2. Rinse in PBS for 5 min at RT twice (*see* **Note 2**).
3. Observe under confocal microscopy. An example of the result is shown in Fig. 2.

3.2 Cleaved Caspase-3 Whole-Mount Immunohistochemistry

This protocol is a modified version of the one reported by Corson et al. [54]. All steps are carried out with gentle rocking in a test tube rocker.

1. Fix the mouse limb in 8 % PFA ON at 4 °C. (Detection of posttranslational modifications may require immediate fixation of embryo samples) (*see* **Note 3**).

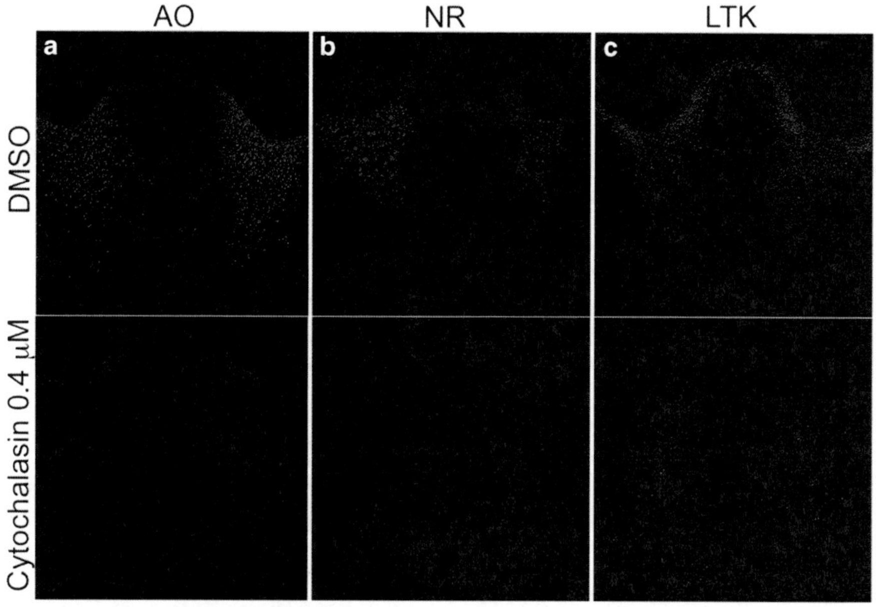

Fig. 2 AO, NR, and LTK detect phagocytosis. Mouse limbs were cultivated for 8 h in presence or absence of 0.4 mM cytochalasin (in DMSO), an inhibitor of phagocytosis. After culture, limbs were sunk in staining solutions of AO (**a**), NR (**b**), and LTK (**c**) and then observed under confocal microscopy

2. Next day, wash the tissues in PBN twice for 10 min each at 4 °C.
3. Dehydrate the tissues by sequential 10 min washes in 25 % v/v methanol–PBN, 50 % v/v methanol–PBN, 75 % v/v methanol–PBN, and absolute methanol at 4 °C.
4. Store in absolute methanol at −20 °C or continue the protocol until end.
5. Next day quench endogenous peroxidase activity by washing 1 h in a 1:1 v/v solution of H_2O_2 in PBN at 4 °C.
6. Wash in cold absolute methanol for 10 min at 4 °C
7. Rehydrate samples by sequential 10 min washes in 75 % v/v methanol–PBN, 50 % v/v methanol–PBN, and 25 % v/v methanol–PBN at 4 °C.
8. Wash twice in PBT for 5 min at 4 °C.
9. Block from unspecific antibody binding with PBSST for 2 h at 4 °C.
10. Incubate with the Cleaved Caspase-3 antibody diluted 1:200 in PBSST ON at 4 °C (*see* **Note 4**).
11. Remove the primary antibody solution and wash twice for 15 min in PBSST at RT.
12. Wash in PBSST five times, 1 h each, at 4 °C.
13. Incubate ON at 4 °C with the secondary antibody (biotinylated anti-rabbit) 1:200 diluted in PBSST (*see* **Note 4**).
14. Remove the secondary antibody solution and wash in PBSST twice for 15 min each at RT.
15. Wash in PBSST five times, 1 h each, at 4 °C.
16. Form the peroxidase–avidin complex by mixing 5 μL of Vector Reagent A plus 5 μL of Vector Reagent B in 1 mL PBSST. Let sit for 15 min at RT (*see* **Note 4**).
17. Incubate samples ON in the previous peroxidase–avidin solution at 4 °C.
18. Wash in PBT for 20 min at RT.
19. Incubate samples in 0.3 mg/mL DAB in PBT for 20 min at RT (*see* **Note 4**).
20. Remove the previous solution and incubate samples in 0.3 mg/mL DAB in PBT plus 0.03 % H_2O_2 until the signal becomes apparent and contrasts with the background; it usually takes less than 5 min to develop.
21. Wash in PBT for 5 min twice at RT.
22. Postfix in PFA for 1 h at RT.
23. Clear first with 50 % v/v glycerol–PBS for 30 min and then with 75 % v/v glycerol–PBS for additional 30 min. An example of the result is shown in Fig. 3.

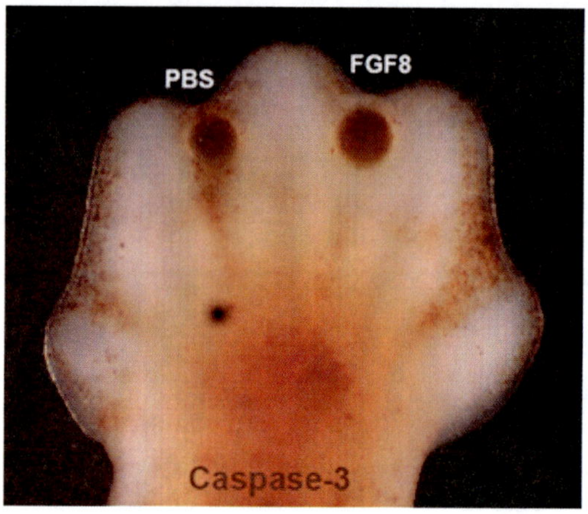

Fig. 3 Caspase-3 immunohistochemistry on whole-mount samples. Mouse limbs were cultured for 8 h with a FGF8- or PBS-soaked bead. The unsectioned samples were processed for caspase-3 immunodetection. Note that an FGF8-soaked bead reduces caspase-3 activation

3.3 Cleaved Caspase-3 Immunohisto-chemistry Combined with TUNEL on Slide Sections

1. Fix the mouse embryo (or fragment of it) with 4 % PFA. Incubate ON at 4 °C.
2. Next day, wash twice in PBS for 5 min each at 4 °C.
3. Dehydrate the tissue by sequential 30 min washes in 50 % v/v ethanol/PBS, 70 % v/v ethanol/PBS, 90 % v/v ethanol/PBS, and absolute ethanol at RT.
4. Wash in 1:1 v/v ethanol–xylene for 30 min at RT.
5. Wash in xylene 30 min at RT.
6. Wash in 1:1 v/v xylene–Paraplast for 30 min at 60 °C.
7. Wash in 1:2 v/v xylene–Paraplast for 30 min at 60 °C.
8. Remove the previous solution and incubate in 100 % Paraplast ON at 60 °C.
9. Transfer the sample to small aluminum pans filled with 100 % Paraplast. Place pans on a warming plate to keep the wax from melting until the tissue is oriented correctly. Once the tissue is in the pan and oriented, set it aside to harden undisturbed for at least 20 min. At this time place the pans at RT for at least 1 h.
10. Cut out blocks of wax containing the embryo or embryo fragment, and mount on wooden dowels that fit into microtome chucks. Section at 8 μm (or more) in a microtome (Leica RM2125).
11. Keep the tissue slides at 37 °C ON.
12. Next day rinse the slides in pre-warmed xylene at 60 °C for 20 min.

13. Rinse the slides in absolute ethanol for 10 min at RT.
14. Rehydrate by sequentially washing samples in 90 % v/v ethanol–PBS, 70 % v/v ethanol–PBS, 50 % v/v ethanol–PBS, 10 min each at RT.
15. Wash with PBS for 10 min at RT.
16. Delineate a hydrophobic contour with ImmEdge pen on each tissue section.
17. Proteinase K solution. Cover each tissue section with enough quantity of Proteinase K solution (10 mg/mL Proteinase K in Tris–HCl pH 8). Keep in a humidified chamber at 37 °C for 20 min.
18. Wash three times in PBS for 5 min at RT.
19. Cover each tissue section with enough quantity of the blocking solution for 1 h at RT.
20. Dilute anti-cleaved caspase-3 antibody in blocking solution 1:50. Incubate in a humidified chamber ON at 4 °C.
21. Next day remove previous solution and wash three times with PBS for 10 min each at RT.
22. Dilute secondary antibody anti-rabbit Alexa fluor 594 on blocking solution 1:1,000. Cover each tissue section with this solution. Keep in a humidified chamber in the dark for 1 h at RT.
23. Wash three times in PBS for 10 min each at RT.
24. Cover each tissue section with the TUNEL reaction mixture. Keep in a humidified chamber in the dark at 37 °C for 1 h.
25. Wash three times in PBS at RT.
26. Add a small drop of 50 % v/v glycerol in PBS on each tissue-section, put a cover glass, and observe under confocal microscopy. An example of the result is shown in Fig. 4.

3.4 Whole-Mount In Situ Hybridization Combined with LysoTracker Detection

3.4.1 After LTK Staining

1. Wash twice in PBT (diluted in DEPC-H_2O) for 10 min each at RT.
2. Fix with 1 mL 4 % PFA ON at 4 °C.
3. Next day remove the PFA and wash in PBT twice in PBS for 10 min at RT.
4. Dehydrate by sequential 10 min washes in 25 % v/v methanol–PBT, 50 % v/v methanol–PBT, 75 % v/v methanol–PBT, and absolute methanol at RT. Store at −20 °C until use or continue with the protocol until the end (*see* **Note 5**).

3.4.2 Preparation of Single-Stranded RNA Probes

1. In an Eppendorf tube mix the following at RT.
   (a) 10 μL of DEPC-H_2O.
   (b) 2 μL 10× Transcription buffer.

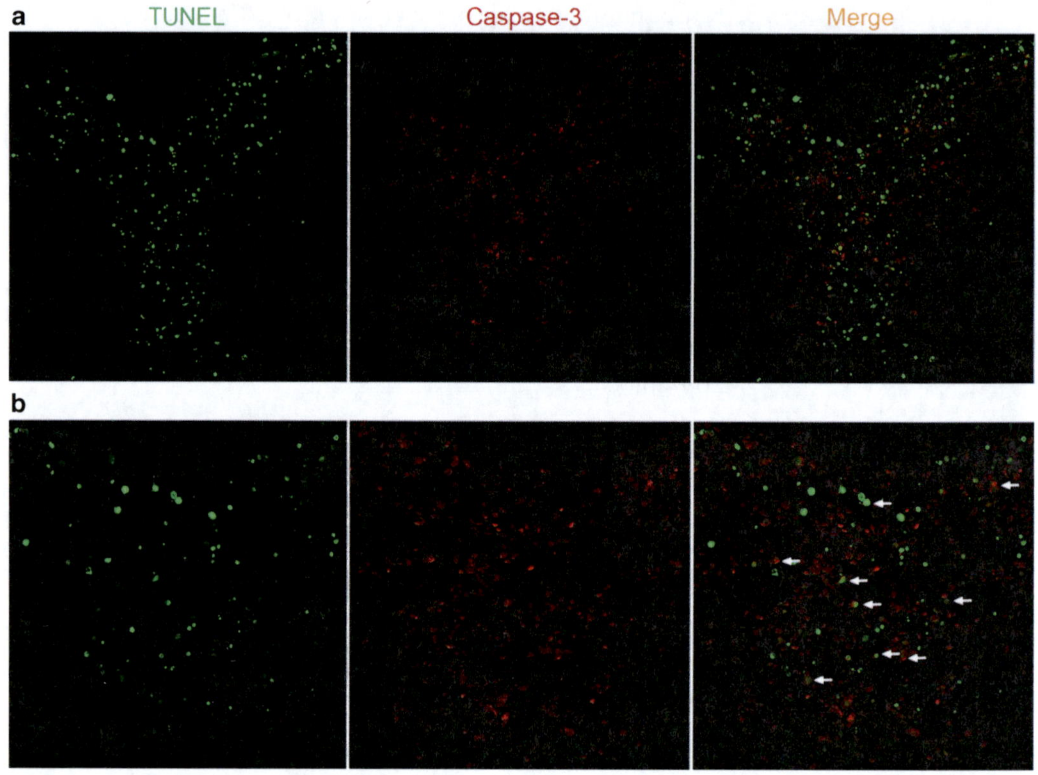

Fig. 4 Capase-3 immunohistochemistry combined with TUNEL. Mouse limbs were processed for TUNEL (**a**) and caspase-3 immunohistochemistry (**b**) on paraffin-sections.Note that many positive cells for TUNEL are also positive for caspase-3 (*arrows*)

   (c) 2 μL of 0.1 M DTT.
   (d) 2 μL of DIG RNA labeling mix.
   (e) 1 μL of the linearized plasmid (1 μg/μL).
   (f) 0.5 μL RNAsin.
   (g) 1 μL of SP6, T3, or T7 RNA polymerase.
2. Incubate at 37 °C for 2 h.
3. Add 100 μL of DEPC-H_2O, 10 μL of 4 M LiCl plus 300 μL of absolute ethanol.
4. Keep at −20 °C for 2 h.
5. Centrifuge at 15,000 × *g* for 10 min at 4 °C.
6. Wash the pellet with 70 % v/v ethanol in DEPC-H_2O at RT.
7. Let dry for 30 min at RT.
8. Dissolve the pellet with 100 μL of DEPC-H_2O. Keep at −20 °C or at −70 °C for a long-term storage. Load 5 μL on a 1 % agarose gel. The RNA band should be at least five times more intense than the DNA template.

3.4.3 Whole-Mount In Situ Hybridization

The protocol is a modified version of the one reported by Wilkinson and Nieto [55]. Carry out all steps in 1.5 mL screw-cap microcentrifuge tubes with gentle shaking (e.g., in a test tube rocker). During the hybridization at 70 °C use a hybridization incubator with a platform rocker (Combi-V12; Daigger, USA). All the steps are carried out at RT unless otherwise indicated. Take the samples required from the stock at −20 °C in absolute methanol. One milliliter of each of the solutions below is enough to cover most of the mouse embryonic samples (*see* **Note 5**).

Hybridization: Day 1

1. Rehydrate samples by sequential 5 min washes in 75 % v/v methanol–PBT, 50 % v/v methanol–PBT, 25 % v/v methanol–PBT, and PBT.
2. Wash four times, 5 min each, in PBT.
3. Treat samples with 10 μg/mL proteinase K in PBT. The time of treatment varies depending on the sample and the probe (range 2–10 min).
4. Treat samples with 2 mg/mL glycine in PBT (fresh) for 5 min.
5. Wash twice in PBT for 5 min.
6. Fix samples with 0.2 % glutaraldehyde/4 % PFA in PBT for 20 min (add 8 μL 25 % glutaraldehyde and 5 μL 20 % Tween per 1 mL 4 % PFA).
7. Wash twice in PBT for 5 min.
8. Incubate in the pre-hybridization solution for 1 h at 70 °C.
9. Remove the previous pre-hybridization solution, and then add pre-hybridization solution + 10 μL probe (hybridization solution). Incubate ON at 70 °C.

Hybridization: Day 2

10. Wash twice in the Washing Solution 1 for 30 min at 70 °C.
11. Wash in a mixture 1:1 v/v of Washing Solution 1:Washing Solution 2 for 10 min at 70 °C.
12. Wash three times in the Washing Solution 2 for 5 min.
13. Wash in the Washing Solution 3 for 5 min.
14. Wash twice in the Washing Solution 3 for 30 min at 65 °C. (Start with **step 18** in parallel now.)
15. Wash three times in 1× TBST for 5 min.
16. Preblock embryos by incubating them with 10 % v/v sheep serum in PBT for 2–3 h. (Inactivate the sheep serum by incubating it 30 min at 70 °C before use.)
17. Preabsorb antibody with "embryo powder" as follows: (1) 3 mg embryo powder in 500 μL TBST, 30 min at 70 °C with shaking (use a shaking heater block); (2) place on ice and allow

to cool for few minutes; (3) add 5 μL sheep serum, 1 μL anti-DIG antibody; (4) rock gently for 1 h at 4 °C; (5) centrifuge for 5 min at 15,000 × *g* and then transfer the supernatant to a fresh tube and dilute it to 2 mL with 1 % v/v sheep serum/TBST.

18. Remove the 10 % v/v sheep serum/TBST solution from embryos.
19. Replace with the preabsorbed antibody (from **step 18**).
20. Rock gently ON at 4 °C.

Hybridization: Day 3

21. Wash three times in 1× TBST for 5 min.
22. Wash four times in 1× TBST for 1 h.
23. Wash in 1× TBST ON.

Hybridization: Day 4

24. Wash three times in NTMT for 10 min.
25. Transfer embryos (or fragment of them) to a 4-well tissue culture dish.
26. Incubate samples with gentle shaking in NTMT + 3.5 μL NBT/mL + 3.5 μL BCIP/mL (you will need 1 mL/well) for few minutes protected from light. It is important to observe the samples each 10 min, because the optimal signal is different for each riboprobe and tissue.
27. Wash twice in PBT for 20 min.
28. Wash in 50 % v/v glycerol–PBT for 20 min.
29. Store in 80 % v/v glycerol–PBT at 4 °C. An example of the result can be found in Hernandez-Martinez et al. [4].

4 Notes

1. TUNEL technique works well with whole-mount samples but we do not recommend it because, due to the amount of reagents needed, the protocol becomes very expensive.
2. During the vital dye staining it is important not to over-stain the sample. If you accidentally leave it more than 15 min in AO or NR staining solution, it is necessary to do more PBS washes.
3. The most important step during the whole-mount immunohistochemistry is the fixation. We already tried 4 % of PFA and we almost never observed a signal or it was very low.
4. Do not incubate more than three limbs in a small volume, caspase-3 positive signal decrease.

5. During the combined in situ hybridization and LTK staining protocols do not use hydrogen peroxide, it will completely remove the LTK signal.

Acknowledgements

We are grateful to Concepción Valencia, Elizabeth Mata, Sergio González, Marcela Ramírez, and Andrés Saralegui for their technical assistance. This work was supported by grants IN218607 and IN225910 from PAPIIT-DGAPA.

References

1. Vaux DL (1993) Toward an understanding of the molecular mechanisms of physiological cell death. Proc Natl Acad Sci U S A 90:786–789
2. Clarke PG (1990) Developmental cell death: morphological diversity and multiple mechanisms. Anat Embryol 181:195–213
3. Cuervo R, Covarrubias L (2004) Death is the major fate of medial edge epithelial cells and the cause of basal lamina degradation during palatogenesis. Development 131:15–24
4. Hernandez-Martinez R, Castro-Obregon S, Covarrubias L (2009) Progressive interdigital cell death: regulation by the antagonistic interaction between fibroblast growth factor 8 and retinoic acid. Development 136: 3669–3678
5. Weil M, Jacobson MD, Raff MC (1997) Is programmed cell death required for neural tube closure? Curr Biol 7:281–284
6. Baehrecke EH (2002) How death shapes life during development. Nat Rev Mol Cell Biol 3:779–787
7. Conradt B (2009) Genetic control of programmed cell death during animal development. Annu Rev Genet 43:493–523
8. Meyer N, Kim SS, Penn LZ (2006) The Oscar-worthy role of Myc in apoptosis. Semin Cancer Biol 16:275–287
9. Matsuzawa A, Ichijo H (2001) Molecular mechanisms of the decision between life and death: regulation of apoptosis by apoptosis signal-regulating kinase 1. J Biochem 130:1–8
10. Finegan KG, Wang X, Lee EJ, Robinson AC, Tournier C (2009) Regulation of neuronal survival by the extracellular signal-regulated protein kinase 5. Cell Death Differ 16:674–683
11. Kumar S (2007) Caspase function in programmed cell death. Cell Death Differ 14: 32–43
12. Kurokawa M, Kornbluth S (2009) Caspases and kinases in a death grip. Cell 138:838–854
13. Ranger AM, Malynn BA, Korsmeyer SJ (2001) Mouse models of cell death. Nat Genet 28: 113–118
14. Riedl SJ, Salvesen GS (2007) The apoptosome: signalling platform of cell death. Nat Rev Mol Cell Biol 8:405–413
15. Chipuk JE, Green DR (2008) How do BCL-2 proteins induce mitochondrial outer membrane permeabilization? Trends Cell Biol 18: 157–164
16. Ekert PG, Vaux DL (2005) The mitochondrial death squad: hardened killers or innocent bystanders? Curr Opin Cell Biol 17:626–630
17. Widlak P, Garrard WT (2009) Roles of the major apoptotic nuclease-DNA fragmentation factor-in biology and disease. Cell Mol Life Sci 66:263–274
18. Ravichandran KS, Lorenz U (2007) Engulfment of apoptotic cells: signals for a good meal. Nat Rev Immunol 7:964–974
19. Bratton DL, Fadok VA, Richter DA, Kailey JM, Guthrie LA, Henson PM (1997) Appearance of phosphatidylserine on apoptotic cells requires calcium-mediated nonspecific flip-flop and is enhanced by loss of the aminophospholipid translocase. J Biol Chem 72:26159–26165
20. Zucker RM, Hunter ES 3rd, Rogers JM (1999) Apoptosis and morphology in mouse embryos by confocal laser scanning microscopy. Methods 18:473–480
21. Zucker RM, Hunter S, Rogers JM (1998) Confocal laser scanning microscopy of apoptosis in organogenesis-stage mouse embryos. Cytometry 33:348–354
22. Salas-Vidal E, Lomelí H, Castro-Obregón S, Cuervo R, Escalante-Alcalde D, Covarrubias L (1998) Reactive oxygen species participate in

the control of mouse embryonic cell death. Exp Cell Res 238:136–147
23. Chipuk JE, Moldoveanu T, Llambi F, Parsons MJ, Green DR (2010) The BCL-2 family reunion. Mol Cell 37:299–310
24. Nicholson DW, Ali A, Thornberry NA, Vaillancourt JP, Ding CK, Gallant M, Gareau Y, Griffin PR, Labelle M, Lazebnik YA et al (1995) Identification and inhibition of the ICE/CED-3 protease necessary for mammalian apoptosis. Nature 376:37–43
25. Niles AL, Moravec RA, Riss TL (2008) Caspase activity assays. Methods Mol Biol 414: 137–150
26. Srinivasan A, Roth KA, Sayers RO, Shindler KS, Wong AM, Fritz LC, Tomaselli KJ (1998) In situ immunodetection of activated caspase-3 in apoptotic neurons in the developing nervous system. Cell Death Differ 5:1004–1016
27. Bardet PL, Kolahgar G, Mynett A, Miguel-Aliaga I, Briscoe J, Meier P, Vincent JP (2008) A fluorescent reporter of caspase activity for live imaging. Proc Natl Acad Sci U S A 105: 13901–13905
28. Kroemer G, Martin SJ (2005) Caspase-independent cell death. Nat Med 11:725–730
29. Hughes FM Jr, Gorospe WC (1991) Biochemical identification of apoptosis (programmed cell death) in granulosa cells: evidence for a potential mechanism underlying follicular atresia. Endocrinology 129: 2415–2422
30. Garcia-Martinez V, Macias D, Ganan Y, Garcia-Lobo JM, Francia MV, Fernandez-Teran MA, Hurle JM (1993) Internucleosomal DNA fragmentation and programmed cell death (apoptosis) in the interdigital tissue of the embryonic chick leg bud. J Cell Sci 106(Pt 1):201–208
31. Zakeri ZF, Quaglino D, Latham T, Lockshin RA (1993) Delayed internucleosomal DNA fragmentation in programmed cell death. FASEB J 7:470–478
32. Gavrieli Y, Sherman Y, Ben-Sasson SA (1992) Identification of programmed cell death in situ via specific labeling of nuclear DNA fragmentation. J Cell Biol 119:493–501
33. Hewitson TD, Bisucci T, Darby IA (2006) Histochemical localization of apoptosis with in situ labeling of fragmented DNA. Methods Mol Biol 326:227–234
34. Bass BP, Tanner EA, Mateos San Martin D, Blute T, Kinser RD, Dolph PJ, McCall K (2009) Cell-autonomous requirement for DNaseII in nonapoptotic cell death. Cell Death Differ 16:1362–1371
35. Sanchez-Carbente MR, Castro-Obregon S, Covarrubias L, Narvaez V (2005) Motoneuronal death during spinal cord development is mediated by oxidative stress. Cell Death Differ 12:279–291
36. van Engeland M, Nieland LJ, Ramaekers FC, Schutte B, Reutelingsperger CP (1998) Annexin V-affinity assay: a review on an apoptosis detection system based on phosphatidylserine exposure. Cytometry 31:1–9
37. Darzynkiewicz Z (1990) Differential staining of DNA and RNA in intact cells and isolated cell nuclei with acridine orange. Methods Cell Biol 33:285–298
38. Gonzalez K, McVey S, Cunnick J, Udovichenko IP, Takemoto DJ (1995) Acridine orange differential staining of total DNA and RNA in normal and galactosemic lens epithelial cells in culture using flow cytometry. Curr Eye Res 14:269–273
39. Hotz MA, Gong J, Traganos F, Darzynkiewicz Z (1994) Flow cytometric detection of apoptosis: comparison of the assays of in situ DNA degradation and chromatin changes. Cytometry 15:237–244
40. Tsujimoto Y, Shimizu S (2005) Another way to die: autophagic programmed cell death. Cell Death Differ 12(Suppl 2):1528–1534
41. Melendez A, Neufeld TP (2008) The cell biology of autophagy in metazoans: a developing story. Development 135:2347–2360
42. Komatsu M, Waguri S, Chiba T, Murata S, Iwata J, Tanida I, Ueno T, Koike M, Uchiyama Y, Kominami E, Tanaka K (2006) Loss of autophagy in the central nervous system causes neurodegeneration in mice. Nature 441:880–884
43. Kuma A, Hatano M, Matsui M, Yamamoto A, Nakaya H, Yoshimori T, Ohsumi Y, Tokuhisa T, Mizushima N (2004) The role of autophagy during the early neonatal starvation period. Nature 432:1032–1036
44. Djehiche B, Segalen J, Chambon Y (1994) Ultrastructure of mullerian and wolffian ducts of fetal rabbit in vivo and in organ culture. Tissue Cell 26:323–332
45. Price JM, Donahoe PK, Ito Y, Hendren WH 3rd (1977) Programmed cell death in the Mullerian duct induced by Mullerian inhibiting substance. Am J Anat 149:353–375
46. Roberts LM, Visser JA, Ingraham HA (2002) Involvement of a matrix metalloproteinase in MIS-induced cell death during urogenital development. Development 129:1487–1496
47. Teng CS (2000) Differential expression of c-Jun proteins during mullerian duct growth

and apoptosis: caspase-related tissue death blocked by diethylstilbestrol. Cell Tissue Res 302:377–385

48. Mizushima N, Yoshimori T, Levine B (2010) Methods in mammalian autophagy research. Cell 140:313–326
49. Mizushima N, Yamamoto A, Matsui M, Yoshimori T, Ohsumi Y (2004) In vivo analysis of autophagy in response to nutrient starvation using transgenic mice expressing a fluorescent autophagosome marker. Mol Biol Cell 15:1101–1111
50. Maiuri MC, Zalckvar E, Kimchi A, Kroemer G (2007) Self-eating and self-killing: crosstalk between autophagy and apoptosis. Nat Rev Mol Cell Biol 8:741–752
51. Zuzarte-Luis V, Hurle JM (2002) Programmed cell death in the developing limb. Int J Dev Biol 46:871–876
52. Rodriguez-Leon J, Merino R, Macias D, Ganan Y, Santesteban E, Hurle JM (1999) Retinoic acid regulates programmed cell death through BMP signalling. Nat Cell Biol 1: 125–126
53. Pajni-Underwood S, Wilson CP, Elder C, Mishina Y, Lewandoski M (2007) BMP signals control limb bud interdigital programmed cell death by regulating FGF signaling. Development 134:2359–2368
54. Corson LB, Yamanaka Y, Lai KM, Rossant J (2003) Spatial and temporal patterns of ERK signaling during mouse embryogenesis. Development 130:4527–4537
55. Wilkinson DG, Nieto MA (1993) Detection of messenger RNA by in situ hybridization to tissue sections and whole mounts. Methods Enzymol 225:361–373

Chapter 19

Microscopic Computed Tomography-Based Virtual Histology of Embryos

Suresh I. Prajapati, David R. Rodriguez, and Charles Keller

Abstract

Advances in imaging technologies and computational capabilities have made possible novel methods for phenotypic assessments and visualization of detailed anatomical structures of whole embryos. We recently reported a rapid and inexpensive technique for achieving high-resolution virtual histology for phenotyping assessment of mouse embryos (Johnson et al., PLoS Genet 2:e61, 2006). By en bloc staining in a solution of electron-dense osmium tetroxide followed by volumetric X-ray computed tomography, whole embryos can be imaged at isometric resolutions as high as 2.5 μm, depending on the size of the specimen. The datasets generated by these techniques are compatible with state-of-the-art computational methods of organ pattern analysis. This method of Microscopic Computed Tomography (microCT)-based Virtual Histology of embryos allows one to rapidly and accurately phenotype transgenic embryos or to engage in developmental and reproductive toxicology studies of investigational drugs at better resolution, less time, and less expense than traditional histology, magnetic resonance microscopy, or the classical Wilson and Staples procedures.

Key words Virtual histology, microCT, High-resolution phenotyping of embryos

1 Introduction

Gene targeting in mice provides unprecedented insight into the function of genes and their roles in patterning the mammalian embryo [1, 2]. A complete understanding of mammalian development by the gene-targeting approach for every one of the approximately 25,000 or more mouse genes may seem like a very daunting task. Yet more than 40 % of known mouse genes have already been disrupted by gene targeting. Moreover, efforts have begun in the USA, Canada, and Europe to create collections of mouse lines with disruption of every known gene [3–6]. Developmental biologists therefore face the challenge of systemically analyzing morphological phenotypes as well as determine the quantitative contribution of each gene towards patterning of the embryo, en masse. Rapid,

Mark Lewandoski (ed.), *Mouse Molecular Embryology*, Methods in Molecular Biology, vol. 1092,
DOI 10.1007/978-1-60327-292-6_19, © Springer Science+Business Media New York 2014

inexpensive, and easily accessible high-throughput methods of high-resolution anatomical imaging (described here), as well as stage-specific statistically averaged wild-type morphological atlases that can be used to discern normal variation from mutant phenotype, will be necessary components of this "phenomic" analysis tool-set [7].

More than a decade ago, magnetic resonance microscopy (MRM) was introduced as a faster technique than classical histology in the screening of E6.5–E19 mouse embryos for mutant morphological phenotypes [8] (reviewed by Schneider and Bhattacharya [9] and Tyszka et al. [10]). However, MRM requires specialized and expensive equipment which are not widely available and scans at useful resolutions (12–43 μm, but generally 25 μm) require 9–14 h of instrument time [11, 12] at a cost of approximately US$200 per hour.

We introduce a new method of obtaining virtual histology using X-ray microscopic computed tomography (microCT). By this technique, mid-gestation embryos (up to mouse embryonic day 13.5) can be scanned at up to 2.5 μm resolution with comparable or less scan time, and at a fraction of the expense of MRM. This new method, which employs osmium tetroxide to differentially stain tissues, enables the quantitative, three-dimensional analysis of mouse developmental defects for researchers across a broad range of institutions at relatively little expense and at resolutions and throughput comparable to or exceeding magnetic resonance methods (*see* **Note 1**).

2 Materials

2.1 Personal Protection Equipment (See Note 2)

2.1.1 Sodium Cacodylate Trihydrate

- Respiratory protection: None needed except when using spray applicators.
- Protective gloves: Neoprene.
- Eye protection: Goggles or face shield.
- Additional clothing and/or equipment: Rubber apron.

2.1.2 Glutaraldehyde, 25 % Aqueous Solution

- Use air-supplied mask in high concentrations.
- Wear polyethylene, butyl, or nitrile protective gloves.
- Wear vapor-proof goggles or face shield.
- Rubber overshoes and chemical apron may be required.

2.1.3 Osmium Tetroxide

- Wear appropriate NIOSH/MSHA approved respirator, chemical resistant gloves, safety goggles, and other protective clothing.

- Under no circumstances should this material be used with open tipped shoes. We also recommend a safety shower be in close vicinity of the work area and this would include also a safety eye bath.
- Use only in a chemical fume hood and under no circumstances, ever breathe in any of the vapors.
- Do not get the material in the eyes, on the skin, or on clothing. Always wash thoroughly after handling.

2.2 2× Cacodylate Buffer (0.2 M)

ddH_2O	100 ml
NaCacodylate trihydrate	21.4 g (0.2 M final)
0.2 M HCl	100 ml
ddH_2O	To 490 ml
Adjust pH to 7.2 with NaOH or HCl	
ddH_2O	To 500 ml

Filter-sterilize the solution. Final solution can be stored at room temp up to several months.

2.3 1× Cacodylate Buffer (0.1 M)

2× Cacodylate buffer	25 ml (0.1 M final)
ddH_2O	25 ml

2.4 Fixing Solution

2× Cacodylate buffer	25 ml (0.1 M final)
Glutaraldehyde	To 3 %
ddH_2O	To 50 ml

2.5 OsO_4 Staining Solution

2× Cacodylate buffer	1.5 ml (0.1 M final)
Glutaraldehyde	To 1 % (v/v)
Osmium Tetroxide (OsO_4)	To 1 % (wt/vol = g/dL)
ddH_2O	To 3 ml

Use fresh solution each time.

2.6 Other Necessary Solutions

Along with above mentioned solutions, 1× PBS, 100 % EtOH, 10 % Buffered Formalin, will also be required.

3 Methods

3.1 Embryo Preparation

1. Dissect embryo in cold PBS after first cutting the umbilical cord cleanly with a *scalpel* (not with scissors or by tearing; this is important to optimize vessel imaging).
2. Carefully remove both the amnion and the inner thin serosa membrane (have as little extraembryonic tissue as possible).
3. Place the cleaned embryo in PBS on ice.
4. When feasible, take one or two photographs of the embryo under a dissecting microscope.
5. Pre-label a glass screw-lock tube for each embryo with a lid-label. All embryos must be individually labeled by a unique identifier number (*see* **Note 3**).

3.2 Embryo Fixation

1. Transfer embryo to the glass screw-lock tube using a wide-mouth plastic transfer pipet.
2. Wash embryo by rocking with PBS for 2–10 min three times before fixing to remove all extraneous membrane and tissue bits (this is important to prevent artifacts).
3. Incubate in either 10 % buffered formalin or Fix Solution rocking overnight at room temp with a fixative volume about ten times the volume of the embryo itself. Well fixed embryos can sit for up to 6 months without adversely affecting the subsequent staining methods.

3.3 Embryo Staining with Osmium Tetroxide (OsO_4)

1. Using a glass Pasteur pipet, aspirate off most (95 %) of the Fix Solution (*see* **Note 4**).
2. Fill the tube with OsO_4 Staining Solution (*see* **Note 5**). Protect OsO_4 from light by wrapping in foil.
   (a) Stain with 1 % OsO_4 staining solution in 4 ml acid-prepped glass vial.
3. Rock 6 h at 4 °C (may require longer if tissue >5 mm thick) (please refer **Note 6**).
4. The next day, do rinses.
   (a) With a Pasteur pipet, aspirate off most (95 %) of the OsO_4 Staining Solution. Dispose of OsO_4 solution in compliance with EHS guidelines.
   (b) Incubate with 1× Cacodylate Buffer twice, rocking for 30 min each time. This solution must be disposed of according to EHS guidelines.
   (c) Rinse in PBS twice for 15 min (treat as OsO_4 containing waste).

3.4 Dehydration and Transition to Ethanol

1. Incubate in 25 % ethanol in PBS for 15–30 min.
2. Incubate in 50 % ethanol in PBS for 15–30 min.
3. Incubate in 75 % ethanol in PBS for 15–30 min.
4. Incubate in 100 % ethanol for 15–30 min.
5. Incubate in 100 % ethanol for 15–30 min.
6. Pack embryo in a loose sponge in ethanol in a microcentrifuge tube.
7. Proceed to Scanning.

3.5 MicroCT Scanning

1. Perform a high-resolution (2.5–27 μm) scan of the specimen.

4 Notes

1. All osmium-containing solutions must be handled in a chemical hood employing gloves, lab coat, and face shield. See below for additional safety information. Done improperly, this protocol can result in severe injury.
2. This procedure is extremely dangerous and should only be undertaken by mature scientists (not students) and only after careful review of MSDS documents, and with the approval of your Institutional Biosafety/Environmental Health & Safety (EHS) committee.
3. Acid-treat the glass vial with 0.1 N HCl and then rinse the vial with copious ddH_2O (use 4 ml tubes for embryos).
4. This solution needs to be discarded in a special container.
5. All incubations are done while rocking.
6. The penetration rate at room temperature is approximately 1 mm/h; however, if poorly fixed prior to OsO_4 staining, staining will be suboptimal. Osmium tetroxide in the presence of light, heat, or organic materials will convert to osmium dioxide, a black compound that is ineffective at fixing tissue.

Acknowledgements

The subject matter of this chapter is disclosed in International Patent Application Publication WO2007/08964, titled "Process and Apparatus for Imaging," which is exclusively licensed to Numira Biosciences Inc., 2151 Michelson Drive Suite 250 Irvine, CA 92612.

Conflict of Interest Statement. C.K. is cofounder of Numira Biosciences.

References

1. Johnson JT, Hansen MS, Wu I, Healy LJ, Johnson CR, Jones GM, Capecchi MR, Keller C (2006) Virtual histology of transgenic mouse embryos for high-throughput phenotyping. PLoS Genet 2:e61
2. Capecchi MR (2005) Gene targeting in mice: functional analysis of the mammalian genome for the twenty-first century. Nat Rev Genet 6:507–512
3. Austin CP, Battey JF, Bradley A, Bucan M, Capecchi M, Collins FS, Dove WF, Duyk G, Dymecki S, Eppig JT, Grieder FB, Heintz N, Hicks G, Insel TR, Joyner A, Koller BH, Lloyd KC, Magnuson T, Moore MW, Nagy A, Pollock JD, Roses AD, Sands AT, Seed B, Skarnes WC, Snoddy J, Soriano P, Stewart DJ, Stewart F, Stillman B, Varmus H, Varticovski L, Verma IM, Vogt TF, von Melchner H, Witkowski J, Woychik RP, Wurst W, Yancopoulos GD, Young SG, Zambrowicz B (2004) The knockout mouse project. Nat Genet 36:921–924
4. Floss T, Schnutgen F (2008) EUCOMM–the European conditional mouse mutagenesis program. Methods Mol Biol 435:127–138
5. Friedel RH, Seisenberger C, Kaloff C, Wurst W (2007) Conditional gene trapping using the FLEx system. Brief Funct Genomic Proteomic 6:180–185
6. Skarnes WC, von Melchner H, Wurst W, Hicks G, Nord AS, Cox T, Young SG, Ruiz P, Soriano P, Tessier-Lavigne M, Conklin BR, Stanford WL, Rossant J (2004) A public gene trap resource for mouse functional genomics. Nat Genet 36:543–544
7. Jacobs RE, Ahrens ET, Dickinson ME, Laidlaw D (1999) Towards a microMRI atlas of mouse development. Comput Med Imaging Graph 23:15–24
8. Smith BR, Johnson GA, Groman EV, Linney E (1994) Magnetic resonance microscopy of mouse embryos. Proc Natl Acad Sci U S A 91:3530–3533
9. Schneider JE, Bhattacharya S (2004) Making the mouse embryo transparent:Identifying developmental malformations using magnetic resonance imaging. Birth Defects Res C Embryo Today 72:241–249
10. Tyszka JM, Fraser SE, Jacobs RE (2005) Magnetic resonance microscopy: recent advances and applications. Curr Opin Biotechnol 16:93–99
11. Schneider JE, Bamforth SD, Farthing CR, Clarke K, Neubauer S, Bhattacharya S (2003) High-resolution imaging of normal anatomy, and neural and adrenal malformations in mouse embryos using magnetic resonance microscopy. J Anat 202:239–247
12. Smith BR, Linney E, Huff DS, Johnson GA (1996) Magnetic resonance microscopy of embryos. Comput Med Imaging Graph 20:483–490

Chapter 20

Collection and Preparation of Rodent Embryonic Samples for Transcriptome Study

Yelena Golubeva and David Symer

Abstract

The need for large-scale collection of rodent embryos and individual embryonic tissues for genomic and proteomic studies requires modification of traditional practices of embryo necropsy. The sample intended for transcriptome study should be rapidly dissected and stabilized to preserve its molecular integrity. The retrieval of high-quality RNA, DNA, and proteins from the target tissue is crucial for informative molecular analysis (e.g., gene profiling on microarray platform). We present a reliable method of collection and preparation of rodent embryos for genomic studies supported by detailed protocols and RNA extraction results for different stages of mouse embryonic development.

Key words Rodent embryo dissection, Rodent embryo staging, RNAlater®, RNA extraction, RNA integrity, Transcriptome study

1 Introduction

The embryological techniques, notoriously laborious and time consuming, have been confined to the description of developmental stages and individual features, and involved a limited number of embryos at a time. In the past, the majority of embryo studies required the stabilization of morphology and antigens of interest by fixation. Most often the stage of development was determined by examination of histological sections. Embryo staging by external features was performed in cold phosphate buffered saline (PBS) under a dissecting microscope prior to fixation.

In modern developmental biology, traditional histological methods have been expanded to laser capture microdissection, in situ hybridization, genomic amplification, gene mutation detection, and protein and gene expression profiling. The success of many downstream molecular applications depends on the sample molecular quality [1–3].

Mark Lewandoski (ed.), *Mouse Molecular Embryology*, Methods in Molecular Biology, vol. 1092,
DOI 10.1007/978-1-60327-292-6_20, © Springer Science+Business Media New York 2014

Transcriptome study of developing embryos presents a particular challenge because a large number of staged embryos must be collected, processed, and analyzed in a timely, standard manner.

We have developed a systematic approach to a large-scale embryo collection to include effective mating, modified necropsy in RNAlater® solution to preserve RNA integrity, embryo staging by appearance of external and internal embryonic features after incubation in RNAlater®, and standard RNA extraction protocols for whole embryo and embryonic organs. The reliability of collection method is demonstrated with RNA quality and yield.

Embryo necropsy procedure and staging tables are supported by photographs of embryos processed in RNAlater® solution.

2 Materials

Materials, instruments, and equipment in Subheadings 2.1 and 2.2 complement the list in Subheadings 2.3 and 2.4.

2.1 Embryo Transcriptome Study Considerations

1. Estrus cycle monitor EC40 Lab Coat Model, Fine Science Tools, Inc., #22500-10.
2. Mouse Probe 3.8 mm diameter, Fine Science Tools, Inc., #22500-11.
3. Rat Probe, Fine Science Tools, Inc., #22500-12.

2.2 Embryo Staging Overview

1. Tissue Collection: RNA Stabilization Solution. Applied Biosystems, # AM 7020 (*see* **Note 1**).

2.3 Embryo Necropsy Overview

1. Paraplast®. McCormick Scientific, St. Louis MO, USA, # 501006.
2. Paraplast Plus® McCormick Scientific, St. Louis MO, USA, # 502004.
3. Parafilm® VWR, # 291-1211.
4. Sylgard®, Silicon Elastomer, two Parts. World Precision Instruments, Inc. # SYLG184.
5. RNase AWAY™, Molecular BioProducts, #7000.
6. UV treated nuclease free water from any pure water system (e.g., Picopure®2UVPlus, HYDRO Inc.).
7. Minuten pins 0.15 mm. BioQuip #1208SA.
8. Ambidextrous Pinning Forceps, BioQuip #4743.
9. Pin Vise. BioQuip #4845.
10. Stainless steel insect pins. BioQuip #1208S00.
11. Sapphire blade, 1 mm wide. World Precision Instruments, Inc. # 500314.

12. Sapphire knife handle, 13 cm long. World Precision Instruments, Inc. # 500317.
13. Camel hairbrush #2. BioQuip #1153A.
14. Disposable Transfer Pipettes. Samco Scientific. # 202-1S.

2.4 Age-Related Embryo Necropsy Protocols

1. Necropsy cork board (Mopec, Oak Park, Michigan, USA), # BC005.
2. Utility Wipes, 12" × 10.25", Wypall L10, Kimberly-Clark®, 34155-30.
3. Cryo tubes (VWR) # 66008-956—2 ml, 66008-957—4 ml.
4. Thermal Labeling System, BRADY TLS PC Link™, #2742-10310-0000.
5. Portable CryoThermal Labels, BRADYTLS2200®/TLS PCLink™, # PTL-32-427.
6. Collection Petri Dish, polystyrene 60 × 15 mm, sterile, Falcon Fisher Scientific, 1007-08772B.
7. Collection Petri Dish, polystyrene 35 × 10 mm, sterile, Falcon Fisher Scientific, 351008.
8. Tissue Culture Plate. Multiwell™ 12 Well. . Falcon Fisher Scientific, 353225.
9. Tissue Culture Plate. Multiwell™ 6 Well. . Falcon Fisher Scientific, 353046.
10. Specimen container with cap. VWR. #25384-146.
11. Tube, conical, 50 ml capacity, polypropylene, sterile, Falcon, #2098.
12. Tube, centrifuge, polypropylene 2.0 ml, Neptune, CLP #3465.S.
13. Re-Freez-R-Brix Foam Refrigerant. Polar Tech Industries, Inc. # RB-15.
14. Koolit®Gel Paks. VWR. # 33500-589.
15. Styrofoam box.
16. Wet ice.
17. Dry ice.
18. Liquid nitrogen.
19. Dewar's flask for liquid nitrogen. Thermo Scientific.
20. Kimwipes, EX-L, 4.5″ × 8.5″, Kimberly-Clark®, #34155.
21. Scalpel, disposable, sterile #20 and #11, Bard Parker, #1620, 1611.
22. Scalpel/single edge razor blades (American Safety Razor, #62-0167).
23. Forceps #5 BioQuip #4524 (*see* **Note 2**).
24. Forceps #5a BioQuip #4525.

25. Forceps #7 BioQuip #4527.
26. McPherson-Vannas Scissors, curved 3 mm blades, 0.1 mm tips. World Precision Instruments, Inc. # 501233.
27. Tissue forceps, 1×2 teeth 5.5″. Roboz Surgical Instrument Co., Inc, # RS-8164.
28. Light operating scissors, $4^3/_4$″, straight sharp-blunt, Roboz # RS-6700.
29. Obrien Angular Scissors. Roboz Surgical Instrument Co., Inc, # RS-5928 (*see* **Note 3**).
30. Shaker.

2.5 Extraction, Quantification, and Quality Control of Embryonic RNA

1. TRIzol® Reagent, #15596-018, Invitrogen™ (*see* **Note 4**).
2. Tube, conical, 50 ml capacity, polypropylene, sterile, Falcon, #2098.
3. Screw Cap Round Bottom Centrifuge Tube 13 ml, PP, Sarstedt, #60.540.500.
4. OMNI Tissue Homogenizer TH-115, OMNI International, Inc., #THP115.
5. OMNI Tips-Rotor Stator Generator Probes, OMNI International, Inc., #30750H-AC (*see* **Note 5**).
6. Timers.
7. Wet ice.
8. Dry ice.
9. 3 M Sodium acetate buffer, pH 5.2, Sigma, S-7899.
10. Isopropanol, Richard-Allen Scientific, 9511.
11. Ethanol USP, Absolute-200 proof.
12. Chloroform, EM Science, GX 1055-6.
13. Linear Polyacrylamide (25 mg/ml), Gen-Elute™-LPA, Sigma, 56575.
14. Tube, centrifuge, polypropylene 1.8 ml.
15. Phase Lock Gel Tube (PLG), 2 ml, Heavy, Eppendorf, 955754045.
16. RNase--free pipette tips with barrier.
17. NanoDrop 1000™ spectrophotometer. Thermo Scientific.
18. 2100 Bioanalyzer, Agilent Technologies Inc. #G2938C.
19. RNA 6000 Pico LabChip kit, Agilent Technologies Inc., #5067-1513.
20. RNA 6000 Nano LabChip kit, Agilent Technologies Inc., #5067-1511.
21. RNA Storage Solution, Applied Biosystems, # AM7001.
22. RNase-free water.

3 Methods

3.1 Embryo Transcriptome Study Considerations

1. Define sample size (number of embryos or organs to be pooled together) to create a single sample for nucleic acid or protein extraction to minimize variation in downstream analysis (e.g., microarray bias) (*see* **Note 6**).
2. Estimate number of animals in the study (*see* **Note** 7) and obtain the females from the same source.
3. House females in groups of up to 3–5 animals per cage and maintain standard colony conditions. This way, the majority of females will be synchronized in the reproductive cycle within 3 days.
4. Use an estrus monitor for mating females for embryonic sample collection [4] (*see* **Notes 8** and **9**).
5. Restrict female mating, estrus measurements and embryo harvest to a particular time of day to minimize the cases when the majority of embryos in the litter do not match the day post conception (dpc) age.
6. Use standard necropsy, euthanasia, and sample preparation and extraction protocol for the entire study (*see* **Note 6**).
7. Use RNase-free precaution for necropsy, sample homogenization, and RNA extraction (*see* **Note 10**).
8. Perform pregnant female and embryo necropsy in accordance with the Guide for the Care and Use of Laboratory animals [5, 6].
9. Inspect pregnant females for lesions, and if present, exclude the litter from the study or process and analyze it separately (*see* **Note 11**).
10. Restrict necropsy time (time between female sacrifice and placement of embryos into RNAlater® solution) to 10 min for 17–18 dpc, to 15 min for 12–16 dpc, and to 30 min for 9–11 dpc embryos (*see* **Note 12**).
11. Stage all the embryos in a litter to create a uniform sample (*see* **Note 13**).
12. Keep notes or cage card with female identification number (ID) and embryo staging table. Save the females ears for ID confirmation, and freeze its tail for genotype (*see* **Note 14**).
13. Avoid contamination of the dissected target with a different tissue type (*see* **Note 15**).

3.2 Embryo Staging Overview

1. For transcriptome samples we stage embryos after placement in RNAlater. In RNAlater® the appearance of the whole embryo and the majority of features is different from the appearance of the embryo in PBS (Figs. 1, 2 and 3a, c).

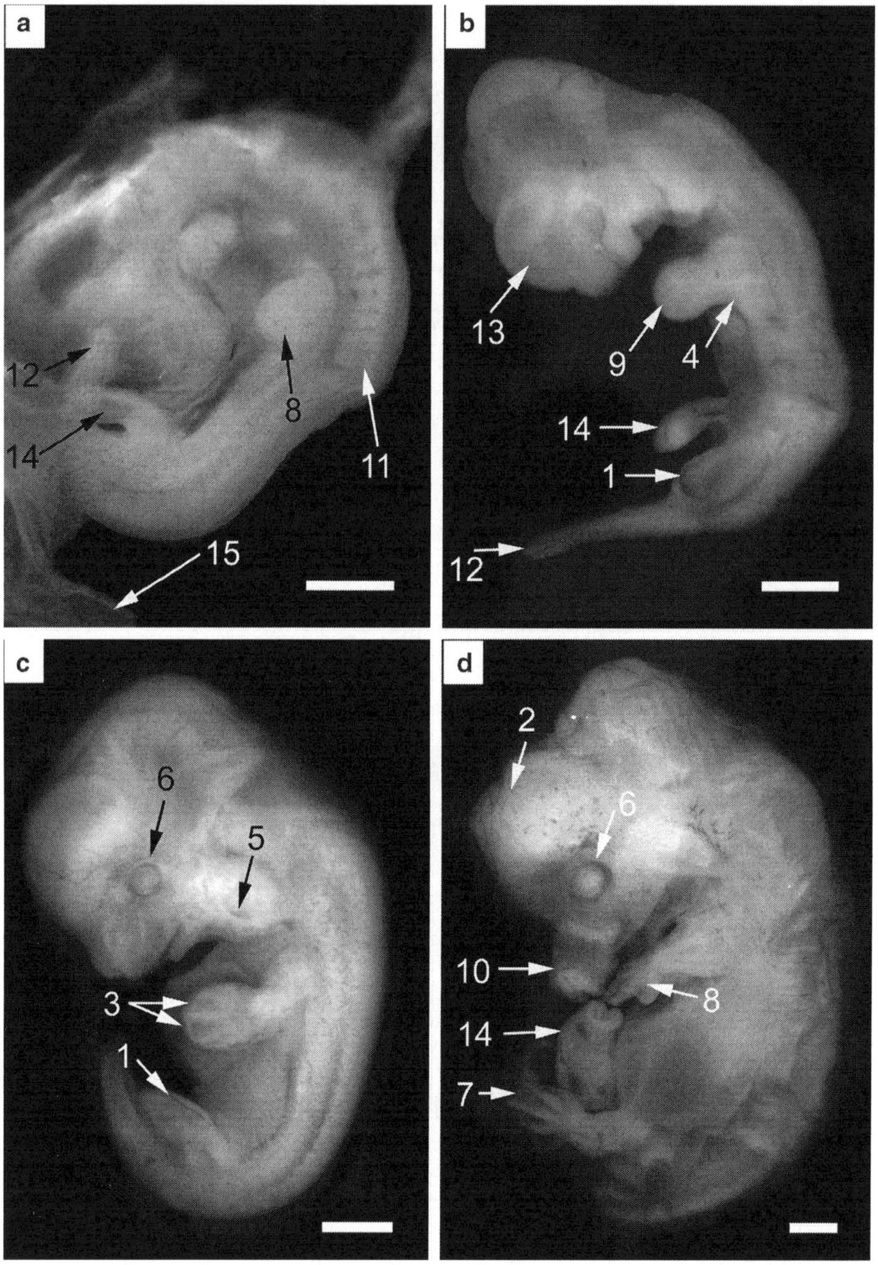

Fig. 1 Effect of RNAlater® on mouse embryo appearance. (**a**) An advanced 10 dpc embryo in natural position (incubation 30 min). (**b**) An 11 dpc embryo (unfolded) (incubation 1 h). (**c**) A 12 dpc embryo (incubation 12 h). (**d**) A 15 dpc embryo (incubation 4 h). (**a**–**d**) Bar = 1,000 μm. (*1*) Apical ridge of the foot plate, (*2*) brain, (*3*) digital interzones of the hand plate, (*4*) distal region of the hand plate, (*5*) external meatus, (*6*) eye, (*7*) foot plate, (*8*) hand plate, (*9*) proximal region of the hand plate, (*10*) rostrum, (*11*) somites, (*12*) tail, (*13*) telencephalic lobes, (*14*) umbilical hernia, (*15*) yolk sac

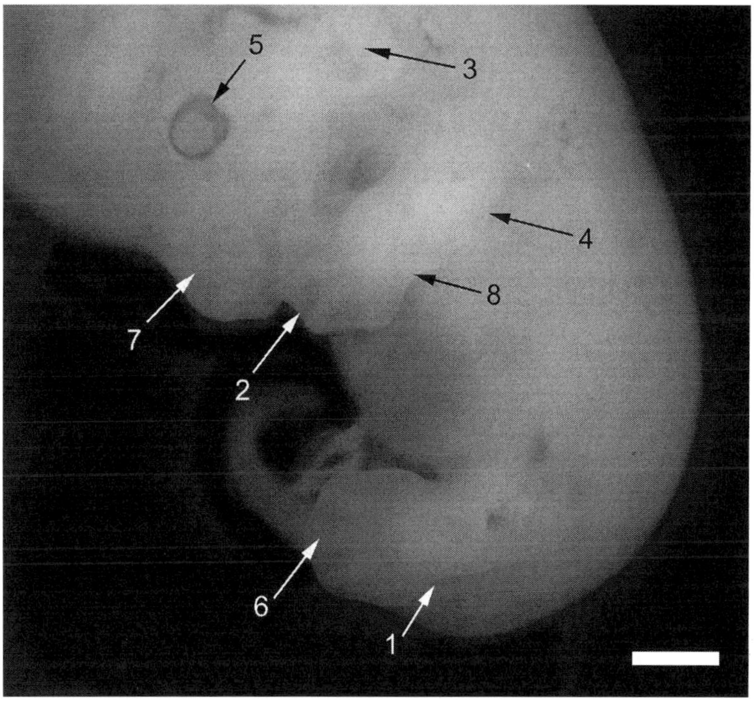

Fig. 2 Appearance of a 13 dpc embryo after a 5-min incubation in RNAlater® (bar = 1,000 μm). (*1*) Ankle, (*2*) digital interzone, (*3*) ear pinna, (*4*) elbow, (*5*) eye pigmentation, (*6*) foot plate, (*7*) vibrissae, (*8*) wrist

Accordingly, we stage embryos by anatomical features readily recognizable after incubation in RNAlater® (*see* Tables 1, 2, 3, 4, 5, 6, 7, 8, 9, and 10).

2. Embryo staging and terminology are those of M. H. Kaufman [7].

3. Since histological sections cannot be used, staging is based on observation of external and internal anatomical features: embryo body, hand and foot plate, eye, head, mouth and tongue, palate, ear, tail, umbilical hernia, hair follicles, liver, lung, heart, thymus, kidney, adrenals, spleen, stomach, pancreas, gonads, salivary gland, brain, ribs, and tracheal cartilaginous rings (*see* **Note 16**) (Figs. 1, 2, 3, 4, 5, 6, 7, 8, 9, 10, 11, 12, 13, 14, 15, 16, 17, 18, 19, 20, 21, 22, and 23).

4. A number of features are specific only for a particular stage of embryonic development [7–9]. We use these specific features as a diagnostic for a particular stage of development (*see* Table 1).

5. Some features gradually go through a transition during a particular stage and are no longer present in the following stage (e.g., webbing between the digits of hand and foot plate).

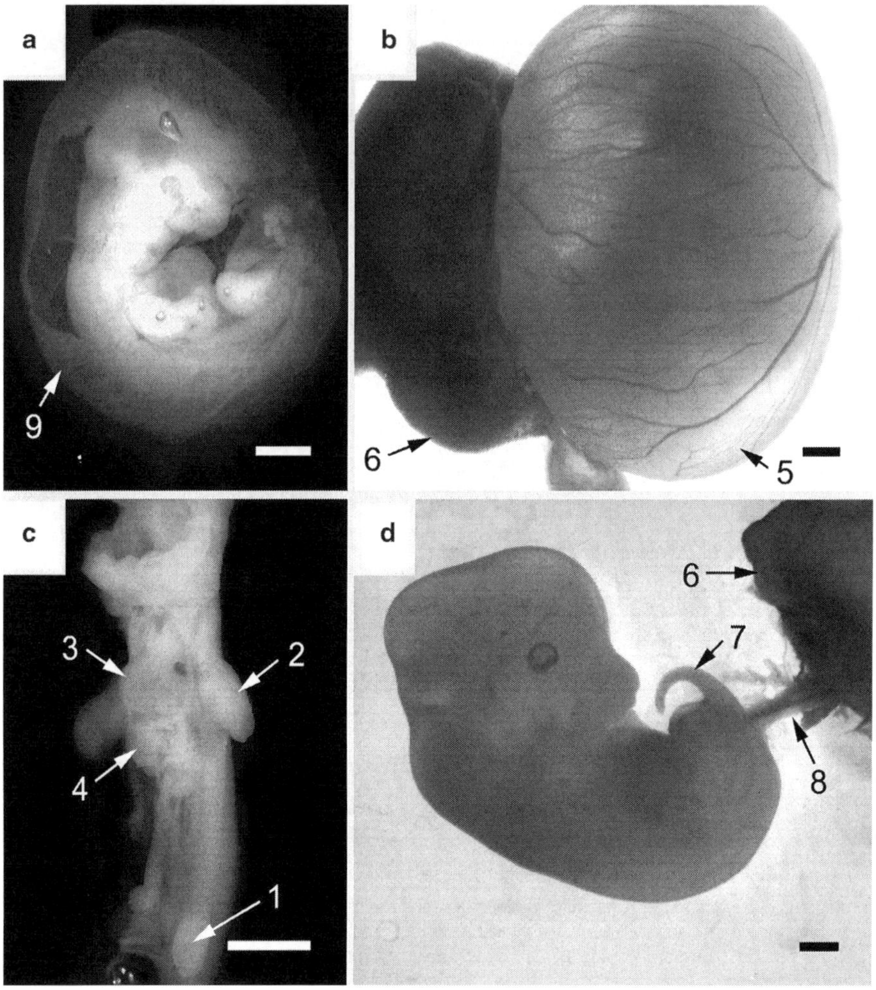

Fig. 3 Mouse embryo necropsy: Collection of embryos in RNAlater®. (**a**) An 11 dpc embryo inside the yolk sac (incubation 3 h). (**b**) A 13 dpc placenta–yolk sac–embryo. (**c**) An 11 dpc embryo unrolled from its natural position prior to organs harvest (incubation 20 min). (**d**) A 13 dpc embryo dissected out of the yolk sac with umbilical cord attached to placenta. (**a**–**d**) Bar = 1,000 μm. (*1*) Foot plate, (*2*) hand plate, (*3*) heart, (*4*) liver, (*5*) location of incision to release the embryo from the yolk sac, (*6*) placenta, (*7*) somites in the tail, (*8*) umbilical cord, (*9*) yolk sac

Some organs gradually change their size and position inside the body during a particular stage, and then only increase in volume during the next stage (e.g., thymus, liver, kidney, gonads). Due to the different growth rates of embryonic organs, the size ratio of internal organs as well as the body appearance changes from stage to stage, and at some stages the change is particularly noticeable (e.g., heart and liver at 14 dpc). Some transitional features might develop faster or slower, and in the embryo staged as 15 dpc it is not unusual to find one or two features of 14 or 16 dpc embryo.

Table 1
Age-related key features readily observable after embryo incubation in RNAlater®

Embryo age	Key features
7 dpc	Ectoplacental cone. No somites
8 dpc	Somites. Embryo is not turned
9 dpc and older	Embryo is turned
10 dpc and older	Both limb buds are present. No otic pit and neuropore. Rathke's pouch
11 dpc and older	Cervical somites are not visible above the hand plate. Caudal part of the tail is slender. Folds along Rathke's pouch. Umbilical hernia. Genital tubercle
12 dpc and older	Rathke's pouch is closed. Anterior and posterior naris. Vibrissa primordia. Eye lids. Hand plate with digital interzones. Multi-loop umbilical hernia. Visible spleen and thymus primordium
13 dpc and older	Somites are not visible in the lower trunk. Tail does not extend beyond the hand plate. Ear pinna. Hand plate digits are splayed out. Foot plate indented. Visible adrenals and thyroid primordium. Right lung lobe demarcation. Pre-cartilaginous condensation in the ribs
14 dpc and older	Ear pinna is turned forward. Hand plate digits are separated. Foot plate digits are splayed out. Superficial veins. Heart is much smaller proportionally than liver. Testes and ovaries are anatomically different
15 dpc and older	Somites are not visible in tail. Anterior part of the tongue has a spatula appearance. Palatal shelves touch in the midline. Foot plate digits are separated. Heart is smaller than lung. Diaphragm is closed. Gall bladder is visible in mouse embryo
16 dpc and older	Back of the embryo is straight. Palatal shelves are fused. Prominent rugae. Nail primordia on hand and foot plate digits. Liver does not occupy the entire peritoneal cavity. Ovaries are much smaller than testes. Discrete ribs and xiphoid process. Repositioning of umbilical hernia
17 dpc and older	Caudal part of the body is long. Eyelids are fused. Upper lip vibrissa erupted. No umbilical hernia. Skin wrinkled. Thymus lobes are close together
18 dpc	Head is proportionally smaller in relation to the body. Tail does not protrude beyond the foot plate when embryo positioned on its back. Eyes are barely visible through the lids. Ear pinna covers the meatus. Testes are inside the pelvis. Ovarian capsule around ovaries

To accommodate transitional changes for transcriptome sample collection, we evaluate embryos within a stage as early, advanced, or late by a combination of stable and transitional features (*see* Tables 1, 2, 3, 4, 5, 6, 7, 8, 9, and 10). In this way, the transcriptome sample can be defined as either a whole range of embryos (early, advanced, and late), or narrowed to an, for example, early, advanced, or late group within a stage. This approach is especially useful for early stages of rodent development (10–15 dpc), and has been developed on 108 litters

Table 2
External features of RNAlater® processed 9–18 dpc embryos (transitional features are marked with*): ear, eye and foot plate

Embryo age	Staging features		
	Ear	Eye	Foot plate
9 dpc		Optic eminence is shallow (Fig.15b)	No hind limb bud. Elevation* at the tail end progressively develops through the stage
10 dpc		Optic eminence* is prominent and turns into optic pit with variable indent depth* through the stage (Figs. 1a and 16a–c)	The sign of hind limb bud* in the middle of the tail with the beginning of proximal and distal differentiation*, volume increase* and rostral ascent* through the stage (Figs. 4a and 16a, c)
11 dpc		Progressive formation of lens vesicle* as elevated optic eminence through the stage. Peripheral margins of the eye are well defined. Beginning of eye pigmentation in non-albino mice (Fig.17a)	Division into proximal and distal part is evident, apical ridge* develops to the end of the stage (Fig. 1b)
12 dpc	External meatus, no pinna (Fig. 1c)	Prominent eye pigmentation (outer layer of the retina) in non-albino strains. Upper and lower eyelids (Fig. 21a)	Progressive development of angular appearance* through the stage (Figs. 1c and 23b)
13 dpc	Ear pinna is formed, and partially covers the meatus. Ear undergoes rostral ascent* (Fig. 2)	Pigmentation is as a wide strip. Eye lids* size increases through the stage (Fig. 2)	Indented, with webbing in digital interzones, smooth angular appearance, smaller than hand plate. Ankle and no knee. Foot plates embrace the tail (Figs. 2 and 23a)
14 dpc	Ear pinna is turned f orward, and covers less than a half of the meatus (Fig. 21d)	Eye is in anterior position. It has wide pigmentation in non-albino strains. In albino strains unpigmented area is well defined. Eyelids* increase in volume through the stage (Fig. 21b, c)	Foot plate digits are splayed out. Progressive separation* of digits through the stage (Fig. 23d)

15 dpc	Ear pinna covers more than a half of a meatus	Eye is in adult position (Fig. 1d)	Foot plate digits are separated and splayed out. Plantar surface developed. Residual webbing* progressively disappears through the stage (Figs. 1d and 23f)
16 dpc	Ear pinna covers two thirds of a meatus	Progressive closing of eyelids* through the stage	Three middle digits are close to each other. The lateral digits are splayed out. No webbing between the digits. Nail primordia on all digits. Pads on plantar (ventral) surface. Foot plate digits are of the same length. Prominent ankle
17 dpc	Ear pinna almost covers a meatus, except anterior part	Eye lids are fused, but eyes are visible through the lids	Foot plate three middle digits are almost parallel, 1st and 5th splayed. Nail primordia are prominent on all digits. Pads are smaller than on hand plate
18 dpc	Ear pinna completely covers the meatus	Eyes are barely visible through the lids	Plantar pads of the foot are well developed. All the digits are parallel

Table 3
External features of RNAlater® processed 9–18 dpc embryos (transitional features are marked with*): genital tubercle, hair follicles and hand plate

Embryo age	Staging features		
	Genital tubercle	Hair follicles	Hand plate
9 dpc			Limb bud stage. Limb bud is visually at the lower heart level (Fig. 15b)
10 dpc			Hand plate* progressively increases in volume through the stage and gradually undergoes proximal and distal differentiation* and rostral ascent* (Figs. 1a, 4a, and 16a, c)
11 dpc	Tubercle* progressively develops through the stage to become prominent (Fig. 11a)		Strong division* into proximal and distal part. Hand plate is well separated from the base* to the end of the stage. Paddle shaped hand plate develops an apical ridge* (thick line of ectoderm, framing the plate) to the end of the stage (Figs. 1b and 17a)
12 dpc	Prominent tubercle		Clear angular shape with digital interzones (Figs. 1d and 15a)
13 dpc	Increase in size (Fig. 5b)		Indented, webbing in digital interzones. Digits are splayed out, but not separated. Elbow and wrist are distinguishable (Fig. 2)
14 dpc		Scarce follicles in cephalic and pelvic region	Hand plate digits are separated, and splayed out. Webbing* progressively disappears through the stage (Fig. 23c)
15 dpc		Moderate number of hair follicles all over the body, except tail, ankle and wrist	Digits splayed out. No webbing (Figs. 1d, 4c, and 23f)
16 dpc		Large number of hair follicles on the head, trunk, proximal limb	Nail primordia in digits 2–5. Pads on palmar (ventral) surface. Four digits are not yet parallel. Elbow is prominent
17 dpc		Hair follicles cover the whole body	Nail primordia on all digits. Four digits are parallel to each other. Pads* become more prominent through the stage. First digit is shorter than the rest of the digits
18 dpc		Hair follicles cover the whole body	Nail primordia on all digits. Four digits are parallel to each other. Pads* increase in size through the stage. First digit is much shorter than the rest of the digits

Table 4
External features of RNAlater® processed 9–18 dpc embryos (transitional features are marked with*): head, mouth and nose

Embryo age	Staging features: Head	Mouth	Nose
9 dpc	The groove in the middle of the head is not prominent. Beginning of two telencephalic lobes in the brain (Fig.15b)	The first branchial arch has a maxillary and mandibular component. Second branchial arch is visible. Midline gap between mandibular processes is wide (Fig.15b)	Olfactory placode (place of future nasal) has rounded profile (Fig. 15b)
10 dpc	Telencephalic lobes* are well defined (gradually more prominent through the stage). Groove between the lobes is obvious (Fig.16)	The 1st branchial arch is C-shaped. Maxillary and mandibular components are easily distinguishable. The mandibular components* are not fused in the midline, but the gap* in between narrows through the stage (Fig. 16b–d)	Olfactory placode* slightly flattened with indentation deepening through the stage and forming olfactory (nasal) pit. Margins around the pit are not fused (Fig. 16b)
11 dpc	Telencephalic lobes are separated from olfactory (nasal) region and midbrain by deep transverse grooves* which tend to disappear to the end of the stage (Fig. 17a)	Mandibular components* undergo progressive midline fusion through the stage. Maxillary and mandibular components move close to each other. Tongue primordium* as two swellings on dorsal side of mandibular component. It progressively increases in volume through the stage (Figs.17 and 18a)	Olfactory pit entrance* progressively narrows through the stage. Margins around the pit are fused. Nasal processes* progressively merge in the midline through the stage (Fig.17b)
12 dpc	All parts of the brain are clearly defined. Smoother outline between brain areas (Fig.1c)	Mouth is formed. The first evidence of vibrissa primordia on upper lips. Tongue is formed (Figs. 1c and 18b)	Primitive anterior naris, and the beginning of posterior naris which becomes obvious at the end of the stage (Figs. 18b and 19b)

(continued)

Table 4 (continued)

Embryo age	Staging features		
	Head	**Mouth**	**Nose**
13 dpc	No demarcation between hindbrain and cervical spinal cord. Subcutaneous dorsal spinal cord bulge in cervical–upper trunk area. The head looks more adult (Figs. 2 and 3d)	Face has more adult features. Five almost parallel rows of vibrissae precursors on upper lips. Volume of tongue increased. Tongue protrudes forward over the lip, and the tip is attached to the lip (Figs. 2, 3d, and 19b)	Anterior nares migrated rostrally. Nasal septum (distance between posterior nares) is wide (Fig. 19a)
14 dpc	Adult features are more prominent	Rows of vibrissae precursors on upper lips become more prominent through the stage. Tongue flattened and detached from the lower lip (Fig. 21c)	Nasal septum is narrow (Fig. 20a)
15 dpc	Rostrum elongates (Fig. 1d)	Tongue has a concave, profile, anterior part has a spatular profile	Progressive fusion* of nasal septum with palatal shelves (Fig. 20c, d)
16 dpc	Adult proportions	Vibrissa hair follicles are well developed. Tongue has a spatula profile	Nasal septum fused with palatal shelves
17 dpc	Adult proportions.	Vibrissa hair follicles of upper lip erupted. Tongue has a spatula profile	
18 dpc	Adult appearance, like in the newborn	Vibrissa hair is long. Tactile and sinus follicle erupted	

Table 5
External features of RNAlater® processed 9–18 dpc embryos (transitional features are marked with*): organ proportions, otic pit, palate and Rathke's pouch

Embryo age	Staging features			
	Organ proportions	Otic pit	Palate	Rathke's pouch
9 dpc	Cephalic region is larger than other regions of the embryo (Fig. 15b)	Otic pit* gradually closing through the stage (Fig. 15)		
10 dpc	Cephalic region enlarged compared to the other parts of the embryo. Heart is larger than liver (Fig. 16c)	Otic pit closed		Rathke's pouch* is wide open, and it progressively narrows through the stage (Fig. 16d)
11 dpc	Cephalic region enlarged compared to the other parts of the embryo. Heart is larger than liver (Figs. 1b and 3b)		The postero-medial vertical folds* along Rathker's pouch become parallel later through the stage (Fig. 18a)	Rathke's pouch* has a narrow entrance. It is barely visible at the end of the stage (Fig. 18a)
12 dpc	Cephalic region enlarged compared to the other parts of the embryo. Heart and liver are close in size (Fig. 1c)		The wide gap between palatal shelves. Tongue tightly fits between the shelves. Medial borders of palatal shelves are vertical (Fig. 18b)	Pouch is closed (Fig. 18b)
13 dpc	Cephalic region enlarged compared to the other parts of the embryo. Heart is slightly smaller than liver (Figs. 3d and 5b)		Anterior part of medial border of the palatal shelve is horizontal and posterior is vertical (Fig. 19a)	

(continued)

Table 5 (continued)

Embryo age	Staging features: Organ proportions	Otic pit	Palate	Rathke's pouch
14 dpc	Heart is definitely smaller than liver. The neck region is defined (elongated and narrowed) (Fig. 5c)		Medial border of palatal shelves* become progressively horizontal and the distance between the shelves* progressively diminishes through the stage (Fig. 20a, b)	
15 dpc	Heart is significantly smaller than liver, and smaller than lungs (Fig. 5d)		Palatal shelves* touch in the midline and progressively fuse through the stage (Fig. 20c, d)	
16 dpc	Back is straight, and liver doesn't occupy the entire peritoneal cavity		Palatal shelves are fused; no entrance to the primitive naris. Entrance to the nasopharynx is shaped. Rugae are prominent	
17 dpc	Liver occupies less then a half of peritoneal cavity, caudal part of the body is long. The face elongates		Palatal shelves are fused, no entrance to the primitive naris. Entrance to the nasopharynx is shaped. Rugae are prominent	
18 dpc	Head is proportionally smaller in relation to the body. The back is straight and long		Palatal shelves are fused, no entrance to the primitive naris. Entrance to the nasopharynx is shaped. Rugae are prominent	

Table 6
External features of RNAlater® processed 9–18 dpc embryos (transitional features are marked with*): skin, somites, tail and umbilical hernia

	Staging features			
Embryo age	**Skin**	**Somites**	**Tail**	**Umbilical hernia**
9 dpc		Somites are visible (Fig. 15a)	Caudal neuropore is wide open on the dorsal part of the tail. Tail is turned to the side (Fig. 15a)	
10 dpc		Cervical somites above the hand plate are visible (Fig. 1a)	Caudal neuropore closed. The end of the tail flattens* (Figs. 1a and 4a)	Progressive development of umbilical hernia* through the stage as the gut curvature (Fig. 1a)
11 dpc		Cervical somites are no longer visible above the hand plate (Fig. 1b)	Caudal part of the tail is slender. Tail gradually narrows to the blunt ended tip and passes on one side of the neck region, close to the nasal pit (Fig. 17)	Umbilical hernia* increases in volume through the stage and forms a prominent single loop to the end of the stage (Figs. 1b and 11a)
12 dpc		Cervical somites are no longer visible above the hand plate (Fig. 1c)	Tail is slender, with blunt tip. Somites are visible in the tip	Hernia is formed by two-three loops
13 dpc		Somites are not visible in lower trunk, but visible in tail after longer incubation in RNA later (Figs. 3d and 22a)	Distal part narrows to a sharp point. Tail is shorter proportionally, positioned in median plane between the foot plates, does not extent beyond the level of the hand plates. (Figs. 3d and 22a)	Hernia is significantly larger in size and has several thick loops (Fig. 22a)

(continued)

Table 6 (continued)

Embryo age	Staging features			
	Skin	Somites	Tail	Umbilical hernia
14 dpc	Smooth. Superficial veins in cephalic region and hand plates	Somites are not visible in lower trunk but visible in the tail (except the tail tip) after longer incubation in RNA later	Distal part narrows to a sharp point. Tail is shorter propotially, positioned in median plane between the foot plates, does not extent beyond the level of the hand plates	Umbilical hernia large (Fig. 22b)
15 dpc	Skin smooth. Superficial veins on the whole body	Somites are not visible in lower trunk and tail (Fig. 1d)	Distal part narrows to a sharp point. Somites are not visible. Tail is shorter propotially, positioned in median plane between the foot plates, does not extent beyond the level of the hand plates	Umbilical hernia is large (Fig. 1d)
16 dpc	Skin smooth. Superficial veins on the whole body	Somites are not visible in lower trunk and tail	Tail is thin and long	Repositioning of umbilical hernia
17 dpc	Skin is wrinkled, particularly on the head, trunk, proximal limb. No wrinkles on the face, distal limb, tail. Superficial veins are visible only on dorsal side of foot and hand plate		Tail is thin and protrudes beyond the foot plates when embryo positioned on its back	No umbilical hernia
18 dpc	Skin is wrinkled everywhere, but distal part of limbs, tail and face		The tail is shorter and wider, the tip is less pointed, doesn't protrude beyond the foot plates when embryo positioned on its back	

Table 7
Internal features of RNAlater® processed 9–18 dpc embryos (transitional features are marked with *): adrenals, diaphragm and gonads

Embryo age	Staging features		
	Adrenals	Diaphragm	Gonads
9 dpc			
10 dpc			Gonadal primordium is visible as elevation in the median part of the urogenital ridge (Fig. 4a)
11 dpc			Elevations* on the median part of urogenital ridges increase in size through the stage (Fig. 11a)
12 dpc	Adrenals are not visible	Diaphragm* progressively develops through the stage (incomplete diaphragm) (Figs. 6b and 13a)	Gonads* become progressively denser and compact, and increase in size through the stage. No anatomical difference between male and female gonads (Fig. 11b–d)
13 dpc	Adrenals are embedded in mesenchymal tissue, and far separated from kidneys (Fig. 12a)	Diaphragm is incomplete (central tendon over liver) (Figs. 5b and 6c)	Ovaries and testes are close in size and appearance (indefinite gonads), and located laterally at upper halve of the kidney (Figs. 11b–d)
14 dpc	Adrenals are well developed, closely associated with kidneys (Figs. 12b and 13a, c)	Diaphragm is closed (Fig. 6d)	Testes are 3–4 times larger than ovaries. They are ovoid, with prominent blood vessel along the ridge, and associated with anterolateral surface of kidney. Ovaries are elongated, without prominent blood vessel along the ridge and associated with inferolateral surface of kidney. They are 3–6 times as long as their width (Figs. 12b and 13b)

(continued)

Table 7 (continued)

Embryo age	Staging features: Adrenals	Diaphragm	Gonads
15 dpc	Adrenals are well developed, and closely associated with kidneys (Fig. 12c, d)	Diaphragm is closed (Fig. 5d)	Testes are partially descended in adult position. Ovaries are in caudal position (Figs. 12c, d)
16 dpc	Adrenals associated with kidneys		Testes have more rounded profile. Ovaries are significantly smaller than testes, with suspensory ligament at the rostral pole
17 dpc	Adrenals associated with kidneys		Testes are rounded, with epididymis, descended to the brim of the pelvis. Bladder is prominent. Ovaries are much smaller than kidneys. Suspensory ligament is present
18 dpc	Adrenals are proportionally smaller		Testes are inside the pelvis as “adult.” Ovaries with completed ovarian capsule located near the lower kidney pole and partially hidden by kidney

Table 8
Internal features of RNAlater® processed 9–18 dpc embryos (transitional features are marked with *): kidneys, liver and lung

Embryo age	Staging features		
	Kidneys	Liver	Lung
9 dpc		Liver is not visible. Heart is visible (Fig. 15b)	
10 dpc		Liver primordium* is diffused and very light in color. It progressively increases in volume through the stage (Figs. 10a and 16c)	Single lung buds* are visible on ventral aspect of foregut late in the stage (Fig. 6a)
11 dpc		Four lobes: two left and two right. The color*intensifies through the stage (Fig. 5a)	Lung is in caudal position under liver. Single buds* progressively increase in size through the stage and undergo demarcation to the end of the stage: three buds on the right lobe and two buds on the left lobe (Figs.6a and 8a)
12 dpc	The elongated metanephros (future kidney) is located in the caudal part of urogenital ridge (Fig. 11b–d)	Liver* progressively increases in size and becomes denser through the stage. Right lower part of the liver is subdivided transversely on upper and lower, the latter is subdivided vertically (Fig. 10b)	Further lung demarcation. Right lung has four buds, left lung has two buds. Lung undergoes rostral ascent (Figs. 6b and 8b)
13 dpc	The top1/3 of a kidney is hidden under the gonad and covered with mesenchyme. Kidneys are enlarged, ascended from the previous pelvic position. Ureter is visible (Fig. 12a)	Liver occupies much of the upper part and (together with the pancreas) the right side of the peritoneal cavity. It is dense. Caudates are visible at the duodenum–stomach junction, and one is twice larger than the other (Figs. 5b and 10c)	Dramatic size increase. Lung is demarcated into lobes. Left lung is smaller than the right and not divided. Right lung has four lobes, one is located close to the midline. Two main bronchi. Lobes separation is incomplete (Figs. 6c and 8c)

(continued)

Table 8 (continued)

Embryo age	Staging features: Kidneys	Liver	Lung
14 dpc	Kidneys are progressively round*, their volume* increases, and they continue to ascend* through the stage (Fig. 12b)	All lobes are developed. Caudates are larger (Fig.5C)	Lung increases in size* through the stage. All the lobes are clearly demarcated and separated. Right bronchus larger than left (Figs. 6d and 8d)
15 dpc	Kidneys are rounded. They undergo volume increase and rostral ascent through the stage (Fig. 12c, d)	Progressive size increase* through the stage. Prominent gall bladder (Fig. 5d)	Progressive size* increase through the stage (Fig. 5d)
16 dpc	Kidneys progressively increase in size* through the stage	Liver is large and dense. All lobes are readily distinguishable	Lung ascended to pleural cavity.
17 dpc	Kidneys progressively ascend* to the adult position through the stage	All liver lobes are prominent. The gall bladder is visible	Volume* increases through the stage
18 dpc	Kidneys are in adult position. Bladder is extremely prominent and muscular	Gall bladder is prominent. Liver is the largest organ of the peritoneal cavity	Dramatic volume* increase through the stage. Lung occupies the entire pleural cavity

Table 9
Internal features of RNAlater® processed 9–18 dpc embryos (transitional features are marked with *): pancreas, salivary gland, skeletal system and spleen

	Staging features			
Embryo age	Pancreas	Salivary gland	Skeletal system	Spleen
9 dpc				
10 dpc				
11 dpc	Two pancreatic primordia			
12 dpc	Pancreas primordia are not fused (Fig. 7b)			Mesenchymal aggregates are embedded in the dorsal mesentery (mesogastrium) of the stomach close to pancreas (short spleen primordium, slightly longer than combined length of two pancreatic primordia) (Fig. 7a, b)
13 dpc	Visible increase in size, dorsal and ventral pancreas fused (Fig. 7d)		Condensations (ossification) for the ribs is visible (Fig. 12a)	More prominent mesenchymal aggregates within the dorsal mesentery of the stomach close to pancreas: elongated spleen primordium ("fuzzy ribbon") (Fig. 7c, d)
14 dpc	Pancreas is large and diffused (Fig.7e)		Pre-cartilaginous condensation in tracheal rings, larynx and ribs.	Spleen increases in size* through the stage. It is ribbon-like, located in close proximity to the left gonad (Figs. 6d and 7e)
15 dpc	Pancreas is diffused, and located close to the left kidney (Fig. 7f)	Salivary gland lobes are prominent structures at the base of the tongue (Fig. 14d)	Ribs and trachea rings are prominent.	Spleen is well defined. It has dense architecture, and can be peeled off the surface of stomach (Fig. 7f)
16 dpc	Pancreas progressively increases in size* through the stage	Size* increase through the stage	Discrete ribs and x iphoid process	Spleen is well defined, denser architecture, can be effortlessly peeled off the surface of stomach
17 dpc	Pancreas is large, located in the left side of peritoneal cavity	Dense architecture	Discrete ribs and xiphoid process	Spleen increases in volume* through the stage and has peppered appearance
18 dpc	Pancreas is large, and located in the left side of peritoneal cavity	Dense architecture	Discrete ribs and xiphoid process	Spleen capsule developed

Table 10
Internal features of RNAlater® processed 9–18 dpc embryos (transitional features are marked with *): stomach, thymus, thyroid and urogenital ridge

	Staging features			
Embryo age	**Stomach**	**Thymus**	**Thyroid**	**Urogenital ridge**
9 dpc				
10 dpc	"'Dilatation of the foregut" stage. Stomach* progressively increases in volume through the stage			Urogenital ridges* progressively increase in volume and density (Fig. 4a)
11 dpc	Straight stomach* progressively enlarges on the left side (Fig. 6a)			The ridges* increase in size and stretch from caudal part of the hand plate to caudal part of the foot plate. The strip underlying gonads (mesonephros and mesentery), is wide and thick (Figs. 3c, 6a and 11a)
12 dpc	Stomach* progressively increases in size and slants to the left (Figs. 6b and 14a)	Visible as two separate white condensations at the base of larynx (Fig. 14a)		The strip underlying gonads (urogenital mesentery) is wide and thin (Fig. 11b–d)
13 dpc	The stomach is large, slanted to the left and occupies most of the left side of the peritoneal cavity (Fig. 6c)	Two separate prominent lobes visually located at the base of larynx (Fig. 14b)	Thyroid is visible from both sides of larynx	Urogenital mesentery, underlaying gonads, is wide and thin (Figs. 12a, 8e, and 13d)
14 dpc	Stomach increases in size, completely slanted to the left, smooth, tight (Fig. 6d).	Thymus moves caudally; prominent lobes are positioned on the arch of aorta on either side of the trachea (Fig. 14c)	Volume increase. Thyroid lobes are prominent from either side of the larynx (Fig. 14c)	Urogenital mesentery, underlying gonads, is narrow (Fig. 13b)

15 dpc	Stomach is large and tight (Fig. 7f)	Thymus is a large structure, descended closer to the heart. Lobes move closer to each other (Figs. 5d and 14d)	Thyroid is well defined from either side of the larynx (Fig. 14d)	
16 dpc	Stomach is large and tight	Thymus is a large structure close to the heart, lobes* are progressively closer to each other through the stage	Thyroid is prominent from either side of the larynx	
17 dpc	Stomach is muscular. Volume* of small intestine increases dramatically through the stage	Thymus is in adult position. Dramatic volume* increase through the stage. Lobes are close together	Prominent from either side of the larynx	
18 dpc	Limiting ridge primordium is visible	Thymus is large with thin fibrous capsule. It descended in adult position	Parathyroid is readily distinguishable	

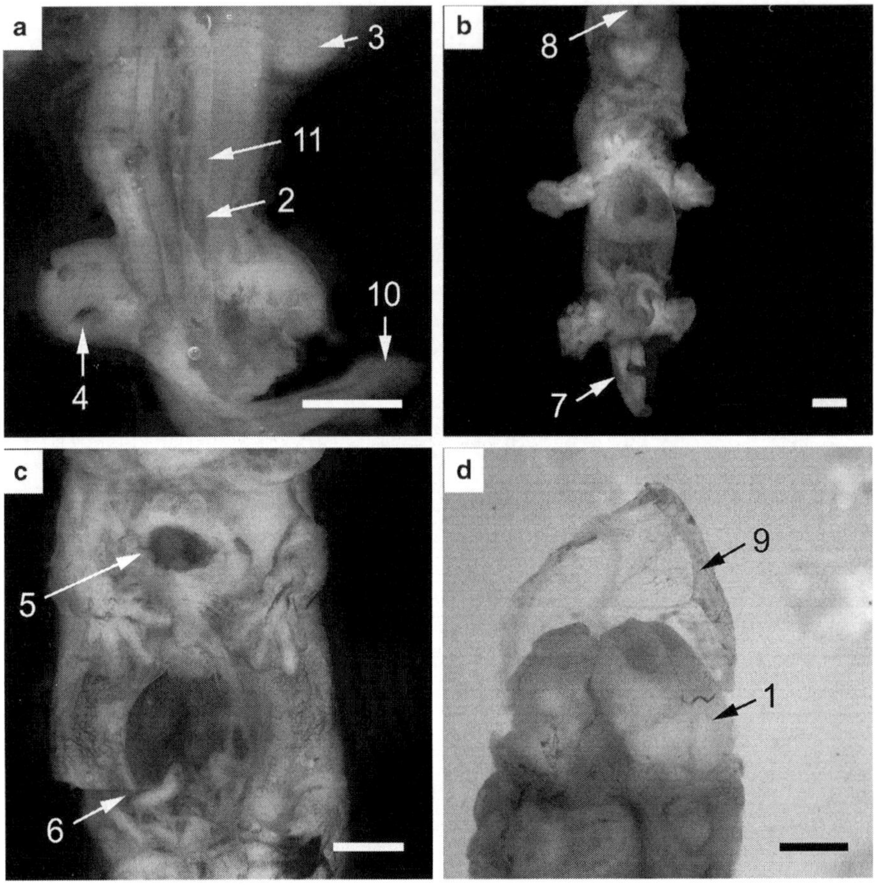

Fig. 4 Mouse embryo necropsy: Preparation of embryos for organ dissection in RNAlater®. (**a**), (**b**) 10 and 13 dpc embryos in pinned down position (incubation 30 min and 12 h, respectively). (**c**, **d**) 15 dpc embryos. Incisions over the heart and under the liver (**c**), and skin peeled off the brain (**d**) for RNAlater® penetration (72 h of storage at −20 °C after 4 h of incubation at RT). (**a**–**d**) Bar = 1,000 μm. (*1*) Brain, (*2*) gonad, (*3*) hand plate, (*4*) hole from a minute pin in the foot plate, (*5*) incision over the heart, (*6*) incision under the liver, (*7*) position of the lower pin, (*8*) position of the upper pin, (*9*) skin peeled off the brain, (*10*) tail, (*11*) urogenital ridge

(around 800 embryos) of four mouse strains and 71 litters (around 400 embryos) of two rat strains (*see* Table 11).

3.3 Embryo Necropsy Overview

1. Rodent embryo necropsy requires good hand coordination, gentle touch and practice for dissecting under magnification.
2. The model of a dissecting microscope is a personal choice. However, a dissecting microscope with quality optics and light source (*see* **Note 17**) is crucial for dissecting 7–8 dpc embryos, and organs from 10 to 12 dpc embryos.
3. We use two types of materials to make dissecting dishes: Paraplast® (*see* **Note 18**) and Sylgard® (*see* **Note 19**). Use Paraplast®-filled dissecting dish to remove 9–12 dpc embryos out of the uterus, to open peritoneal cavity of 12–16 dpc, to

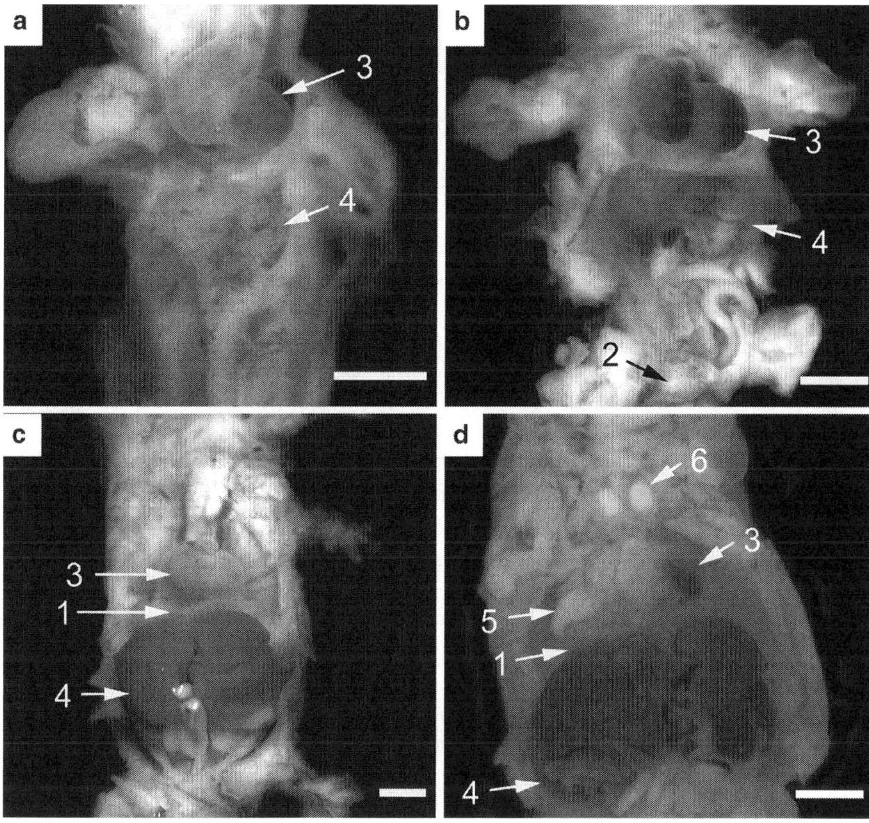

Fig. 5 Mouse embryo necropsy in RNAlater®: Ventral skin removed. (**a**) An 11 dpc embryo (incubation 3 h). (**b**) A 13 dpc embryo (72 h of storage at −20 °C after 4 h of RT incubation). (**c**), (**d**) 14 and 15 dpc embryos, respectively (incubation 4 h). (**a–d**) Bar = 1,000 μm. (*1*) Diaphragm, (*2*) genital tubercle, (*3*) heart, (*4*) liver, (*5*) lung, (*6*) thymus

observe external features and determine sex of 14–16 dpc embryos, and for necropsy of 17–18 dpc embryos. Use Sylgard® dish for staging of 7–11 dpc embryos and organ dissections from 9 to 16 dpc embryos.

4. We found a sapphire knife to be very useful for organ dissection. Its diamond shape with double edge blade allows to cut and, at the same time, to roll the organ off the adjacent tissue with the flat surface of the blade. Dissected organs, picked up from the floating position in RNAlater, readily stick to the flat part of the blade, and readily come off once the blade placed back in RNAlater®. The blade size does not obscure the organs during dissection. Minuten entomological pin inserted in a pin vise is also a good dissecting tool.
5. For transcriptome study, embryos of all ages, except 7–9 dpc (*see* **Note 20**), should be dissected in RNAlater® to preserve RNA integrity in target organs.
6. 9–11 dpc embryos should be dissected out of the uterus with intact yolk sac around them (*see* **Note 21**) (Fig.3a).

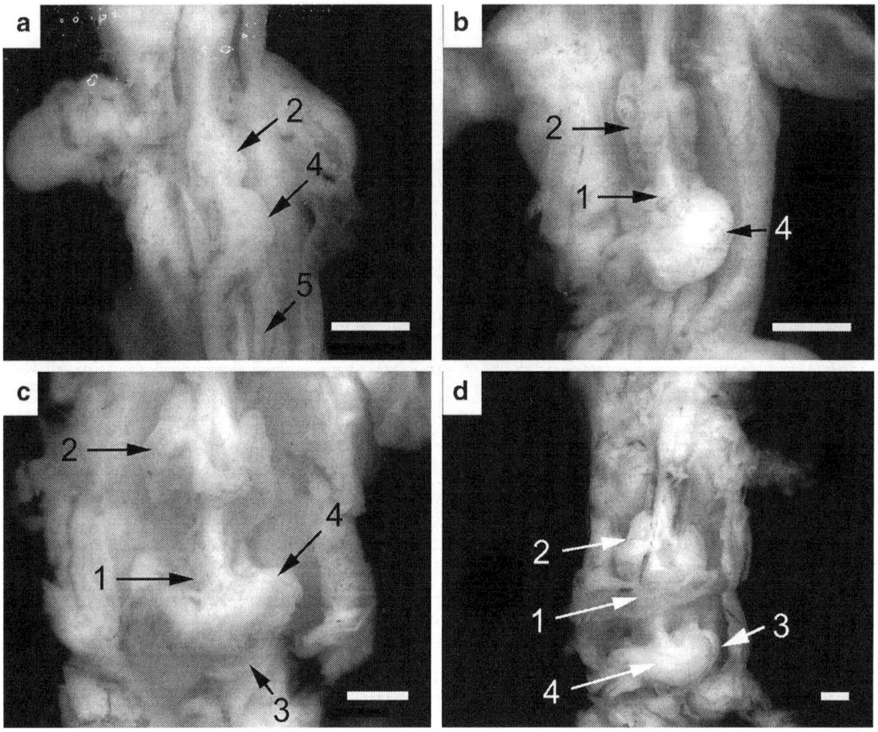

Fig. 6 Mouse embryo necropsy in RNAlater®: Liver and heart removed. (**a**) An early 11 dpc embryo (incubation 2 h). (**b–d**) 12, 13, and 14 dpc embryos, respectively (incubation 3 h). (**a–d**) Bar = 1,000 μm. (*1*) Diaphragm, (*2*) lung, (*3*) spleen, (*4*) stomach, (*5*) urogenital ridge

7. For 12–14 dpc embryos the yolk sac should be cut open on the side opposite to placenta (dorsal side of the embryo) (Fig. 3b) to keep umbilical cord and hernia intact. Umbilical cord should be severed at the placenta (Fig. 3d). The embryos bleed out through the intact umbilical cord during incubation in RNAlater®. Effective blood removal from the embryo body ensures high quality of RNA in dissected organs (*see* **Note 22**). 15–18 dpc embryos bleed out in RNAlater through the cut in cervical area after cervical dislocation.
8. Allow embryos to bleed out for 10 min on a shaker with gentle agitation. Use fresh RNAlater for embryo sexing.
9. Before organ dissection it is important to unroll the embryo from its natural position inside the yolk sac and pin it down to the surface of the dissecting dish (Figs. 1b, 3c, and 4a, b). We use thin, RNase-free treated cactus needles or minuten entomological pins for 9–12 dpc embryos, and thin entomological needles for older ages.
10. After external staging and sexing, follow with a 30-min incubation of embryos in RNAlater® to harden up organs for their better visualization and clean removal.

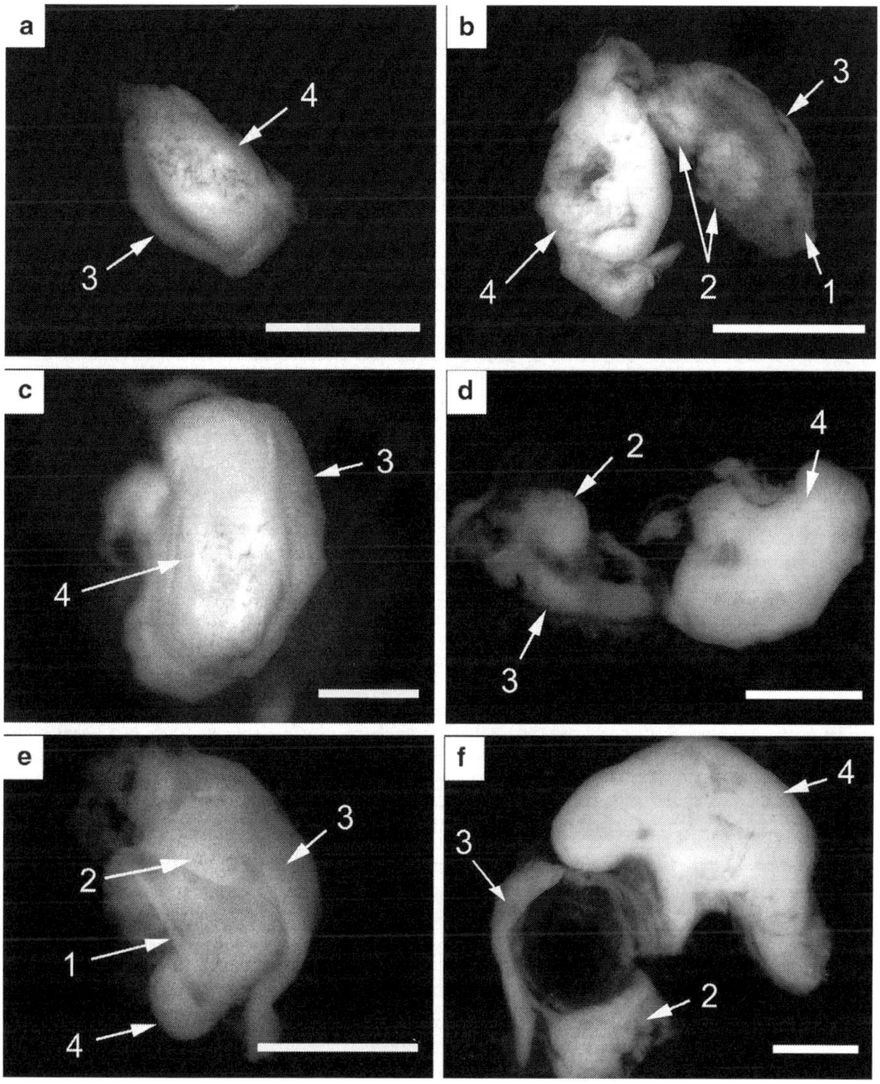

Fig. 7 Mouse embryo necropsy in RNAlater®: Stomach, spleen, and pancreas. (**a**), (**c**), (**e**) Stomach with spleen and pancreas of 12, 13, and 14 dpc embryos, respectively. (**b**), (**d**), (**f**). Spleen and pancreas with mesenchymal tissue separated from the dorsal surface of stomach of 12, 13, and 15 dpc embryos, respectively. (**a**–**f**) Incubation 3 h. (**a**–**d**, **f**) Bar = 1,000 μm. (**e**) Bar = 500 μm. (*1*) Dorsal mesentery, (*2*) pancreas, (*3*) spleen, (*4*) stomach

11. The procedure of embryo collection and organ dissection in RNAlater can be divided into two parts. Part 1 is the removal of embryos from the uterus, sex determination, and observation of external anatomical features for staging. Part 2 is the observation of internal anatomical features for staging (*see* **Note 23**), and dissection and preparation of organs for RNA extraction. After completion of Part 1, embryos can be incubated in RNAlater for 3 h to overnight at RT (*see* **Note 24**) and then

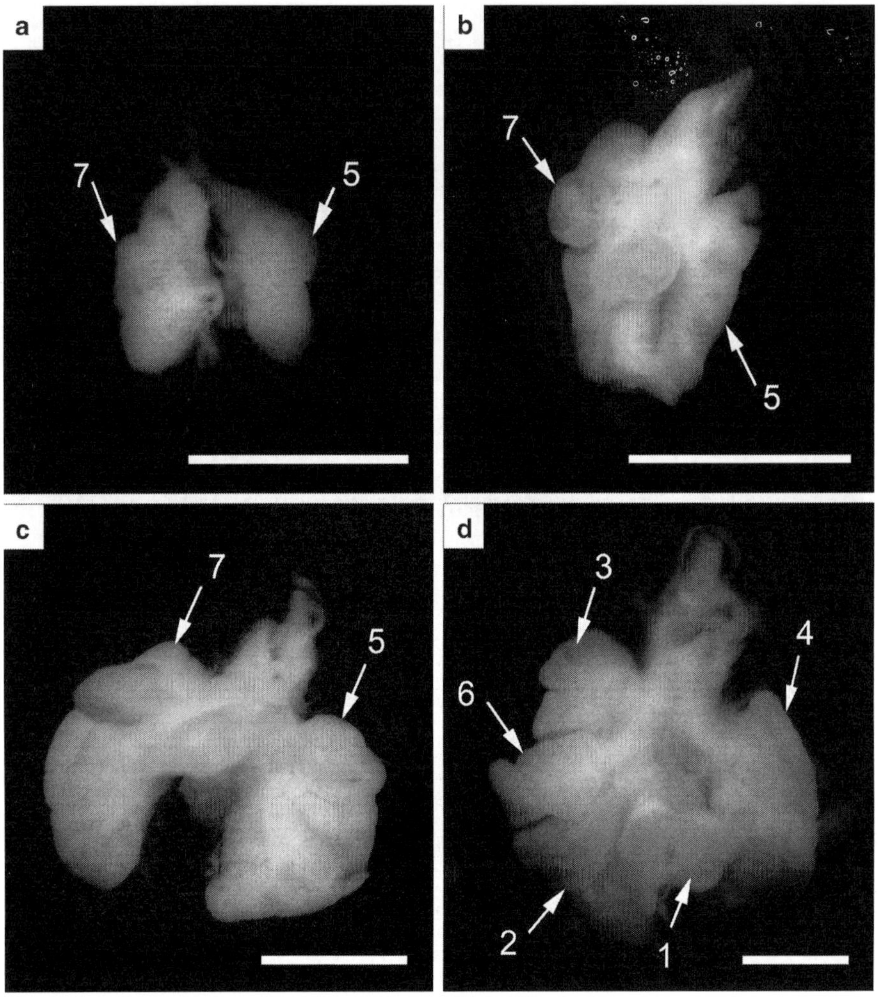

Fig. 8 Mouse embryo necropsy in RNAlater®: Lung. (**a–d**) 11, 12, 13, and 14 dpc embryos lungs, respectively (incubation 4 h, bar = 1,000 μm). (*1*) Accessory lobe, (*2*) caudal lobe, (*3*) cranial lobe, (*4*) left lobe, (*5*) left lung, (*6*) middle lobe, (*7*) right lung

transferred to –20 °C. Part 2 can be performed any time up to 6 weeks, when embryos are stored at –20 °C (*see* Subheading 3.4) (*see* **Note 25**).

12. Use hairbrush (cleaned for RNase-free conditions) or flat surface of sapphire knife for picking up a tiny floating organ (e.g., 14 dpc thymus, spleen, ovaries, testes, and kidney) and transferring it in the collecting well (*see* **Note 26**).
13. Before homogenization in lysis buffer, RNAlater® should be blotted off the tissue (*see* **Note 27**). 9–11 dpc embryos and larger organs (e.g., heart, brain, liver) can be blotted with Kimwipes®. Tiny organs should be transferred from the collection well on Kimwipes® and let drain off (*see* **Note 28**).

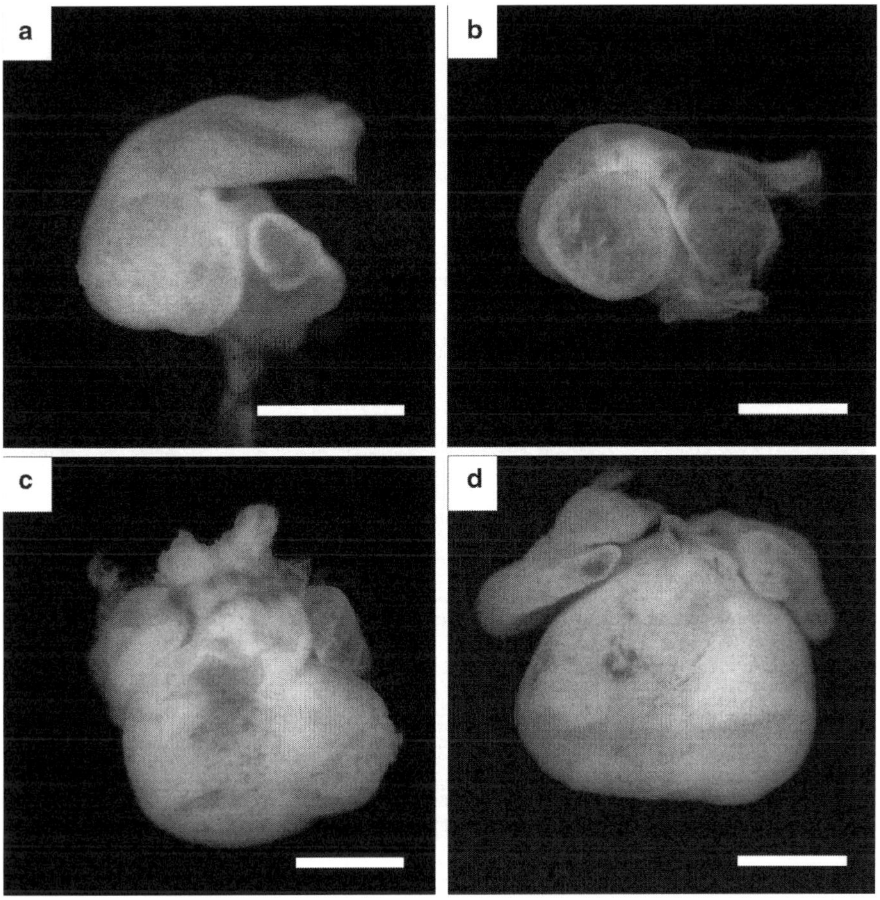

Fig. 9 Mouse embryo necropsy in RNAlater®: Heart. (**a**–**d**) 11, 12, 13, and 14 dpc embryos heart, respectively (incubation 4 h, bar = 1,000 μm)

14. We have been able to dissect a variety of organs from embryos of different ages (Table 12). The feasibility of organ dissection depends on embryo age and size (*see* **Note 29**).
15. After incubation in RNAlater®, embryos can be fixed in 10 % neutral buffered formalin (NBF) for further histological evaluation (*see* **Note 30**).

3.4 Age-Related Embryo Necropsy Protocols for RNA Retrieval from Whole Embryos and Embryonic Organs

3.4.1 Necropsy Protocol of 9–11 dpc Whole Embryo Collection

1. Follow RNase-free precautions for the entire procedure.
2. Perform cervical dislocation on a pregnant female mouse.
3. Dissect the pregnant uterus together with the bladder (for orientation) (*see* **Note 31**).
4. Place uterus in sterile petri dish on wet ice in the styrofoam box.
5. Diagram embryo positions in the uterine horns on embryo necropsy report form (*see* **Note 32**) and place the lid on the styrofoam box. Keep a lid on the box at all times (*see* **Note 33**).

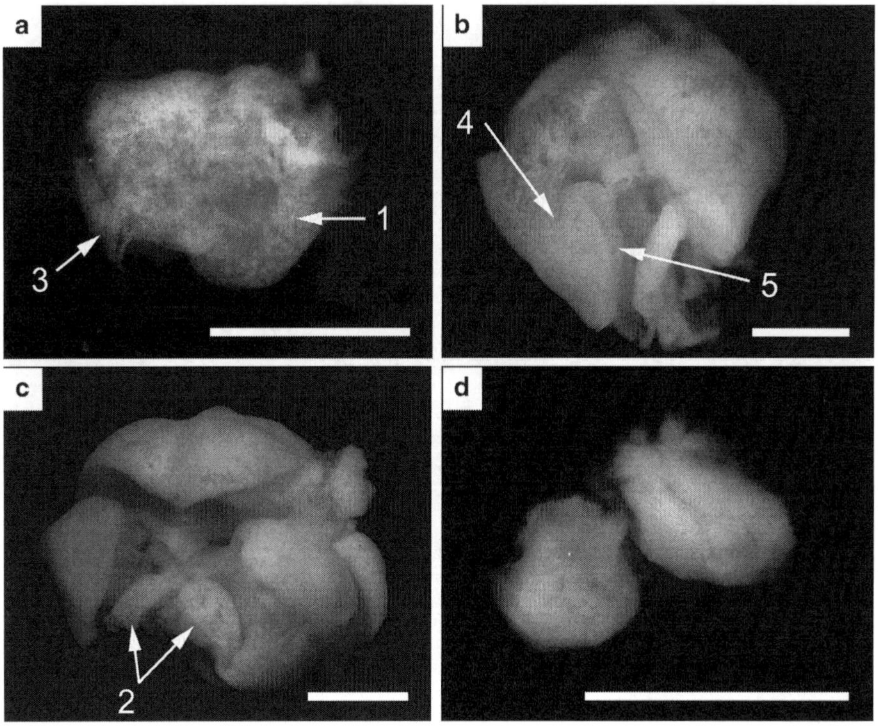

Fig. 10 Mouse embryo necropsy in RNAlater®: Liver and thymus. (**a**–**c**) 10, 12, and 13 dpc embryos liver, respectively. (**d**) A 13 dpc embryo thymus. (**a**–**d**) Incubation 4 h, bar = 1,000 μm. (*1*) Left liver lobes, (*2*) caudate, (*3*) right liver lobes, (*4*) transverse subdivision, (*5*) vertical subdivision

6. Freeze dam's ears and tail (*see* **Note 14**).
7. For each embryo prepare two tubes with embryo ID, labeled as "yolk sac" and "embryo."
8. Prepare Sylgard® filled 6-multiwell plate with embryo IDs written on the outside bottom of each well.
9. Pour 1 ml of RNAlater® in each designated well.
10. With clean instruments, dissect embryos out of fetal compartments (*see* **Note 34**).
11. Cut right uterine horn off and transfer it to an RNase-free paraffin-filled petri dish with 3–4 ml of RNAlater.
12. Pin the uterine horn down at ovary and cervix side with reproductive fat facing the bottom of the petri dish.
13. Dissect embryos in RNAlater® one by one, according to the uterus diagram of fetal compartments.
14. Make a surface incision in the uterine swelling at the border with adjacent fetal compartment, and carefully cut the uterus open along the placental side.
15. Gently peel the decidua off with blunt forceps and expose the yolk sac.

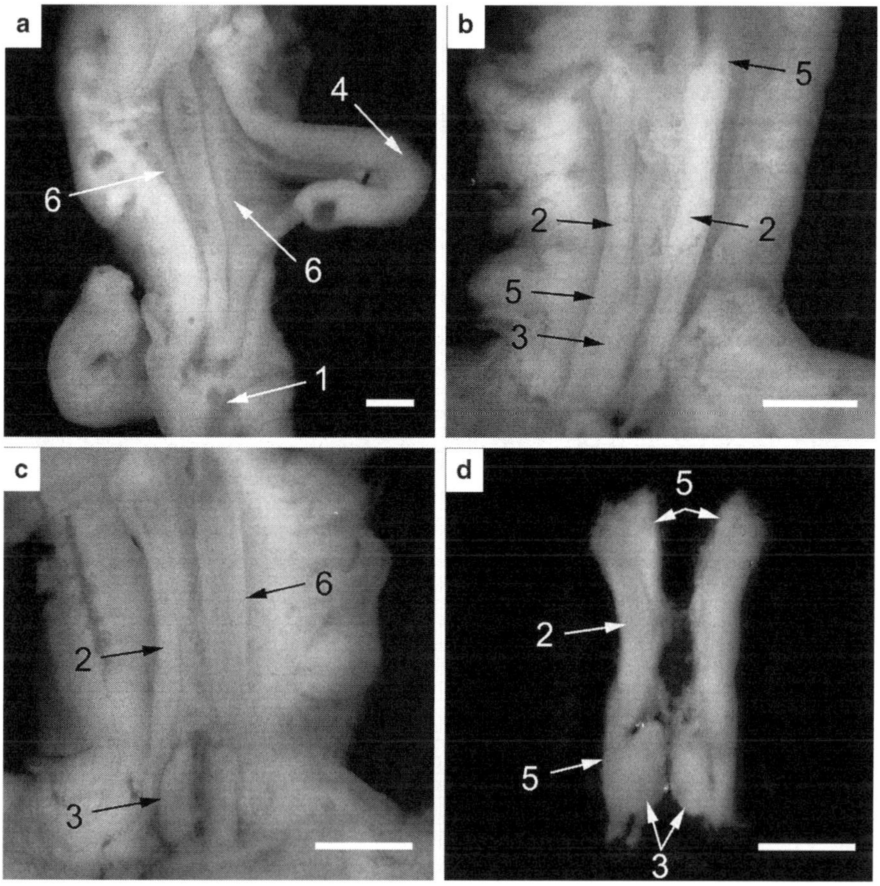

Fig. 11 Mouse embryo necropsy in RNAlater®: Stomach, lung and intestine removed (early stages). (**a**) Urogenital ridges with gonadal primordial of an 11 dpc embryo. (**b–d**) A 12 dpc embryo. (**b**) Caudal position of indefinite gonads and metanephros (definite kidney). (**c**) Kidneys exposed out of mesenchymal tissue. (**d**) Dissected urogenital ridges with gonads and kidneys. (**a–d**) Incubation 4 h, bar = 1,000 μm. (*1*) Genital tubercle, (*2*) gonad, (*3*) kidney, (*4*) umbilical hernia, (*5*) urogenital mesentery, (*6*) urogenital ridge

16. Roll the embryo with the yolk sac out of the decidua and transfer it into the corresponding well with RNAlater (*see* **Note 35**).
17. Discard the dissected horn and transfer the left horn in the same RNAlater® petri dish.
18. Repeat **steps 12** through **16**.
19. Peel the yolk sac of the embryo (*see* **Notes 36** and **37**). Blot the yolk sac with Kimwipes® and snap-freeze in the corresponding cryotube for genotyping.
20. Inspect embryo for normal development, stage it (*see* 9–12 dpc in Tables 2, 3, 4, 5, 6, 7, 8, 9, and 10) and make a record on necropsy report form.
21. Blot embryo with new Kimwipes® and snap-freeze in a corresponding cryotube.
22. Transfer samples to −80 °C storage on dry ice.

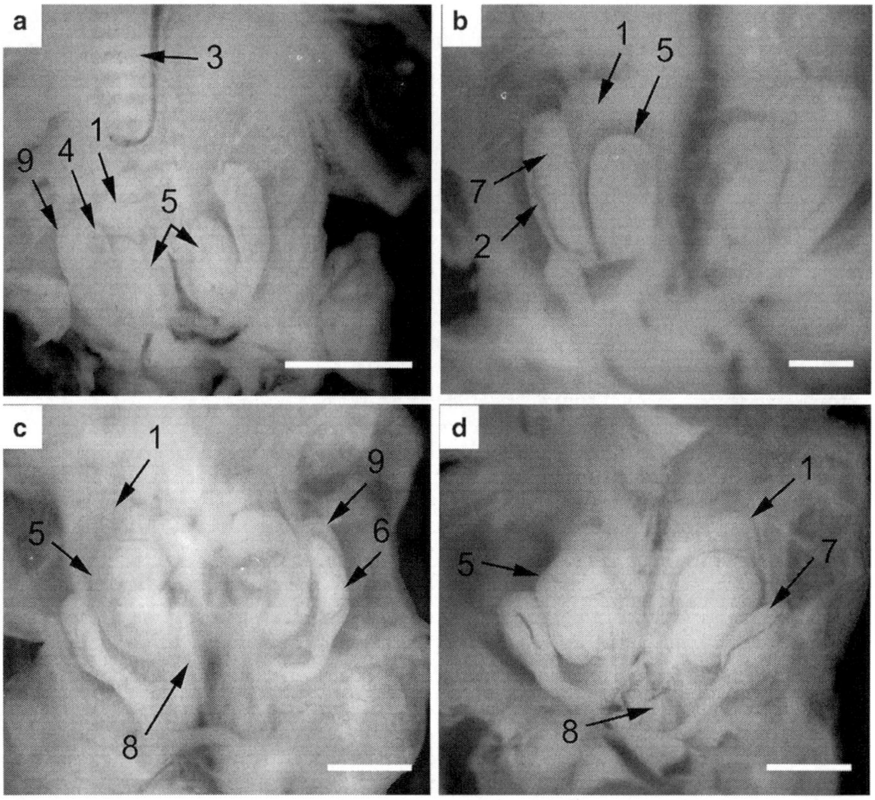

Fig. 12 Mouse embryo necropsy in RNAlater®: Stomach, lung, and intestine removed (late stages). (**a**) Kidneys and indefinite gonads of a 13 dpc embryo. (**b**, **d**) Position of testes in relation to kidneys of a 14 and 15 dpc embryos, respectively. (**c**) Position of ovaries in relation to kidneys of a 15 dpc embryo. (**a**–**d**) Incubation 4 h, bar = 1,000 μm. (*1*) Adrenal, (*2*) blood vessel, (*3*) future ribs, (*4*) gonad, (*5*) kidney, (*6*) ovary, (*7*) testicle, (*8*) ureter, (*9*) urogenital mesentery

3.4.2 Necropsy Protocol of 9–11 dpc Embryo Organ Collection

Part 1

1. Follow **steps 1** through **6** of Subheading 3.4.1.
2. For each embryo prepare a cryotube with embryo ID labeled as "yolk sac."
3. Prepare 12-multiwell plate with embryo IDs written on the outside bottom of each well.
4. Follow **steps 9** through **20** of Subheading 3.4.1.
5. Unfold the embryo out of its natural position (Figs. 1b, 3c, and 4a).
6. Seal the multiwell plate inside the plastic bag and keep at −20 °C until genotyping is completed (*see* **Note 38**).

Part 2

7. Prepare separate 12-multiwell plate for male and female sample with organ names written on the outside bottom of the well and 0.5 ml of RNAlater® per well. Cool it at −20 °C at least for

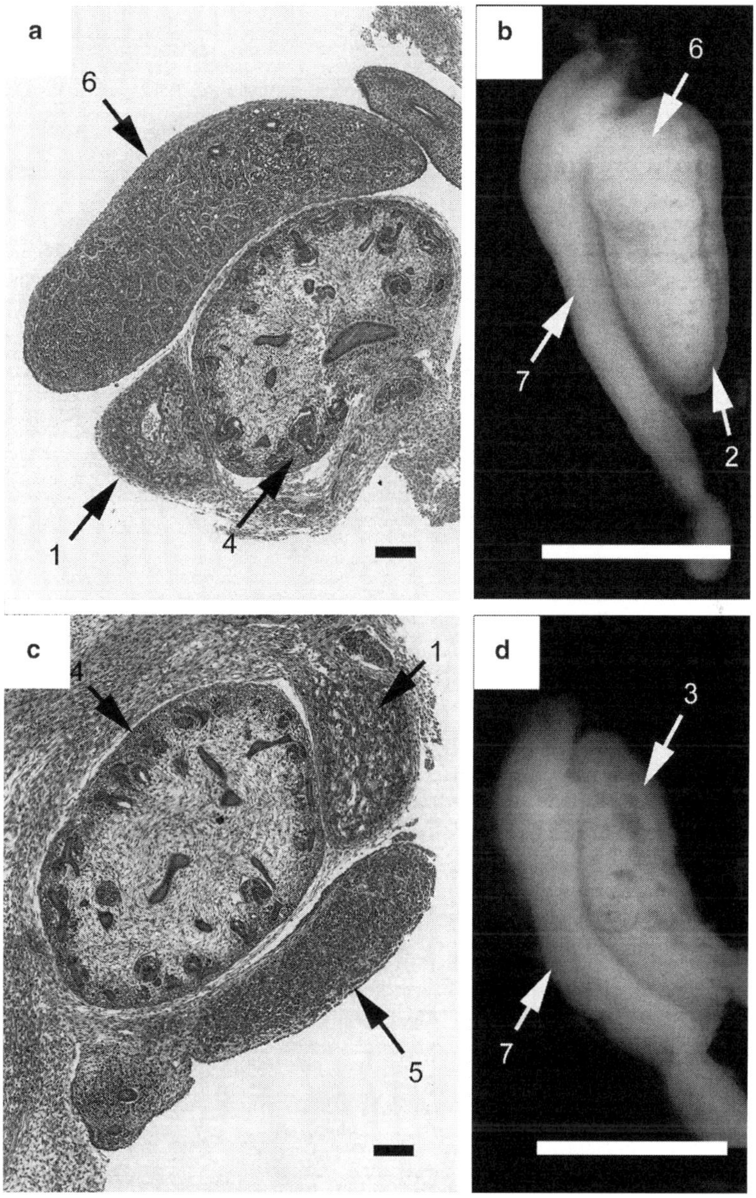

Fig. 13 Mouse embryo necropsy in RNAlater®: Gonads. (**a**, **c**) H&E section of a 14 dpc embryo testicle and ovary (respectively). Gonads were dissected with kidneys "in block" after 4 h incubation in RNAlater®, then fixed in 10 % NBF, processed and embedded in paraffin (Bar = 100 μm). (**b**) A 14 dpc embryo testicle (definite gonad) (Bar = 1,000 μm). (**d**) A 13 dpc embryo indefinite gonad (Bar = 500 μm). (**b**, **d**) Incubation 4 h. (*1*) Adrenal, (*2*) blood vessel, (*3*) gonad, (*4*) kidney, (*5*) ovary, (*6*) testicle, (*7*) urogenital mesentery

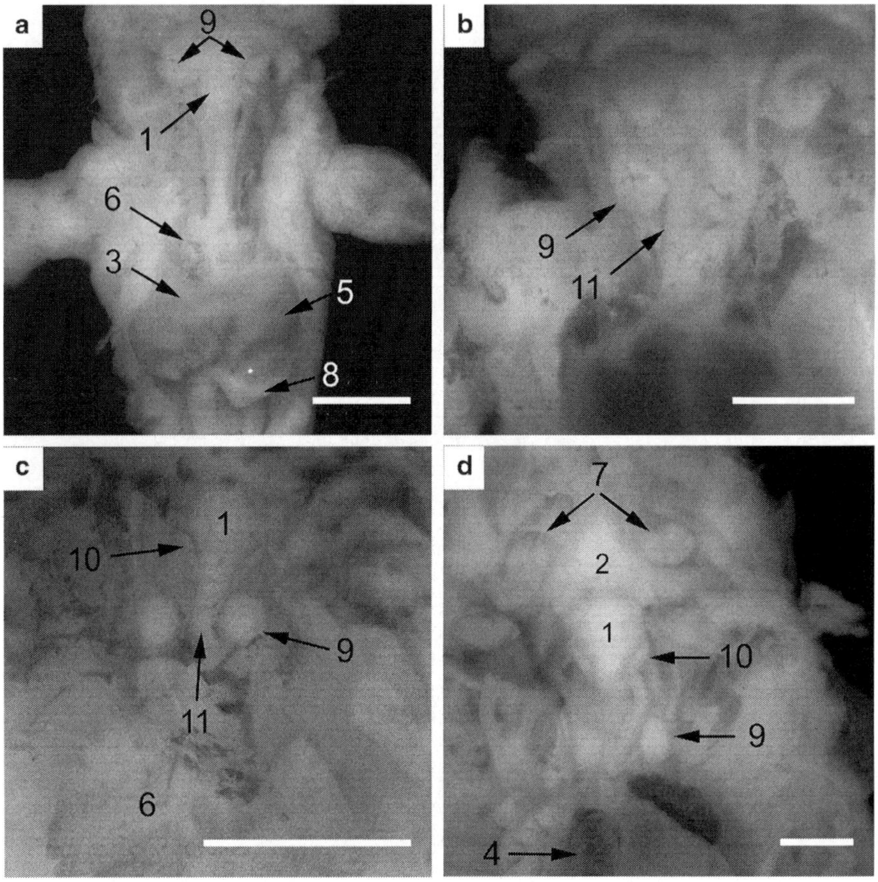

Fig. 14 Mouse embryo necropsy in RNAlater®: Age-related appearance and position of thymus. (**a**) Rostral position of thymus at the base of larynx of a 12 dpc embryo. (**b–d**) Caudal descent of thymus of 13 dpc (**b**), 14 dpc (**c**) and 15 dpc (**d**) embryos. (**a–d**) Incubation 4 h, bar = 1,000 μm. (*1*) Base of larynx, (*2*) base of tongue, (*3*) diaphragm, (*4*) heart, (*5*) liver, (*6*) lung, (*7*) salivary gland, (*8*) stomach, (*9*) thymus, (*10*) thyroid, (*11*) trachea

30 min and place it in a styroform box on top of a frozen cool pack (*see* **Note 39**).

8. Prepare 2.0 ml tubes (CLP, Neptune) with the cryolabel for each target organ.
9. Transfer embryos from –20 °C storage to a styrofoam box with wet ice.
10. One at a time, transfer embryos to a 35 mm Sylgard®-filled petri dish together with RNAlater® and place it on a frozen cool pack under a dissecting microscope (*see* **Note 40**).
11. Pin the embryo down with minuten pins through the head at the mandibular level, tail, and/or hand and foot plate (Figs. 3c and 4a).
12. Peel off the abdominal skin from over the heart towards the tail (Fig. 5a).

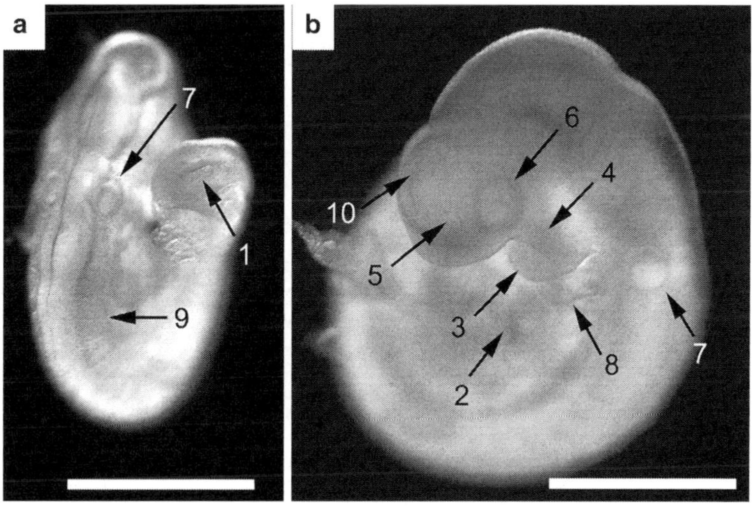

Fig. 15 Rodent embryo staging: Features of 9 dpc embryos readily visible in RNAlater®. (**a**) Dorsal view. (**b**) Sagittal view. Fixation—4 % paraformaldehyde, glycerol (*see* **Note 53**) (Courtesy of Kristin K. Biris and Terry P. Yamaguchi). (**a**), (**b**) Bar = 1,000 μm. (*1*) Caudal neuropore at the end of the tail, (*2*) heart, (*3*) mandibular component of the first branchial arch, (*4*) maxillary component of the first branchial arch, (*5*) olfactory placode, (*6*) optic eminence, (*7*) otic pit, (*8*) second branchial arch, (*9*) somites, (*10*) telencephalic lobes

13. Cut heart off moving the blade from the tail up and place it in a corresponding well (*see* **Note 41**) (Figs. 5a and 9a).
14. Roll liver (Figs. 5a and 10a) off the body and place it in a corresponding well.
15. Remove intestine together with lung, pancreas, and stomach and move it to the bottom of the petri dish (Fig. 6a).
16. Separate lung buds (Figs. 6a, 8a, and 11a), pancreas, and stomach in the above order with the blade and place in corresponding wells.
17. Roll gonads (Figs. 6a and 11a) off the urogenital ridges (*see* **Note 42**) and place them in a corresponding well.
18. Position the embryo on its side (Fig. 1a, b) and separate the brain vesicles cutting under the telencephalic lobes and gradually moving the blade down along the brain. Place it in a corresponding well.
19. One at a time, transfer collected organs from the well on Kimwipes® to draw RNAlater® off. Place organs in a corresponding tube and put it on dry ice inside the styrofoam box (*see* **Note 28**).
20. Put male and female sample in separate plastic bags and transfer samples to the –80 °C storage on dry ice (*see* **Note 43**).

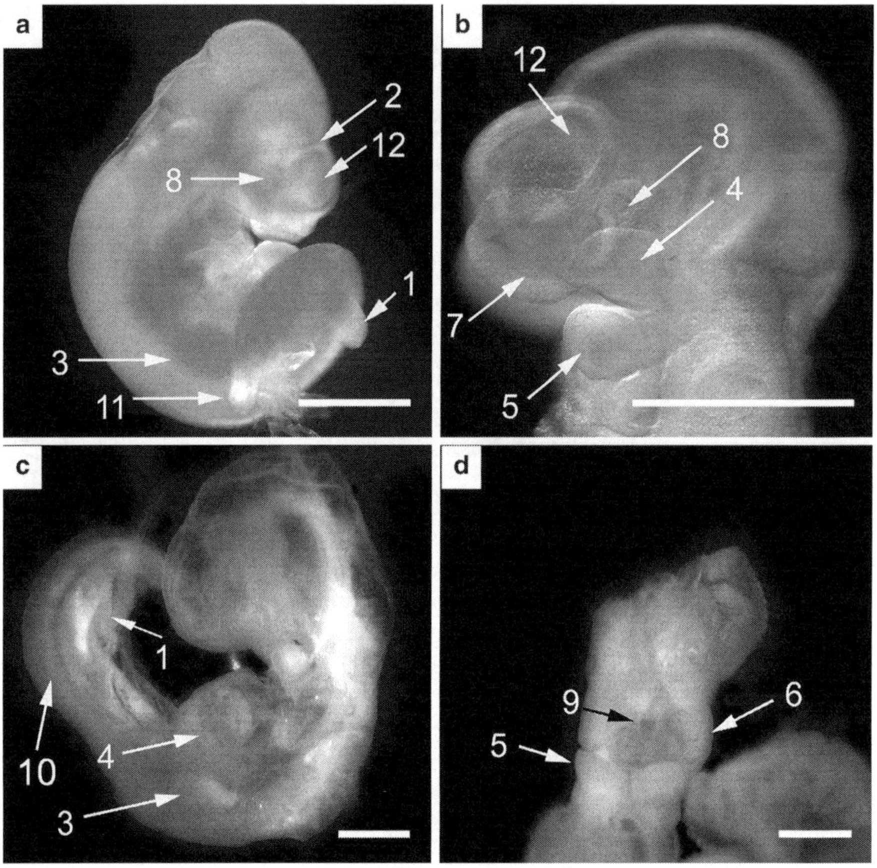

Fig. 16 Rodent embryo staging: Features of 10 dpc embryos readily visible in RNAlater®. (**a**), (**b**) Advanced 10 dpc embryos. Fixation—4 % paraformaldehyde, glycerol (*see* **Note 53**) (Courtesy of Kristin K. Biris and Terry P. Yamaguchi). (**c**) An early 10 dpc embryo (incubation 1 h). (**d**) Frontal view with Rathke's pouch exposed of an advanced 10 dpc embryo (24 h of storage at −20 °C after 4 h of RT incubation). (**a**–**d**) Bar = 1,000 μm. (*1*) Foot plate, (*2*) groove, (*3*) hand plate, (*4*) heart, (*5*) mandibular component of the first branchial arch, (*6*) maxillary component of the first branchial arch, (*7*) olfactory (nasal) pit, (*8*) optic pit, (*9*) Rathke's pouch, (*10*) somites, (*11*) tail, (*12*) telencephalic lobes

3.4.3 Necropsy Protocol of 12–13 dpc Embryo Organ Collection

Part 1

1. Follow **steps 1** through **3** of Subheading 3.4.2.
2. Pour 1 ml of RNAlater® in each designated well.
3. Inside the styrofoam box, in order, dissect embryos out of the uterus keeping umbilical cord intact (Fig. 3b, d). Snap-freeze the yolk sac and place the embryo into the corresponding well with RNAlater®.
4. Carefully agitate the multiwell plate each time an embryo is placed in it, to wash blood off.
5. Inspect embryo for normal development, observe the external features and inspect the palate and umbilical hernia for staging (*see* 11–14 dpc in Tables 2, 3, 4, 5, 6, 7, 8, 9, and 10).

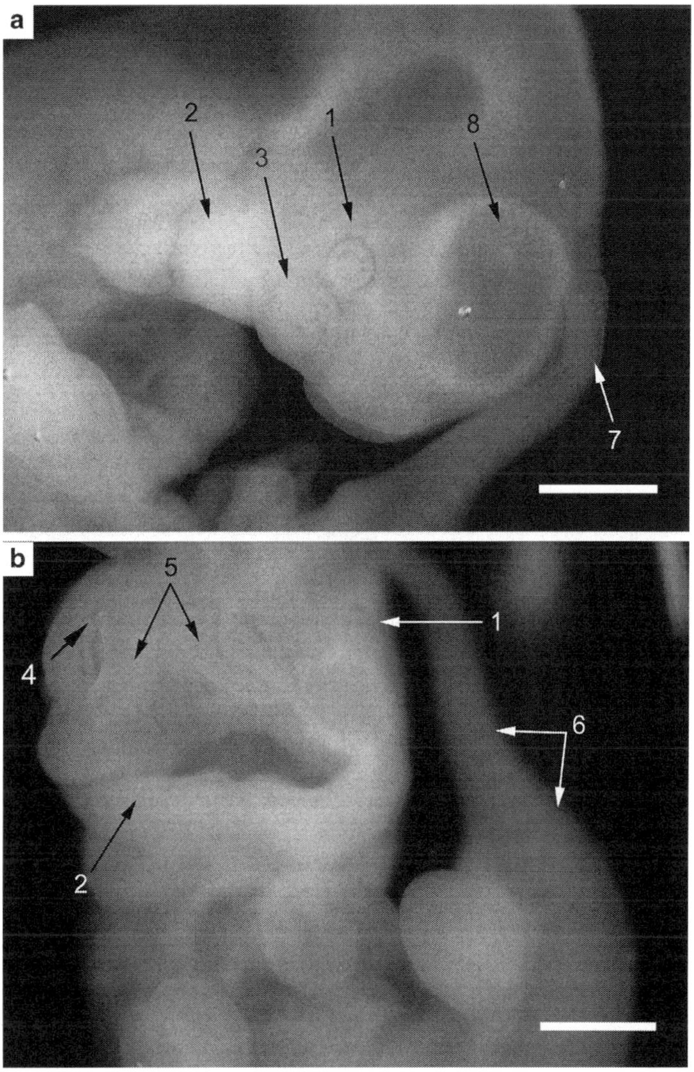

Fig. 17 Rodent embryo staging: Features of 11 dpc embryos readily visible in RNAlater®. (**a**) Sagittal view. (**b**) Frontal view. (**a**), (**b**) Incubation 20 min, bar = 1,000 μm. (*1*) eye, (*2*) mandibular component of the first branchial arch, (*3*) maxillary component of the first branchial arch, (*4*) nasal pit, (*5*) nasal process, (*6*) somites, (*7*) tail, (*8*) telencephalic lobes

6. Unfold the embryo from its natural position (Fig. 4b).
7. Place the plate on the shaker for 10 min to bleed the embryos out.
8. Make incisions under the brain, heart, and liver and peel skin off the brain (*see* **Note 44**) (Fig. 4c, d).
9. Transfer embryos to a multiwell plate with fresh RNAlater®.

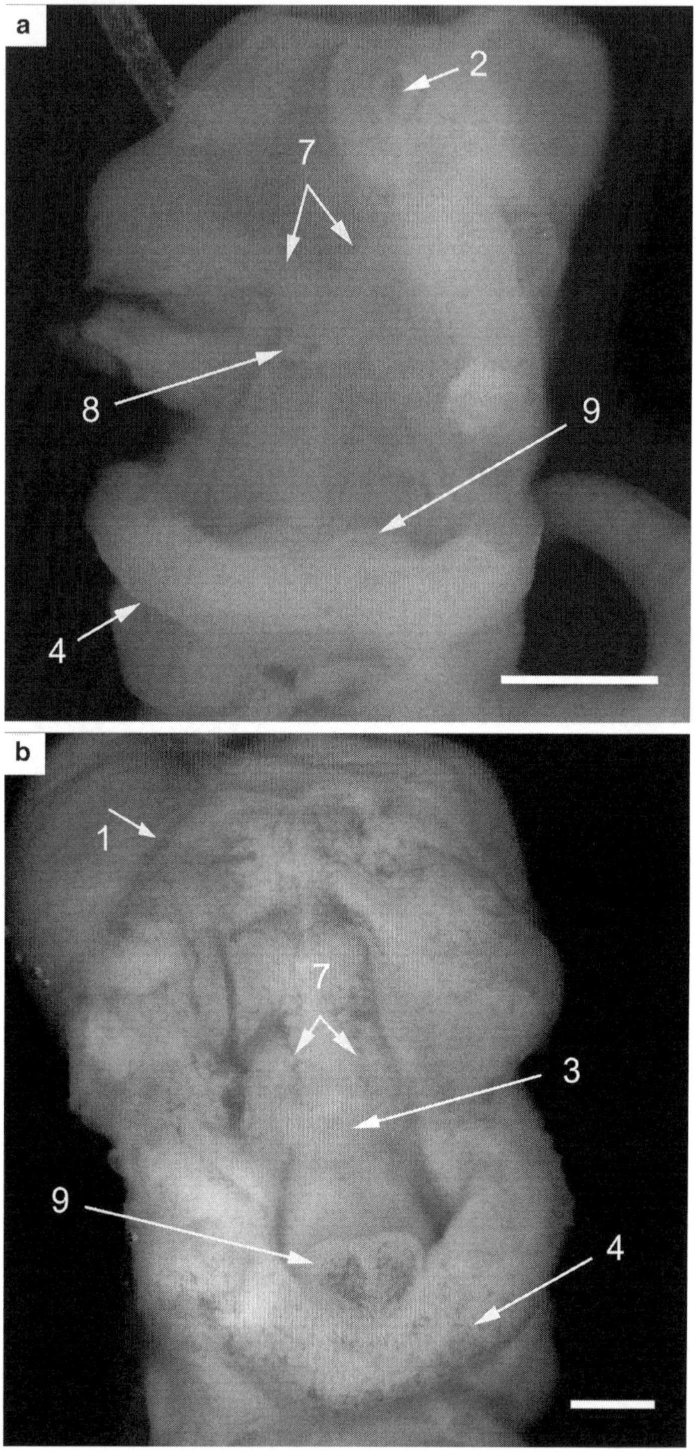

Fig. 18 Rodent embryo staging in RNAlater®: 11 and 12 dpc embryos palate. (**a**) Frontal view of an 11 dpc embryo head with Rathke's pouch exposed. (**b**) Frontal view of a 12 dpc embryo head with palate and former Rathke's pouch location exposed (incubation 4 h). (**a**), (**b**) Bar = 1,000 μm. (*1*) Anterior naris, (*2*) nasal pit, (*3*) location of closed Rathke's pouch, (*4*) lower lip, (*5*) mandibular component of the first branchial arch, (*6*) palatal shelves, (*7*) postero-medial vertical folds, (*8*) Rathke's pouch, (*9*) tongue

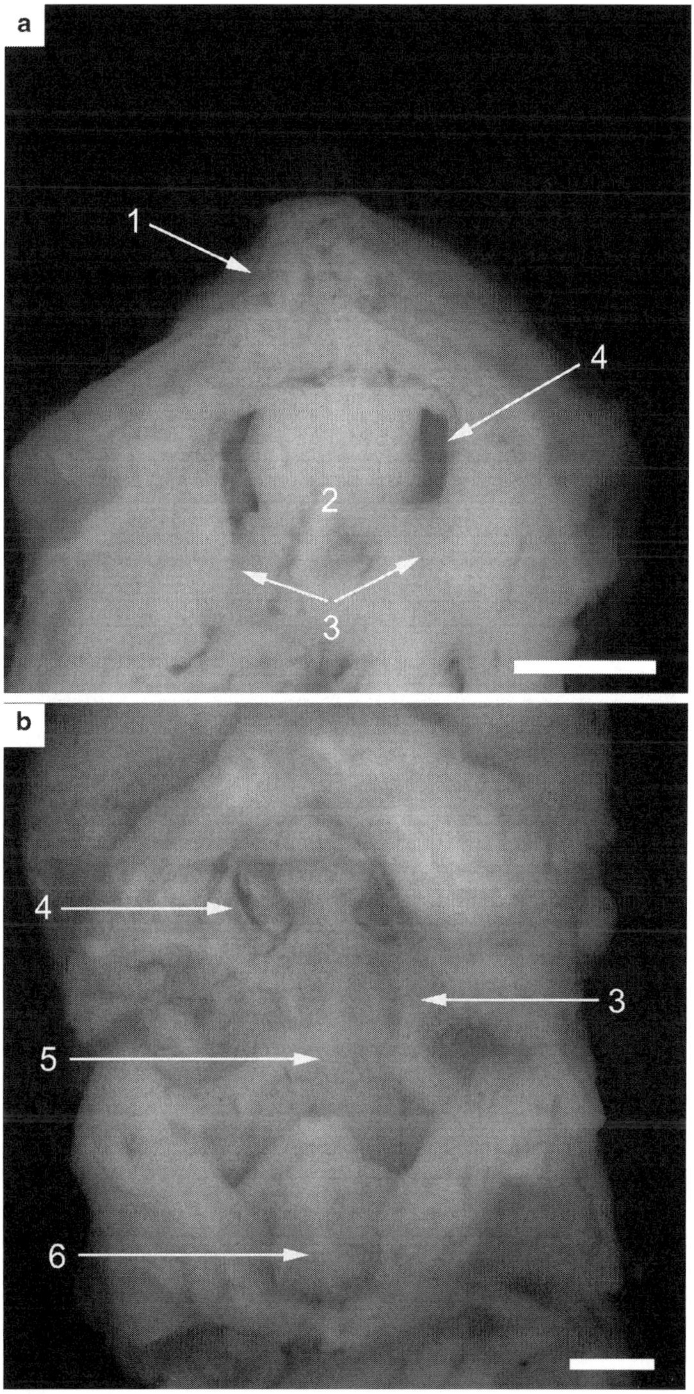

Fig. 19 Rodent embryo staging in RNAlater®: 13 dpc embryos palate. (**a**) An early 13 dpc embryo palate. (**b**) Frontal view of an advanced 13dpc embryo head with palate and former Rathke's pouch location exposed. (**a**), (**b**) (4 weeks storage at −20 °C after 4 h of RT incubation, bar = 1,000 μm). (*1*) Anterior naris, (*2*) nasal septum, (*3*) palatal shelves, (*4*) posterior naris, (*5*) location of closed Rathke's pouch, (*6*) tongue

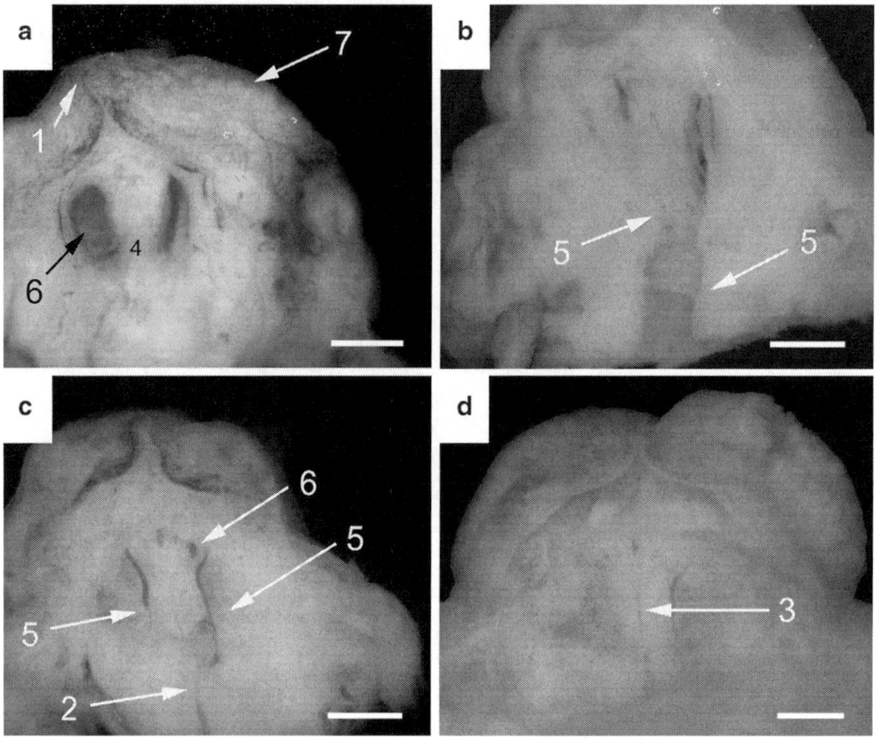

Fig. 20 Rodent embryo staging in RNAlater®: 14 and 15 dpc embryos palate. (**a**), (**b**) Early and late 14 dpc embryos, respectively. (**c**), (**d**) Early and late 15 dpc embryos, respectively. (**a**)–(**d**) 4 weeks storage at −20 °C after 4 h of RT incubation, bar = 1,000 μm. (*1*) Anterior naris, (*2*) beginning of midline fusion of palatal shelves, (*3*) line of fusion, (*4*) nasal septum, (*5*) palatal shelves, (*6*) posterior naris, (*7*) vibrissae

10. Put the plate with all the embryos on a shaker for 4 h or overnight at RT.
11. Seal the multiwell plate inside the plastic bag and keep at −20 °C until genotyping is completed (*see* **Note 38**).

Part 2

12. Follow **steps** 7 through **10** of Subheading 3.4.2.
13. Pin the embryo down with minuten pins through the head just under the brain between the eyes, and through the tail (Figs. 4b and 22b).
14. Cut open the intact part of the peritoneal cavity and peel the skin off (Fig. 5b).
15. Separate spleen and pancreas with supporting mesenchymal tissue from stomach and move them to the bottom of the dish (Fig. 7a–d).
16. Clean pancreas and spleen from mesenchymal tissue and transfer organs to their corresponding wells.
17. Dissect gonads out and transfer them to their corresponding wells (*see* **Note 42**) (Figs. 11b–d and 13d).

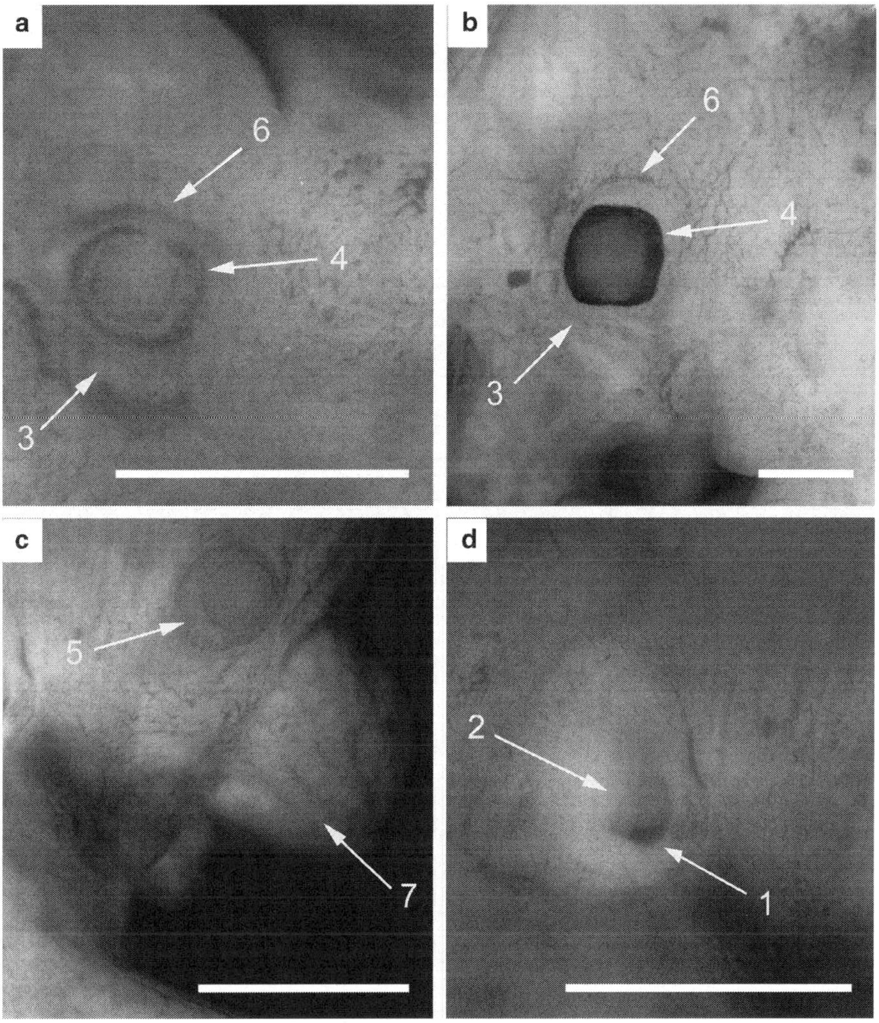

Fig. 21 Rodent embryo staging in RNAlater®: Eye and ear. (**a**), (**b**) 12 and 14 dpc embryos eye (non-albino strain). (**c**) A 14 dpc embryo eye (albino strain). (**d**) A 14 dpc embryo ear. (**a**)–(**d**) Incubation 4 h, bar = 1,000 μm. (*1*) Ear meatus, (*2*) ear pinna, (*3*) lower eye lid, (*4*) pigmented area, (*5*) unpigmented area, (*6*) upper eye lid, (*7*) vibrissae

18. Dissect kidneys out and transfer them to their corresponding well (Figs. 11b–d and 12a).
19. Dissect adrenals out and transfer them to their corresponding well (Fig. 12a).
20. Dissect liver out, carefully "rolling" it off the stomach, and transfer it to its corresponding well (Figs. 5b, 10b, c, and 14a).
21. Dissect stomach out and transfer it to its corresponding well.
22. Dissect intestine out together with umbilical hernia and transfer it to its corresponding well (Fig. 22a).
23. Dissect heart out and transfer it to its corresponding well (Figs. 5b and 9b, c).

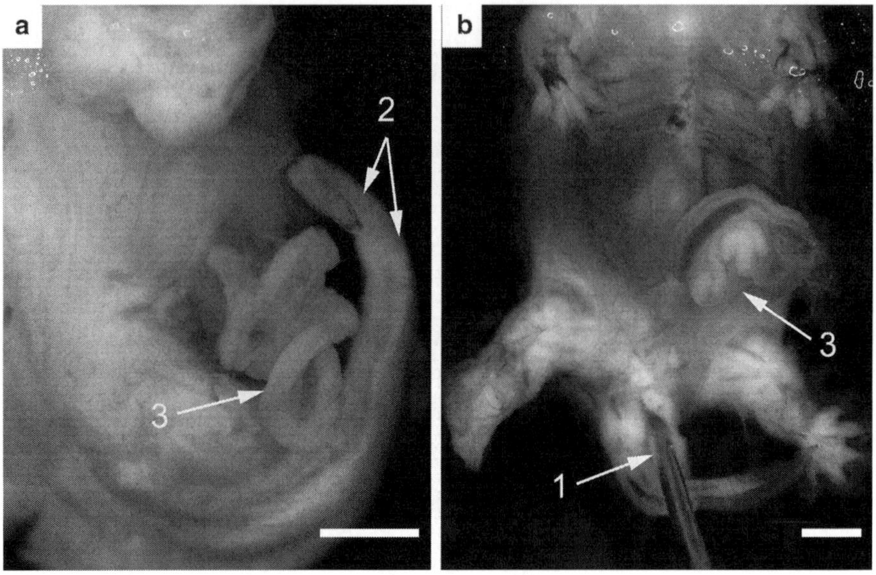

Fig. 22 Rodent embryo staging in RNAlater®: Umbilical hernia. (**a**), (**b**) 13 and 14 dpc embryos hernia, respectively (incubation 4 h, bar = 1,000 μm). (*1*) needle, pinning embryo down through the tail, (*2*) somites in the tail, (*3*) umbilical hernia

24. Dissect thymus lobes out and transfer them to their corresponding well (Figs. 10d and 14a, b).
25. Dissect lung lobes out and transfer them to their corresponding well (Figs. 6b, c and 8b, c).
26. Peel the rest of the skin layer off the brain, dissect brain out and transfer it to the corresponding well.
27. One at a time, transfer collected organs from the well on the Kimwipe to draw RNAlater® off (large organs e.g., liver should be blotted with Kimwipe). Place them in a corresponding labeled tube (*see* **Note 28**) and put on dry ice inside the styrofoam box.
28. Follow **step 20** of the Subheading 3.4.2.

3.4.4 Necropsy Protocol of 14–16 dpc Embryo Organ Collection

Part 1

1. Follow **step 1** through **6** of Subheading 3.4.1.
2. Fill 60 mm sterile petri dish with 5 ml of RNAlater®.
3. Dissect 14 dpc embryos out of the uterus inside the styrofoam box and move them into the petri dish with RNAlater. Euthanize 15 and 16 dpc embryos by cervical dislocation with scissors or scalpel before placement in RNAlater®. Carefully agitate the dish each time the embryo placed in, to wash blood off the embryo.
4. Incubate the embryos on a shaker at RT for 10 min to bleed them out.

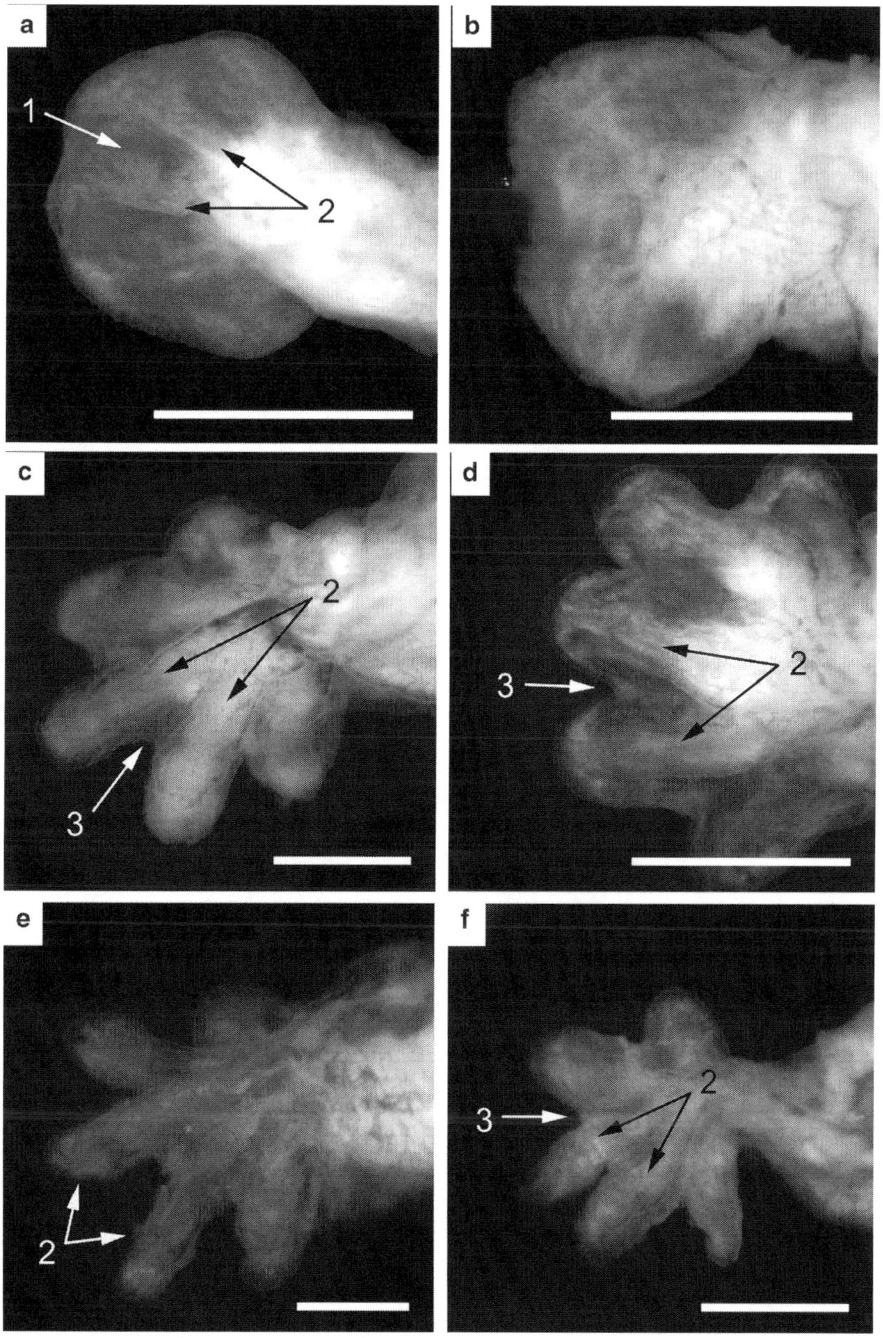

Fig. 23 Rodent embryo staging in RNAlater®: Hand and foot plates. (**a**), (**c**), (**e**) 13, 14, and 15 dpc embryos hand plates, respectively. (**b**), (**d**), (**f**) 13, 14, and 15 dpc embryos foot plates, respectively. (**a**–**f**) Four week storage at −20 °C after 4 h of RT incubation, bar = 1,000 μm. (*1*) Digital interzones, (*2*) digits, (*3*) webbing

Table 11
Breeding and rate of embryo development of different rodent strains

Strain	Pregnancy (%) among females plugged in estrus	Average number of vital embryos per litter	Tendency of development (% of litters with different age[a] embryos)					
			E	O	E and O mix	Y	E and Y mix	E, O, Y mix
C57BL/6J (M)	91	7.4 ± 2.1 (*n* = 27)	39	29	25	4	3	0
129×1/SVJ (M)	75	4.7 ± 2.4 (*n* = 27)	60	8	11	12	12	1
129S1/SVIMJ (M)	87	6.3 ± 2.5 (*n* = 28)	41	24	21	7	7	0
A/J (M)	94	7.3 ± 1.8 (*n* = 26)	67	0	0	0	33	0
BN/Mcwi (R)	78	5.1 ± 1.4 (*n* = 57)	84	0	0	12	4	0
F344/NCr (R)	70	8.0 ± 2.7 (*n* = 14)	72	21	7	0	0	0

[a]Litter age: *E* expected age, *O* older than expected, *Y* younger than expected
(M)—mouse, (R)—rat

5. Pour 3–4 ml of RNAlater® into the 60 mm paraffin dissecting dish.
6. Fill two petri dishes with 5 ml of RNAlater and label them with embryo litter ID, and sex.
7. One at a time, transfer embryos in a paraffin dissecting petri dish for staging, sexing, and preparation for organ dissection.
8. Observe the external features and inspect the palate and umbilical hernia for staging before the embryo is pinned down (*see* 13–17 dpc in Tables 2, 3, 4, 5, 6, 7, 8, 9, and 10).
9. Pin the embryo down through the head just under the brain between the eyes and through the tail (Fig. 4b).
10. Make incisions under the brain, heart, and liver and peel the skin off the brain (*see* **Note 44**) (Fig. 4c, d).
11. Carefully expose a gonad (Fig. 12).
12. Record the sex of each embryo and place it in the corresponding petri dish.
13. Put the petri dish on a shaker at RT for 20 min for immediate organ harvest or for 4 h to overnight for delayed harvest.
14. After the incubation dissect embryos at RT (*see* **Note 45**), or transfer embryos into the labeled sample cup, and add up to 15 ml of RNAlater® for 10 embryos before moving the cup in –20 °C storage.

Table 12
Age-related feasibility of organ dissection from mouse and rat embryos for RNA retrieval

Organ	9 dpc	10 dpc	11 dpc	12 dpc	13 dpc	14 dpc	15 dpc	16 dpc	17 dpc	18 dpc
Adrenal					M	M	M, R	M, R	M, R	M, R
Bladder				M	M	M, R	M, R	M, R	M, R	M, R
Brain			M	M, R	M, R	M, R	M, R	M, R	M, R	M, R
Eye				M	M, R	M, R	M, R	M, R	M, R	M, R
Epididymis									M	M, R
Gonads (in different)			M	M, R	M, R	R				
Gall bladder							M	M	M, R	M, R
Heart	M	M	M, R	M, R	M, R	M, R	M, R	M, R	M, R	M, R
Intestine		M	M, R	M, R	M, R	M, R	M, R	M, R	M, R	M, R
Kidney				M	M	M, R	M, R	M, R	M, R	M,
Liver		M	M, R	M, R	M, R	M, R	M, R	M, R	M, R	M, R
Lung			M	M, R	M, R	M, R	M, R	M, R	M, R	M, R
Lymph node									M	M, R
Ovary						M	M	M, R	M, R	M, R
Pancreas			M	M, R	M, R	M, R	M, R	M, R	M, R	M, R
Pituitary gland										M
Salivary gland						M	M	M, R	M, R	M, R
Spleen				M	M	M, R	M, R	M, R	M, R	M, R
Stomach		M	M	M, R	M, R	M, R	M, R	M, R	M, R	M, R
Testes						M	M	M, R	M, R	M, R
Thymus					M	M, R	M, R	M, R	M, R	M, R
Thyroid						M	M, R	M, R	M, R	M, R
Tongue			M	M	M, R	M, R	M, R	M, R	M, R	M, R
Uterus								M	M, R	M, R

(*M* mouse. *R* rat. Numbers 9–18 dpc embryos age). Note that rat development extends to 20 dpc

Part 2

15. Follow **step** 7 through **9** of Subheading 3.4.2.
16. One at a time, transfer embryos in a 35 mm Sylgard®-filled petri dish with 2–3 ml of cold RNAlater® (*see* **Note 46**), and place it on a frozen cool pack under a dissecting microscope.
17. Pin down the transferred embryo through the existing holes (head and tail) (Fig. 4b).
18. Cut open the intact part of the peritoneal cavity and separate spleen and pancreas with supporting mesenchymal tissue from stomach (Fig.7e, f).
19. Clean pancreas and spleen from mesenchymal tissue and transfer organs in their corresponding wells.
20. Dissect gonads (Figs. 12 and 13b) out and transfer them in the corresponding well (*see* **Note 42**).
21. Dissect kidneys and adrenals (Fig. 12) out and transfer them to their corresponding wells.
22. Dissect liver (Fig. 5c, d) out, carefully "rolling" it off the stomach, and transfer it to its corresponding well.
23. Dissect stomach out and transfer it to its corresponding well.
24. Dissect intestine out together with umbilical hernia and transfer it to its corresponding well.
25. Peel skin over the heart apart to open the pleural cavity.
26. Dissect heart (Figs. 5c, d and 9d) out and transfer it to its corresponding well.
27. Dissect thymus lobes (Figs. 5d and 14c, d) out and transfer them to their corresponding well.
28. Dissect lung lobes (Figs. 6d and 8d) out and transfer them to their corresponding well.
29. Peel the skin off the neck area towards the mouth to expose salivary glands (Fig. 14d).
30. Dissect salivary glands out and transfer them to their corresponding well.
31. Follow **steps 26** through **28** of Subheading 3.4.3.

3.4.5 Necropsy Protocol of 17–18 dpc Embryo Organ Collection

Part 1

1. Follow **steps 1** and **2** of Subheading 3.4.4.
2. Dissect embryos from the uterus inside a styrofoam box with wet ice one by one. Euthanize embryos by cervical dislocation with scissors or scalpel and move them into the petri dish with RNAlater. Carefully agitate the dish each time an embryo is added, to wash blood off.
3. Follow **steps 4** through 7 of Subheading 3.4.4.

4. Observe the external features for staging before the embryo is pinned down (*see* 16–18 dpc in Tables 2, 3, 4, 5, 6, 7, 8, 9, and 10).
5. Pin the embryo down through the front and hind limbs.
6. Make incision under the brain and peel the scull off the cerebrum (front lobes of the brain).
7. Cut open peritoneal and plural cavities.
8. Follow **steps 12** through **14** of Subheading 3.4.4.

Part 2

9. Follow **step 15** of the Subheading 3.4.4.
10. Pin down the transferred embryo through the existing holes in the limbs.
11. Peel spleen off the stomach and transfer it to its corresponding well (*see* **Note 47**).
12. Separate pancreas from the duodenum and transfer it to its corresponding well.
13. Dissect the following organs in the listed order and transfer them to their corresponding wells: gonads, bladder, kidneys, adrenals, liver, stomach, intestine, mesenteric lymph node, heart, thymus, and lungs.
14. Peel the skin off the neck area towards the mouth to expose salivary glands.
15. Dissect salivary glands out and transfer them to their corresponding well.
16. Dissect thyroid lobes out and transfer them to their corresponding well.
17. Dissect brain and pituitary gland out and transfer them to their corresponding wells.
18. Dissect eyes out and transfer them to their corresponding well.
19. Follow **steps 26** through **28** of Subheading 3.4.3.

3.5 Extraction, Quantification, and Quality Control of Embryonic RNA

RNA extraction from embryos and embryonic organs was performed with TRIzol® protocol, provided by the manufacturer, with the following modifications.

3.5.1 RNA Extraction

1. 1 μl of LPA was added to each sample after 30 s of homogenization and mixed with the sample by 30 s of vortexing before the 5-min incubation at RT (*see* **Note 48**).
2. After the addition of chloroform, each sample was incubated for 3 min at RT.
3. PLG tubes were used for aqueous phase recovery (*see* **Note 49**) according to the manufacturer's instructions.

4. In the precipitation step, to each volume of aqueous phase, we added 0.1 volume of 3 M sodium acetate and 1 volume of isopropanol.
5. The precipitation step was carried at −20 °C overnight.
6. Each RNA pellet was dissolved in RNA Storage Solution, incubated at +65 °C for 3 min, cooled on wet ice for 1 min, vortexed, and spun down.
7. Purified RNA was stored at −80 °C (*see* **Note 50**).

3.5.2 Quantification and Quality Control of Extracted Embryonic RNA

1. RNA concentration was determined with NanoDrop spectrophotometer using 1.5 μl of the sample per measurement.
2. The RNA quality was evaluated by Agilent Bioanalyzer NanoChip (*see* **Note 51**).
3. We accepted RIN # as a measure of RNA quality [1, 10].
4. Selected results of extraction and preparation of embryonic RNA for microarray analysis are summarized in Table 13 (*see* **Note 52**).

4 Notes

1. RNAlater® solution has been found to be a reliable method of tissue RNA preservation for transcriptome study [11, 12].
2. Use item 23–26 for embryo and items 27–28 for pregnant female necropsy. Fine instruments should be kept protected from accidental falls and bends. During dissection do not hold instruments with the hand you use for adjusting the microscope focus, or pinning the embryo to the dissecting dish.
3. Use these scissors for opening the uterus around the fetal compartments.
4. TRIzol® is considered to provide a balanced recovery of all transcripts in a sample [13, 14].
5. Disposable plastic rods are effective for homogenization of embryonic tissues. They eliminate the need for cleaning between samples and risk of sample cross-contamination. To reduce the loss of RNA during homogenization of embryonic organs, lift the spinning rod out of the fluid and briefly touch the side of the tube to draw liquid on the rod down.
6. To have enough RNA for direct (without amplification) hybridization to microarrays, the organs from a number of embryos should be combined for a single extraction. Also, many factors can introduce alterations in gene expression profile during tissue harvesting (e.g., the method of euthanasia, prolonged period of tissue harvest, individual differences in the

Table 13
Selected results of extraction, quantification, and quality control of rodent embryonic RNA (quality and quantity assessment before DNAse treatment)*-paired organ

Sample	Embryo age (dpc)	Strain	Sex	Sample size (*n*)	Incubation in RNAlater (h)	Extraction protocol	Extraction volume of Trizol (ml)	RNA volume (µl)	RNA yield (µg)	RNA quality (RIN)
Embryo	10	C57BL/6J	Male	10	1	Without PLG tube	5	100	815	10
Embryo	10	C57BL/6J	Female	8	1	Without PLG tube	5	100	629	10
Embryo	9	BN/Mcwi	Mixed	37	1	Without PLG tube	0.5	23	9.3	10
Brain	14	C57BL/6J	Male	10	4	Without PLG tube	5	100	1,680	9.9
Brain	14	A/J	Male	10	4	Without PLG tube	5	100	580	9.8
Liver	14	C57BL/6J	Male	10	4	Without PLG tube	5	200	2,140	9.9
Liver	14	A/J	Male	10	4	Without PLG tube	5	200	1,481	9.9
Testes*	14	C57BL/6J	Male	10	4	With PLG tube	1	24	10	9.8
Testes*	14	C57BL/6J	Male	13	672	With PLG tube	1	24	17.6	9.6
Testes*	14	A/J	Male	10	4	With PLG tube	1	24	5	9.8
Testes*	14	129x1	Male	17	1,008	With PLG tube	0.5	24	19.5	9.8
Thymus	14	C57BL/6J	Male	10	4	With PLG tube	1	24	11	9.8
Spleen	14	C57BL/6J	Male	10	4	With PLG tube	1	24	5	9.6
Spleen	13	BN/Mcwi	Mixed	35	4	With PLG tube	0.5	23	8.8	9.7
Kidneys*	13	BN/Mcwi	Mixed	35	4	With PLG tube	0.5	23	37	10
kidneys*	14	C57BL/6J	Male	10	4	With PLG tube	1	24	46	9.7

(continued)

Table 13 (continued)

Sample	Embryo age (dpc)	Strain	Sex	Sample size (n)	Incubation in RNAlater (h)	Extraction protocol	Extraction volume of Trizol (ml)	RNA volume (µl)	RNA yield (µg)	RNA quality (RIN)
Ovaries*	14	C57BL/6J	Female	5	4	With PLG tube	1	24	8	9.9
Ovaries*	14	C57BL/6J	Female	9	672	With PLG tube	1	24	10	9.8
Ovaries*	14	A/J	Female	10	4	With PLG tube	1	24	4.9	9.7
Ovaries*	14	A/J	Female	10	1,008	With PLG tube	1	24	4.5	9.6
Ovaries*	14	129×1	Female	7	4	With PLG tube	1	24	4	9.8
Heart	13	BN/Mcwi	Mixed	15	2	With PLG tube	0.5	33	45	10
Heart	11	BN/Mcwi	Mixed	6	2	With PLG tube	0.2	14	6	10
Lung	11	BN/Mcwi	Mixed	6	2	With PLG tube	0.1	14	0.35	9.7
Lung	12	BN/Mcwi	Mixed	7	2	With PLG tube	0.5	14	2.1	9.8
Lung	13	BN/Mcwi	Mixed	21	2	With PLG tube	0.5	23	20	9.9
Lung	14	BN/Mcwi	Mixed	7	2	With PLG tube	0.2	14	5.4	9.8

dissecting technique of the same tissue, introduced by the necropsy technician, variable ratio of early, advanced, and late embryos in different litters, method of RNA extraction). Methods of euthanasia vary by institution but should be in compliance with local Institutional Animal Care and Use Committee guidelines. A combined sample also ensures against stochastic errors in case of a low copy gene.

7. The number of animals in the study depends on the required amount of RNA for the downstream applications and average RNA content of the target organ. RNA content of embryonic organs depends on extraction and homogenization method, embryo size, age, sex, and animal strain. The number of required animals will also depend on such strain characteristics as litter size, pregnancy rate, and the degree of deviation from the required age among the embryos in the litter (*see* Tables 11 and 13).

8. Based on vaginal smear evaluation, mouse estrus corresponds to the range of numbers from 5 to 11 (data for C57BL/J6), and rat estrus (Sprague-Dawley) from 4 to 10. Mating in estrus dramatically increases the number of pregnancies for plugged females. After the plug has been found, taking consecutive measurements during the next expected cycle helps to detect pregnancy at an early stage of embryo development when palpation is not effective [4].

9. We have established the following protocol for breeding females:
   (a) Mice used for timed pregnancies should be between 2 and 6 months old.
   (b) Females should have their estrus measurement taken at the same time each day.
   (c) On the day of mating, females with an estrus reading between 5 and 11 should be put with a good male. If the estrus reading is less than 5, females should not be used for this round of mating.
   (d) Females should be checked for plugs the day after they are put together with the male.
   (e) If a female in estrus is not plugged the next morning, it should be separated from the male and saved for use at a later date. Females with no plugs for three consecutive matings should be excluded from the mating pool, as well as females with a plug and no pregnancy after two consecutive matings.
   (f) Plugged females should have their estrus measurement taken for the next 5 days. We consider the day of plug formation as day "0." Typically mice with low estrus readings (ideally below 2) during 5 days after plug formation are

pregnant, and should be sacrificed at the correct stage of pregnancy. From day 13, females should also be palpated.

(g) Mice whose estrus reading sharply increase (3, 4, 5, and greater) during 5 consecutive days after plug formation, are most likely not pregnant. They should be palpated at 14 days to confirm the absence of pregnancy and returned to the mating pool.

10. A necropsy station should be wiped down with RNase AWAY™ and lined with disposable utility wipes. All the materials in Subheading 2 were tested for RNase-free conditions. Use a new package for RNase-free techniques. The materials continue to be suitable if the package was sealed and used only for RNase-free setup. If you take out Kimwipes® from the opened box, discard the first one and use the following for the RNase-free conditions. Wear gloves at all times. Aliquot large quantities of reagents. Use RNase-free instruments and clean them for RNase-free conditions (spray with RNAse AWAY™, wipe with Kimwipes® and thoroughly rinse with RNase-free water).

11. The presence of lesions in an individual animal changes the genetic status of many organs, not only of the organ with the lesion.

12. These recommendations are based on our regular protocol. Reported times allow to retrieve high-quality RNA from embryonic tissues. We have not tested the effect of longer dissection times on RNA integrity in harvested organs.

13. Deviations in developmental stage among the embryos of the same litter are normal in mouse and rat, especially during early development [7, 8, 15–17]. Embryos positioned closer to the ovaries might be delayed in development by 2 days compared with embryos positioned at the base of the uterine horn. Different strains develop at different rates [18], and males develop faster than females [19, 20]. A litter may to be older or younger than the age calculated by the plug date, (*see* Table 11) due to delays in implantation, missing a plug during a checkup, or fast implantation after mating.

14. It allows rechecking a female's identification and genotype in case of bias in downstream analysis of embryonic RNA or unexpected phenotypes in genetically engineered strains [21].

15. Contamination of the sample can dramatically affect downstream analysis. Instruments should be treated for RNAse-conditions after removal of such RNAse-rich tissues as pancreas, liver, heart, and adrenals. Instruments and dissecting dishes should be treated between male and female samples.

16. For staging of 7–8 day embryos by morphological landmarks, *see* refs. [22] and [23].

17. Combination of ring light with directed illumination from "goose neck" light source reduces eye fatigue and facilitates photography.
18. Melted Paraplast® should be poured into a sterile petri dish half way. Close the lid and place the petri dish on a cold plate until Paraplast® solidifies. Paraplast®-filled dishes are not transparent for microscope base lights, and the white background makes it difficult to see early embryos and embryonic organs. Paraplast® flakes off and cracks after embryos are repeatedly pinned to it with dissecting needles. However, it is an easy and fast way to make a disposable dissecting dish suitable for a variety of tasks.
19. Sylgard® is an inert silicon material. Silicon is widely used in a range of molecular biology products (e.g., tubes, films, laser capture collection tubes). Cured Sylgard® can be frozen down to −50 °C or heated up to 200 °C without damage. Prepare the mixture according to manufacturer's instructions and slowly pour it into sterile petri dishes avoiding bubbles. Close the lids and leave overnight at RT until it solidifies. Solidified Sylgard® is transparent to light from the microscope base. Sylgard® firmly holds minuten dissecting needles and allows tissue cutting without breaking sapphire knives and damaging forceps tips. The surface of the dissecting dishes can be cleaned for RNase-free conditions (wipe with Kimwipes® soaked in RNase AWAY™ and thoroughly rinse with RNase-free water). Sylgard® can be used with different ratios of components to prepare sticky surfaces. Early age embryos or embryonic organs, gently pressed against such surfaces, stay in place during observation. Sylgard® dishes can be also used for fixation of pinned embryos or organs with any fixative, or in whole mount staining procedures.
20. RNA extracted from 9 dpc embryo heart, dissected in RNase-free cold PBS, showed high-quality RNA (RIN = 9.9), but RNA of 10 dpc and 11 dpc lungs and heart dissected under the same conditions showed some degradation (RIN = 8.5, 8.0, and 8.2 respectively).
21. Gently rolling the yolk sac out of the decidua helps to preserve embryos intact for staging. The yolk sacs can be easily peeled off when embryos become harder during incubation in RNAlater®. Avoid "popping" embryos out by applying pressure to the deciduas. Instead, cut the uterus wide open and peel the decidua off to expose intact yolk sac. Pressure on the decidua dramatically changes the appearance of 7–8 dpc embryos and damages 9–12 dpc embryos. "Popping" embryos out may destroy urogenital ridges, the end of the tail, and the developing umbilical hernia (an important feature for staging 10–11 dpc embryos).

22. Blood is rich with RNases. Contamination of dissected organs with blood contributes to RNA degradation.
23. We stage early embryos (9–11 dpc) primarily by external features (Tables 2, 3, 4, 5, and 6). Internal features described in Tables 7, 8, 9, and 10 are used for confirmation of the ages determined by external features, if organ harvest is required.
24. It has been shown by microarray experiments that incubation of the tissue in cold RNAlater® (+4 °C) immediately after dissection impacts its gene profile [11, 24–26].
25. We have retrieved high-quality RNA from 14 dpc organs dissected from embryos stored at −20 °C up to 6 weeks. Longer storage has not been tested.
26. The practice of picking up tiny organs with a glass pipette often results in a loss or damage of samples because of their remarkable ability to stick to the outer or inner walls of the glass pipette or plastic pipette tip. It is difficult to see and detach them.
27. Blotting the tissues with Kimwipes® prevents RNAlater® carryover which can change the salt content of reagents during RNA extraction. Moderate carryover will be shown as low 260/230 ratios on NanoDrop reports.
28. Do not let tiny organs over dry on Kimwipes® because they become static and tend to fly away. As soon as the last transferred fluid drains off, collect them with forceps by sticking the organs together in a lump. Open a 2 ml collection tube and, holding it upside down, and press the tube lid against the bench surface. Transfer organs to the lid and press the tube against the lid. Flip it to the up position and tap to deposit the tissue on the tube bottom. Homogenize the sample, or put it on dry ice and transfer to −80 °C until RNA extraction.
29. Embryos of different strains and species differ in size at identical gestation ages. For instance, mouse embryos of A/J strain at 11 dpc are the size of 9 dpc C57BL/6 J embryos; dissection of gonads, lung, and pancreas from 11 dpc A/J embryos is not feasible.
30. Transfer an embryo or organs (from one embryo) out of RNAlater® to 200 ml of 10 % NBF for 48 h. Fix at RT on a shaker before processing and paraffin embedding. Change to fresh fixative at the half time of incubation. We have fixed and routinely processed to paraffin block 13 dpc embryos, as well as stomach with spleen and pancreas, and gonads after 4 h of incubation in RNAlater®. Slides stained with hematoxylin and eosin (H&E) were suitable for evaluation by a veterinary pathologist (Fig. 13a, c). Similar results have been reported for a range of human tissues [26], as well as successful immunostaining of RNAlater® preserved esophageal human tumors [27].

31. A bladder in the "up" position allows distinguishing the left and right uterine horns for labeling fetal compartments in a pregnant uterus. It is crucial for embryo genotyping and documentation of developmental abnormalities.
32. Necropsy report form should contain the following: a cage card with female information, an expected gestation day, a sketch of gravid uterus with labeled fetal compartments, and any observations about the female and embryos.
33. This procedure guaranteed high-quality RNA retrieval when the gravid uterus was on wet ice for 40 min (from the time of necropsy to the placement of embryos into RNAlater®) A different approach has not been tested.
34. We use the following cleaning procedure for the instruments. Prepare a 50 ml Falcon tube of RNase AWAY™ and two tubes of RNase-free water. Dip the instruments into RNase AWAY™ and wipe with new Kimwipes®, and follow with two rinses in water. Wipe the instrument with new Kimwipes® after each rinse.
35. A yolk sac with embryo should be picked up by the loose part of a yolk sac, or transferred with the transfer pipette with the cutoff end (to form the wide opening).
36. It is easier to observe an embryo for staging inside a partially open yolk sac (Fig. 3a). Pin the yolk sac with a minuten pin to the bottom of the well and open the yolk sac. After staging, completely remove the yolk sac.
37. RNAlater® crystallizes fast, so while working with one embryo, cover the rest of the wells with Parafilm®.
38. If genotyping is not required, proceed to the Part 2 and dissect organs in RNAlater® at RT.
39. The multiwell plate and a 50 ml Falcon tube with RNAlater® can be stored in the freezer. In this case, pour 0.5 ml of cold RNAlater in a cold multiwell plate before dissections.
40. Rinse the dissecting dish with RNase-free water and wipe with Kimwipes® after each embryo necropsy to get rid of debris. After dissection of male embryos, clean the instruments and the dissecting dish for RNase-free conditions for female samples.
41. After each organ removal, rinse the tools in RNase-free water and wipe with Kimwipes® prior to the next organ dissection. Inspect the tips of the dissecting tools under the microscope for presence of debris.
42. Turn the dissecting dish 90° so that the gonads' long axis (Figs. 11b–d and 12a) are parallel to you. Cut under both ends of the gonad and roll it off the urogenital ridge with the scraping motion of the sapphire blade.

43. Keep female and male samples separate during sample collection, preparation, and extraction. Mixed up (failed) samples prepared for microarray are costly.
44. The incision under the liver should be made on the dissector's left side to keep the spleen and stomach intact. The location where you make an incision for RNAlater® penetration depends on the target organs of your necropsy protocol.
45. For immediate dissection skip to **step 17**.
46. Volume of RNAlater® solution depends on embryo size. The solution should cover an embryo completely.
47. All the organs of 17 and 18 dpc embryos, except the gonads, look similar to the organs of an adult rodent.
48. LPA as an RNA carrier maximizes the recovery of RNA from small samples. On Agilent Pico Chip electropherogram such samples will show a broad peak represented by recovered transcripts <200 bp in the 5S low molecular weight area. Do not add LPA to homogenates if you are going to freeze the samples at −80 °C and proceed with extraction later, because LPA fragments in these conditions lose their carrier efficiency. In this case, add LPA in a thawed out homogenates before RNA extraction.
49. The use of PLG tubes for small extraction volume increases the recovery of aqueous phase by 30–40 % compared to the routine procedure (*see* also ref. [28]). Do not touch the wax inside the tube during pipetting. Contamination of the RNA sample will affect the accuracy of NanoDrop quantification and will lower the RIN number for PicoChip analysis.
50. A novel RNA storage medium, RNAstable® (Biomatrica, San Diego, CA) can be used to store RNA samples at room temperature as a dried pellet for extended period of time [29].
51. Samples with NanoDrop concentrations below 60–100 ng/μl should be evaluated on Agilent Pico Chip for reproducible results (*see* also [1, 10]).
52. The protocol with PLG tubes was used for all samples with projected RNA recoveries less than 100 μg. For larger samples we used a protocol without PLG tubes.
53. 8–10 dpc embryos fixed in 4 % paraformaldehyde for 24 h look similar to the same age embryos incubated in RNAlater® for 10 min.

Acknowledgments

The authors would like to thank Anna Trivet of NCI-Frederick for her technical expertise and Keith Rogers, currently of BSF–Singapore for his continuous support during many different phases of this study, also Scott Lawrence of NCI-FCRDC, DCTD, DTP, PADIS for generous help with photography and Tamara Morgan of Histotechnology Laboratory, NCI–Frederick for technical assistance. This work was funded by NCI Contract HHSN261200800001E.

References

1. Strand C, Enell J, Hedenfalk I, Ferno M (2007) RNA quality in frozen breast cancer samples and the influence on gene expression analysis–a comparison of three evaluation methods using microcapillary electrophoresis traces. BMC Mol Biol 22:8–38
2. Coposis V, Bibeau F, Bascoul-Mollevi C, Salvertat N, Chalbos P, Bareil C et al (2007) Impact of RNA degradation on gene expression profiles: assessment of different methods to reliably determine RNA quality. J Biotechnol 127:549–559
3. Ibberson D, Benes V, Muckenthaler MU, Mirco C (2009) RNA degradation compromises the reliability of microRNA expression profiling. BMC Biotechnol 9:102
4. Ramos SD, Lee JM, Peuler JD (2001) An inexpensive meter to measure differences in electrical resistance in the rat vagina during the ovarian cycle. J Appl Physiol 91:667–670
5. Institute of Laboratory Animal Resources, Commission on Life Sciences, National Research Council (1996) Guide for the care and use of laboratory animals, 7th edn. National Academy Press, Washington, DC, p 125
6. Phifer CB, Terry LM (1986) Use of hypothermia for general anesthesia in preweaning rodent. Physiol Behav 38:887–890
7. Kaufman MH (1992) The atlas of mouse development. Academic, London, 525 pp
8. Witschi E (1962) Development: rat. In: Altman PL, Dittmer DS (eds) Growth, including reproduction and morphological development. Fed. Soc. Exp. Biol, Washington, DC, p 608
9. Theiler K (1972) The house mouse. Development and normal stages from fertilization to 4 weeks of age. Springer, New York, 178 pp
10. Imbeaud S, Graudens E, Boulanger V, Barlet X, Zaborski P, Eveno E et al (2005) Towards standartization of RNA quality assessment using user–independent classifiers of microcapillary electrophoresis traces. Nucleic Acids Res 33:e56
11. Mutter GL, Zahrieh D, Liu C, Neuberg D, Finkelstein D, Baker HE et al (2004) Comparison of frozen and RNAlater® solid tissue storage methods for use in RNA expression microarrays. BMC Genomics 5:88
12. Kihara AH, Moriscot AS, Ferreira PJ, Hamassaki DE (2005) Protecting RNA in fixed tissue: an alternative method for LCM users. J Neurosci Methods 148:103–107
13. Ojaniemi H, Evengard B, Lee DR, Unger ER, Vernon SD (2003) Impact of RNA extraction from limited samples on microarray results. Biotechniques 35:1–5
14. Chomczynski P, Sacchi N (2006) The single-step method of RNA isolation by acid guanidine thiocyanate–phenol–chloroform extraction: twenty–something years on. Nat Protoc 1: 581–585
15. Goedbloed JF (1976) The embryonic and postnatal growth of rat and mouse. IV. Prenatal growth of organs and tissues: age determination and general growth pattern. Acta Anat 95:8–33
16. Goedbloed JF (1980) The embryonic and postnatal growth of rat and mouse. VI. Prenatal growth of organs and tissues: individual organs; final remarks on parts I–VI, phase transitions. Acta Anat 106:108–128
17. Fujinaga M, Baden FM (1992) Variation in development of rat embryos at the presomite period. Teratology 45:661–670
18. Byrne MJ, Newmark JA, Warner CM (2006) Analysis of the sex ratio in preimplantation embryos from B6.K1 and B6.K2 Ped gene congenic mice. J Assist Reprod Genet 23: 321–328
19. Perez-Crespo M, Ramirez MA, Fernandez-Gonzalez R, Rizos D, Lonergan P, Pintado B et al (2005) Differencial sensitivity of male and female mouse embryos to oxidative induced heat stress is mediated by glucose-6-phosphate dehydrogenase gene expression. In Gary M. Wessel, John Willey & Sons, Inc., (ed.) Providence, RI. Mol Reprod Dev 72: 502–510
20. Seller MJ, Perkins-Cole KJ (1987) Sex difference in mouse embryonic development at neurulation. J Reprod Fertil 79:159–161
21. Ward JM, Devor-Hanneman DE (2000) Gestational mortality in genetically engineered mice: evaluating the extraembryonal embryonic placenta and membranes. In: Pathology of genetically engineered mice, p 103–120
22. Fujinaga M, Brown NA, Baden JM (1992) Comparison of staging systems for the gastrulation and early neurulation period in rodents: a proposed new system. Teratology 46:183–190

23. Downs KM, Davies T (1993) Staging of gastrulating mouse embryos by morphological landmarks in the dissecting microscope. Development 118:1255–1266
24. Micke P, Ohshima M, Tahmasebpoor S, Ren Z, Ostman A, Ponten F et al (2006) Biobanking of fresh frozen tissue: RNA is stable in nonfixed surgical specimens. Lab Investig 86:202–211
25. Ellis M, Davis N, Coop A, Liu M, Schumaker L, Lee RY et al (2002) Development and validation of a method for using breast core needle biopsies for gene expression microarray analyses. Clin Cancer Res 8:1155–1166
26. Florell SR, Coffin SM, Holden JA, Zimmermann JW, Gerwels JW, Summers BK (2001) Preservation of RNA for functional genomic studies: a multidisciplinary tumor bank protocol. Mod Pathol 14:116–128
27. Smith E, De Young NJ, Tian Z, Caruso M, Ruszkiewicz AR, Liu J et al (2008) Methylation of TIMP3 in esophageal squamous cell carcinoma. World J Gastroenterol 14:203–210
28. Murphy NR, Hellwig RJ (1996) Improved nucleic acid organic extraction through use of a unique gel barrier material. Biotechniques 21:934–939
29. Hernandez GE, Mondala TS, Steven R (2009) Assessing a novel room-temperature RNA storage medium for compatibility in microarray gene expression analysis. Biotechniques 47:667–670

Chapter 21

The Latest Improvements in the Mouse Sperm Preservation

Takehito Kaneko

Abstract

Sperm preservation is an important technique for maintaining valuable genetic resources in biomedical research and wildlife. In the mouse, the sperm cryopreservation method has been established and adopted by large-scale sperm preservation projects in cryobanks. Recently, a new sperm preservation method using freeze-drying has been studied in various mammals. Freeze-drying is the ultimate method by which sperm can be preserved long term in a refrigerator (4 °C). And it is possible to realize easy and safe transportation of sperm at an ambient temperature that requires neither liquid nitrogen nor dry ice. Furthermore, it has been demonstrated that the fertilizing ability of sperm cryopreserved or freeze-dried by the methods described in this chapter is well maintained during long-term preservation. This chapter introduces the latest protocols for cryopreservation and freeze-drying of mouse sperm, and the anticipated results of the fertilizing ability of these sperm preserved long-term.

Key words Sperm, Cryopreservation, Freeze-drying, Long-term preservation, In vitro fertilization, Intracytoplasmic sperm injection

1 Introduction

Successful cryopreservation of mouse sperm was reported in the 1990's [1–6]. Subsequent improvements led to high fertilization rates using sperm cryopreserved in a solution containing 18 % raffinose and 3 % skim milk [7–10]. Presently, the number of novel transgenic mouse strains has expanded rapidly, and the maintenance of these strains by standard breeding colonies is becoming increasingly difficult [11]. The cryopreservation method overcame this problem and has contributed to large-scale sperm preservation projects in cryobanks. Furthermore, cryopreservation of sperm is relatively simple, rapid, and economical because a large number of sperm ($1\text{–}3 \times 10^7$) can be preserved immediately after collection from each male, and many animals can be produced from one aliquot of preserved sperm [12].

Freeze-drying is the ultimate method by which sperm can be preserved long term in a refrigerator (4 °C). Furthermore, it is possible to realize easy and safe transportation of sperm at an

Mark Lewandoski (ed.), *Mouse Molecular Embryology*, Methods in Molecular Biology, vol. 1092,
DOI 10.1007/978-1-60327-292-6_21, © Springer Science+Business Media New York 2014

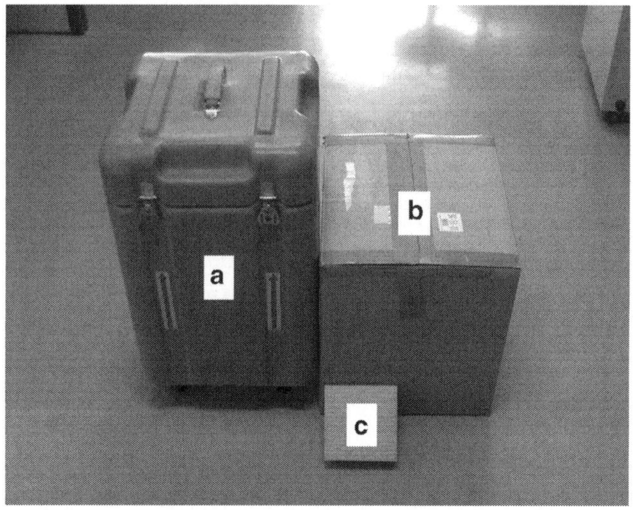

Fig. 1 (**a**) A dryshipper for transportation of cryopreserved samples in the gas phase of liquid nitrogen, (**b**) a box for transportation of cryopreserved samples in the dry ice, (**c**) a box for transportation of freeze-dried samples at ambient temperature

ambient temperature that requires neither liquid nitrogen nor dry ice (Fig. 1). Although attempts to freeze-dry sperm have been reported previously, it has been studied in various mammals as a new sperm preservation method [13–17] since the first offspring were obtained from oocytes fertilized with mouse freeze-dried sperm [18, 19]. Freeze-drying methods for mouse sperm have been improved for efficient long-term preservation [20–22], and a simple solution containing Tris and ethylenediaminetetraacetic acid (EDTA) has allowed long-term preservation of sperm while maintaining fertilizing ability [23].

In sperm preservation, the most important problem is the decrease in the fertilizing ability (deterioration) of sperm during handling and subsequent preservation, and phenotypic differences in the offspring derived from these sperm. Using the sperm preserved by the methods described in this chapter, a high proportion of offspring without phenotypic differences were obtained from embryos fertilized with sperm cryopreserved for 10 years [24] or stored at 4 °C for 1–1.5 years after freeze-drying [22, 23, 25].

This chapter introduces the latest protocols for cryopreservation and freeze-drying of mouse sperm and anticipated results of the fertilizing ability of these long-term preserved sperm.

2 Materials

2.1 The Sperm Cryopreservation Solution

1. Dissolve 18 % of d (+)-raffinose and 3 % of powdered skim milk in 10 ml of distilled water at 60 °C.
2. Divide the aliquots of the solution into 1.5 ml microcentrifuge tubes.

3. Centrifuge the tubes at 10,000 × *g* for 15 min at room temperature (*see* **Note 1**).
4. Sterilize the supernatant using a 0.45 μm disposable filter.
5. Store the filter-sterilized solution in glass ampoules at room temperature.

2.2 The Sperm Freeze-Drying Solution

1. Dissolve 10 mM Tris–HCl and 1 mM EDTA in 10 ml of distilled water, and adjusted to pH 8.0 (*see* **Note 2**).
2. Store the freeze-drying solution at room temperature.
3. Sterilize the freeze-drying solution using a 0.22 μm disposable filter before use.

3 Methods

3.1 Cryopreservation of Sperm

1. Add 100 μl of cryopreservation solution into two of the four wells of a 4-well disposable multidish (Nunc A/S, Roskilde, Denmark).
2. Sacrifice matured male by cervical dislocation, and remove two cauda epididymides using a pair of small scissors. Note that methods of euthanasia vary by institution but should be in compliance with local Institutional Animal Care and Use Committee guidelines.
3. Remove blood vessels and adipose tissue, rinse the two cauda epididymides in one of the wells containing cryopreservation solution, and then transfer them to another well containing cryopreservation solution.
4. Cut the epididymides finely into the cryopreservation solution using micro-spring scissors, and then leave for 3 min at room temperature to disperse spermatozoa.
5. Connect a 0.25 ml plastic straw (IMV Technologies, L'Aigle, France) to a 1 ml syringe with a silicone cap (Fig. 2).

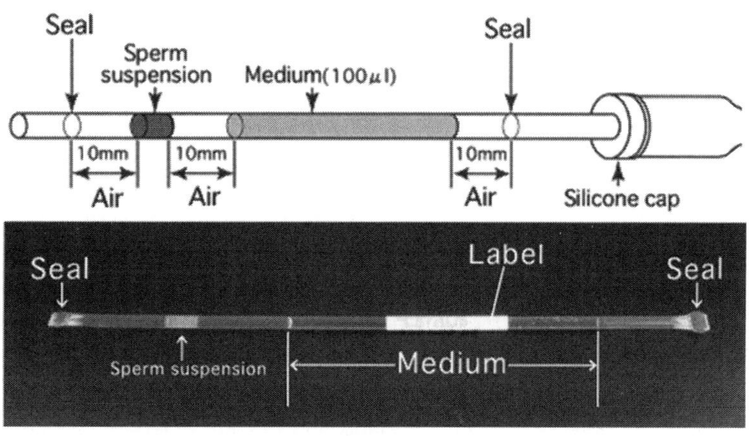

Fig. 2 The straw loaded sperm suspension

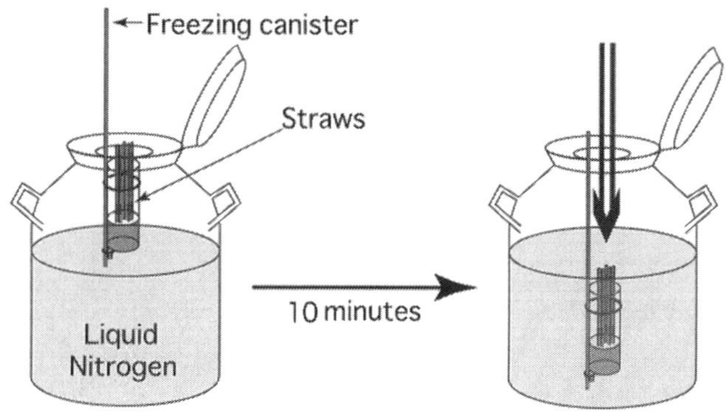

Fig. 3 Cooling and subsequent freezing of straws using a freezing container in a cryobiological container

6. Carefully load 100 μl of HTF medium [26], 10 mm of air, 10 μl of the sperm suspension, and another 10 mm of air, successively into the straw using the syringe (Fig. 2) (*see* **Note 3**).
7. Heat-seal both ends of the straw using a polysealer.
8. Make up 8–9 straws per mouse in the same manner.
9. Put the straws into a freezing canister and cool in the gas phase by floating on liquid nitrogen in a cryobiological container (Fig. 3).
10. After 10 min, quickly immerse the freezing canister in the liquid nitrogen (Fig. 3).
11. Keep straws in the liquid nitrogen (*see* **Note 4**).

3.2 Thawing of Cryopreserved Sperm

1. Immerse a cryopreserved straw in a water bath maintained at 37 °C.
2. After 10 min, wipe any water from the straw.
3. Cut the sealed end of the straw furthest from the sperm, and connect the cut end to a 1 ml syringe with a silicone cap.
4. Cut the other sealed end of the straw and transfer only the sperm suspension to a drop of HTF medium (90 μl) in the culture dish.
5. Place the culture dish at 37 °C under 5 % CO_2 and 95 % air for 1 h.
6. For in vitro fertilization, add 10 μl of cultured sperm suspension to each drop of HTF medium containing oocytes collected from females (*see* **Note 5**).

3.3 Freeze-Drying of Sperm

1. Prepare 1 ml of freeze-drying solution in a 1.5 ml microcentrifuge tube.
2. Sacrifice matured male by cervical dislocation and remove the two cauda epididymides using a pair of small scissors.

Fig. 4 Lyophilizing of the sperm suspension in ampoules connected to the manifold of a freeze-drying machine

3. Remove blood vessels and adipose tissue, and squeeze a dense mass of sperm out of the two cauda epididymides using sharply pointed forceps.
4. Gently place the sperm mass on the bottom of a 1.5 ml microcentrifuge tube containing 1 ml of freeze-drying solution.
5. Leave for 10 min at room temperature to allow the sperm to disperse into the freeze-drying solution.
6. Collect 800 µl of supernatant in another tube and transfer 100 µl aliquots of the sperm suspension into a 2 ml long-necked glass ampoule for freeze-drying (Wheaton, Millville, NJ, USA).
7. Make up 8 ampoules per mouse in the same manner.
8. Plunge ampoules into liquid nitrogen for 20 s.
9. Connect ampoules to the manifold of a freeze-drying machine (Freeze-drying systems 77530, Labconco Corporation, Kansas City, MO, USA) and lyophilize for 4 h at a pressure of 0.03–0.05 hPa (Fig. 4) (*see* **Note 6**).
10. Flame-seal each ampoule using a gas burner with the inside pressure of the ampoules kept at 0.03–0.05 hPa (Fig. 5) (*see* **Note** 7).
11. Store ampoules at 4 °C (*see* **Note 8**).

3.4 Rehydration of Freeze-Dried Sperm

1. Open an ampoule and rehydrate the sperm by adding 100 µl of sterile distilled water.
2. Use sperm that show a morphologically normal shape for ICSI (*see* **Notes 9** and **10**).

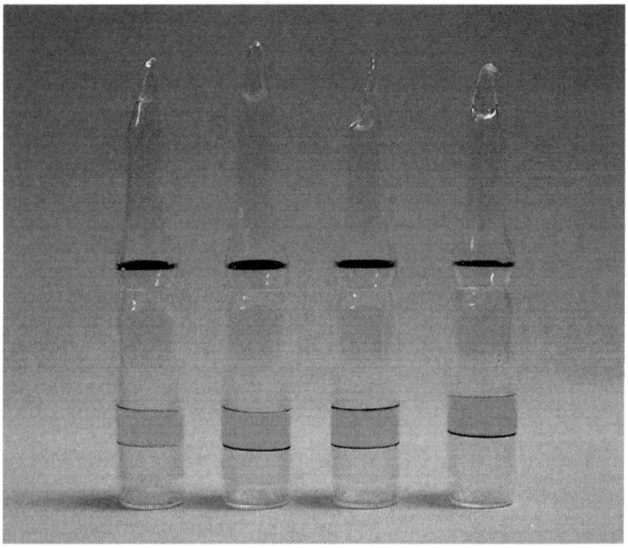

Fig. 5 Flame-sealed ampoules after freeze-drying

4 Notes

1. If the supernatant is not clear after centrifuging, centrifuge again until the supernatant becomes clear.
2. Use a commercial product (Ambion, An Applied Biosystems Business, Austin, TX, USA).
3. HTF medium acts as a weight to sink the straw into the liquid nitrogen.
4. All manipulations are done punctually and carefully to highly maintain the motility and fertilizing ability of the sperm after cryopreservation because they are sensitive to physical shock.
5. Anticipated results of the fertilizing ability of sperm cryopreserved for 10 years by the methods described in this chapter are shown in Table 1 [24]. In the mouse sperm after thawing, however, fertilizing ability showed strain-dependent susceptibility to damage by freezing. We established a simple method of using oocytes laser-microdissected zona pellucida to increase the fertilizing ability of these mouse sperm with low fertility [27]. Moreover, when sperm are immotile after thawing, oocytes can be fertilized using intracytoplasmic sperm injection (ICSI) [28, 29].
6. The pressure during lyophilizing is important to highly maintain the fertilizing ability of the sperm.
7. Do not make a hole by flame-sealing in order to maintain the inside pressure of the ampoules.
8. Freeze-dried sperm can be stored for up to 3 months at room temperature (24 °C) [22].

Table 1
Fertilizing ability of sperm cryopreserved for 10 years and subsequent development of oocytes fertilized in vitro with these sperm

Strain of origin	No. of oocytes examined	No. of oocytes developed to the 2-cell stage (%)	No. of 2-cell embryos transferred	No. of offspring (%)
C57BL/6J	140	93 (66)	60	23 (38)
DBA/2N	366	338 (92)	60	9 (15)
BALB/cA	243	177 (73)	60	26 (43)
C3H/HeJ	140	46 (33)	46	12 (26)
B6D2F1	436	263 (60)	60	23 (38)
B6C3F1	363	195 (54)	60	10 (17)

The data are reproduced from ref. [24]

Table 2
Fertilizing ability of freeze-dried B6D2F1 sperm preserved at 4 °C for 1 year and subsequent development of oocytes injected with these sperm

Storage term	No. of oocytes injected	No. of oocytes survived (%)	No. of oocytes fertilized	No. of 2-cell embryos transferred	No. of offspring (%)
1 month	46	24 (52)	23	20	11 (55)
3 month	100	55 (55)	51	47	27 (58)
1 year	104	79 (76)	76	63	41 (65)

The data are reproduced from ref. 23

Table 3
Development of oocytes injected with freeze-dried B6D2F1 sperm after international air transportation at ambient temperature

Transportation	No. of oocytes fertilized	No. of 2-cell embryos transferred	No. of offspring (%)
Before	25	25	15 (60)
After	31	28	15 (54)

The data were produced by Kaneko et al. (unpublished previously)
Ampoules were transported between the USA and Japan (6 days)

9. After rehydration, some sperm with separated heads and broken tails are observed.
10. Anticipated results of the fertilizing ability of sperm stored at 4 °C for 1 year after freeze-drying by the methods described in this chapter are shown in Table 2 [23]. Freeze-dried spermatozoa can be transported worldwide by air at an ambient temperature (Table 3, *see* **Note 8**). Unfortunately, the sperm after

freeze-drying are completely immotile, and ICSI is necessary to fertilize the oocytes and recover live offspring from these sperm. However, the ICSI technique is no longer difficult as the birth of normal offspring using ICSI has already been widely reported in various mammals [30]. Moreover, the advantage of ICSI is that only one spermatozoon is required to fertilize an oocyte, and the waste of valuable preserved gametes can be avoided compared with IVF and other reproductive techniques.

References

1. Okuyama M, Isogai S, Saga M, Hamada H, Ogawa S (1990) In vitro fertilization (IVF) and artificial insemination (AI) by cryopreserved spermatozoa in mouse. J Fertil Implant 7:116–119
2. Tada N, Sato M, Yamanoi J, Mizorogi T, Kasai K, Ogawa S (1990) Cryopreservation of mouse spermatozoa in the presence of raffinose and glycerol. J Reprod Fertil 89:511–516
3. Yokoyama M, Akiba H, Katsuki M, Nomura T (1990) Production of normal young following transfer of mouse embryos obtained by in vitro fertilization using cryopreserved spermatozoa. Exp Anim 39:125–128
4. Takeshima T, Nakagata N, Ogawa S (1991) Cryopreservation of mouse spermatozoa. Exp Anim 40:493–497
5. Penfold LM, Moore HD (1993) A new method for cryopreservation of mouse spermatozoa. J Reprod Fertil 99:131–134
6. Songsasen N, Betteridge KJ, Leibo SP (1997) Birth of live mice resulting from oocytes fertilized in vitro with cryopreserved spermatozoa. Biol Reprod 56:143–152
7. Nakagata N, Takeshima T (1992) High fertilizing ability of mouse spermatozoa diluted slowly after cryopreservation. Theriogenology 37:1283–1291
8. Nakagata N, Takeshima T (1993) Cryopreservation of mouse spermatozoa from inbred and F1 hybrid strains. Exp Anim 42: 317–320
9. Nakagata N, Ueda S, Yamanouchi K, Okamoto M, Matsuda Y, Tsuchiya K, Nishimura M, Oda S, Koyasu K, Azuma S, Toyoda Y (1995) Cryopreservation of wild mouse spermatozoa. Theriogenology 43:635–643
10. Nakagata N (1996) Use of cryopreservation techniques of embryos and spermatozoa for production of transgenic (Tg) mice and for maintenance of Tg mouse lines. Lab Anim Sci 46:236–238
11. Knight J, Abbott A (2002) Full house. Nature 417:785–786
12. Nakagata N (2000) Cryopreservation of mouse spermatozoa. Mamm Genome 11:572–576
13. Keskintepe L, Pacholczyk G, Machnicka A, Norris K, Curuk MA, Khan I, Brackett BG (2002) Bovine blastocyst development from oocytes injected with freeze-dried spermatozoa. Biol Reprod 67:409–415
14. Kwon IK, Park KE, Niwa K (2004) Activation, pronuclear formation, and development in vitro of pig oocytes following intracytoplasmic injection of freeze-dried spermatozoa. Biol Reprod 71:1430–1436
15. Liu JL, Kusakabe H, Chang CC, Suzuki H, Schmidt DW, Julian M, Pfeffer R, Bormann CL, Tian XC, Yanagimachi R, Yang X (2004) Freeze-dried sperm fertilization leads to full-term development in rabbits. Biol Reprod 70:1776–1781
16. Hirabayashi M, Kato M, Ito J, Hochi S (2005) Viable rat offspring derived from oocytes intracytoplasmically injected with freeze-dried sperm heads. Zygote 13:79–85
17. Kaneko T, Kimura S, Nakagata N (2007) Offspring derived from oocytes injected with rat sperm, frozen or freeze-dried without cryoprotection. Theriogenology 68:1017–1021
18. Wakayama T, Yanagimachi R (1998) Development of normal mice from oocytes injected with freeze-dried spermatozoa. Nat Biotechnol 16:639–641
19. Kusakabe H, Szczygiel MA, Whittingham DG, Yanagimachi R (2001) Maintenance of genetic integrity in frozen and freeze-dried mouse spermatozoa. Proc Natl Acad Sci U S A 98:13501–13506
20. Kaneko T, Whittingham DG, Yanagimachi R (2003) Effect of pH value of freeze-drying solution on the chromosome integrity and developmental ability of mouse spermatozoa. Biol Reprod 68:136–139

21. Kaneko T, Whittingham DG, Overstreet JW, Yanagimachi R (2003) Tolerance of the mouse sperm nuclei to freeze-drying depends on their disulfide status. Biol Reprod 69:1859–1862
22. Kaneko T, Nakagata N (2005) Relation between storage temperature and fertilizing ability of freeze-dried mouse spermatozoa. Comp Med 55:140–144
23. Kaneko T, Nakagata N (2006) Improvement in the long-term stability of freeze-dried mouse spermatozoa by adding of a chelating agent. Cryobiology 53:279–282
24. Kaneko T, Yamamura A, Ide Y, Ogi M, Yanagita T, Nakagata N (2006) Long-term cryopreservation of mouse sperm. Theriogenology 66: 1098–1101
25. Ward MA, Kaneko T, Kusakabe H, Biggers JD, Whittingham DG, Yanagimachi R (2003) Long-term preservation of mouse spermatozoa after freeze-drying and freezing without cryoprotection. Biol Reprod 69:2100–2108
26. Quinn P, Kerin JF, Warnes GM (1985) Improved pregnancy rate in human in vitro fertilization with the use of a medium based on the composition of human tubal fluid. Fertil Steril 44:493–498
27. Kaneko T, Yanagi M, Nakashima T, Nakagata N (2006) The improvement in fertilizing ability of cryopreserved mouse spermatozoa using laser-microdissected oocytes. Reprod Med Biol 5:249–253
28. Szczygiel MA, Kusakabe H, Yanagimachi R, Whittingham DG (2002) Intracytoplasmic sperm injection is more efficient than in vitro fertilization for generating mouse embryos from cryopreserved spermatozoa. Biol Reprod 67:1278–12784
29. Sakamoto W, Kaneko T, Nakagata N (2005) Use of frozen-thawed oocytes for efficient production of normal offspring from cryopreserved mouse spermatozoa showing low fertility. Comp Med 55:136–139
30. Yanagimachi R (2005) Intracytoplasmic injection of spermatozoa and spermatogenic cells: its biology and applications in humans and animals. Reprod Biomed Online 10:247–288

Chapter 22

Analyzing Gene Function in Whole Mouse Embryo and Fetal Organ In Vitro

Satomi S. Tanaka, Yasuka L. Yamaguchi, Vanessa J. Jones, and Patrick P.L. Tam

Abstract

A well-established experimental paradigm to analyze gene function in development is to elucidate the impact of gain and loss of gene activity on cell differentiation, tissue modelling, organogenesis, and morphogenesis. This chapter describes the experimental protocols to study gene function by means of electroporation and lipofection to manipulate genetic activity in whole embryos and fetal organs in vitro. These techniques allow for more precise control of the timing, with reference to developmental age or stage, and the cell/tissue-specificity of the changes in gene activity. They provide an alternative strategy that can expedite the analysis of gene function before resorting to the conventional means of transgenesis and gene targeting in the whole organism.

Key words Mouse embryo, Fetal ovary, In vitro culture, Electroporation, Lipofection, siRNA

1 Introduction

A time-tested experimental approach to analyze gene function in mammalian development is to assess the consequence of changes to gene activity on cell proliferation and differentiation, tissue growth and patterning, and organogenesis. Gene activity may be modified to produce a gain or a loss of gene function. For gain-of-function, the specific gene is expressed at a level over and above the normal activity, or at a developmental stage/age of the embryo and in cells/tissues different from the normal schedule and pattern. The outcome of the gain-of-function modification will inform us about the full impact of the gene activity, albeit with the caveat that the effect may reach beyond its normal function. For loss-of-function, the activity of the specific gene may be abrogated by altering or ablating the critical coding sequences and/or regulatory elements of the locus, or to express a functionally counteracting factor (dominant negative counterpart, RNA interference, antisense/morpholino inhibition). The resultant phenotype may reveal

Mark Lewandoski (ed.), *Mouse Molecular Embryology*, Methods in Molecular Biology, vol. 1092,
DOI 10.1007/978-1-60327-292-6_22, © Springer Science+Business Media New York 2014

the developmental process which most critically requires the function of the gene. However, the effect of compensatory activity of the rest of the genome and the epigenetic processes in response to the loss of specific gene function may have to be taken into consideration in the interpretation of the loss-of-function effect.

It is also feasible to introduce genetic modifications to express in situ reporters which are functionally neutral. This capability has enabled the visualization of gene expression patterns or the downstream effect of other gene activity, and the testing for molecular regulation of gene activity. Incorporation of a reporter gene may facilitate tracing cell fate and movement in the developing embryo for a longer time than by the dye-labelling method [1], because incorporated reporter genes are not diluted by cell divisions like the exogenous and non-replenishable labels. This approach is also effective for investigating the requirement of regulatory elements (promoters and enhancers) for controlling the tissue- or developmental stage-specific gene expression. In the developing chick embryo, for example, regulatory modules for the tissue-specific *Sox2* expression were identified using this means, in conjunction with comparative analysis of avian and mammalian genomic sequences [2].

To perform the gain-of-function or loss-of-function study in the mice, the established approaches are transgenesis and targeted mutagenesis of the whole organism. For example, transgenic mice can be generated by the incorporation of transgenic vectors into the host genome. This is commonly achieved by microinjection of a DNA construct into fertilized oocytes, viral infection of early embryos or transfection of a vector to generate random integration into the genome of embryonic stem (ES) cells, which can then be used to generate mice. Both gain and loss of function modifications can be made through targeted mutagenesis by homologous recombination at the specific gene locus using the ES cell-chimera production technology. However, generation of these genetically modified mice requires lengthy and laborious effort of animal production before the targeted gene function can be studied. In this chapter, we describe an alternative experimental approach which may expedite the analysis of the impact of the modification of genetic activity in whole embryos and organ explants using a combination of techniques of genetic manipulation and in vitro embryo/organ culture. In essence, this is accomplished by introducing plasmid DNA or short interference RNA into the mouse embryo and fetal organs, followed by in vitro culture. A number of recent studies utilizing this approach are summarized in Table 1.

1.1 Molecular Tools for the Modification of Gene Function

1.1.1 Gain of Function

Plasmid DNA or bacteria artificial chromosome (BAC) constructs are used for the over-expression of transgenes containing the coding sequence of the gene of interest under the control of a constitutively active regulatory element, resulting in widespread and sustained expression. Alternatively, the transgene may be

Table 1
Examples of genetic manipulation experiments performed on embryos and fetal organs using viral transduction, electroporation, and lipofection

Genetic manipulation				
Method	**Reagents**	**Target**	**Outcome**	**Reference**
Lentiviral transduction: preimplantation embryo in vitro followed by embryo transfer and intrauterine development	Expression constructs of trophoblast-specific genes	Trophectoderm of mouse blastocyst	Expression of genes ameliorates placental defects and rescues embryonic lethality of mutant	[29]
Electroporation: whole embryo in vitro	CMV-regulated expression construct, tagged by GFP	Posterior endoderm of late-streak and early-head-fold stage mouse embryo	Ectopic expression of *Ifitm1* and *Ifitm3*, revealing gene function in navigating cell movement	[4]
Electroporation: whole embryo in vitro	dsDNA and morpholino	Node of late-streak stage mouse embryo	Knockdown of *Otx2* and *Foxa2* activity	[30]
Electroporation: whole embryo in vitro	Reporter constructs for expression of EGFP and lacZ	Germ layers of late-streak stage mouse embryo	Expressing the reporter in specific populations of cells in the germ layers of the gastrulating embryo for tracking developmental fates	[3]
Electroporation: whole embryo in vitro	DNA fragments of chicken *Sox2* genes	Embryonic tissues of Stage 4 chick embryo	Testing for tissue-specific *Sox2* enhancer activity	[2]
Electroporation: embryo in utero	Dominant negative construct and shRNA	Embryonic cerebral cortex of 13- and 14-day embryos	Loss of function of Trk receptor activity for testing effect on cell proliferation and differentiation	[31]
Electroporation: organ in situ	siRNA	Neural tube of 10.0-day mouse embryo	Knockdown of co-transfected β-galactosidase activity	[32]

(continued)

Table 1
(continued)

Genetic manipulation				
Method	**Reagents**	**Target**	**Outcome**	**Reference**
Lipofection: organ in vitro	siRNA	Explants of embryonic kidney	Knockdown of *WT1* to test to phenocopy the effect of loss of gene function in vivo	[25]
Lipofection: organ in vitro	siRNA	Explants of fragments of fetal ovary	Knockdown of *Ipo13* to reveal its role in regulating meiotic differentiation of oocyts	[16]
Lipofection: embryo in vitro	Expression of GFP constructs with pegylation of the lipofectamine-based DOSPA/DOPe lipoplex	Left outflow tract of the heart of HH Stage 16 chick embryo	Efficient transfection of yolk sac vasculature and embryonic tissues	[15]
Lipofection: whole embryo in vitro	CMV- regulated expression construct, tagged by GFP	Visceral endoderm of pre-gastrulation stage mouse embryo	Ectopic expression of *Nodal* and its antagonist for testing effect on cell movement	[33]
Lipofection: whole embryo in vitro	CMV- regulated expression construct, tagged by GFP	Edoderm layer of 7.5-day mouse embryo	Broad expression of *Ifitm1* for revealing gene function in navigating cell movement	[34]

driven by stage- or tissue-specific regulatory elements to produce a distinctive pattern of transgenic activity. Similarly, these reagents may be used for the expression of a reporter gene which reveals the activity of the endogenous gene of which the regulatory elements is coupled with the reporter gene.

1.1.2 Loss of Function

This may be accomplished by expressing plasmid DNA vector encoding (1) an antagonist for the target gene product, (2) a dominant-negative form of the target gene product that competitively interferes with the gene function, or (3) short-hairpin RNA (shRNA) that induces the degradation of target mRNA resulting in the "knockdown" of gene activity. In addition, introduction of antisense oligonucleotide (morpholino DNA) or a decoy-double-strand-DNA (comprising DNA binding sequences of the specific transcription factor) that interferes with the binding of the factor to the promoter of the target gene can suppress expression (Table 1). Instead of using DNA reagents, siRNA against specific target sequences or double-strand small RNA (dsRNA) can also be used to knockdown gene function. A combination of vectors/ RNAs, each targeted to specific genes, may also be used to elucidate the interaction of genetic activities. However, several factors such as the efficiency and duration of the knockdown effects and off-target effects, may have significant influence on the outcome of the experiment.

1.2 Methods and Reagents for Modifying Gene Function

1.2.1 Transduction Using Viral Vectors

Viruses are highly evolved biological machines that efficiently gain access to host cells and exploit the cellular machinery to facilitate their replication. By exploiting these properties, several types of modified viral vectors have been used for delivering foreign genes, such as adenovirus, retrovirus and lentivirus derived vectors. By viral transduction, the genome of adenovirus vectors persists in the cell nucleus predominantly as extrachromosomal DNA, whereas that of retrovirus and lentivirus vector integrates into host genome. Adenovirus vector is therefore useful to induce a transient expression of transgenes, whereas retrovirus and lentivirus vector can sustain long-term gene expression. In particular, lentiviral vector, which could integrate efficiently into the host genomic DNA, enables transduction in cells that are not actively dividing. However, the efficacy of the viral transduction system may be compromised by the lack of site-specific integration, and the necessity to incorporate the gene into the viral genome and to generate a sufficiently high titer for transduction (Table 2). There is a time lag (up to 12 h) between transduction and the expression of gene activity, compared to the faster effects (after 3–4 h) of other delivery techniques such as electroporation and lipofection. In view of the rapid rate of embryonic development from gastrulation to early organogensis during the immediate post-implantation period, the latter two techniques are preferred, and will be the focus of this chapter.

Table 2
Comparison of three methods for gene-delivery to the mouse embryo

	Electroporation	Lipofection	Viral transduction
Specificity of targeting sites	Yes, via electrode positioning and direction of current flow	Yes, locally at the site of application, with some diffusion	Not easily controlled unless contained within a confined space
Reagents	DNA and RNA	DNA and RNA	DNA and virus-based shRNA
Transfection efficiency	High	Moderate to low	Very high, applicable in nondividing cells
Immediacy of expression	3–4 h after electroporation	3–4 h after electroporation	More than 12 h after infection
Damage or cytotoxicity	Potential tissue damage	Not significant	Not significant
Instrumentation	Electroporation apparatus and micromanipulator (special case)	Micromanipulation (special case)	None

1.2.2 Electroporation of Nucleic Acids

Introduction of DNA or RNA molecules into cells can be achieved by permeating the cell membrane using electroporation. When a cell is exposed to a strong electric field that alters the transmembrane potential, rearrangement of the molecular architecture of the membrane occurs. This leads to formation of pores in the membrane and consequently an increase in the cell membrane permeability to ions, small molecules and even macromolecules. Electroporation can be performed on specific groups of cells or cell layers by varying the direction of the electric field so that the negatively charged DNA or RNA molecules are delivered into the cells as they are driven towards the positive electrode. Therefore, electroporation has a great advantage in the directional introduction of DNA or RNA molecules.

Many kinds of plasmid DNA vectors and RNAs can be effectively introduced by electroporation into the embryo or fetal organs to examine the impact of modulating gene function (Table 1). Preferably, the vectors also contain a reporter such as an internal ribosomal entry site (IRES)-GFP cassette (Clontech), which allows both the transgene and GFP to be expressed from a single mRNA transcript. This bi-cistronic construct may give a more consistent indication of the site and level of expression of the vector than that offered by the co-transfection of the transgene vector and a separate reporter vector. Fluorescent and chromogenic reporter genes, driven by a CMV promoter, have been used for this purpose and the signals are detected 2–3 h after electroporation. CRE-mediated activation of reporters has also been used but the signals are detected after a longer time lag of 8–9 h after the introduction of *cre*-expressing vector [3]. However, by using a CAGGS (CMV-IE enhancer and beta-actin promoter), reporter gene activity is visible sooner at 5–6 h after electroporation [4].

Electroporation is also effective for introducing large genomic DNA constructs, including yeast artificial chromosome (YAC) or BAC clones [5]. The use of large genomic DNA constructs enables the expression of the target gene largely under the control of the intrinsic regulatory elements. This approach is also particularly useful to drive expression transgenes and reporters by integrating the expression sequences into the gene locus which in many cases provides a higher fidelity of the expression pattern and level.

Knockdown of gene function can be achieved by vector-based expression of shRNA, which are double-stranded RNAs with a fold-back stem-loop structure that are processed into a siRNA duplex in the transfected cell. Expression of the shRNAs is usually driven by RNA Polymerase III promoter which is presumed to be universally active. However, it is not uncommon to find a wide range of knockdown effects among the transfected cell clones [6].

Recent improvement of the electroporation technique means it could be adapted for the introduction of transgenes into tissues or organs in vivo (for reviews, 5, 7, 8), as well as into mouse embryos in utero [9–12], and followed by in vitro culture [3, 13, 14].

1.2.3 Lipofection of Nucleic Acids

Lipofection is a method of introducing DNA/RNA into cells using specially designed cationic lipids as carriers or neutral liposomes whereby nucleic acids are encapsulated within the liposomes. The preparation of the DNA/liposome complex should take into consideration the ratio of DNA and liposome to optimize the efficiency. Transfection efficiency is also influenced by the quality of the DNA which is usually purified by using the anion-exchange resin or the silica membrane columns that are commercially available (i.e., Qiagen, Sigma, Macherey-Nagel). Presence of residual materials, such as proteins or excess bacteria culture liquid, may affect the purification procedure, interfere with the formation of DNA–liposome complexes, the interaction of the liposome with the cell membrane, and may cause undesirable cytotoxic effects following lipofection.

Recent improvements in the lipofection reagents show that cationic lipid reagent-mediated transfection engenders higher efficiency and lower cell toxicity than the conventional means of using neutral liposomes, DEAE-dextran and polybrene, and calcium phosphate co-precipitation. The cationic lipid contains a positively charged head group, which mediates the interaction between the lipid and the phosphate backbone of the nucleic acid, and facilitates the DNA–cationic lipid complex formation (liposome/nucleic acid transfection complex). The positive charge of the liposomes also facilitates the interaction of the nucleic acid and the cell membrane, enabling the fusion of the liposome/nucleic acid transfection complex with the negatively charged cell membrane, followed by entry into the cell via endocytosis. Once inside the cell, the complex is imported to the nucleus for activating expression. Improvement of cationic lipid-based gene delivery methods, together with the conjugation of polyethylene glycerol chain to the lipid (peglation of the transfection reagent) can reduce the recognition and clearance of the reagents by the host cells, which leads to enhanced transfection efficiency [15].

1.3 Strategies for Delivering the Reagents

1.3.1 Electroporation

Reagents for manipulating gene activity can be delivered to cell populations over a broad area or at a localized site in the fetal organ or whole mouse embryo. The strategy for tissue-specific delivery may be illustrated by the electroporation of cell markers into the three germ layers (ectoderm lining an internal cavity, endoderm on the outside and the mesoderm in between the two layers) by a combination of site-specific delivery of the reagent and focal electroporation [3]. For electroporation of an epithelial cell layer such as the endoderm of the embryo, the reagent is adsorbed onto the apical surface of the cells by soaking the embryo in the DNA solution prior to electroporation. The embryo is then electroporated with the negative electrode positioned near the surface of endoderm to drive the reagent into this cell layer. For the ectoderm layer that lines the amniotic cavity, reagents are microinjected into the cavity to gain access to the cells in this enclosed environment. Electroporation of this cell layer is then achieved by positioning the positive electrode near the surface of the embryo, such that the negatively charge nucleic acid is driven into the ectoderm cells through their luminal surface. For electroporating the mesoderm cells between the ectoderm and endoderm, the reagent is microinjected into the extracellular space to bathe the cells before electroporation, with the polarity of the electrodes set for that of either ectoderm or endoderm, the vector will go into cells in this layer that is sandwiched between the ectoderm and the endoderm. For fetal organs, the reagent may be microinjected prior to electroporation into the extracellular space of tissues, such as the stroma of the kidney, lung and gonad; or into a confined space, such as the lumen of a tubular-like structure.

1.3.2 Lipofection

Lipofection can be accomplished either by exposing the cells, tissue fragments, fetal organs and whole embryos to a lipofection solution containing the liposome–DNA complex, which gains entry to the cells through the interaction of liposome with the cell membrane. However, especially for organs and whole embryos, a localized transfection may be more desirable. This can be achieved, as for electroporation, by delivering the lipofection reagents to a cavity (e.g., amniotic cavity, heart chambers, neural tube lumen and brain vesicles, intratubular space of gland and ducts, etc) lined by the cells of interest or into the extracellular space between cells.

In these situations, only cells exposed to the reagents will be lipofected. Alternatively, the fetal organ may be dissected to expose the cells intended for lipofection. For example, to achieve efficient lipofection of fetal oocytes, the ovaries are sliced open and immersed in a liposome solution [16]. Sliced ovaries recovered during subsequent culture. Oocytes also grow normally, entering meiotic division under the culture conditions, like their counterparts in vivo [16]. This technique can achieve high transfection efficiency (60–95 %) in a wide area of the fetal ovary, whereas electroporation produces transfected cells only at localized regions [16]. Lipofection causes less damage to cells than electroporation and it can be used for introducting transgene in the E6.5 or younger stage mouse embryos, although the transfection efficiency in a specific population of cells appeared to be lower than that of the electroporation method (Table 2).

siRNA can also be introduced into cells by lipofection. Although siRNA can be tagged with a fluorescent dye to monitor the transfection efficiency, the signal is usually too low to detect the siRNA transfected cells in the whole embryos cultured in vitro. Therefore, to assess the knockdown effect in cells in an embryo, specific antibodies are required to visualize the down-regulation of the target gene product.

To introduce the reagents into the embryo or fetal organs cultured in vitro, lipofection may have to be performed in a serum-containing culture media. Caution should be taken to test the transfection efficiency of the reagents under conditions that the embryo or fetal organs will be cultured, since some lipofection reagents show poor efficiency in the presence of serum supplement. During lipofection, antibiotics are commonly omitted from the medium to avoid their toxic effects on the transfected cells.

1.4 Adjunct Techniques: Isolation and Culture of Fetal Organs and Early Post-implantation Embryo

Experimental manipulation of genetic activity on explanted mouse embryos and fetal organs is a powerful and expeditious approach to analyze gene function in mouse development. To accomplish this goal, two additional techniques are required. First is to devise the appropriate methods to isolate the mouse embryo and fetal organs for experimentation. In this chapter, we will describe the dissection of gastrula- to early-organogenesis stage mouse embryos and the fetal ovaries to showcase the basic principles. Second is the establishment of an appropriate in vitro-culture system that enables the assessment of the impact of manipulating gene function. Current post-implantation embryo culture methods can support extensive growth and morphogenesis of the whole mouse embryo from the E6.5 gastrulating stage to as far as E10.5 late organogenesis stage, and a detailed procedure for whole embryo culture has been described [17–19]. In the organogenesis stage, the organ culture systems for several fetal organs are well established. They can support an appropriate growth and development of fetal organs, such as the genital ridge (gonad), kidney, liver, lung and retina, and allow for the examination of the target gene function in the organs cultured in vitro. Generally, fetal organs are cultured at the interface of medium and air on a filter membrane using a floating filter or a membrane culture insert, or are cultivated in a three-dimensional matrix gel (e.g., Matrigel), hanging medium drop or rotating culture for maintaining the three-dimensional structures of the organs to minimize the artifact of in vitro culture. General procedure for the organ culture system using a floating filter membrane is described in this chapter.

2 Materials

2.1 Dissection of Embryos and Fetal Organs

1. PB1 medium is prepared according to Table 3. Ensure the chemical components are added in the correct order as documented. The pH should be between 7.3 and 7.4. The solution is sterilized by passing through a 0.22 μm millipore filter and stored in 50 mL aliquots at 4 °C. The solution should be used within 2 weeks.
2. Dissecting instruments: iridectomy scissors (cat. No. 131011: Fine Scientific Tools, Germany), fine scissors (BC 50; Aesculap, Canada), standard scissors (BC 374; Aesculap, Canada), fine

Table 3
The composition of PB1

Components	Concentration (g/L)
NaCl	8.0
KCl	0.2
$Na_2HPO_4 \cdot 12H_2O$	2.86
KH_2PO_4	0.2
$CaCl_2 \cdot 2H_2O$	0.13
$MgCl_2 \cdot 6H_2O$	0.1
Sodium pyruvate[a]	0.036
Phenol red[a]	0.01
Penicillin[a]	0.06
Glucose	1.0
Bovine serum albumin	4.0
pH	7.3–7.4
Osmolarity	286–292 mOsmol/L

[a]See *Reagents* for method of preparation

forceps (BD 331; Aesculap, Canada), and watchmaker's forceps (Fine Scientific Tools, Germany).

3. Dissecting microscope (Leica Microsystems, Deerfield, IL).
4. Petri dishes (60-mm, Falcon).
5. 9-in. Pasteur pipettes (cat. No. 93; Chase Instruments).

2.2 Manipulation and Electroporation of Embryos

1. De Fonbrune syringe (cat. No. 4095; Alcatel, Malakoff, France), an oil-filled system, used to apply suction and expulsion actions to the injection micropipette to deliver the fluid into an embryo. (Alternative equipment is the Eppendorf Oil-Tram injector Order No. 920002030).
2. Electroporation apparatus components: 0.2 mm platinum wire, 0.1 mm tungsten wire, 60- and 150-mm petri dish lids, banana socket, banana plugs.
3. Fluorescence stereomicroscope system (Leica MZ FLIII) with excitation filter GFP-1 filter set, 425 nm.
4. Injection pipettes (thin-walled glass capillaries; outer diameter (od): 1.0 mm, inner diameter (id): 0.75 mm, cat. No. 900 021 61 Drummond) and holding pipettes (thick-walled glass capillaries; od: 1.0 mm, id: approx. 0.60 mm, cat. No. 520-119; Leica Microsystems, Deerfield, IL).
5. Microforge (Narishige Scientific Instrument Laboratory, Tokyo, Japan, MF79) (Alternative equipment is the MF-900 microforge, Narishige).
6. Micromanipulation apparatus (Leica): base plate with fixture points to hold two manipulators, left and right manipulators instrument holders, instrument sleeves.
7. Micrometer syringe (e.g., IM5A/5B, IM88, Narishige, Tokyo, Japan) an oil-filled system, used to apply suction to the holding pipette (Alternative equipment is the Eppendorf Air-Tram injector Order No. 920002021).
8. Positioner (Taurus-R, World Precision Instruments Inc, USA), used to hold the tungsten wire in place.
9. Square wave electroporator (ECM 830, BTX, San Diego, CA, USA).
10. Tyrode–Ringer saline (see chemical ingredients in Table 4), stored at 4 °C.

2.3 Lipofection Reagents

1. Lipofectamine 2000 (Invitrogen, cat. No.11668-019).
2. Highly purified Plasmid DNA prepared by using QIAGEN Plasmid Midi Kit (Qiagen, cat. No. 12143).
3. RNAiFect Transfection Reagent (Qiagen, cat. No. 301605).
4. Opti-MEM (Gibco, cat. No. 31985-070).
5. siRNA, commercially supplied from Qiagen (HP Genomewide siRNA).

Table 4
The composition of Tyrode–Ringer saline for electroporation of embryos

Compound	g/500 mL
NaCl	4.0
KCl	0.15
$NaH_2PO_4 \cdot 2H_2O$	0.0465
KH_2PO_4	0.0125
$NaHCO_3$	0.5
Glucose	1.0

2.4 Culturing Embryos and Fetal Organs

1. 4- and 8-well chamber slides (Lab-Tek™ Chamber Slides™, Nunc, Rochester, NY, USA).
2. Penicillin (Sigma). Add 599 mg to 100 mL of 0.9 % (w/v) NaCl. Store at −20 °C.
3. DMEM (high glucose) (12100-061, Gibco Labs, Invitrogen, USA). Before DMEM is used for culture, add 10 mL each of 200 mM stock glutamine (Gibco Labs, Invitrogen, USA) and 5 mg/mL penicillin/streptomycin stock solution (Gibco Labs, Invitrogen, USA) to 1 L of DMEM.
4. 100 % Rat serum (RS): the RS is thawed and inactivated by heating at 56 °C for 30 min immediately before use. Preparation of RS has been described previously [19].
5. Fetal calf serum (FCS) (Trace Biosciences): the FCS is thawed and inactivated by heating at 56 °C for 30 min immediately before use.
6. 24-well culture plate (Falcon).
7. 0.3 μm of isopore membrane filters (Millipore) or insert Transwell (0.4 μm; Corning).
8. Water-jacketed CO_2 incubator (Forma Scientific, Marietta, OH, Model 3336).
9. Razor blade (Feather).

3 Methods

3.1 Electroporation of Expression Vector into the Embryo

Electroporation is an effective method for introducing plasmid DNA vector, as well as siRNA into the mouse embryo. Detailed methods for the electroporation of mouse embryos have been described previously [3, 19]. Here, we introduce the electroporation method for delivering DNAs/RNAs into the embryo.

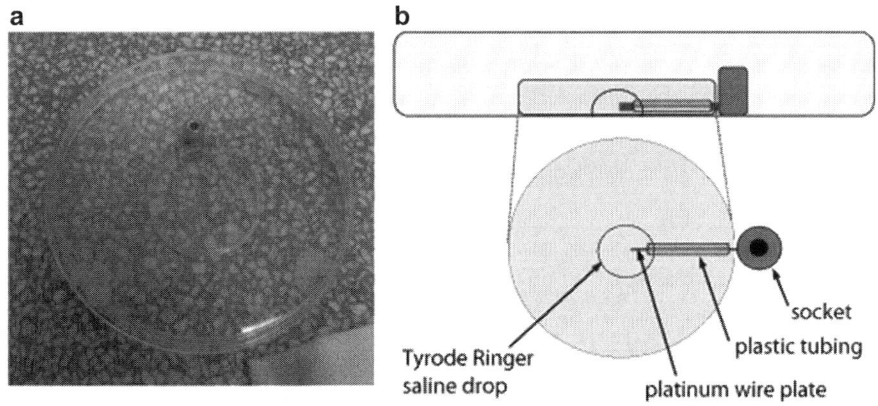

Fig. 1 An electroporation chamber for the mouse embryo. (**a**) The electroporation chamber is made from a 100-mm petri dish (outside) with a socket screwed onto it. Platinum wire-plate is then fused to the socket and covered by plastic tubing, only leaving a small amount of the wire free at the end for electroporation. A 35-mm petri dish is then used on the inside with a 5-mm hole on the side wall to allow entry of the platinum wire plate. (**b**) Schematic illustration of the electroporation chamber. A drop of Tyrode–Ringer saline is placed over the platinum wire-plate for the electroporation

3.1.1 Making Pipettes for Embryo Manipulation

1. Holding pipettes are made from thick-walled glass capillary tubing, being pulled once melted in a flame, and a microforge is then used to break the glass giving an internal diameter of 10–50 μm, depending on the size of the embryo used for the experiment (ideally the diameter should be half the size of the embryo).
2. Injection pipettes are made from thin-walled glass capillary tubing being pulled on a horizontal pipette puller, and a microforge is used to break the glass giving an internal diameter of roughly 10 μm.
3. Bend the injection and holding pipettes 1–2 cm from the tip using the heat from a small flame from a Bunsen burner to produce a bend of ~100 °. Make a similar bend, in opposite direction,at about 1 cm proximal to the first bend, to allow the pipettes to reach over the lip of the petri dish to the embryos.
4. Set up the micromanipulation apparatus. Attach the holding pipette to the instrument holder on the left manipulator—the holding pipette is then controlled using the micrometer syringe on the far right.

3.1.2 Setting the Electroporation Apparatus

1. The plate electrode for electroporation is made from a platinum wire, and the plate platinum electrode is attached to the lid of a petri dish using a banana socket (Fig. 1a, b).
2. The point electrode is made from a tungsten wire, which is held in place by a positioner.
3. Connect the plate electrode and point electrode to an Electro Square Porator. The plate should be set as the positive pole and the point as the negative pole for electroporation (*see* **Note 1**).

4. Place a drop of Tyrode–Ringer saline over the platinum plate and tungsten wire until both are completely immersed.
5. Place a 10–20 μL drop of concentrated plasmid DNA solution (1 to 5 mg/mL in water) on the petri dish (*see* **Note 2**), next to the Tyrode–Ringer Saline drop.

3.1.3 Dissection of the Embryos for Culture

1. Embryos are dissected in the PB1 dissected medium. Detailed protocols for dissection of the early implanted embryos have been described in [20].
2. Trim away the Reichardt's membrane, leaving only a small skirt around the ectoplacental cone (*see* **Note 3**).
3. All experimental procedures should be done within 1–2 h of collecting the embryos.
4. Do not place the embryos on ice especially at the heart-beating-stage embryos (E8.5–9.5), as cold shock can adversely affect the viability of the embryos.

3.1.4 Electroporation of the Surface Germ Layer of the Embryo

1. Put 1.5 mL 100 % rat serum (or DMEM with 50 % or 75 % rat serum) in one well of a 4-Well Chamber Slide (Nunc) and pre-incubate it in 5 % CO_2 in air at 37 °C for at least 1 h before culture.
2. Place dissected embryos into the drop of plasmid DNA solution on the electroporation dish.
3. Incubate the embryos for 2–5 min in the DNA solution (*see* **Note 4**).
4. Transfer embryos to the drop of Tyrode–Ringer saline in the electroporation dish.
5. Using the holding pipette, hold one embryo between the platinum plate and tungsten wire, without touching either electrode (Fig. 2a; *see* **Note 5**).
6. Using the Electro Square Porator, deliver 5 pulses at 15 V, 50 ms each and 1-s gap between each pulse (*see* **Note 6**).
7. Transfer the embryos to the culture medium in the 4-Well Chamber and place in the 5 % CO_2 in air at 37 °C for 4 h of static culture. This allows sufficient time for the production of GFP driven by a CMV promoter (Fig. 2b).
8. Culture the embryos for 24 h (*see* **Note 7**) (Fig. 2c).

3.1.5 Injection of DNA or siRNA Solution into the Embryo Followed by Electroporation

1. Using the holding pipette, hold the embryo (*see* **Note 8**).
2. Aspire a small volume of the plasmid DNA (1–5 μg/μL) or siRNA (0.1–0.5 μg/μL) solution into a micropipette and inject about 0.5 μL of the solution into the proamniotic cavity of the embryo (for targeting the ectoderm) or 0.1 μL of the solution into the space between the ectoderm and endoderm (for targeting the mesoderm) (Fig. 3) (*see* **Note 9**).

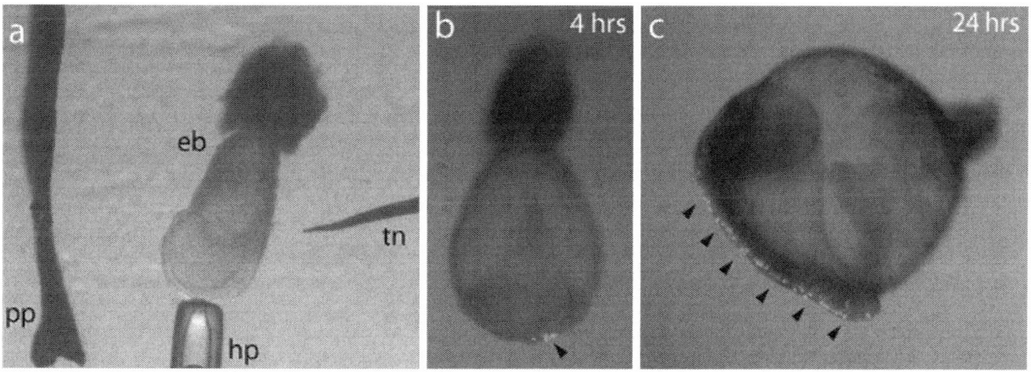

Fig. 2 (**a**) A view under the dissection microscope of the arrangement of the E7.5 embryo (eb), holding pipette (hp), platinum plate electrode (pp) and tungsten needle electrode (tn) during electroporation. The platinum plate is set as the anode and the tungsten wire-needle is set as the cathode, so that the DNA will enter via the apical surface of the cells driven towards the platinum plate (*yellow arrow*). Photos taken at (**b**) 4 h and (**c**) 24 h after electroporation of a CMV-GFP vector into the distal endoderm of E7.5 embryo cultured in vitro. Although the posterior region of this embryo is not perfectly normal, this specimen is shown to illustrate the extensive distribution of the descendants of the electroporated cells (*arrowheads*) in the endoderm (Color figure Online)

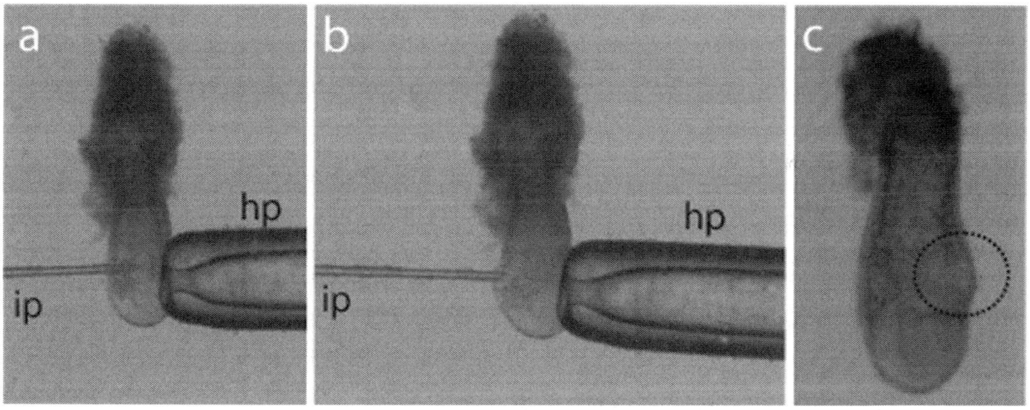

Fig. 3 A view under the dissection microscope showing the arrangement of the injection pipette (ip), holding pipette (hp), and the E6.75 embryo injected with DNA and lipofection solution. The solution can be injected into the amniotic cavity (**a**), or injected into the site between the endoderm and the epiblast (**b**). After the injection, a small bulge at the targeted site can be observed (**c**)

3. Perform electroporation using Electro Square Porator to deliver 5 pulses at 15 V, 50 ms each and 1-s gap between each pulse, with the polarity of the electrodes adjusted for the desired direction of electroporation (*see* **Note 10**).
4. Culture the embryos for 24 h in 4-Well Chamber Slides with static culture medium (*see* **Note 7**).

5. When transfecting siRNA, examine the knockdown effect by monitoring the target gene/protein expression at an appropriate time (usually 24–72 h after the transfection, depending on the cell types).
6. After the culture, embryos can be examined by fluorescence imaging, if a GFP reporter gene is incorporated into the construct; prior to fixation for histological analyses (*see* **Note 11**).

3.2 Electroporation of the Fetal Organs

The procedure for introduction of DNAs/RNAs into fetal organs by electroporation is essentially the same as that for the whole embryo. Localized introduction can be performed using plate and needle electrodes, in combination with the DNA or siRNA solution injection. Alternatively, a chamber with both plate type electrodes could be used for electroporating fetal organs, as well as late-organogenesis stage embryos. Depending on the type of target organ and features such as its size and structure, the electroporation conditions (i.e., voltage of electric pulses) should be optimized for transfection efficiency. Here, we introduce the procedure for the introduction of DNA/RNA into E12.5 genital ridges by electroporation using plate and needle electrodes (Fig. 4a–b″). For other organs, such as retina, kidney, and palate, detailed procedures have been described elsewhere [21–23].

1. Put 0.5 mL DMEM with 10 % FCS in one well of a 24-well culture plate. Float 0.3 μm of isopore membrane filters (Millipore) in the well and pre-incubate the 24-well plate in 5 % CO_2 in air at 37 °C for at least 1 h before culture.
2. Isolate genital ridges from E12.5 embryos. Dissection procedures for genital ridges cultured in vitro have been described elsewhere [24].
3. Transfer the dissected tissues to a drop of PB1 medium on a 60 mm petri dish.
4. DNA (1–5 μg/μL) or siRNA (0.1–0.5 μg/μL) solution is injected into the genital ridge by using an injection pipette. Adding the dye (1 % Fast Green) to the DNA solution enables the monitoring of the injection procedure. In this case the tissue fragment is large enough to be held by forceps. No holding pipette is required.
5. Using the Electro Square Porator, deliver 5 pulses at 15 V, 50 ms each and 1-s gap between each pulse, where the solution was injected.
6. The electroporated genital ridge is then placed onto the filter and cultured in 5 % CO_2 in air at 37 °C.

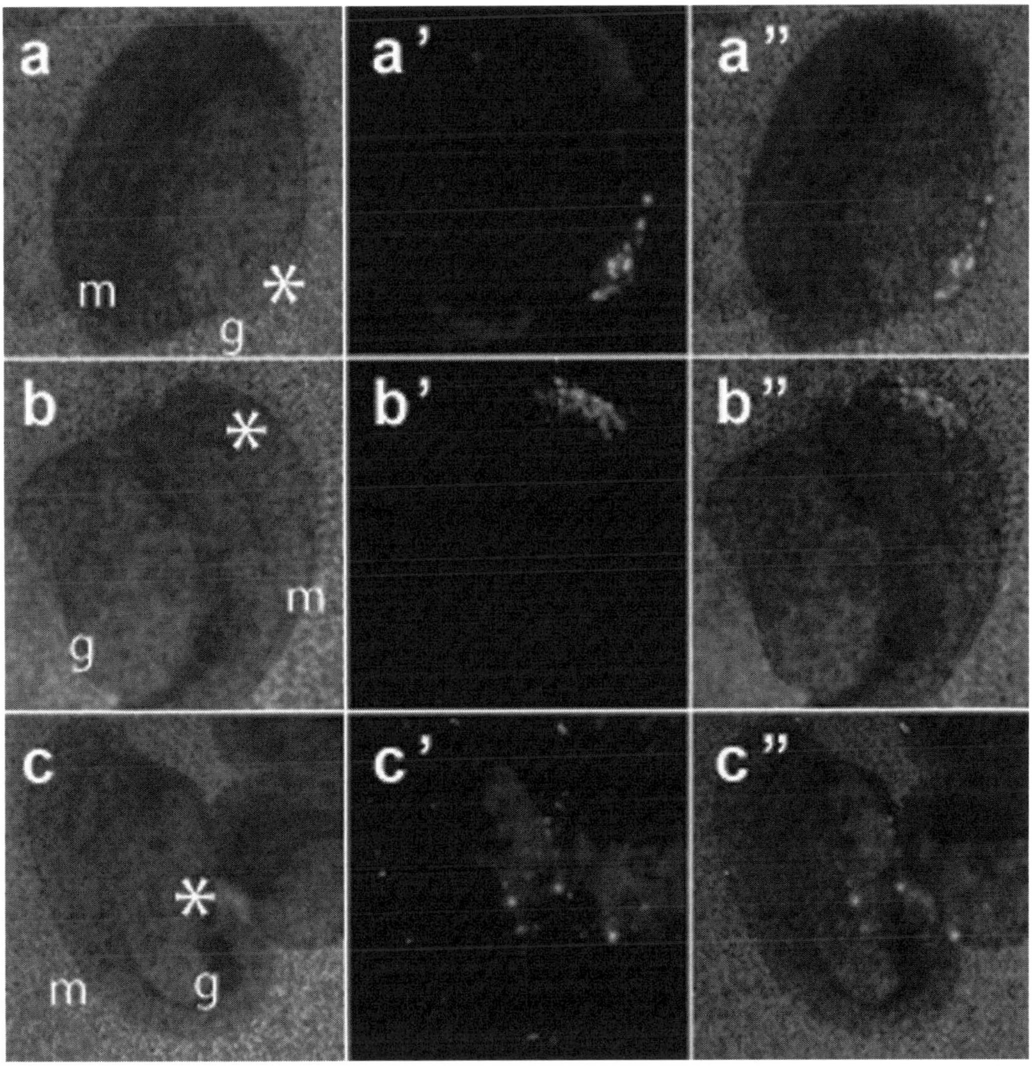

Fig. 4 Whole-mount view of cultured E12.5 genital ridges transfected with CMV-GFP by electroporation into the gonad (**a**) and mesonephroi (**b**), and by lipofection of the stroma of the gonad (**c**). Note that electroporation produced more restricted expression of CMV-GFP in the specimen. *Asterisks* indicate the site where DNA or lipofection solution was injected. (**a**, **b**, **c**) Bright-field images. (**a**′, **b**′, **c**′) Fluorescent images. (**a**″, **b**″, **c**″) Merged images. *m* mesonephroi, *g* gonad

3.3 Lipofection of Plasmid DNA Vector or siRNA into Embryos

3.3.1 Preparation of Lipofection Solution for Plasmid DNA

Preparation of the lipofection working solution follows the manufacture's instruction. Several kinds of lipofection reagents are provided by manufacturers, and each of them may have specific advantages, such that the reagent may transfect efficiently depending on the conditions for instance if there is serum-containing or antibiotic-containing media, low toxicity, long-time transgene expression, etc. (*see* **Note 12**). Here, we outline the lipofection method using Lipofectamine 2000 (Invitrogen). Other types of reagents may also be used, and you should be optimized for the specific experiment to achieve the best efficiency.

1. Mix 25 μL Opti-MEM and 1 μL Lipofectamine 2000 (Invitrogen) in a 1.5-mL microcentrifuge tube. Incubate for 5 min at room temperature.
2. Mix 25 μL Opti-MEM and 1 μL expression vector (0.5–1.0 mg/mL; *see* **Note 13**).
3. Mix the liposome and DNA solutions from above (*see* **Note 14**). Incubate for a further 20 min at room temperature.

3.3.2 Preparation of Lipofection Solution for siRNA

Preparation of the lipofection working solution for siRNA transfection follows the manufacturer's instructions. In this study, siRNAs were purchased from Qiagen, and we used the QIAGEN RNAiFect transfection reagent (Qiagen) according to the manufacturer's recommendation. Other lipofection reagents may also be used for siRNA transfection (*see* **Note 15**).

1. Mix 50 μL Opti-MEM and 1 μL siRNA (0.2–1.0 μg/mL, HP genomewide siRNA, Qiagen) in 1.5-mL microcentrifuge tube (*see* **Note 16**).
2. Add 3 μL of QIAGEN RNAiFect transfection reagent (Qiagen) and mix by vortexing for 10 s.
3. Incubate for 10–15 min at room temperature to allow formation of transfection complex.

3.3.3 Transfection to the Surface Layer of the Embryo

Introduction of transgenes by lipofection usually takes place by culturing the cells in the lipofection solution containing media. When we introduce the transgene into the explanted E6.5 epiblast, soaking the epiblast in the lipofection solution for a few minutes before the culture, and then culturing the epiblast in the liposome containing culture media provided an efficient transgene transfection at the surface of the epiblast (data not shown). If the plan is to examine the modulating gene activity on the surface layer of the embryo, such as a broad area of the endoderm layer, culturing the embryo in the liposome containing culture media and/or soaking the embryo in the lipofection solution before the culture will allow transgene integration into the outside tissue of the embryo [34]. Here we describe the method for delivering transgene to the superficial endoderm layer of the E7.5 embryo by lipofection (*see* **Note 17**). The procedures are essentially the same for embryos at other developmental stage, except for the in vitro culture conditions. Details of the culture conditions are described elsewhere [17].

1. Put 0.25 mL DMEM with 50 % rat serum in one well of an 8-Well Chamber Slide (Nunc) and pre-incubate it in 5 % CO_2 in air at 37 °C for at least 1 h before culture.

2. Isolate E7.5 embryos with extraembryonic tissue in the PB1 dissection medium. Detailed procedures for the dissection of early implanted embryos were described in [20].
3. Dissected embryos are kept in Opti-MEM (Gibco).
4. Transfer the embryos into the lipofection solution (50 μL), and immerse the embryos for 3 min in the liposome solution.
5. Transfer the embryos from the liposome solution (50 μL) into the well of the 8-Well Chamber Slides containing rat serum.
6. Culture the embryos for 24 h in the 8-Well Chamber Slide (Nunc) (*see* **Note 18**).
7. If the transgene contains a GFP reporter cassette, fluorescence imaging of the transfected embryos should be performed before fixation (*see* **Note 11**).

3.3.4 Injection of Lipofection Solution

To examine the modulating gene activity on the cells inside the embryo or at a localized region, the lipofection solution has to be injected into the embryo or the site. Here we present the procedures for injecting the lipofection solution into E6.5–E7.5 embryos with the aid of micromanipulators for targeting the transfection of transgenes to the specific tissue layers. Handling of later stage embryos will be easier, and may not require the use of holding pipettes. The embryos could be immobilized by lodging against a mold of agarose or held by forceps during the injection. Moreover, injection of the lipofection solution could be done by using a mouth-controlled injection pipette rather than an oil-operated micropipette.

1. Set up the micromanipulation apparatus, such that the holding pipette is controlled using the micrometer syringe on the right and the injection pipette is controlled using the de Fonbrune syringe on the left.
2. Back-fill the holding with heavy paraffin oil by adjusting the syringes. Back-fill the injection pipette but not all the way to the end.
3. Place a drop of PB1 in the center of a 60-mm petri dish.
4. Place a drop of lipofection solution in the same petri dish 1–2 cm from the PB1 drop.
5. Using a transfer pipette, made by pulling a Pasteur pipette in a flame, transfer 4–7 embryos into the drop of PB1.
6. Position the manipulation dish on the microscope to focus on the medium drop in the center of the dish. Position one of the embryos (using a steel needle to roll the embryo to the desired position), and realign the microinjection pipettes so that the embryos and instruments are all at the same focal plane.

7. Move the holding pipette close to the embryo and apply suction on the micrometer syringe so that a small piece of the yolk sac membrane is drawn into the pipette (*see* Fig. 3a). Move the embryo gently back and forth to ensure that it is held tightly.
8. Move the injection pipette into the lipofection solution drop and draw a small volume of the solution into the injection pipette (*see* **Note 19**).
9. Insert the injection pipette by a sharp jabbing action through the endoderm layer into the amniotic cavity or to the space between the endoderm and the epiblast/ectoderm (*see* Fig. 3c).
10. Inject a small amount of the lipofection solution by applying a positive pressure to the de Fonbrune syringe (*see* **Note 20**).
11. Slowly withdraw the injection pipette and try to minimize the escape of lipofection solution.
12. Culture the embryos for 24 h in 4-Well Chamber Slides (*see* **Note 7**).

3.4 Lipofection of Expression Vector or siRNA into Fetal Organs Cultured In Vitro

By culturing in the liposome containing media, transgenes can be efficiently introduced into the cells of fetal organs, such as an explanted embryonic kidney [25] and embryonic lung [26]. When transgenes are to be introduced into organs covered by a tunica tissue layer, such as the fetal gonad and kidney, incubating in the liposome containing medium alone is unlikely to results in efficient transfection, because lipofection complexes cannot access cells in the core of the organ. Microinjection of the lipofection solution into the organs may help to deliver the reagent to the core tissues, but it would be restricted to a localized region where the reagent is injected. To achieve more efficient transfection of the core tissues, the organ may be sliced open to allow better access by the liposome reagent during incubation [16]. Results of a preliminary study on the efficiency of introducing EGFP expression construct by either electroporation (Fig. 5a) or microinjection of liposome reagent can achieve localized expression of the vector in the E15.5 fetal ovary (Fig. 5b). However, transfection of vector is more widespread in the stromal tissues of fetal ovary that were dissected before lipofection (Fig. 5c) than electroporation and microinjection of the intact organ. In the E15.5 fetal ovary that were treated with control vector tagged by Alexa488 at and cultured for 3 days in vitro, 60–95 % of oocytes express the fluorescent tag and react positively to anti-synaptonemal complex protein 3 antibody, indicating normal meiotic differentiation [16]. This result highlights that the transfection and culture procedures are efficient and have no discernible adverse effect on the embryonic tissue.

In the following sections, we describe the procedures for lipofection of expression vector and siRNAs into the explanted fetal organ by culturing in a liposome containing media (Subheading 3.4.3) (*see* **Note 21**), and for lipofection of cells of

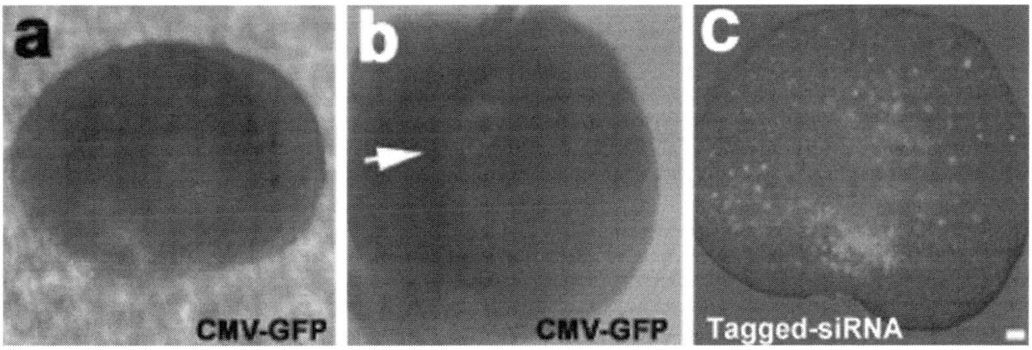

Fig. 5 Introduction of transgenes in an ovary culture system. Whole mount view of cultured fetal ovaries with CMV-GFP transfected either by (**a**) electroporation, or (**b**) microinjection of the lipofection solution, the *arrow* in (**b**) indicates GFP expression. (**c**) A whole mount view of an E15.5 fetal ovary, cultured after lipofection with Alexa488 tagged siRNA that emits green fluorescence. Scale bar: 10 μm

the fetal gonads by microinjection for a localized transfection (Subheading 3.4.4) and by microsurgical manipulation to achieve wider transfection of the tissue (Subheading 3.4.5).

3.4.1 Preparation of DNA–Liposome Lipofection Solution

Procedures for the preparation and optimization of DNA–liposome lipofection reagents are described in the Subheading 3.3.1.

3.4.2 Preparation of siRNA–Liposome Lipofection Solution

Procedures for the preparation of the siRNA–liposome lipofection working solution are described in Subheading 3.3.2. To monitor the transfection efficiency, the 3′ end of the sense sequence of each siRNA construct is tagged with Alexa488 green Fluorescent dye (Qiagen). After 4–8 h transfection, uptake of fluorescent siRNA can be detected, although the fluorescent signal may become weaker over time.

3.4.3 Transfection by Culturing with Lipofection Solution into Explanted Fetal Kidney

1. Put 0.3 mL DMEM with 10 % FCS in a well of a 24-well culture plate (*see* **Note 22**).
2. Float 0.3 μm of isopore membrane filters (Millipore) or insert Transwell (0.4 μm; Corning) into the well and pre-incubate the 24-well plate in 5 % CO_2 in air at 37 °C for at least 1 h before culture.
3. Isolate a metanephroi from E11.5 embryos. Dissection procedures for metanephros that are cultured in vitro were described elsewhere [25, 27].
4. 100 μL of lipofection solution is added into the culture medium according to the methods for the transfection into cultured cells provided by the manufacturer.
5. Move the metanephroi onto the filter and culture it in 5 % CO_2 in air at 37 °C.

6. After 24 h, the medium is changed to fresh DMEM containing 10 % FCS medium (*see* **Note 21**).
7. Examine the knockdown effect by monitoring the target gene/protein expression at an appropriate time (usually 24–72 h after the transfection, depending on the cell types).

3.4.4 Transfection by Microinjection of Lipofection Solution into the E12.5 Genital Ridges

1. Put 0.5 mL DMEM with 10 % FCS in one well of a 24-well culture plate.
2. Float 0.3 μm of isopore membrane filters (Millipore) in the well and pre-incubate the 24-well plate in 5 % CO_2 in air at 37 °C at least 1 h before culture.
3. Isolate genital ridges from E12.5 embryos. Dissection procedures for genital ridges that are cultured in vitro were described elsewhere [24].
4. Move the dissected tissues to a drop of PB1 medium on a 60-mm petri dish.
5. Lipofection solution is injected into the loose connective compartment of the genital ridge by using an injection pipette.
6. Transfer the genital ridge onto the filter and culture in 5 % CO_2 in air at 37 °C.
7. Observe the reporter gene activity after 24 h to determine the site where lipofection is acheived (Fig. 5c).

3.4.5 Transfection of Plasmid DNA or siRNA to Oocytes of the E15.5 Fetal Ovary by Lipofection

1. Put 0.5 mL DMEM with 20 % FCS in a well of a 24-well culture plate.
2. Float 0.3 μm of isopore membrane filters (Millipore) in the well and pre-incubate the 24-well plate in 5 % CO_2 in air at 37 °C for at least 1 h before culture.
3. Remove fetal ovaries with oviduct and uterine tissues from the E15.5 mouse embryo in the 60 mm petri dish filled with dissection medium (PB1).
4. Move the dissected tissues to fresh PB1 medium.
5. Working under a dissecting microscope and using watchmaker's forceps and/or fine scissors, dissect the ovary out from the oviduct.
6. Dissected ovary is sliced open by using sterilized razor blade (Fig. 6a–c), and then immerse in 50 μL liposome solution for 4 min.
7. Bring the ovary with the liposome solution (50 μL) onto the 0.3 μm isopore membrane filter (Millipore) (Fig. 6d).
8. Culture the treated ovaries for 3 days on the membrane filters floating on 500 μL DMEM with 20 % FCS in 5 % CO_2 in air at 37 °C (Fig. 6e [28]).

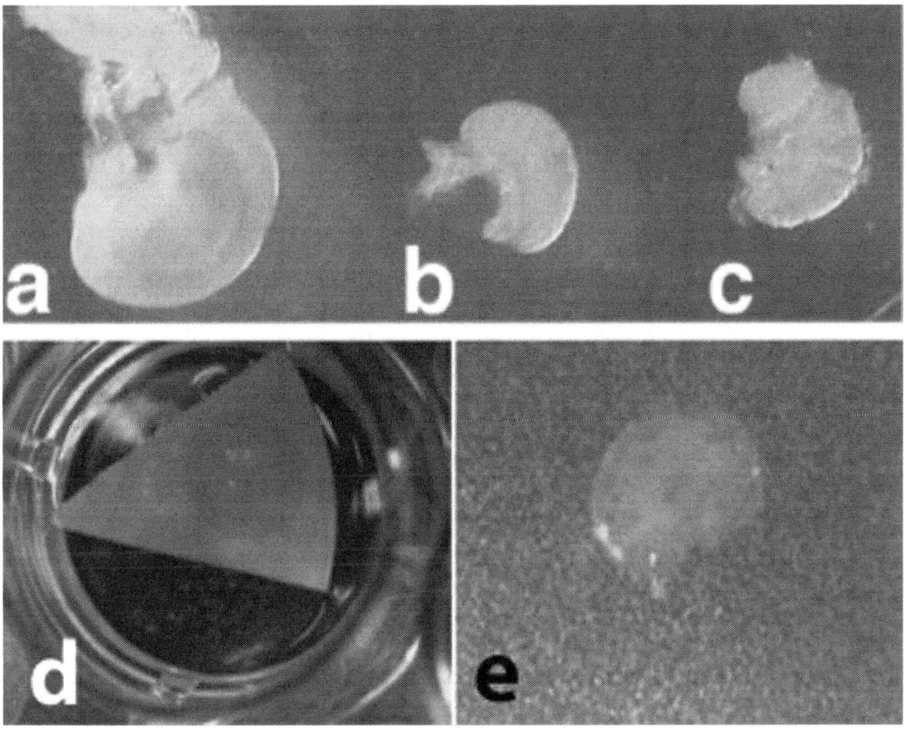

Fig. 6 The ovary is sliced open to facilitate access of the lipofection solution to the inside of the ovary. The dissection of E15.5 fetal ovaries for lipofection: Ovaries are (**a**) dissected out from the embryo, then (**b**) the Müllerian ductal tissues removed, and (**c**) are sliced open prior to immersion in the lipofection solution. (**d**) The lipofected ovaries were cultured on the floating membrane filter and (**e**) regained the morphology after 24 h in vitro

4 Notes

1. As the DNA is negatively charged, it will be driven toward the positive electrode. Depending on the setting of the polarity of the electrodes, the position of the embryo and where the DNA is delivered in the preparatory step, the appropriated germ layer can be targeted.
2. The quality of the plasmid reagent is critical for good transfection efficiency. Purification of plasmid DNA is performed by an anion-exchange resin or silica membrane column, such as a QIAGEN-tip plasmid purification kit (Qiagen). The quality of DNA should be monitored by UV spectrophotometry and agarose gel analysis, and by quantitative analysis of reporter gene activity in the specimens.
3. Do not dissect away the ectoplacental cone because it is required for the normal growth of the embryo in vitro.

4. If the transfection efficiency is low, increase the DNA concentration and incubation period.
5. It is essential that the embryo is properly positioned between the electrode and that the point electrode is as close as possible to the site of electroporation while not causing burn damage to the tissue. Air bubbles will emit from the tungsten wire when electric current is generated. It is important to leave space for the air bubbles to escape from the electrode without disturbing the embryo.
6. Electroporation has to be done swiftly to minimize the time spent by the embryo in Tyrodes solution.
7. For later stages embryos, such as E8.5 and E9.5 embryos, embryos were cultured in a roller bottle with continuous gassing. Detailed culture conditions have been previously described [17].
8. Hold the embryo at a point away from the embryo proper whenever this is feasible (e.g., ectoplacental cone or extraembryonic region) and at the best orientation relative to the point electrodes for targeting the electroporation. If you wish to transfect DNA/RNA into the later stage embryos (E8.5–E9.5), embryos are immobilized in a mold of agarose for the injection and electropration instead of holding the embryo by the pipette.
9. Since the mesoderm is a relatively spacious layer, the DNA should be injected close to the endoderm layer, so electroporation will be targeted to the majority of the mesoderm cells further away from the endoderm.
10. Electroporation of the ectoderm is done by reversing the polarity of the electrodes set for the mesoderm or the endoderm, so that the DNA will be driven towards the luminal surface of the ectoderm from within the proamniotic or amniotic cavity.
11. After fixation, for histochemical analysis, transfected embryos/specimens will have a diminished GFP signal. Avoid using ethanol and methanol as fixative, which may abolish the GFP signals.
12. For transfection into the embryo or fetal organs cultured in vitro, a lipofection reagent that allows efficient transfection in the serum-containing media is preferred.
13. The quality of the plasmid DNA is critical for good transfection efficiency. Ascertain the purity of isolated DNA by UV spectrophotometry and agarose gel analysis and perform a quantitative analysis of reporter gene activity of the specimens.
14. For efficient transfection, the lipofection reagents may be optimized by changing the ratio of the mixing volume of DNA and the lipofection solutions. Titration of the concentration of DNA solution may also help the optimization.

15. Similar transfection efficiency was obtained by using either QIAGEN RNAiFect transfection reagent (Qiagen) or Lipofectamine 2000 (Invitrogen) for siRNA transfection into the mouse embryo (data not shown).
16. The concentration of siRNA solution may be titrated for optimizing the transfection efficiency.
17. Our preliminary results showed that transgenes were transfected into the broad area of the endoderm of E7.5 embryos by this method but strong reporter activity was not observed. Further optimization will be required.
18. Do not add antibiotics to the medium. Antibiotics sometimes decrease the transfection efficiency depending on the lipofection reagent, since antibiotics have toxic effects on the cells transfected with liposome.
19. Be careful not to let the lipofection solution touch the paraffin oil in the injection pipette, as it will adhere to the oil molecules.
20. Monitor the solution–air meniscus to control the amount of lipofection solution to be injected. A small bulge at the injection site is a good visual indication of successful injection (*see* Fig. 3c).
21. Long-time culture with liposome-containing media sometimes may have cytotoxic effects on the cultured organs.
22. For siRNA knockdown in the explanted kidney by lipofection, culturing with Richter's Modified Improved MEM (Gibco) supplemented with 10 μg/mL iron-loaded human transferrin without serum would enhance the transfection efficiency and kidney development [24].

References

1. Tam PP, Khoo PL, Lewis SL, Bildsoe H, Wong N, Tsang TE, Gad JM, Robb L (2007) Sequential allocation and global pattern of movement of the definitive endoderm in the mouse embryo during gastrulation. Development 134:251–260
2. Uchikawa M, Ishida Y, Takemoto T, Kamachi Y, Kondoh H (2003) Functional analysis of chicken Sox2 enhancers highlights an array of diverse regulatory elements that are conserved in mammals. Dev Cell 4:509–519
3. Davidson BP, Tsang TE, Khoo PL, Gad JM, Tam PP (2003) Introduction of cell markers into germ layer tissues of the mouse gastrula by whole embryo electroporation. Genesis 35:57–62
4. Tanaka SS, Yamaguchi YL, Tsoi B, Lickert H, Tam PP (2005) IFITM/Mil/fragilis family proteins IFITM1 and IFITM3 play distinct roles in mouse primordial germ cell homing and repulsion. Dev Cell 9:745–756
5. Ogura T (2002) In vivo electroporation: a new frontier for gene delivery and embryology. Differentiation 70:163–171
6. Lickert H, Cox B, Wehrle C, Taketo MM, Kemler R, Rossant J (2005) Dissecting Wnt/beta-catenin signaling during gastrulation using RNA interference in mouse embryos. Development 132:2599–2609
7. Itasaki N, Bel-Vialar S, Krumlauf R (1999) 'Shocking' developments in chick embryology: electroporation and in ovo gene expression. Nat Cell Biol 1:E203–E207
8. Swartz M, Eberhart J, Mastick GS, Krull CE (2001) Sparking new frontiers: using in vivo electroporation for genetic manipulations. Dev Biol 233:13–21

9. Saito T, Nakatsuji N (2001) Efficient gene transfer into the embryonic mouse brain using in vivo electroporation. Dev Biol 240:237–246
10. Tabata H, Nakajima K (2001) Efficient in utero gene transfer system to the developing mouse brain using electroporation: visualization of neuronal migration in the developing cortex. Neuroscience 103:865–872
11. Fukuchi-Shimogori T, Grove EA (2001) Neocortex patterning by the secreted signaling molecule FGF8. Science 294:1071–1074
12. Fukuchi-Shimogori T, Grove EA (2003) Emx2 patterns the neocortex by regulating FGF positional signaling. Nat Neurosci 6:825–831
13. Osumi N, Inoue T (2001) Gene transfer into cultured mammalian embryos by electroporation. Methods 24:35–42
14. Takahashi M, Sato K, Nomura T, Osumi N (2002) Manipulating gene expressions by electroporation in the developing brain of mammalian embryos. Differentiation 70:155–162
15. deCastro M, Saijoh Y, Schoenwolf GC (2006) Optimized cationic lipid-based gene delivery reagents for use in developing vertebrate embryos. Dev Dyn 235:2210–2219
16. Yamaguchi YL, Tanaka SS, Yasuda K, Matsui Y, Tam PP (2006) Stage-specific Importin13 activity influences meiosis of germ cells in the mouse. Dev Biol 297:350–360
17. Tam PP (1998) Postimplantation mouse development: whole embryo culture and micromanipulation. Int J Dev Biol 42:895–902
18. Franklin V, Bildsoe H, Tam PP (2007) Fate-mapping technique: grafting fluorescent cells into gastrula stage mouse embryos at 7–7.5 days post coitum. CSH Protoc. doi:10.1101/pdb.prot4892
19. Khoo PL, Franklin VJ, Tam PP (2007) Fate-mapping technique: targeted whole embryo electroporation of DNA constructs into the germ layers of 7–7.5 dpc mouse embryos. CSH Protoc. doi:10.1101/pdb.prot4893
20. Nagy A, Gertenstein M, Vintersten K, Behringer R (2003) Isolation, culture and manipulation of postimplantation embryos, Chapter 5 "Manipulating the Mouse Embryo", 3rd edn. Cold Spring Harbor Laboratory Press, Cold Spring Harbor, NY
21. Donovan SL, Dyer MA (2006) Preparation and square wave electroporation of retinal explant cultures. Nat Protoc 1:2710–2718
22. Alie TM, Vrljicak PJ, Myburgh DB, Gupta IR (2007) Microinjection and electroporation of embryonic kidney explants: an improved method. Kidney Int 72:121–125
23. Lee JM, Kim JY, Cho KW, Lee MJ, Cho SW, Kwak S, Cai J, Jung HS (2008) Wnt11/Fgfr1b cross-talk modulates the fate of cells in palate development. Dev Biol 314:341–350
24. Hiramatsu R, Kanai Y, Mizukami T, Ishii M, Matoba S, Kanai-Azuma M, Kurohmaru M, Kawakami H, Hayashi Y (2003) Regionally distinct potencies of mouse XY genital ridge to initiate testis differentiation dependent on anteroposterior axis. Dev Dyn 228:247–253
25. Davies JA, Ladomery M, Hohenstein P, Michael L, Shafe A, Spraggon L, Hastie N (2004) Development of an siRNA-based method for repressing specific genes in renal organ culture and its use to show that the Wt1 tumour suppressor is required for nephron differentiation. Hum Mol Genet 13:235–246
26. Liu J, Beqaj S, Yang Y, Honore B, Schuger L (2001) Heterogeneous nuclear ribonucleoprotein-H plays a suppressive role in visceral myogenesis. Mech Dev 104:79–87
27. Nishinakamura R, Matsumoto Y, Nakao K, Nakamura K, Sato A, Copeland NG, Gilbert DJ, Jenkins NA, Scully S, Lacey DL, Katsuki M, Asashima M, Yokota T (2001) Murine homolog of SALL1 is essential for ureteric bud invasion in kidney development. Development 128:3105–3115
28. Park EH, Taketo T (2003) Onset and progress of meiotic prophase in the oocytes in the B6.YTIR sex-reversed mouse ovary. Biol Reprod 69:1879–1889
29. Okada Y, Ueshin Y, Isotani A, Saito-Fujita T, Nakashima H, Kimura K, Mizoguchi A, Oh-Hora M, Mori Y, Ogata M, Oshima RG, Okabe M, Ikawa M (2007) Complementation of placental defects and embryonic lethality by trophoblast-specific lentiviral gene transfer. Nat Biotechnol 25:233–237
30. Mellitzer G, Hallonet M, Chen L, Ang SL (2002) Spatial and temporal 'knock down' of gene expression by electroporation of double-stranded RNA and morpholinos into early postimplantation mouse embryos. Mech Dev 118:57–63
31. Bartkowska K, Paquin A, Gauthier AS, Kaplan DR, Miller FD (2007) Trk signaling regulates neural precursor cell proliferation and differentiation during cortical development. Development 134:4369–4380
32. Calegari F, Haubensak W, Yang D, Huttner WB, Buchholz F (2002) Tissue-specific RNA interference in postimplantation mouse embryos with endoribonuclease-prepared short interfering RNA. Proc Natl Acad Sci U S A 99:14236–14240

33. Yamamoto M, Saijoh Y, Perea-Gomez A, Shawlot W, Behringer RR, Ang SL, Hamada H, Meno C (2004) Nodal antagonists regulate formation of the anteroposterior axis of the mouse embryo. Nature 428:387–392

34. Tanaka SS, Yamaguchi YL, Steiner KA, Nakano T, Nishinakamura R, Kwan KM, Behringer RR, Tam PP (2010) Loss of *Lhx1* activity impacts on the localization of primordial germ cells in the mouse. Dev Dyn 239:2851–2859

Chapter 23

Using the Textpresso Site-Specific Recombinases Web Server to Identify Cre Expressing Mouse Strains and Floxed Alleles

Brian G. Condie and William M. Urbanski

Abstract

Effective tools for searching the biomedical literature are essential for identifying reagents or mouse strains as well as for effective experimental design and informed interpretation of experimental results. We have built the Textpresso Site Specific Recombinases (Textpresso SSR) Web server to enable researchers who use mice to perform in-depth searches of a rapidly growing and complex part of the mouse literature. Our Textpresso Web server provides an interface for searching the full text of most of the peer-reviewed publications that report the characterization or use of mouse strains that express Cre or Flp recombinase. The database also contains most of the publications that describe the characterization or analysis of strains carrying conditional alleles or transgenes that can be inactivated or activated by site-specific recombinases such as Cre or Flp. Textpresso SSR complements the existing online databases that catalog Cre and Flp expression patterns by providing a unique online interface for the in-depth text mining of the site specific recombinase literature.

Key words Cre, loxP, RMCE, Flp, Frt, Floxed, Mouse, PhiC31, Database, Text mining

1 Introduction

Many powerful genetic approaches in the mouse use site-specific recombinase (SSR) technology. SSRs are used in a range of approaches including tissue specific or temporal specific gene knockouts or gene activation, lineage tracing, cell ablation, the generation of genetic mosaics, and the production of precise chromosomal rearrangements [1–16]. This is not an exhaustive list of methods and new SSR based research tools are being continuously developed. Along with the rapid expansion in experimental approaches, there has been a dramatic increase in the number of mouse strains that can be used in SSR-based experimental strategies. Furthermore, the number of publications that describe the generation, validation and genetic analysis of these mice has burgeoned. The volume of information makes it very challenging to identify

Mark Lewandoski (ed.), *Mouse Molecular Embryology*, Methods in Molecular Biology, vol. 1092,
DOI 10.1007/978-1-60327-292-6_23, © Springer Science+Business Media New York 2014

appropriate research resources for a particular project. It is also difficult to identify publications of interest that report the results of research that use SSR based approaches. This is a rapidly growing portion of the mouse genetics literature and it is of particular interest because of the wealth of detailed information it contains concerning the construction and validation of recombinase expressing mouse strains as well as the research results obtained in the phenotypic analysis of these mouse strains.

To assist investigators in retrieving information from this portion of the literature we have built a text-mining Web server that allows researchers to search a full text database of over 10,000 publications. Our text-mining server, Textpresso Site Specific Recombinases (Textpresso SSR) uses the Textpresso information retrieval system [17]. A diverse group of Textpresso Web servers have been generated and they are accessible via the main Textpresso Web site (http://www.textpresso.org/). Textpresso Web servers allow users to search a publication database. This database consists of the information typically retrieved in a PubMed search as well as the full text of the publication when available. In the case of Textpresso SSR 95 % of the publication entries in the database consist of the full text of the paper plus the abstract and citation information normally found via PubMed. In a Textpresso search, the Keyword query is searched against all of the information within the database (depending on user specifications). The ability to search the full text of publications greatly increases the effectiveness of information retrieval, especially in searches that are designed to retrieve technical details or specific experimental results that tend to be embedded in the text of publications. In addition, the Textpresso interface presents the retrieved information in a format that is easily scanned and analyzed by researchers.

In the case of Textpresso SSR our goal was to build a publication database that would include papers describing or validating site specific recombinase technologies as well as papers reporting research results obtained using the SSR-based research tools. We included a wide range of fields and topics in the database [18]. We have also included most of the publications that describe mouse strains that express site-specific recombinases such as Cre as well as mouse strains that carry alleles that are activated or inactivated in the presence of SSR activity [18]. Table 1 summarizes the topic areas relevant to mouse researchers within the publication database. Textpresso SSR complements existing online resources such as the Cre expression pattern databases that can be viewed via the Mouse Genome Informatics Web site and other Web servers [19–22]. These additional online resources are listed in Table 2. An important and unique feature of the Textpresso SSR text database is that it exclusively contains peer-reviewed published literature. This distinguishes it from the other existing recombinase databases and Web servers.

Table 1
Topic areas relevant to mouse genetics that are represented in the Textpresso SSR text database

General topic area	Specific topics within general topic area
SSR expressing mice	Characterization of mouse strains expressing Cre, Flp, Dre, or PhiC31 Mouse strains expressing tamoxifen inducible recombinases
Conditional knockout mice	Characterization and validation of conditional alleles Phenotype analysis of tissue specific or stage specific knockout mice
Conditional gene activation	Recombinase regulated overexpression alleles and transgenes Rosa26 knockin alleles Cell lineage and cell fate studies in mice Brainbow, MADM, STARS and related approaches Recombinase activity reporter transgenes or alleles
Emerging SSR tools	Studies of Dre recombinase and PhiC31 integrase in mice

Note: The categories in this table are not an exhaustive list of topics contained within the database. Performing a search may reveal other information categories of interest

Table 2
Other recombinase related databases for mouse research

Resource	URL	Contents	Reference
MGI-recombinase	http://www.creportal.org/	Images[a]	
Cre-X-mice	http://nagy.mshri.on.ca/cre_new/index.php	Text description[b]	[21]
Allen Institute	http://transgenicmouse.alleninstitute.org/	Images[a]	[19]
Cre driver mice database	http://www.credrivermice.org/database	Images[a]	
Gensat Cre mice	http://www.gensat.org/cre.jsp	Images[a]	[22]

[a]These resources contain images of recombinase expression patterns
[b]This database uses text and icons to describe expression pattern of the recombinase

Textpresso SSR was built to help researchers deal with the large number and rapidly increasing volume of peer reviewed published information about SSR-based research tools and the experiments that use them. In general, the information overload caused by the steadily increasing volume of published papers and review articles in biomedical fields isolates researchers from the scientific literature. Text mining and information retrieval systems like Textpresso help researchers find key information within this unmanageable corpus of information. In fact, text mining and other forms of information management are becoming an essential part of every researcher's work. Several reviews have been published that give a good introduction or overview of text mining in biomedical research [23–28]. In addition to tools such as Textpresso, PDF management software is available to assist researchers and students in managing the growing collections of PDF files we all have on our own computers. We use one of the PDF managers, Mekentosj Papers (http://mekentosj.com/papers/) to maintain a backup publication database for Textpresso SSR and to perform regular updates. An excellent recent review article provides an overview of software and online tools for researchers who are interested managing and searching through their own digital libraries of publications [29].

2 Materials

The Textpresso Site Specific Recombinases Web server is accessed at http://ssrc.genetics.uga.edu/. The site does not require any registration or login. The Web site has been tested using several commonly used Web browsers on multiple operating systems (Windows, Mac, and Linux).

An understanding of how to use other mouse information resources will maximize the user's ability to effectively utilize Textpresso SSR. The primary resource for information regarding mouse genetics and mouse research resources is the Mouse Genome Informatics (MGI) portal maintained by the Jackson Laboratory (http://www.informatics.jax.org/). A basic familiarity with MGI will assist the user in formulating complex queries for extensive searches of Textpresso SSR.

3 Methods

Subheading 3 of this chapter outlines basic search strategies as well as our suggestions for optimizing the effectiveness of your search using Textpresso SSR. A version of the methods outlined here is available on the FAQ and User Guide page on the Textpresso SSR Web site. Most researchers working with mouse strains will be searching for information related to a particular gene or for specific mouse strains. In this section, we outline strategies for maximizing the retrieval of information from Textpresso SSR for these types of research resources.

In general, anyone performing an extensive information search using any search or text mining tool should keep good notes. The notes should include details about the queries/keywords used and the results obtained with specific queries. This will help you to reconstruct what you have done and will allow you to avoid duplicating previous searches. Keywords that have been used should be saved and Web pages with search results can be saved in various formats by Web browsers. This will allow you to create an archive of your searches. Treat the search process and the results of literature/text mining as though it is a bench experiment, and keep complete records of what you do and what results you obtain. The literature in many areas is quite extensive and several searches will be necessary to retrieve a large sample of the relevant information. Saving your queries/keywords and search results will make it easy for you to reexamine the results of multiple searches.

We have written the detailed methods with the assumption that the reader will be viewing the Web site (and the other Web sites mentioned in the methods such as MGI) while reading the protocols. Our intent is to provide a brief tour of the major features of Textpresso SSR and to discuss specific search strategies that will be of greatest use to most mouse researchers. Additional details about user options and detailed descriptions of the home, search and ontology browser pages are available on the FAQ and User Guide page.

3.1 An Introduction to the Textpresso SSR Web Server

1. The URL for the Web server is http://ssrc.genetics.uga.edu/. There is no registration or other login requirements to use Textpresso SSR.
2. The first page you will land on is the home page. We recommend that new users read the FAQ (frequently asked questions) and User Guide page. Click on the "FAQ & User Guide" button located at the top of the home page to access the FAQ/User Guide page. You can print this page to provide a reference while using the site.
3. A second link to the FAQ/User Guide page is located below the Keyword textbox and is labeled "How do I use this Website?"
4. A search can be initiated from the home page or you can click on the "Search" link at the top of the page to enter the search page. There are a few additional user options on the search page.
5. Keyword(s) are entered in the "Keywords" textbox on the left side of the page on the home page or the search page.
6. If you enter multiple keywords please note that Textpresso translates queries consisting of keywords separated by single spaces into the Boolean operator "AND" while queries consisting of commas with no spaces between keywords are translated into the Boolean operator "OR". In our tests the best results are obtained when only one Boolean operator is used at a time. You can use several keywords in a query as long as they are all linked by the same Boolean operator.
7. You have the option of selecting "Exact match" and/or "Case sensitive" to adjust the parameters used to match your keyword(s) to words in the text database.
8. On the search page you can also select the fields within the database that will be searched (abstract, author, body, title, or year). The default setting is to search abstract, body and title

fields of the publications in the database excluding the author and year fields. The user can select any combination of these fields for the search. Please note the specific instructions on the FAQ for performing a search for a particular author name without retrieving author names listed in the reference lists of the retrieved publications.

9. *See* Subheading 3.3 for a discussion of search results display options and Subheading 3.4 for a discussion of how to search within the Textpresso search results.

3.2 Searches Using Search Categories in the Query

1. On the home and search pages the user can select up to four search categories. The names of the categories are on the lists under the "Category" heading on the home or search pages.
2. The categories are collections of terms that are relevant to the subject area denoted by the name of the category. The publications in a Textpresso database are indexed so that terms within each category are tagged with the category name. This allows the terms listed in the categories to be found during searches.
3. You can determine which category contains a particular term or terms by searching the category lists using the ontology browser page. The ontology browser is accessed by clicking on the "Ontology" button at the top of any Textpresso SSR page. You can search the category lists with multiple terms at one time if desired.
4. As an example, enter the term "gastrulation" into the text box of the Ontology Browser. The browser returns a results table listing the category name "developmental process" as the category in which gastrulation is listed.
5. The user can perform a keyword search with or without selecting categories. You can also search the database by selecting one to four categories without entering any keywords.
6. To perform a keyword + category search, enter a keyword or list of keywords and then select 1–4 categories from the category menu. If you select multiple categories they will automatically be linked with the Boolean operator "AND". The search results returned each contain terms from each category.

3.3 Adjusting the Display of the Search Results

1. The search results are displayed below the keyword textbox and category pull-down menu display. In cases where multiple pages of search results are returned, each page is displayed in the same format. For each page scroll down to examine the search results.
2. The search results include the title, author list, citation information, abstract, and matching sentences for each retrieved publication.

3. Immediately above the search results there are two display options for controlling the display of the search results.
4. The "matching sentences" display option controls the number of sentences displayed flanking each keyword or category match.
5. If you want to scan through the search results without examining the matching sentences for each publication select the "none" option under matching sentences. This allows you to view the search results without viewing the matching sentences.
6. If your search returns a large number of results, it is helpful to select the 10, 20, or 50 entries per page display option.
7. Textpresso displays a great deal of text in the search results and some users find it helpful to increase the size of the text. This can be done using the "zoom" control of your Web browser.

3.4 Retrieving Publications Using a Query that Includes Synonymous Gene Symbols for a Gene of Interest

1. Any given gene can have several alternative gene symbols. Synonymous gene symbols often are different from each other and are unrelated to the standardized gene symbol. For example the mouse transcription factor Egr1 is also known as A530045N19Rik, Egr-1, ETR103, Krox-1, Krox-24, Krox24, NGF1-A, NGFIA, TIS8, Zenk, Zfp-6, and Zif268. Currently, Textpresso SSR does not translate a gene symbol entered as a keyword into a query containing of all known synonymous gene symbols. However, a user can easily craft a keyword string of gene names.
2. One good source for gene symbol synonyms is the Mouse Genome Informatics (MGI) Web site http://www.informatics.jax.org/. On the MGI home page click on the "Genes" link and then click on the "Genes and Markers Query" link. On the next page enter the gene name in the appropriate text box and go to the page for that gene. The synonymous symbols for your gene of interest are listed on the Gene Detail page underneath the standardized gene name.
3. In your Web browser copy the gene symbol synonyms and paste them directly into the Textpresso keyword text box. If you need to edit a large number of symbols copy them into a text file or word processing file. Modify them as described below and then copy and paste them into the keyword text box.
4. Retain the commas between the gene names but remove the spaces so that Textpresso will insert the Boolean operator OR between the keywords when it processes the query. Perform the search using the keyword string.

3.5 Parsing the Search Results

1. Textpresso presents a great deal of information in the search results. You have a couple of options for searching within the results.

2. A very straightforward option is to use the search or find function of your Web browser to perform a search of the search results. If you use a very broad keyword to retrieve a large number of publications this can allow you to identify publications within the search results that contain specific terms or phrases. This is a simple way to quickly search through results for specific items.
3. Alternatively, you can narrow the search results obtained with a keyword and/or category search by filtering the search results with the filter function that is built into the Textpresso system. The textbox for the filter can be found on the search results page immediately under the "Narrow your search results with filter" heading.
4. Save your search results as HTML or PDF files for future reference.

References

1. Brown SD, Wurst W, Kuhn R, Hancock JM (2009) The functional annotation of mammalian genomes: the challenge of phenotyping. Annu Rev Genet 43:305–333
2. Wang SZ, Liu BH, Tao HW, Xia K, Zhang LI (2009) A genetic strategy for stochastic gene activation with regulated sparseness (STARS). PLoS One 4:e4200
3. Dymecki SM, Kim JC (2007) Molecular neuroanatomy's "Three Gs": a primer. Neuron 54:17–34
4. Lewandoski M (2007) Analysis of mouse development with conditional mutagenesis. Handb Exp Pharmacol 178:235–262
5. Brault V, Besson V, Magnol L, Duchon A, Herault Y (2007) Cre/loxP-mediated chromosome engineering of the mouse genome. Handb Exp Pharmacol 178:29–48
6. Feil R (2007) Conditional somatic mutagenesis in the mouse using site-specific recombinases. Handb Exp Pharmacol 178:3–28
7. Brault V, Pereira P, Duchon A, Herault Y (2006) Modeling chromosomes in mouse to explore the function of genes, genomic disorders, and chromosomal organization. PLoS Genet 2:e86
8. Zong H, Espinosa JS, Su HH, Muzumdar MD, Luo L (2005) Mosaic analysis with double markers in mice. Cell 121:479–492
9. Branda CS, Dymecki SM (2004) Talking about a revolution: the impact of site-specific recombinases on genetic analyses in mice. Dev Cell 6:7–28
10. Dymecki SM, Rodriguez CI, Awatramani RB (2002) Switching on lineage tracers using site-specific recombination. Methods Mol Biol 185:309–334
11. Lewandoski M (2001) Conditional control of gene expression in the mouse. Nat Rev Genet 2:743–755
12. Nagy A (2000) Cre recombinase: the universal reagent for genome tailoring. Genesis 26:99–109
13. Grieshammer U, Lewandoski M, Prevette D, Oppenheim RW, Martin GR (1998) Muscle-specific cell ablation conditional upon Cre-mediated DNA recombination in transgenic mice leads to massive spinal and cranial motoneuron loss. Dev Biol 197:234–247
14. Brocard J, Warot X, Wendling O, Messaddeq N, Vonesch JL, Chambon P, Metzger D (1997) Spatio-temporally controlled site-specific somatic mutagenesis in the mouse. Proc Natl Acad Sci U S A 94:14559–14563
15. Feil R, Brocard J, Mascrez B, LeMeur M, Metzger D, Chambon P (1996) Ligand-activated site-specific recombination in mice. Proc Natl Acad Sci U S A 93:10887–10890
16. Ramirez-Solis R, Liu P, Bradley A (1995) Chromosome engineering in mice. Nature 378:720–724
17. Muller HM, Kenny EE, Sternberg PW (2004) Textpresso: an ontology-based information retrieval and extraction system for biological literature. PLoS Biol 2:e309
18. Urbanski WM, Condie BG (2009) Textpresso site-specific recombinases: a text-mining

server for the recombinase literature including Cre mice and conditional alleles. Genesis 47:842–846

19. Madisen L, Zwingman TA, Sunkin SM, Oh SW, Zariwala HA, Gu H, Ng LL, Palmiter RD, Hawrylycz MJ, Jones AR, Lein ES, Zeng H (2009) A robust and high-throughput Cre reporting and characterization system for the whole mouse brain. Nat Neurosci 13:133–140
20. Bult CJ, Kadin JA, Richardson JE, Blake JA, Eppig JT (2010) The mouse genome database: enhancements and updates. Nucleic Acids Res 38:D586–D592
21. Nagy A, Mar L, Watts G (2009) Creation and use of a cre recombinase transgenic database. Methods Mol Biol 530:365–378
22. Gong S, Doughty M, Harbaugh CR, Cummins A, Hatten ME, Heintz N, Gerfen CR (2007) Targeting Cre recombinase to specific neuron populations with bacterial artificial chromosome constructs. J Neurosci 27:9817–9823
23. Rodriguez-Esteban R (2009) Biomedical text mining and its applications. PLoS Comput Biol 5:e1000597
24. Rzhetsky A, Seringhaus M, Gerstein M (2008) Seeking a new biology through text mining. Cell 134:9–13
25. Cohen KB, Hunter L (2008) Getting started in text mining. PLoS Comput Biol 4:e20
26. Ananiadou S, Kell DB, Tsujii J (2006) Text mining and its potential applications in systems biology. Trends Biotechnol 24:571–579
27. Hunter L, Cohen KB (2006) Biomedical language processing: what's beyond PubMed? Mol Cell 21:589–594
28. Spasic I, Ananiadou S, McNaught J, Kumar A (2005) Text mining and ontologies in biomedicine: making sense of raw text. Brief Bioinform 6:239–251
29. Hull D, Pettifer SR, Kell DB (2008) Defrosting the digital library: bibliographic tools for the next generation web. PLoS Comput Biol 4:e1000204

Chapter 24

Live Imaging Mouse Embryonic Development: Seeing Is Believing and Revealing

Sonja Nowotschin and Anna-Katerina Hadjantonakis

Abstract

The use of genetically encoded fluorescent proteins has revolutionized the fields of cell and developmental biology and redefined our understanding of the dynamic morphogenetic processes that work to shape the embryo. Fluorescent proteins are routinely used as vital reporters to label tissues, cells, cellular organelles, or proteins of interest and in doing so provide contrasting agents enabling the acquisition of high-resolution quantitative image data. With the advent of more accessible and sophisticated imaging technologies and abundance of fluorescent proteins with different spectral characteristics, the dynamic processes taking place in situ in living embryos can now be probed. Here, we provide an overview of some recent advances in this rapidly evolving field.

Key words Mouse embryo, Transgenic, Gene targeted, Reporter, Genetically encoded fluorescent protein, Green fluorescent protein, GFP, Red fluorescent protein, RFP, Live imaging, 3D, Time-lapse

1 Introduction

A goal of developmental biology is to formulate an understanding of the carefully orchestrated stereotypical cell behaviors, as well as the molecular and physical mechanisms that underlie the complex morphogenetic processes that shape a multicellular organism. Even though genetic manipulations and the analysis of mutants have identified many genes important for regulating key developmental events in the mouse embryo, we are still a long way from understanding how many key morphogenetic events are regulated.

The details of critical events in the development of the mammalian embryo, such as, early lineage commitment and segregation at preimplantation, the establishment of the germ layers and elaboration of the axes at early postimplantation, and the subsequent morphogenesis of individual organ systems remain obscure. Intrinsic cell behavior is not only determined by gene expression but also by intercellular interactions and physical forces generated

Mark Lewandoski (ed.), *Mouse Molecular Embryology*, Methods in Molecular Biology, vol. 1092,
DOI 10.1007/978-1-60327-292-6_24, © Springer Science+Business Media New York 2014

between cells, as well as between cells and their substrata. It is therefore critical to move analyses of sequentially staged dead embryos forward and place them into a dynamic context so as to reveal the cell behaviors underlying normal embryonic development. Only with such information at hand can we begin to appreciate how mutations can result in an aberrant course of events.

Microscopy is a central tool in developmental biology, both for determining the normal course of developmental events and for investigating the consequences of experimental perturbations. Traditionally (dead) embryos have been visualized at single time points either in whole mount, to reveal gross morphological features, or after sectioning and histological staining, to provide cellular resolution. Dynamics and the sequence of events are usually inferred by analyzing multiple sequentially staged embryos. The ability to observe a single specimen continuously and at cellular resolution holds great promise and represents the evolution of classical histology into digitized multidimensional information documenting actual developmental progression. Recent advances in microscopy techniques for data acquisition as well as in computational methods for data processing in combination with the ever-increasing repertoire of reporter strains expressing spectrally distinct genetically encoded fluorescent proteins (FPs) provide powerful tools for the generation of 3D time-lapse (i.e., 4D) information. Moreover such data promises to provide insight not only to the field of developmental biology but also beyond, by furthering our understanding of homeostasis and disease progression.

2 Live Imaging Mouse Embryonic Development

A fertile period of methodological innovation has spearheaded the rapid evolution of the live imaging field. This rapid progress is a reflection of technical improvements and the increased affordability of digital microscopy techniques in parallel with the advent of genetically encoded fluorescent proteins with different spectral characteristics. By exploiting on-stage whole embryo or explant cultures and utilizing appropriate environmental climate control, short periods of normal development can be fostered ex utero [1–4]. While various imaging modalities can be used to study the tissue architecture and dynamic processes taking place in biological samples at high resolution, there is no single approach that is ideal for all applications, and a discussion of the various modalities is beyond the scope of this review. Certain applications, for example, determining gross tissue organization, are better served by imaging of large samples (e.g., whole embryos) at lower resolution, while others require investigations at the subcellular level. As a consequence while imaging modalities such as optical projection tomography (OPT), optical coherence tomography (OCT), and magnetic resonance imaging (MRI) provide insight into embryogenesis and the global organization of structures, they lack the cellular or subcellular resolution of laser-scanning microscopy (LSM) [5].

LSM, including confocal point-scanning, multi-point and slit scanning modalities in addition to multi-photon excitation, has found widespread use for many live high-resolution imaging applications. Data is usually acquired as *x–y* images in the *z* dimension (3D) and in 3D over time (3D time-lapse or 4D). Thus fluorescently tagged objects can be analyzed quantitatively in situ using 3D time-lapse imaging with a high spatial and temporal resolution. LSM can provide information at single-cell resolution by revealing information on cell position, morphology, division, and death in situ. The main limitations of LSM are limited imaging depth in the tissue being sampled, and cell death resulting from phototoxicity. New developments in light sheet microscopy have partially addressed some of these issues, but have yet to be applied in the mouse [6, 7].

Once image data is collected computational methods are used to quantify and segment data. This generates high-resolution information on, for example, cellular organelles, which can be used as descriptors of cell position (e.g., nuclei), and cell morphology (e.g., plasma membrane labels). A detailed discussion of the software available is beyond the scope of this introduction; however, a short list is provided for the interested reader to follow up on. Commercially available, general purpose image analysis software packages include Amira from Mercury Computer Systems (http://3dviz.mc.com/), Imaris from Bitplane (http://www.bitplane.com), MetaMorph from Molecular devices (http://www.moleculardevices.com/pages/software/metamorph.htm), and Volocity from Perkin-Elmer (http://www.cellularimaging.com/products/volocity/). While open source academic packages include the general purpose ImageJ which is based on the NIH Image (http://rsb.info.nih.gov/ij/ rashband et al. 2006), as well as software for image segmentation methods used for the identification and tracking of individual cells which include 3D-DIAS [8], STARRYNIGHT [9, 10], and cell division analysis [11].

Irrespective of imaging modality used, observation of tissues or individual cells of interest and their functional properties is not often possible without molecular tagging. Typical requirements for tagging agents include developmental neutrality, fluorescence intensity and stability, and the possibility of multiplexing. Numerous agents for molecular labeling are available. They range from fluorescent organic dyes, to quantum dots and molecular beacons to genetically encoded reporters [12–14]. Genetically encoded reporters comprise two main categories: chromogenic enzyme-based reporters (e.g., beta-galactosidase and alkaline phosphatase) and vital autofluorescent proteins (e.g., the Green Fluorescent Protein, GFP). In this chapter we focus on the latter as it is currently the most prevalent reporter system used in the mouse.

Strains of mice expressing genetically encoded reporters are routinely generated by gene targeting or gene trapping in embryonic stem (ES) cells and germline transmission through chimeras or by transgenesis (in ES cells or by pronuclear injection of DNA into zygotes) through modification of large insert genomic DNA containing vectors (e.g., Bacterial Artificial Chromosomes—BACs) or by using defined *cis*-regulatory elements. A key advantage of using autofluorescent proteins (FPs) for cell labeling is the lack of exogenous compounds necessary for reporter visualization and the attractive possibility to image noninvasively. However, a disadvantage is that since there is no enzyme-based amplification, what you see is what you get, and therefore low levels of reporter expression may not be readily detectable, and cannot easily be improved upon.

FPs can either be used in their native form or they can be incorporated into protein fusions to follow protein dynamics or interactions or to label specific subcellular compartments or organelles, such as the nucleus, plasma membrane, cytoskeleton, or mitochondria. The increasing number of FPs provides an assortment of probes with increased fluorescence intensities to mix and match for multiplexing and simultaneous detection. As a consequence, such variety can also represent a confusing and sometimes overwhelming choice. Importantly, choice is often best dictated by application. Since no review can exhaustively cover the panoply of available genetically encoded fluorescent proteins, we discuss some of the more commonly used ones and present some of the rationale behind their use.

Given how quickly the field of FPs is evolving, the information presented here will likely be obsolete as soon as it sees the light of publication. With this in mind, the discussion provided is intended for use as a guide, rather than a definitive collection.

3 Reporters for Live Imaging: A Color-Palette of Genetically Encoded Fluorescent Proteins

3.1 Green Fluorescent Protein (GFP) and Its Variants

The cloning of the green fluorescent protein, wtGFP from *Aequorea victoria* [15] the first genetically encoded fluorescent protein, spearheaded a revolution in the use of fluorescent markers as autofluorescent protein tags to study cell function, morphology, and protein–protein interactions. GFP protein and its spectral variants are the favored choice in many applications and are used in many different organisms. GFP and its variants possess a unique structure, which consists of 11 beta sheets and an internal alpha helix. The fluorophore, which is located in the center of the barrel, is formed by three amino acid residues of the alpha helix which form a cyclic tripeptide, Ser65-Tyr66-Gly67, during a process of maturation [16, 17].

Its stability over a wide range of pH and temperatures, as well as the lack of a need for a cofactor to fluoresce, made wtGFP attractive for use in molecular and biochemical applications [18, 19]. Improved versions of GFP with increased levels of fluorescence and photostability, and include enhanced GFP (EGFP) which contains a point mutation that leads to a S65T substitution in the cyclic tripeptide [20], Emerald GFP (EmGFP) [21], Superfolder GFP [22], TagGFP2 [23], and new green FPs such as mWasabi [24], Azami Green (AG) [25] have been discovered or developed through mutagenesis. EGFP was one of the first improved GFPs and has so far been the green FP of choice for most applications in mice. It remains to be seen if any of the newly identified green FPs will perform comparatively to EGFP in mice.

Spectral GFP variants include the blue fluorescent proteins (BFP), cyan fluorescent proteins (CFP), and yellow fluorescent proteins (YFP). These GFP-based variants exhibit different spectral characteristics and were obtain through site-directed mutagenesis of GFP [26]. To date improved versions of all color variants have become available exhibiting improved quantum yields, higher extinction coefficients, and brighter fluorescence than their predecessors. Of the YFPs, Venus [27], Citrine [28] and TagYFP (Evrogen) are less acid sensitive and brighter than YFP, and so are better suited for labeling proteins in secretory organelles. Cerulean [29], the successor of CFP, as well as TagCFP (Evrogen) are characterized through their increased brightness and faster maturation, and so are preferable for use in FRET experiments together with YFPs. The availability of spectrally distinct FPs allows simultaneous visualization of two or more FPs expressed in different tissues, or even in the same cell, and can be used for quantitative dual imaging applications such a fluorescence resonance energy transfer (FRET) and FRET-based sensors [27, 29–31].

The advent of spectrally distinct FPs has afforded multiplexing for simultaneous detection in different tissues while under the control of different *cis*-regulatory elements or when targeted to different locations within the same cell (e.g., labeling nucleus vs. plasma membrane) when placed under the control of the same *cis-regulatory* elements. However, it should be noted that, when using newly isolated fluorescent proteins, new variants of established fluorescent proteins or newly generated fluorescent protein fusions, their biological compatibility, particularly their developmental neutrality must first be carefully evaluated. This is usually achieved by generating strains of mice exhibiting widespread expression of an FP, or FP fusion, of interest. For example, even though DsRed and its early variants were extensively used in cells, only the monomer mRFP1 was proven to be developmentally neutral when widely expressed in mice [32]. Consideration of existing fluorescent reporter strains, which are mainly green, should also be taken into account when generating new strains that might be used in combination [33].

Of great promise are genetically encoded fluorescent protein reporters that can be used as sensors to monitor the activity of proteins or provide a readout and so act as sensors of a functional change in proteins acting within an intracellular network. These reporters have several advantages over widely used synthetic indicators. They are noninvasive and allow monitoring of a specific subset of cells and specific subcellular compartments, respectively. The probes are constantly produced hence permanently label the cell of interest, which makes them suitable for long-term imaging in vivo. Pioneering work has been achieved regarding using indicators for calcium (cameleons, camgaroos, pericams, and G-CaMP) in invertebrates [34–36] as well as in zebrafish [37, 38]. In zebrafish a G-CaMP transgene, expressed under a cardiac specific promoter, was used to live image (optical mapping) the development of the cardiac conduction system. The same transgenic line was also used to identify mutants of an ENU screen by optical methods [38]. Developing similar transgenes in mice would be desirable. Indeed, use of the improved genetically encoded calcium indicator, G-CaMP2, has been reported in mice. When expressed in pyramidal neurons action potentials lead to changes in fluorescent intensity facilitating the detection of bursts of high-frequency action potentials and synaptic currents in vivo [39].

3.2 Orange and Red Fluorescent Proteins

Bright orange and red fluorescent proteins are much sought after for live cell or whole animal imaging. These red-emitting FPs, due to their long wavelength emission, are less toxic for the cell and possess a greater depth penetration. Approaches to mutagenize GFP to obtain FPs with farther red-shifted spectra than YFP have been unsuccessful. To date only few FPs emitting in the orange and red wavelengths (560–650 nm) have been isolated or engineered, with most far-red FPs having been isolated from different sea anemone and coral species.

DsRed from the sea anemone *Discosoma sp.* was the first RFP that was discovered [40]. Since then many variants have been engineered to obtain brighter more photostable and monomeric forms. One of them, mRFP1, was the first true monomeric variant of DsRed [41]. mRFP1 has been successfully used in mice [32], as has as one of its brighter successors, mCherry [42]. Widespread expression of the native form of mRFP1 or mCherry does not affect cell morphology, developmental potential, or animal viability. A tandem dimer variant of mRFP, (td)RFP has been used to generate a Cre recombinase reporter strain of mice for lineage tracing studies [43]. It should be noted that attempts at widespread expression of several mRFP1 fusions in mice, including a myristoylated-mRFP1 and a human histone H2B-mRFP1 fusion that labels active chromatin, have not been successful, suggesting that some mRFP1 fusion reporters might not be neutral (our unpublished observations).

Additional orange and red FPs have been reported, but to date these not been used to generate mouse strains. Kusabira Orange, a tetrameric protein was isolated from the coral *Fungia concinna*. The monomeric versions, mKO [44] and its enhanced version mKO2 [45], have been generated through site-directed mutagenesis. Their brightness is similar to that of EGFP, and they exhibit high photostability making them particularly suitable for live imaging applications [44]. Variants of mKO are commercially available from MBL International (http://www.mblintl.com/). Another variant resulting from the site-directed mutagensis of DsRed was mOrange, its successor mOrange2, and the tandem dimer (td) tomato. Compared to mOrange, (td)tomato has greater photostability and brightness [46]. Other orange–red FPs that are commonly available include TagRFP [47], its enhanced version TagRFP-T [48] and variants including the dimer TurboRFP [47], as well as the monomeric form of TagRFP exhibits high brightness and photostability. In contrast to TurboRFP, the monomeric TagRFP is well tolerated in a variety of fusion proteins [47]. TagRFP and TurboRFP (as well their cyan, green, and yellow couterparts) are commercially available from Evrogen (http://www.evrogen.com/).

Such observations underscore the need for alternative red FPs or variants of existing ones. A series of spectral variants, named after fruits, have been engineered through a large-scale mutagenesis of mRFP1. These variants exhibit excitation/emission maxima between 537 and 610 nm and range from a green FP (mHoneydew) to several reds (mCherry, mStrawberry) through to a far-red (mPlum). For live imaging the tandem dimer tdTomato, and the monomers mCherry and mStrawberry are considered as the most applicable red FPs of this series [46]. mRuby, a recently engineered monomeric RFP, exhibits increased brightness as compared mStrawberry and mCherry, but it is not as photostable. However, its stability at extreme pH values makes it ideal for labeling subcellular organelles for live imaging [49]. Given its photostability, mCherry is currently one of the better choices for time-lapse imaging applications in the mouse. mCherry has also been shown to be incorporated in protein fusions that are correctly targeted, labeling the plasma membrane and the nucleus without affecting the developmental potential [50, 51]. But, as with mRFP1, constitutive widespread expression of an H2B-mCherry fusion is always not compatible with normal embryonic development [52]. However this does not necessarily preclude the use of H2B-mCherry fusions from use a conditional or lineage-specific reporters.

3.3 Far-Red Fluorescent Proteins

Given their physical properties, very bright far-red FPs will be difficult to generate, however, FPs with bright near-infrared spectra are eagerly anticipated as they should provide greater tissue penetration, reduced autofluorescence and less phototoxicity. The fare-red FPs, mPLum and mRasberry, were the first genetically engineered monomeric proteins generated through iterative somatic hypermutation [53]. However, their brightness is only small fraction of that of GFP. Over the recent years additional far-red proteins have become available and include the dimeric Katushka (available from Evrogen as TurboFP635), its monomeric version mKate (available from Evrogen as TagFP635) [54], and its successor mKate2 [55], as well as the unrelated mNeptune [56]. The monomer mKate2 is currently considered to be a preferable far-red FPS for most applications due to its considerable brightness and high photostability making it ideal for cell labeling or protein tagging in the far-red spectrum [55].

A recent report has introduced two new near-infrared dimeric FPs, eqFP650, and eqFP670, where the former has been reported as the brightest far-red FP so far, and the latter is characterized as exhibiting the most red-shifted emission spectrum and high photostability, thereby making them attractive for whole body imaging experiments [57].

3.4 Photomodulatable Fluorescent Proteins

Photomodulatable proteins represent novel tools for cell lineage tracing and fate mapping cells in vivo. Up until now pulse-chase labeling or fate mapping cells are predominantly performed using invasive techniques like the injection of dyes, grafting tissues, or genetically using the binary Cre/*lox*P system. Photomodulatable FPs facilitate the labeling and tracking of single or cohorts of cells noninvasively. In addition, fused to a protein of interest they provide information on localization, turnover and tracking of a protein in specific cell types.

Photomodulatable FPs change their spectral characteristics when exposed to short wavelength illumination. This class of fluorescent proteins comprises two main groups: (1) photoactivatable proteins that change their state from nonfluorescent to a fluorescent upon irradiation with short wavelength light (referred to as photoactivation) and (2) photoconvertible proteins that can be converted from one fluorescent state (i.e., color) to another when irradiated with short wavelength light (referred to as photoconversion). With the growing numbers of proteins that belong to each particular group one has powerful tools at hand to noninvasively label cohorts of cells, single cells, subcellular organelles, or individual proteins in vivo and track them over time. The choice of most appropriate PMFP for the planned experiment depends on the specific characteristics of the PMFP, as well as on the imaging modalities available. The structure of the photomodulatable protein must also be considered for its application. Monomeric ones can be used in fusion proteins to label intracellular molecules. Whereas tetrameric photomodulatable proteins by their nature are not ideal for use in fusion proteins since they could potentially disrupt the function and localization of the protein [58]. Below we briefly review some of the most common PMFPs.

3.5 Photoactivatable Proteins

PA-GFP, a variant of GFP, is a photoactivatable protein that changes irreversibly from a weak green fluorescent state (excitation peak at 400 nm, emission peak at 515 nm) to a bright green fluorescent state (excitation peak at 504, emission peak at 517) with a 100-fold increase in green fluorescence upon irradiation with short wavelength light [59]. Red variants, including PA-RFP, of irreversible photoactivable proteins that change from a nonfluorescent to a red fluorescent state were obtained from the red FP DsRed through mutagenesis; however first generation variants were dim [60]. Their successors PA-mCherry1, -2, and -3 are more promising for broader spectrum of applications including use in the mouse due to their enhanced brightness, photostability, and faster photoactivation [61].

3.6 Reversible Photoactivatable Fluorescent Proteins

This group encompasses the proteins KFP1 [62] and Dronpa [63, 64]. KFP1, kindling fluorescent protein, is derived from the chromoprotein asFP595 of the Anthozoa *Anemonia sulcata* [65]. It converts from a nonfluorescent to a green fluorescent state upon exposure to green light. Depending on the intensity and duration of the activating light this photoactivation can be reversible or irreversible [62, 66]. Dronpa from the coral *Pectiniidae sp.* can be activated from a green to a nonfluorescent state upon activation with blue light [63, 64]. Such reversible proteins have been used in a number of applications for protein tracking [67, 68] and super-resolution imaging [69].

3.7 Photoconvertible Proteins

Another variant of GFP, PS-CFP, fluoresces cyan in its neutral state and changes, upon activation with short wavelength light, irreversibly to a green fluorescing state. The recently engineered PS-CFP2 succeeds PS-CFP due to its increase in brightness [70, 71] and has been proven useful for live cell tracking for several hours up to days in the chick due to its photostability [72].

A large group of photoconvertible proteins comprise FPs that change from a green to a red fluorescent state upon irradiation with short wave length light. Among them are Dendra and its successor Dendra2 [70], Kaede [73, 74], the tandem dimer EosFP and its monomeric variant mEos2 [75] as well as KikGR [76] and its monomeric variant mKikGR [77].

In the mouse, Dendra has been used in an actin fusion to investigate the role of actin in axonal transport in axonogenesis in hippocampal neurons [78]. Dendra2, along with mEos2, finds widespread use in super-resolution microscopy [79, 80], as well as in protein tracking studies [81].

Kaede and KikGR have been shown to be attractive photomodulatable markers for efficiently labeling and tracking single cells or subpopulations of cells in the chick embryo [72, 82], as well as in the mouse embryo [83–85]. Transgenic mouse strains expressing KikGR in a widespread fashion in combination with 3D time-lapse imaging have been shown to be useful tools for cell tracking in various morphogenetic processes [84]. Such strains have been used to demonstrate that the specification of the embryonic–abembryonic axis in the mouse blastocyst is independent of early cell lineage [85]. Since both, Kaede and KikGR, form tetramers, neutral protein fusions comprising these photoactivatable proteins have not been reported in the chick or in mice [58], with the exception of a recent study using electroporation of a Kaede-centrin1 fusion to elegantly visualize mother vs. daughter centrosomes in mammalian neocortical neural progenitors [86]. Even so, the recently developed monomer of KikGR, mKikGR [77], represents a promising new variant that should be tested in protein fusions, and if successful will likely find a frequent application.

A unique recently reported FP is IrisFP. IrisFP is the only FP that can undergo irreversible photoconversion from a green to a red fluorescent state, as well as a reversible photoswitching from a fluorescent to a nonfluorescent form in both the green and the red fluorescent states [87]. The monomeric variant, mIrisFP has already been successfully applied in pulse-chase experiments in combination dual color photoactivation localization microscopy (PALM) [88].

4 Visualizing Mouse Embryonic Development

Live imaging cell behaviors together with mouse genetics represent a powerful approach to gain a deeper understanding of the mechanisms that play a role embryonic development and disease. Engineered mouse strains that express genetically encoded FPs are an essential step in towards fulfilling this goal.

Widespread expressed native fluorescent proteins often do not provide the desired cellular resolution in imaging. Therefore, cells are marked with a tag so one can track and identify individual cells in 3D space. The most prominent of such applications are the incorporation of FPs into protein fusions that serve as vital labels for subcellular compartments [89]. The most commonly used protein fusions are the histone fusions that mark the cell nucleus, signal sequence tagged fusions for the plasma membrane of the cell, cytoskeletal fusion, i.e., actin or tubulin FP fusion proteins and mitochondrial fusions. Besides its low autofluorescence the nucleus is an ideal organelle to be labeled with a fluorescent tag since it is optimal for tracking cells. Histone fusion proteins are bound to chromatin and stay bound during cell divisions. This makes them suitable to visualize dividing cells over time [90–92]. Cells in mitosis can be easily distinguished from cells in interphase. Individual cells can be tracked and identified in a subpopulation of cells. In addition, cells that undergo apoptosis can be visualized. To gain information on cell morphology, lipid modified fusions act as tags for the plasma membrane and secretory pathway. These tags can be fused to different spectrally distinct FPs used to gain information on both, cell morphology and cell position, simultaneously [90, 91, 93–95]. Moreover, a recently published global double Cre reporter expressing membrane tagged (td)Tomato prior to Cre-mediated excision and membrane-tagged GFP after excision is a useful reporter to trace cell lineages and gain information on cell morphology in the context of recombined and nonrecombined cells [95].

Increasingly, sophisticated approaches exploiting the copious FPs are being used to label and trace cells in mice. The combinatorial expression of FPs to label individual neurons and trace their origin in the mouse brain was demonstrated using a Cre/*lox*P strategy, called Brainbow. Brainbow results in the Cre-mediated stochastic expression multiple FPs from a single transgene and simultaneously labels neurons in as many as 90 different colors [96]. A different strategy employing CFP, GFP, YFP, and RFP was chosen to analyze the contributions of clonal progenitors to yolk sac blood islands, the initial sites of hematopoetic and endothelial cells, to reveal that these cell lineages do not arise from a single clonal precursor [97].

Such reporter fusions have already been successfully used in other diverse organisms including nematode worms, fruit flies, zebrafish, and chick embryos [94, 98–101]. Several transgenic and gene-targeted mouse strains expressing histone fusions and various membrane fusions in a variety of colors have been made and are used to perform high-resolution imaging of cell function and behavior. Besides widespread expression, these fusions are also used in a tissue-specific context when they are expressed under defined *cis*-regulatory elements to provide further understanding of specific gene activity or protein function, for example, the H2B-GFP knock-in into the *Pdgfrα* locus, which recapitulates PDGFRα expression and can be used as a marker for PDGFRα expression as well as the tissues in which is expressed *Pdgfrα* is in refs. [102, 103]. The visualization of live cell behaviors in these models will pave the way for an in-depth knowledge of the mechanisms that control embryonic development, homeostasis, and disease progression.

5 Imag(in)ing the Future

Imaging is unique in its ability to capture quantitative data at single cell resolution in living embryos. Improvements in both optical imaging modalities and the characteristics of fluorescent proteins are spearheading the increased popularity of live imaging approaches to investigate mouse embryonic development. Novel reporter strains are continually being developed for diverse live imaging applications. It is likely that the future holds more color variants, brighter, faster maturing, and more photostable proteins. Such FPs will be better suited, for longer time-lapse imaging of mouse embryos or explants in culture. Photomodulatable FPs represent a new class of genetically encoded reporter and further engineering will produce brighter and monomeric versions, more suitable for protein fusions and less toxic for cells. In addition, photoconversion at wavelengths that are less toxic and shifting emission wavelengths to far-red or near infrared regions are desirable and likely will be achieved. One can predict with certainty, is that whatever the future holds, live imaging using genetically encoded fluorescent reporters will be center stage.

Acknowledgments

We apologize to the many authors whose work we were unable to cite due to space constraints. Work in our lab is supported by the NIH (RO1-HD052115 and RO1-DK084391), HFSP, and NYSTEM and the Muscular Dystrophy Association (186552 to S.N.).

References

1. Jones EA, Baron MH, Fraser SE, Dickinson ME (2005) Dynamic in vivo imaging of mammalian hematovascular development using whole embryo culture. Methods Mol Med 105:381–394
2. Jones EAV, Hadjantonakis A-K, Dickinson ME (2005) Imaging mouse embryonic development. In: Yuste R (ed) Imaging in neuroscience and development: a laboratory manual. Cold Spring Harbor Laboratory, New York
3. Nowotschin S, Ferrer-Vaquer A, Hadjantonakis AK (2010) Imaging mouse development with confocal time-lapse microscopy. Methods Enzymol 476:351–377
4. Udan RS, Dickinson ME (2010) Imaging mouse embryonic development. Methods Enzymol 476:329–349
5. Dickinson ME (2006) Multimodal imaging of mouse development: tools for the postgenomic era. Dev Dyn 235:2386–2400
6. Keller PJ, Schmidt AD, Santella A, Khairy K, Bao Z, Wittbrodt J, Stelzer EH (2010) Fast, high-contrast imaging of animal development with scanned light sheet-based structured-illumination microscopy. Nat Methods 7:637–642
7. Keller PJ, Schmidt AD, Wittbrodt J, Stelzer EH (2008) Reconstruction of zebrafish early embryonic development by scanned light sheet microscopy. Science 322:1065–1069
8. Wessels D, Kuhl S, Soll DR (2006) Application of 2D and 3D DIAS to motion analysis of live cells in transmission and confocal microscopy imaging. Methods Mol Biol 346:261–279
9. Bao Z, Murray JI, Boyle T, Ooi SL, Sandel MJ, Waterston RH (2006) Automated cell lineage tracing in *Caenorhabditis elegans*. Proc Natl Acad Sci U S A 103:2707–2712
10. Santella A, Du Z, Nowotschin S, Hadjantonakis AK, Bao Z (2010) A hybrid blob-slice model for accurate and efficient detection of fluorescence labeled nuclei in 3D. BMC Bioinformatics 11:580
11. Quesada-Hernandez E, Caneparo L, Schneider S, Winkler S, Liebling M, Fraser SE, Heisenberg CP (2010) Stereotypical cell division orientation controls neural rod midline formation in zebrafish. Curr Biol 20: 1966–1972
12. Bhattacharyya S, Kulesa PM, Fraser SE (2008) Vital labeling of embryonic cells using fluorescent dyes and proteins. Methods Cell Biol 87:187–210
13. Zhang J, Campbell RE, Ting AY, Tsien RY (2002) Creating new fluorescent probes for cell biology. Nat Rev Mol Cell Biol 3:906–918

14. Slotkin JR, Chakrabarti L, Dai HN, Carney RS, Hirata T, Bregman BS, Gallicano GI, Corbin JG, Haydar TF (2007) In vivo quantum dot labeling of mammalian stem and progenitor cells. Dev Dyn 236:3393–3401
15. Prasher DC, Eckenrode VK, Ward WW, Prendergast FG, Cormier MJ (1992) Primary structure of the *Aequorea victoria* green-fluorescent protein. Gene 111:229–233
16. Ormo M, Cubitt AB, Kallio K, Gross LA, Tsien RY, Remington SJ (1996) Crystal structure of the *Aequorea victoria* green fluorescent protein. Science 273:1392–1395
17. Yang F, Moss LG, Phillips GN Jr (1996) The molecular structure of green fluorescent protein. Nat Biotechnol 14:1246–1251
18. Bokman SH, Ward WW (1981) Renaturation of Aequorea gree-fluorescent protein. Biochem Biophys Res Commun 101: 1372–1380
19. Ward WW, Bokman SH (1982) Reversible denaturation of Aequorea green-fluorescent protein: physical separation and characterization of the renatured protein. Biochemistry 21:4535–4540
20. Heim R, Cubitt AB, Tsien RY (1995) Improved green fluorescence. Nature 373: 663–664
21. Cubitt AB, Heim R, Adams SR, Boyd AE, Gross LA, Tsien RY (1995) Understanding, improving and using green fluorescent proteins. Trends Biochem Sci 20:448–455
22. Pedelacq JD, Cabantous S, Tran T, Terwilliger TC, Waldo GS (2006) Engineering and characterization of a superfolder green fluorescent protein. Nat Biotechnol 24:79–88
23. Subach OM, Gundorov IS, Yoshimura M, Subach FV, Zhang J, Gruenwald D, Souslova EA, Chudakov DM, Verkhusha VV (2008) Conversion of red fluorescent protein into a bright blue probe. Chem Biol 15:1116–1124
24. Ai HW, Olenych SG, Wong P, Davidson MW, Campbell RE (2008) Hue-shifted monomeric variants of Clavularia cyan fluorescent protein: identification of the molecular determinants of color and applications in fluorescence imaging. BMC Biol 6:13
25. Karasawa S, Araki T, Yamamoto-Hino M, Miyawaki A (2003) A green-emitting fluorescent protein from Galaxeidae coral and its monomeric version for use in fluorescent labeling. J Biol Chem 278:34167–34171
26. Tsien RY (1998) The green fluorescent protein. Annu Rev Biochem 67:509–544
27. Nagai T, Ibata K, Park ES, Kubota M, Mikoshiba K, Miyawaki A (2002) A variant of yellow fluorescent protein with fast and efficient maturation for cell-biological applications. Nat Biotechnol 20:87–90
28. Griesbeck O, Baird GS, Campbell RE, Zacharias DA, Tsien RY (2001) Reducing the environmental sensitivity of yellow fluorescent protein. Mechanism and applications. J Biol Chem 276:29188–29194
29. Rizzo MA, Springer GH, Granada B, Piston DW (2004) An improved cyan fluorescent protein variant useful for FRET. Nat Biotechnol 22:445–449
30. Heim R, Prasher DC, Tsien RY (1994) Wavelength mutations and posttranslational autoxidation of green fluorescent protein. Proc Natl Acad Sci U S A 91:12501–12504
31. Heim R, Tsien RY (1996) Engineering green fluorescent protein for improved brightness, longer wavelengths and fluorescence resonance energy transfer. Curr Biol 6:178–182
32. Long JZ, Lackan CS, Hadjantonakis AK (2005) Genetic and spectrally distinct in vivo imaging: embryonic stem cells and mice with widespread expression of a monomeric red fluorescent protein. BMC Biotechnol 5:20
33. Viotti M, Nowotschin S, Hadjantonakis A-K (2011) Afp::mCherry, a red fluorescent transgenic reporter of the mouse visceral endoderm. Genesis 49:124–133
34. Fiala A, Spall T, Diegelmann S, Eisermann B, Sachse S, Devaud JM, Buchner E, Galizia CG (2002) Genetically expressed cameleon in Drosophila melanogaster is used to visualize olfactory information in projection neurons. Curr Biol 12:1877–1884
35. Wang JW, Wong AM, Flores J, Vosshall LB, Axel R (2003) Two-photon calcium imaging reveals an odor-evoked map of activity in the fly brain. Cell 112:271–282
36. Yu D, Baird GS, Tsien RY, Davis RL (2003) Detection of calcium transients in Drosophila mushroom body neurons with camgaroo reporters. J Neurosci 23:64–72
37. Higashijima S, Masino MA, Mandel G, Fetcho JR (2003) Imaging neuronal activity during zebrafish behavior with a genetically encoded calcium indicator. J Neurophysiol 90:3986–3997
38. Chi NC, Shaw RM, Jungblut B, Huisken J, Ferrer T, Arnaout R, Scott I, Beis D, Xiao T, Baier H, Jan LY, Tristani-Firouzi M, Stainier DY (2008) Genetic and physiologic dissection of the vertebrate cardiac conduction system. PLoS Biol 6:e109
39. Mao T, O'Connor DH, Scheuss V, Nakai J, Svoboda K (2008) Characterization and subcellular targeting of GCaMP-type genetically-encoded calcium indicators. PLoS One 3:e1796

40. Matz MV, Fradkov AF, Labas YA, Savitsky AP, Zaraisky AG, Markelov ML, Lukyanov SA (1999) Fluorescent proteins from nonbioluminescent Anthozoa species. Nat Biotechnol 17:969–973
41. Campbell RE, Tour O, Palmer AE, Steinbach PA, Baird GS, Zacharias DA, Tsien RY (2002) A monomeric red fluorescent protein. Proc Natl Acad Sci U S A 99:7877–7882
42. Fink D, Wohrer S, Pfeffer M, Tombe T, Ong CJ, Sorensen PH (2010) Ubiquitous expression of the monomeric red fluorescent protein mcherry in transgenic mice. Genesis 48:723–729
43. Luche H, Weber O, Nageswara Rao T, Blum C, Fehling HJ (2007) Faithful activation of an extra-bright red fluorescent protein in "knock-in" Cre-reporter mice ideally suited for lineage tracing studies. Eur J Immunol 37:43–53
44. Karasawa S, Araki T, Nagai T, Mizuno H, Miyawaki A (2004) Cyan-emitting and orange-emitting fluorescent proteins as a donor/acceptor pair for fluorescence resonance energy transfer. Biochem J 381:307–312
45. Sakaue-Sawano A, Kurokawa H, Morimura T, Hanyu A, Hama H, Osawa H, Kashiwagi S, Fukami K, Miyata T, Miyoshi H, Imamura T, Ogawa M, Masai H, Miyawaki A (2008) Visualizing spatiotemporal dynamics of multicellular cell-cycle progression. Cell 132:487–498
46. Shaner NC, Campbell RE, Steinbach PA, Giepmans BN, Palmer AE, Tsien RY (2004) Improved monomeric red, orange and yellow fluorescent proteins derived from Discosoma sp. red fluorescent protein. Nat Biotechnol 22:1567–1572
47. Merzlyak EM, Goedhart J, Shcherbo D, Bulina ME, Shcheglov AS, Fradkov AF, Gaintzeva A, Lukyanov KA, Lukyanov S, Gadella TW, Chudakov DM (2007) Bright monomeric red fluorescent protein with an extended fluorescence lifetime. Nat Methods 4:555–557
48. Shaner NC, Lin MZ, McKeown MR, Steinbach PA, Hazelwood KL, Davidson MW, Tsien RY (2008) Improving the photostability of bright monomeric orange and red fluorescent proteins. Nat Methods 5:545–551
49. Kredel S, Oswald F, Nienhaus K, Deuschle K, Rocker C, Wolff M, Heilker R, Nienhaus GU, Wiedenmann J (2009) mRuby, a bright monomeric red fluorescent protein for labeling of subcellular structures. PLoS One 4:e4391
50. Egli D, Rosains J, Birkhoff G, Eggan K (2007) Developmental reprogramming after chromosome transfer into mitotic mouse zygotes. Nature 447:679–685
51. Provost E, Rhee J, Leach SD (2007) Viral 2A peptides allow expression of multiple proteins from a single ORF in transgenic zebrafish embryos. Genesis 45:625–629
52. Nowotschin S, Eakin GS, Hadjantonakis AK (2009) Dual transgene strategy for live visualization of chromatin and plasma membrane dynamics in murine embryonic stem cells and embryonic tissues. Genesis 47:330–336
53. Wang L, Jackson WC, Steinbach PA, Tsien RY (2004) Evolution of new nonantibody proteins via iterative somatic hypermutation. Proc Natl Acad Sci U S A 101:16745–16749
54. Shcherbo D, Merzlyak EM, Chepurnykh TV, Fradkov AF, Ermakova GV, Solovieva EA, Lukyanov KA, Bogdanova EA, Zaraisky AG, Lukyanov S, Chudakov DM (2007) Bright far-red fluorescent protein for whole-body imaging. Nat Methods 4:741–746
55. Shcherbo D, Murphy CS, Ermakova GV, Solovieva EA, Chepurnykh TV, Shcheglov AS, Verkhusha VV, Pletnev VZ, Hazelwood KL, Roche PM, Lukyanov S, Zaraisky AG, Davidson MW, Chudakov DM (2009) Far-red fluorescent tags for protein imaging in living tissues. Biochem J 418:567–574
56. Lin MZ, McKeown MR, Ng HL, Aguilera TA, Shaner NC, Campbell RE, Adams SR, Gross LA, Ma W, Alber T, Tsien RY (2009) Autofluorescent proteins with excitation in the optical window for intravital imaging in mammals. Chem Biol 16:1169–1179
57. Shcherbo D, Shemiakina II, Ryabova AV, Luker KE, Schmidt BT, Souslova EA, Gorodnicheva TV, Strukova L, Shidlovskiy KM, Britanova OV, Zaraisky AG, Lukyanov KA, Loschenov VB, Luker GD, Chudakov DM (2010) Near-infrared fluorescent proteins. Nat Methods 7:827–829
58. Nowotschin S, Hadjantonakis AK (2009) Photomodulatable fluorescent proteins for imaging cell dynamics and cell fate. Organogenesis 5:217–226
59. Patterson GH, Lippincott-Schwartz J (2002) A photoactivatable GFP for selective photolabeling of proteins and cells. Science 297: 1873–1877
60. Verkhusha VV, Sorkin A (2005) Conversion of the monomeric red fluorescent protein into a photoactivatable probe. Chem Biol 12:279–285
61. Subach FV, Patterson GH, Manley S, Gillette JM, Lippincott-Schwartz J, Verkhusha VV (2009) Photoactivatable mCherry for high-resolution two-color fluorescence microscopy. Nat Methods 6:153–159
62. Chudakov DM, Belousov VV, Zaraisky AG, Novoselov VV, Staroverov DB, Zorov DB, Lukyanov S, Lukyanov KA (2003) Kindling

fluorescent proteins for precise in vivo photolabeling. Nat Biotechnol 21:191–194

63. Habuchi S, Ando R, Dedecker P, Verheijen W, Mizuno H, Miyawaki A, Hofkens J (2005) Reversible single-molecule photoswitching in the GFP-like fluorescent protein Dronpa. Proc Natl Acad Sci U S A 102:9511–9516
64. Habuchi S, Dedecker P, Hotta J, Flors C, Ando R, Mizuno H, Miyawaki A, Hofkens J (2006) Photo-induced protonation/deprotonation in the GFP-like fluorescent protein Dronpa: mechanism responsible for the reversible photoswitching. Photochem Photobiol Sci 5:567–576
65. Lukyanov KA, Fradkov AF, Gurskaya NG, Matz MV, Labas YA, Savitsky AP, Markelov ML, Zaraisky AG, Zhao X, Fang Y, Tan W, Lukyanov SA (2000) Natural animal coloration can be determined by a nonfluorescent green fluorescent protein homolog. J Biol Chem 275:25879–25882
66. Chudakov DM, Feofanov AV, Mudrik NN, Lukyanov S, Lukyanov KA (2003) Chromophore environment provides clue to "kindling fluorescent protein" riddle. J Biol Chem 278:7215–7219
67. Ando R, Mizuno H, Miyawaki A (2004) Regulated fast nucleocytoplasmic shuttling observed by reversible protein highlighting. Science 306:1370–1373
68. Chudakov DM, Chepurnykh TV, Belousov VV, Lukyanov S, Lukyanov KA (2006) Fast and precise protein tracking using repeated reversible photoactivation. Traffic 7:1304–1310
69. Hofmann M, Eggeling C, Jakobs S, Hell SW (2005) Breaking the diffraction barrier in fluorescence microscopy at low light intensities by using reversibly photoswitchable proteins. Proc Natl Acad Sci U S A 102:17565–17569
70. Chudakov DM, Lukyanov S, Lukyanov KA (2007) Tracking intracellular protein movements using photoswitchable fluorescent proteins PS-CFP2 and Dendra2. Nat Protoc 2:2024–2032
71. Chudakov DM, Verkhusha VV, Staroverov DB, Souslova EA, Lukyanov S, Lukyanov KA (2004) Photoswitchable cyan fluorescent protein for protein tracking. Nat Biotechnol 22:1435–1439
72. Stark DA, Kulesa PM (2007) An in vivo comparison of photoactivatable fluorescent proteins in an avian embryo model. Dev Dyn 236:1583–1594
73. Ando R, Hama H, Yamamoto-Hino M, Mizuno H, Miyawaki A (2002) An optical marker based on the UV-induced green-to-red photoconversion of a fluorescent protein. Proc Natl Acad Sci U S A 99:12651–12656
74. Mizuno H, Mal TK, Tong KI, Ando R, Furuta T, Ikura M, Miyawaki A (2003) Photo-induced peptide cleavage in the green-to-red conversion of a fluorescent protein. Mol Cell 12:1051–1058
75. Wacker SA, Oswald F, Wiedenmann J, Knochel W (2007) A green to red photoconvertible protein as an analyzing tool for early vertebrate development. Dev Dyn 236:473–480
76. Tsutsui H, Karasawa S, Shimizu H, Nukina N, Miyawaki A (2005) Semi-rational engineering of a coral fluorescent protein into an efficient highlighter. EMBO Rep 6:233–238
77. Habuchi S, Tsutsui H, Kochaniak AB, Miyawaki A, van Oijen AM (2008) mKikGR, a monomeric photoswitchable fluorescent protein. PLoS One 3:e3944
78. Flynn KC, Pak CW, Shaw AE, Bradke F, Bamburg JR (2009) Growth cone-like waves transport actin and promote axonogenesis and neurite branching. Dev Neurobiol 69: 761–779
79. Fernandez-Suarez M, Ting AY (2008) Fluorescent probes for super-resolution imaging in living cells. Nat Rev Mol Cell Biol 9:929–943
80. McKinney SA, Murphy CS, Hazelwood KL, Davidson MW, Looger LL (2009) A bright and photostable photoconvertible fluorescent protein. Nat Methods 6:131–133
81. Falk MM, Baker SM, Gumpert AM, Segretain D, Buckheit RW 3rd (2009) Gap junction turnover is achieved by the internalization of small endocytic double-membrane vesicles. Mol Biol Cell 20:3342–3352
82. Kulesa PM, Teddy JM, Stark DA, Smith SE, McLennan R (2008) Neural crest invasion is a spatially-ordered progression into the head with higher cell proliferation at the migratory front as revealed by the photoactivatable protein, KikGR. Dev Biol 316:275–287
83. Tomura M, Yoshida N, Tanaka J, Karasawa S, Miwa Y, Miyawaki A, Kanagawa O (2008) Monitoring cellular movement in vivo with photoconvertible fluorescence protein "Kaede" transgenic mice. Proc Natl Acad Sci U S A 105:10871–10876
84. Nowotschin S, Hadjantonakis AK (2009) Use of KikGR a photoconvertible green-to-red fluorescent protein for cell labeling and lineage analysis in ES cells and mouse embryos. BMC Dev Biol 9:49
85. Kurotaki Y, Hatta K, Nakao K, Nabeshima Y, Fujimori T (2007) Blastocyst axis is specified independently of early cell lineage but aligns with the ZP shape. Science 316:719–723
86. Imai JH, Wang X, Shi SH (2010) Kaede-centrin1 labeling of mother and daughter

centrosomes in mammalian neocortical neural progenitors. Curr Protoc Stem Cell Biol Chapter 5 Unit 5A 5

87. Adam V, Lelimousin M, Boehme S, Desfonds G, Nienhaus K, Field MJ, Wiedenmann J, McSweeney S, Nienhaus GU, Bourgeois D (2008) Structural characterization of IrisFP, an optical highlighter undergoing multiple photo-induced transformations. Proc Natl Acad Sci U S A 105:18343–18348
88. Fuchs J, Bohme S, Oswald F, Hedde PN, Krause M, Wiedenmann J, Nienhaus GU (2010) A photoactivatable marker protein for pulse-chase imaging with superresolution. Nat Methods 7:627–630
89. Tsien RY, Miyawaki A (1998) Seeing the machinery of live cells. Science 280: 1954–1955
90. Hadjantonakis AK, Papaioannou VE (2004) Dynamic in vivo imaging and cell tracking using a histone fluorescent protein fusion in mice. BMC Biotechnol 4:33
91. Fraser ST, Hadjantonakis AK, Sahr KE, Willey S, Kelly OG, Jones EA, Dickinson ME, Baron MH (2005) Using a histone yellow fluorescent protein fusion for tagging and tracking endothelial cells in ES cells and mice. Genesis 42:162–171
92. Kanda T, Sullivan KF, Wahl GM (1998) Histone-GFP fusion protein enables sensitive analysis of chromosome dynamics in living mammalian cells. Curr Biol 8:377–385
93. Rhee JM, Pirity MK, Lackan CS, Long JZ, Kondoh G, Takeda J, Hadjantonakis AK (2006) In vivo imaging and differential localization of lipid-modified GFP-variant fusions in embryonic stem cells and mice. Genesis 44:202–218
94. Teddy JM, Lansford R, Kulesa PM (2005) Four-color, 4-D time-lapse confocal imaging of chick embryos. Biotechniques 39:703–710
95. Muzumdar MD, Tasic B, Miyamichi K, Li L, Luo L (2007) A global double-fluorescent Cre reporter mouse. Genesis 45:593–605
96. Livet J, Weissman TA, Kang H, Draft RW, Lu J, Bennis RA, Sanes JR, Lichtman JW (2007) Transgenic strategies for combinatorial expression of fluorescent proteins in the nervous system. Nature 450:56–62
97. Ueno H, Weissman IL (2006) Clonal analysis of mouse development reveals a polyclonal origin for yolk sac blood islands. Dev Cell 11:519–533
98. Cooper MS, Szeto DP, Sommers-Herivel G, Topczewski J, Solnica-Krezel L, Kang HC, Johnson I, Kimelman D (2005) Visualizing morphogenesis in transgenic zebrafish embryos using BODIPY TR methyl ester dye as a vital counterstain for GFP. Dev Dyn 232:359–368
99. Jacinto A, Wood W, Woolner S, Hiley C, Turner L, Wilson C, Martinez-Arias A, Martin P (2002) Dynamic analysis of actin cable function during Drosophila dorsal closure. Curr Biol 12:1245–1250
100. Langenberg T, Brand M (2005) Lineage restriction maintains a stable organizer cell population at the zebrafish midbrain-hindbrain boundary. Development 132:3209–3216
101. Ribeiro C, Neumann M, Affolter M (2004) Genetic control of cell intercalation during tracheal morphogenesis in Drosophila. Curr Biol 14:2197–2207
102. Hamilton TG, Klinghoffer RA, Corrin PD, Soriano P (2003) Evolutionary divergence of platelet-derived growth factor alpha receptor signaling mechanisms. Mol Cell Biol 23: 4013–4025
103. Plusa B, Piliszek A, Frankenberg S, Artus J, Hadjantonakis AK (2008) Distinct sequential cell behaviours direct primitive endoderm formation in the mouse blastocyst. Development 135:3081–3091

Chapter 25

Genetic Cell Ablation

Damien Grégoire and Marie Kmita

Abstract

Targeted cell ablation has proven to be a valuable approach to study in vivo cell functions during organogenesis, tissue homeostasis, and regeneration. Over the last two decades, various approaches have been developed to refine the control of cell ablation. In this review, we give an overview of the distinct genetic tools available for targeted cell ablation, with a particular emphasis on their respective specificity.

Key words Ablation, Cytotoxic proteins, Diphtheria toxin, Suicide transgenes, Apoptosis, TRIP, Cell death

1 Introduction

Initially, cell/tissue ablations have been generated either by microdissection, aspiration, or laser-based techniques. However, the difficulty of distinguishing neighboring but genetically distinct cell types has restricted the application of surgery-based cell ablation and has led to the emergence of alternative approaches. These latter involve antibodies targeting surface molecules, the use of chemicals that interfere with the survival of particular cell types or transgenes that will directly or indirectly trigger cell death. In this review, we focus on genetic tools, which are the most versatile approaches. Most strategies are based on the expression of a cytotoxin or a protein that renders cells sensitive to cytotoxic products, but methods relying on alternative mechanisms of cell death induction have also been developed. Each procedure has specific characteristics that are either advantages or limitations depending on the aim of the study.

2 The Genetic Cell Ablation Approaches: 20 Years of Technical Innovations

Cells are characterized by the genes they express and have expressed, the activity of which underlies cell fate decision, cell–cell interaction, and eventually their contribution to tissue and organ biology.

Mark Lewandoski (ed.), *Mouse Molecular Embryology*, Methods in Molecular Biology, vol. 1092,
DOI 10.1007/978-1-60327-292-6_25, © Springer Science+Business Media New York 2014

In this respect, specific depletion of genetically defined cell populations (referred to as genetic cell lineages hereafter) is valuable in studies aimed at understanding development, homeostasis, and physiology of tissues and organs in the context of a whole organism. Initial attempts of genetic cell ablation, published 20 years ago, were based on the targeted expression of the diphtheria toxin. The obvious potential of genetic cell ablation has fostered the development of various tools involving either alternative cell death inducers or a refined control of the toxin-based cell ablation.

3 Targeted Expression of Cytotoxic Proteins

In 1987, the laboratories of Ralf Brinster and Alan Bernstein independently developed similar approaches to perform in vivo cell ablation, by expressing a cytotoxic gene product in a cell-specific manner [1, 2] (Fig. 1a). Both procedures were based on the expression of the A chain of the diphtheria toxin (DT-A) from *Corynebacteriun diphtheriae* under the control of cell-specific promoters. In one report, the elastase I promoter was used to target pancreatic ascinar cells, while in the other experiment eye lens cells were targeted using the promoter of the γ2-cristallin gene. In the resulting transgenic animals, they observed that cell death was specifically induced in the targeted tissues and subsequent impaired development of the pancreas and microphtalmia, respectively. These pioneer experiments provided a proof of principle for genetic cell ablation. Since the initial reports on DT-A mediated cell ablation, this approach has been used in a variety of different systems (Table 1).

DT-A is highly cytotoxic. It triggers apoptosis through efficient inhibition of protein synthesis mediated by ADP-ribosylation of elongation factor 2. Once inside the cell, a single molecule of DT-A is sufficient to induce apoptosis [3]. In this respect, DT-A is a very potent means to trigger complete ablation of cells expressing it. However, because of its drastic toxicity, it is essential that the DT-A transgene be totally silent in nontargeted cells. Any leakage in transgene expression, even at low levels, would result in nonspecific cell ablation. To limit this drawback, a less cytotoxic form of DT-A has been generated (referred to as DT-A176 [4]). The main constraint of this approach is that if ablation of the cell population of interest is lethal, it becomes impossible to establish the relevant mutant line. However, this limitation can be circumvented with the use of a conditional DT-A transgene (*see* Subheading 4.1).

A similar approach has been designed with a modified form of the ricin A subunit, a lectin that inhibits ribosome function [5]. However, to date, DT-A remains the most widely used toxin-based approach.

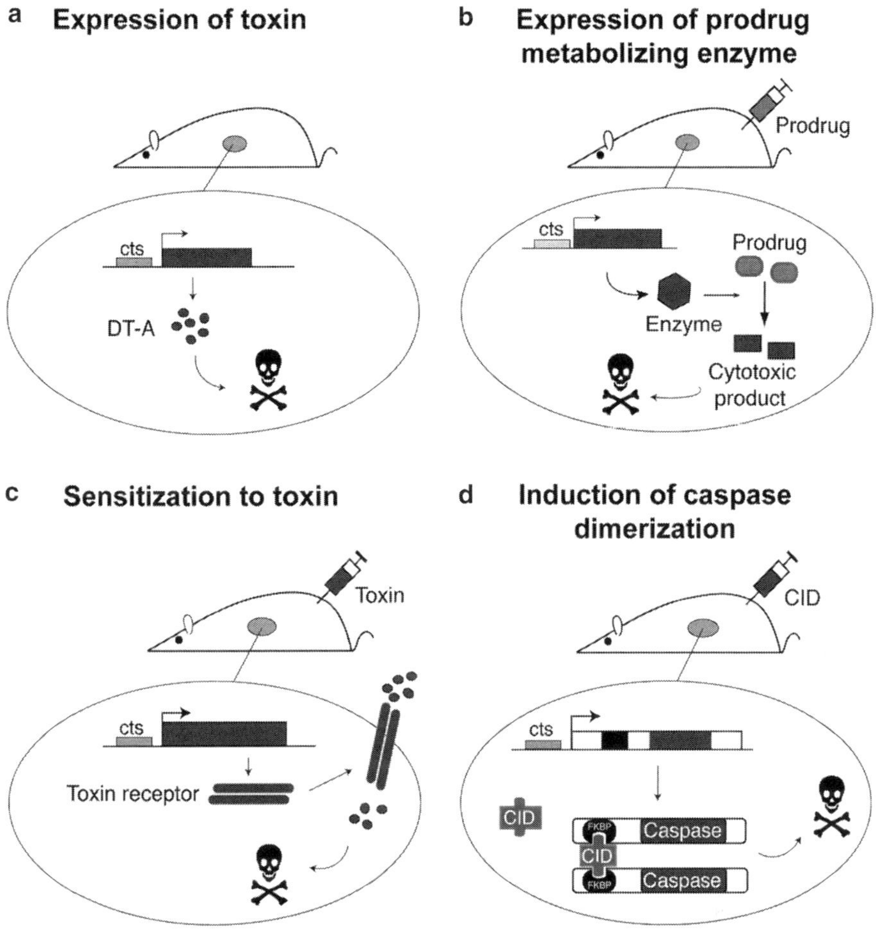

Fig. 1 Monogenic approaches for targeted cell ablation. Scheme of the distinct mechanisms of cell death induction. (**a**) Mutant mice carrying the DT-A encoding gene under the control of a tissue-specific promoter. The cytotoxic effect of DT-A is restricted to cells expressing the transgene. (**b**) Mutant mice expressing a prodrug-metabolizing enzyme in a targeted manner. Upon prodrug injection, apoptosis is induced in cell expressing the prodrug-metabolizing enzyme. (**c**) Targeted expression of a modified dimerizable caspase, which induces the endogenous apoptotic pathway upon injection of the dimerizable catalyst. (**d**) Targeted cytotoxic effect through the controlled expression of the toxin receptor coding sequences

4 Inducible Approaches

4.1 Targeted Expression of Prodrug-Metabolizing Enzymes

Prodrug-metabolizing enzymes are not detrimental for cell survival per se but metabolize prodrugs into cytotoxic products (Fig. 1b). For instance, the Herpes Simplex Virus thymidine kinase (HSV-*tk*) phosphorylates nucleoside analogs such as ganciclovir (GCV) into toxic metabolites that are inserted into replicating DNA, inhibiting replication and subsequently triggering cell death. In 1988, Ronald Evans and colleagues generated a transgene carrying the HSV-*tk*

Table 1
List of genetic cell lineages investigated

Technique	Targeted cell type	Refs.
DT-A expression (1987)[a]	Pancreatic acinar cells	[2]
	Eye lens fiber cells	[1, 46]
	Somatotrope cells (GH+ cells)	[47]
	PNMT-producing cells	[48]
	Paneth cells	[49]
	Cone photoreceptors	[50, 51]
	Natural killer (NK) cells	[52]
	Intestinal goblet cells	[53]
	Roof plate cells	[54]
	Male germ cells	[55]
	Langerhans cells	[56]
	Lens fiber cells	[57]
	K cells (GIP producing cells)	[58]
Conditional DT-A expression (2004)[a]	Myelinating glia cells	[28]
	Germ line cells	[29]
	Nkx2.5-expressing cells	[30]
	Wnt1-expressing cells	[30]
	Oligodendrocytes	[31, 43]
	CD4+ T Cells	[41]
	Myogenic cell lineage	[39, 59]
	Intestinal stem cells	[36]
	TRPV1 expressing neurons	[60]
	Notochord and floorplate	[61]
	Retinal pigment epithelial cells	[62]
DTR expression (2001)[a]	Hepatocytes	[19]
	Dendritic cells	[42–89]
	Cardiomyocytes	[63]
	Langerhans cells	[44, 64, 65]
	Hypothalamic AgRP+ neurons	[66]
	Olfactory sensory neurons	[67]
	Vascular smooth muscle cells	[68]
	Macrophages	[69, 70]
	Fox3P+ T cells	[71]
	NK cells	[45]
	Osteocytes	[72]
	CD11c+ macrophages	[73]
	β-Cells	[20]
	Epidermal Langherans cells	[90]
Conditional DTR expression (2005)[a]	Oligodendrocytes	[32]
	Retinal ganglion cells	[74]
	CD11+ dentritic cells	[88]
hCD59 expression + ILY (2008)[a]	Erythrocytes	[21]
	Endothelial cells	[21]

(continued)

Table 1
(continued)

Technique	Targeted cell type	Refs.
HSV thymidine kinase expression (1988)[a]	B and T cells	[6, 7]
	Thyroid follicle cells	[8]
	IL-4 producing cells	[75]
	Dendritic cells	[76]
	GFAP+ Glia cells	[77, 78]
	Thyrocytes	[8, 9]
	Parietal cells	[10]
	Secretin cells	[79]
	Osteoblasts	[80]
	T cells	[81]
	Bulge epithelial stem cells	[82]
Conditional tk expression (2004)[a]	Chondrocytes	[34]
Nitroreductase (1997)[a]	Luminal cells of mammary gland	[11]
	T cells	[12]
	Astrocytes	[83]
	Olfactory neurons	[84]
	Adipocytes	[85]
	Prostate notch1 expressing cells	[86]
	Neuronal stem cells	[13]
Caspase dimerization (2002)[a]	Hepatocytes	[22]
	Microvascular endothelial cells	[23]
	Myocytes	[24]
	Adipocytes	[25]
Targeted recombination between inverted *loxP* sites (2008)[a]	Limb mesenchymal cells	[35]
	Wnt1-expressing cells	[35]

[a]Year of the first report of cell ablation using the technique

under the control of the immunoglobulin promoter to induce cell death in the lymphoid system [6]. Upon GCV injections they obtained the depletion of B and T lymphocytes [6, 7]. Based on its mechanism of action, it was initially proposed that HSV-*tk*-mediated cell death specifically targets proliferating cells. However, HSV-*tk* ablation of distinct post-mitotic cells has also been reported raising concerns regarding its restriction to cycling cells [8–10].

As far as prodrug-metabolizing enzymes are concerned, an alternative to HSV-*tk* is the nitroreductase (NTR) enzyme from *Escherichia coli*. Upon injection of CB1954 (5-aziridin-1-yl-2-4-dinitrobenzamide), cell death is induced in all NTR expressing cells [11–13]. However, for both prodrug-metabolizing enzymes, bystander effects have been reported in several studies [14–16], suggesting that cell ablation could also encompass nontargeted, neighboring, cells.

4.2 Making Cells Responsive to Exogenous Toxins

Exogenous toxins have to be internalized in the cell to exert their deleterious effect. Therefore the sensitivity to a given toxin depends on the existence of its specific receptor at the cell surface. For instance, the uptake of diphtheria toxin is mediated by the precursor form of the heparin binding epidermal growth factor, *Hbegf* [17]. The affinity of the murine receptor for DT-A is 10^5 lower than that of the human receptor due to a three amino acid sequence variation between the human and murine protein [17, 18]. Thus, except when it is directly expressed within cells, low doses of DT-A is not toxic for mice. In turn, expressing the human DT-A receptor (DTR) can be used in mice to target DT-A cytotoxicity (Fig. 1c). This approach was first used by Saito and colleagues, who generated transgenic mice expressing the human toxin receptor specifically in hepatocytes [19]. Upon DT-A injections, they were able to induce massive ablation of the targeted cell population [19]. DTR-mediated cell ablation has proven to be highly efficient in a wide variety of cell populations (*see* Table 1). For instance, Thorel et al. used this approach to ablate β-cells in mice and showed a near-complete ablation of this cell population, which allowed them to elegantly reveal that β-cells can regenerate in adults [20]. However, it should be mentioned that the DTR approach is inappropriate for cell ablation in developing embryos since DT-A cannot cross the placental barrier.

Recently, a similar approach has been developed with intermedilysin toxin (ILY) from *Streptococcus intermedius*. ILY forms large pores in the cellular membrane, causing very rapid cell lysis [21]. This cytolytic effect relies upon ILY binding to the human membrane-bound CD59 [22]. ILY does not bind to murine CD59 such that targeted cell ablation can be obtained through targeted expression of the human CD59 [22]. Cell death is more rapidly induced following ILY injection than DT-A injection as ILY-mediated cell death is due to cell lysis and not apoptosis.

Addendum: These procedures involve manipulation of toxins that are highly cytotoxic for humans. Therefore it should be emphasized that special safety measures are required.

4.3 Targeted Activation of Apoptotic Machinery

The majority of cell ablation procedures involve exogenous cytotoxic products. It is also possible to induce cell death by triggering the endogenous apoptotic pathway. Mallet and colleagues showed that apoptosis can be induced in targeted cells by inducing the dimerization of the pro-apoptotic form of Caspase 3 [23] (Fig. 1d). They generated a transgenic mouse line driving hepatocyte-specific expression of the human Caspase 3 fused to a modified FKBP domain. Subsequent injection of the FK1012 homolog, AP20187, which binds simultaneously two FKBP domains, triggered homodimerization of the FKBP-Caspase 3 fusion protein thereby inducing the Caspase 3 cell death pathway. Cell ablation driven by the direct activation of the apoptotic pathway was also obtained with the fusion protein FKBP-Caspase 8 [24–26] and FKBP-Fas [27], respectively.

4.4 Specificities of Inducible Approaches

The particularity of the inducible approaches is that they require injection of the appropriate substrate/catalyst to trigger cell death, thereby allowing for both temporal and spatial control of cell ablation. Since the suicide effect relies on the availability of the substrate/catalyst, the induction of cell death is transient. This aspect is of particular interest when studying, for instance, mechanisms of homeostasis. Noteworthy, inducible approaches make feasible the establishment of transgenic lines even if the cell population of interest is critical for mice survival. The injection step being a key parameter for inducible cell ablation, it is important to ensure that injections are performed with minimal variations. To date, a number of distinct cell populations have been successfully targeted using inducible approaches (*see* Table 1). However, the efficiency can vary depending on the cell population of interest. For instance, ablation of embryonic tissues cannot be achieved using DT-A injection (*see* Subheading 4.2).

5 Versatile Tools for Cell Ablation: The Bi-allelic Approach

The major constraint regarding the procedures described above resides in the requirement of generating a specific transgene for each distinct cell population. Recently, efforts have been dedicated to generate more versatile mutant strains. The procedure takes advantage of the *loxP–Cre* recombination system such that the activity of the suicide transgene is under the control of the Cre activity. Thereby, the same transgene can be used to target distinct cell populations simply by combining it with distinct tissue-specific Cre strains.

5.1 Conditional Expression of Suicide Transgenes

The first bi-allelic system was developed in 1998 in Gail Martin's laboratory to establish a conditional DT-A transgenic line, which otherwise was embryonic lethal. They generated a transgene in which the open reading frame of the DT-A encoding gene was disrupted by the insertion of a "stop" cassette [28]. This "stop" cassette was flanked with *loxP* sites in the same relative orientation such that it could be deleted upon Cre-mediated recombination. They subsequently crossed these transgenic mice with mutants expressing the Cre recombinase ubiquitously. In the resulting double heterozygous embryos, the recombinase activity triggered the deletion of the "stop" cassette thereby restoring the coding sequence of the diphtheria toxin and activating the suicide function of the transgene [28].

The use of the bi-allelic approach has been fostered by the availability of numerous tissue-specific Cre transgenes. Versatile conditional transgenes have been engineered for DT-A [29–33], DTR [34], or HSV-*tk* [35]. These transgenes were designed such that the "stop" cassette flanked with *loxP* sites is inserted in between a ubiquitous promoter and the coding sequences of the suicide gene (Fig. 2a, b).

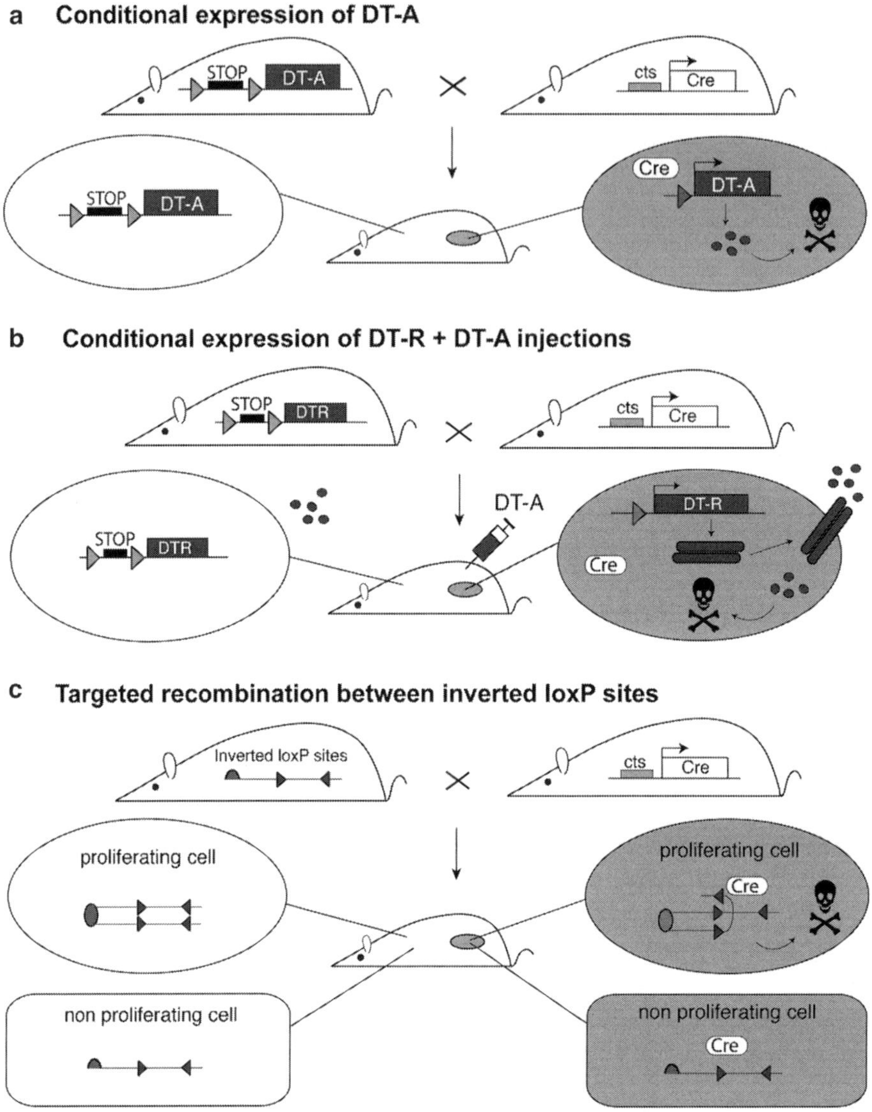

Fig. 2 Bi-allelic strategies for targeted cell ablation. Conditional expression of the diphtheria toxin (**a**) or its receptor (**b**) using the *loxP–Cre* recombination system. A stop cassette flanked with *loxP* sites in the same relative orientation is inserted between the promoter and the coding sequences of the transgene. Upon Cre recombination, the stop cassette is deleted leading to the expression of the transgene. The cell specificity is driven by the selected Cre strain. (**c**) Targeted cell ablation based on the cytotoxicity of Cre-mediated recombination between *loxP* sites in inverted orientation. Recombination between inverted *loxP* sites is cytotoxic only in proliferating cells. After DNA replication, recombination results in unequal exchanges between sister chromatids which ultimately triggers the apoptotic pathway

5.2 Targeted Recombination Between Inverted loxP Sites

Recently, we discovered that targeted recombination between inverted *loxP* sites (referred as to TRIP) is cytotoxic for proliferating cells [36] (Fig. 2c). In contrast to the other bi-allelic systems, TRIP does not activate a suicide transgene but generates a deleterious genomic rearrangement. After DNA replication and before mitosis, Cre-mediated recombination between *loxP* sites in inverted orientation results in unequal crossover between sister-chromatids giving rise to an acentric and a dicentric chromosome. It is most likely that cells with such abnormal karyotype are detected by the cell cycle checkpoint machinery, which in turn activates the endogenous apoptotic cascade. Indeed, we found that cells that have lost the chromosome carrying inverted *loxP* sites were eliminated before entering in metaphase. The extent of cell death following TRIP indicates that it is an efficient means for targeted ablation of proliferating cells within any genetic cell lineage of interest [36]. HSV-*tk*-based cell ablation was initially described as been restricted to proliferating cells; however, it turned out that distinct nonproliferative cells were also affected [8–10]. To our knowledge, TRIP is the only ablation procedure specific for proliferating cells.

5.3 Specificities of Bi-allelic Approaches

Initially developed for conditional gene inactivation, the *loxP–Cre* recombination system is now widely used for cell ablation (Fig. 2). It is particularly appealing as the repertoire of available Cre strains allows for a large variety of genetic cell lineages to be targeted through a simple breeding step, without the need of engineering novel mutant mouse strains. Importantly, the cell ablation experiment can be readily repeated even if the cell population of interest is crucial for mice viability. Furthermore, a temporal control of cell ablation can be added by using tamoxifen-inducible Cre [37], the recombinase activity of which is induced upon intraperitoneal tamoxifen injections [38]. To date, the bi-allelic strategy has been implemented for cell ablation based either on toxins or targeted recombination between inverted *loxP* sites but could be applied to all genetic cell ablation procedures.

6 Technical Aspects to Keep in Mind

6.1 Efficiency of Cell Ablation

The extent of cell ablation is a key parameter to take into account for the interpretation of the ablation outcome, which has sometimes been neglected in previous reports. Even when the efficiency of a given approach is reported, it can vary from one cell population to another. For instance, following injection, the spreading of the prodrug/toxin within the organism is not homogeneous [13, 39] which might impact on the efficiency of cell ablation. How to establish the extent of cell death? The most obvious means is to directly measure cell death through specific labeling of apoptotic cells (e.g., TUNEL assay). *Cre–loxP*-based approaches offer the possibility to label the Cre cell lineage by combining a Cre reporter transgene with the conditional killer transgene chosen. Notably, even if a small percentage of targeted cells survive, it becomes possible to follow their fate and allows for a more reliable analysis of the cell ablation outcome. Any Cre reporter strain could be used to label the Cre lineage [36, 40], but a dual reporter strain such as mT/mG [41] is preferable as it differentially labels cells in which recombination has occurred (GFP positive cells) or not (Tomato red positive cells). Taking into account that nonrecombined alleles can occasionally be found within the Cre-expression domain [34, 42], the use of a dual reporter is recommended.

Even if genetic cell ablation procedures were originally developed to achieve maximum efficiency, partial ablation is valuable for generating models of injury or neurodegenerative diseases. In this respect, inducible approaches (either mono- or bi-allelic) offer the possibility to reduce the ablation efficiency simply by lowering the dosage of the inducer that is injected. The specific ablation of proliferating cells within the genetic cell lineage of interest could also be valuable.

6.2 The Lag Between the Activation of the Genetic System and Cell Ablation

Another factor to take into account when analyzing the outcome of targeted cell ablation is the delay between the activation of the genetic killing system and the actual cell death. This aspect is particularly important when studying developmental processes. Obviously, the least time lag is obtained with direct systems, i.e., nonconditional toxin expressing transgene. However transgene leakiness or lethality could complicate the experiment. The delay in bipartite systems is inherent to the mechanism used to activate cell death. For targeted expression of DT receptor, cell death begins within a day after the first toxin injection [43–45] and should be complete within 3 days. The induction of cell death by intermedilysin injection seems to be faster as targeted ablation of erythrocytes and endothelia initiates within less than an hour following intravenous injection [22]. However, the time lag for less accessible cell population remains to be established. As far as the bi-allelic approach is concerned, there is a time lag between the initial expression of the Cre transgene and the actual cell death. DT-A-induced apoptosis has been reported to begin 12 h after the onset of Cre expression and is completed within 2–3 days [31, 46]. A similar delay was reported for TRIP-mediated cell ablation [36].

7 Conclusion

Genetic cell ablation is to date the most reliable means to perform targeted ablation experiments. Efforts have been dedicated to refine this technique and there are now various tools available to the scientific community. Each procedure has specific characteristics that are either advantages or limitations depending on the aim of the study (*see* Table 2). Yet, conditional cell ablation, based on a bi-allelic strategy, is to date the most powerful approach.

Table 2
Key features of the principal genetic cell ablation procedures

	DT-A	Conditional DT-A	HSV-tk	Conditional puΔ-tk	Nitroreductase	DTR	Conditional DTR	ILY	Caspase dimerization	TRIP
Embryo	+	+	+	+	+	–	–	ND[a]	ND[a]	+
Restricted to proliferating cells	–	–	?	–	–	–	–	–	–	+
Constitutive ablation	+	+	–	–	–	–	–	–	–	+
Inducible/transient ablation	–	iCre[b]	+	+	+	+	+	+	+	iCre[b]
Quantitative control	–	iCre[b]	+	+	+	+	+	+	+	iCre[b]
Concomitant lineage labeling	–	+	–	+	–	–	+	–	–	+

[a] *ND* not determined
[b] iCre, possible if a tamoxifen-inducible Cre is used

Acknowledgments

We thank D. Hipfner and members of the lab for critical reading of the manuscript. This work was supported by the Canadian Institutes for Health Research (CIHR-82880), the Canada Research Chair (to M.K.), and by a postdoctoral fellowship from the Fond de la Recherche en Santé du Québec (to D.G.).

References

1. Breitman ML, Clapoff S, Rossant J, Tsui LC, Glode LM, Maxwell IH, Bernstein A (1987) Genetic ablation: targeted expression of a toxin gene causes microphthalmia in transgenic mice. Science 238:1563–1565
2. Palmiter RD, Behringer RR, Quaife CJ, Maxwell F, Maxwell IH, Brinster RL (1987) Cell lineage ablation in transgenic mice by cell-specific expression of a toxin gene. Cell 50:435–443
3. Yamaizumi M, Mekada E, Uchida T, Okada Y (1978) One molecule of diphtheria toxin fragment A introduced into a cell can kill the cell. Cell 15:245–250
4. Maxwell F, Maxwell IH, Glode LM (1987) Cloning, sequence determination, and expression in transfected cells of the coding sequence for the tox 176 attenuated diphtheria toxin A chain. Mol Cell Biol 7:1576–1579
5. Landel CP, Zhao J, Bok D, Evans GA (1988) Lens-specific expression of recombinant ricin induces developmental defects in the eyes of transgenic mice. Genes Dev 2:1168–1178
6. Borrelli E, Heyman R, Hsi M, Evans RM (1988) Targeting of an inducible toxic phenotype in animal cells. Proc Natl Acad Sci U S A 85:7572–7576
7. Heyman RA, Borrelli E, Lesley J, Anderson D, Richman DD, Baird SM, Hyman R, Evans RM (1989) Thymidine kinase obliteration: creation of transgenic mice with controlled immune deficiency. Proc Natl Acad Sci U S A 86:2698–2702
8. Wallace H, Ledent C, Vassart G, Bishop JO, al-Shawi R (1991) Specific ablation of thyroid follicle cells in adult transgenic mice. Endocrinology 129:3217–3226
9. Wallace H, Clarke AR, Harrison DJ, Hooper ML, Bishop JO (1996) Ganciclovir-induced ablation non-proliferating thyrocytes expressing herpesvirus thymidine kinase occurs by p53-independent apoptosis. Oncogene 13:55–61
10. Canfield V, West AB, Goldenring JR, Levenson R (1996) Genetic ablation of parietal cells in transgenic mice: a new model for analyzing cell lineage relationships in the gastric mucosa. Proc Natl Acad Sci U S A 93:2431–2435
11. Clark AJ, Iwobi M, Cui W, Crompton M, Harold G, Hobbs S, Kamalati T, Knox R, Neil C, Yull F, Gusterson B (1997) Selective cell ablation in transgenic mice expression E. coli nitroreductase. Gene Ther 4:101–110
12. Drabek D, Guy J, Craig R, Grosveld F (1997) The expression of bacterial nitroreductase in transgenic mice results in specific cell killing by the prodrug CB1954. Gene Ther 4:93–100
13. Kwak SP, Malberg JE, Howland DS, Cheng KY, Su J, She Y, Fennell M, Ghavami A (2007) Ablation of central nervous system progenitor cells in transgenic rats using bacterial nitroreductase system. J Neurosci Res 85:1183–1193
14. Hamel W, Magnelli L, Chiarugi VP, Israel MA (1996) Herpes simplex virus thymidine kinase/ganciclovir-mediated apoptotic death of bystander cells. Cancer Res 56:2697–2702
15. Freeman SM, Abboud CN, Whartenby KA, Packman CH, Koeplin DS, Moolten FL, Abraham GN (1993) The "bystander effect": tumor regression when a fraction of the tumor mass is genetically modified. Cancer Res 53:5274–5283
16. Ahtiainen M, Toppari J, Poutanen M, Huhtaniemi I (2004) Indirect Sertoli cell-mediated ablation of germ cells in mice expressing the inhibin-alpha promoter/herpes simplex virus thymidine kinase transgene. Biol Reprod 71:1545–1550
17. Naglich JG, Metherall JE, Russell DW, Eidels L (1992) Expression cloning of a diphtheria toxin receptor: identity with a heparin-binding EGF-like growth factor precursor. Cell 69: 1051–1061
18. Mitamura T, Umata T, Nakano F, Shishido Y, Toyoda T, Itai A, Kimura H, Mekada E (1997) Structure-function analysis of the diphtheria toxin receptor toxin binding site by site-directed mutagenesis. J Biol Chem 272:27084–27090
19. Saito M, Iwawaki T, Taya C, Yonekawa H, Noda M, Inui Y, Mekada E, Kimata Y, Tsuru A,

Kohno K (2001) Diphtheria toxin receptor-mediated conditional and targeted cell ablation in transgenic mice. Nat Biotechnol 19: 746–750
20. Thorel F, Népote V, Avril I, Kohno K, Desgraz R, Chera S, Herrera PL (2010) Conversion of adult pancreatic alpha-cells to beta-cells after extreme beta-cell loss. Nature 464:1149–1154
21. Giddings KS, Zhao J, Sims PJ, Tweten RK (2004) Human CD59 is a receptor for the cholesterol-dependent cytolysin intermedilysin. Nat Struct Mol Biol 11:1173–1178
22. Hu W, Ferris SP, Tweten RK, Wu G, Radaeva S, Gao B, Bronson RT, Halperin JA, Qin X (2008) Rapid conditional targeted ablation of cells expressing human CD59 in transgenic mice by intermedilysin. Nat Med 14:98–103
23. Mallet VO, Mitchell C, Guidotti JE, Jaffray P, Fabre M, Spencer D, Arnoult D, Kahn A, Gilgenkrantz H (2002) Conditional cell ablation by tight control of caspase-3 dimerization in transgenic mice. Nat Biotechnol 20: 1234–1239
24. Wencker D, Chandra M, Nguyen K, Miao W, Garantziotis S, Factor SM, Shirani J, Armstrong RC, Kitsis RN (2003) A mechanistic role for cardiac myocyte apoptosis in heart failure. J Clin Invest 111:1497–1504
25. Pajvani UB, Trujillo ME, Combs TP, Iyengar P, Jelicks L, Roth KA, Kitsis RN, Scherer PE (2005) Fat apoptosis through targeted activation of caspase 8: a new mouse model of inducible and reversible lipoatrophy. Nat Med 11:797–803
26. Wang ZV, Mu J, Schraw TD, Gautron L, Elmquist JK, Zhang BB, Brownlee M, Scherer PE (2008) PANIC-ATTAC: a mouse model for inducible and reversible beta-cell ablation. Diabetes 57:2137–2148
27. Burnett SH, Kershen EJ, Zhang J, Zeng L, Straley SC, Kaplan AM, Cohen DA (2004) Conditional macrophage ablation in transgenic mice expressing a Fas-based suicide gene. J Leukoc Biol 75:612–623
28. Grieshammer U, Lewandoski M, Prevette D, Oppenheim RW, Martin GR (1998) Muscle-specific cell ablation conditional upon Cre-mediated DNA recombination in transgenic mice leads to massive spinal and cranial motoneuron loss. Dev Biol 197:234–247
29. Brockschnieder D, Pechmann Y, Sonnenberg-Riethmacher E, Riethmacher D (2006) An improved mouse line for Cre-induced cell ablation due to diphtheria toxin A, expressed from the Rosa26 locus. Genesis 44:322–327
30. Matsumura H, Hasuwa H, Inoue N, Ikawa M, Okabe M (2004) Lineage-specific cell disruption in living mice by Cre-mediated expression of diphtheria toxin A chain. Biochem Biophys Res Commun 321:275–279
31. Brockschnieder D, Lappe-Siefke C, Goebbels S, Boesl MR, Nave KA, Riethmacher D (2004) Cell depletion due to diphtheria toxin fragment A after Cre-mediated recombination. Mol Cell Biol 24:7636–7642
32. Sato M, Tanigawa M (2005) Production of CETD transgenic mouse line allowing ablation of any type of specific cell population. Mol Reprod Dev 72:54–67
33. Ivanova A, Signore M, Caro N, Greene ND, Copp AJ, Martinez-Barbera JP (2005) In vivo genetic ablation by Cre-mediated expression of diphtheria toxin fragment A. Genesis 43: 129–135
34. Buch T, Heppner FL, Tertilt C, Heinen TJ, Kremer M, Wunderlich FT, Jung S, Waisman A (2005) A Cre-inducible diphtheria toxin receptor mediates cell lineage ablation after toxin administration. Nat Methods 2:419–426
35. Chen YT, Levasseur R, Vaishnav S, Karsenty G, Bradley A (2004) Bigenic Cre/loxP, puDeltatk conditional genetic ablation. Nucleic Acids Res 32:e161
36. Gregoire D, Kmita M (2008) Recombination between inverted loxP sites is cytotoxic for proliferating cells and provides a simple tool for conditional cell ablation. Proc Natl Acad Sci U S A 105:14492–14496
37. Sangiorgi E, Capecchi MR (2008) Bmi1 is expressed in vivo in intestinal stem cells. Nat Genet 40:915–920
38. Danielian PS, Muccino D, Rowitch DH, Michael SK, McMahon AP (1998) Modification of gene activity in mouse embryos in utero by a tamoxifen-inducible form of Cre recombinase. Curr Biol 8(24):1323–1326
39. Bennett CL, Clausen BE (2007) Modification of gene activity in mouse embryos in utero by a tamoxifen-inducible form of Cre recombinase. Trends Immunol 28:525–531
40. Haldar M, Karan G, Tvrdik P, Capecchi MR (2008) Two cell lineages, myf5 and myf5-independent, participate in mouse skeletal myogenesis. Dev Cell 14:437–445
41. Muzumdar MD, Tasic B, Miyamichi K, Li L, Luo L (2007) A global double-fluorescent Cre reporter mouse. Genesis 45:593–605
42. Voehringer D, Liang HE, Locksley RM (2008) Homeostasis and effector function of lymphopenia-induced “memory-like” T cells in constitutively T cell-depleted mice. J Immunol 180:4742–4753
43. Walzer T, Blery M, Chaix J, Fuseri N, Chasson L, Robbins SH, Jaeger S, Andre P, Gauthier L, Daniel L, Chemin K, Morel Y, Dalod M, Imbert J, Pierres M, Moretta A, Romagne F,

Vivier E (2007) Identification, activation, and selective in vivo ablation of mouse NK cells via NKp46. Proc Natl Acad Sci U S A 104: 3384–3389

44. Jung S, Unutmaz D, Wong P, Sano G, De los Santos K, Sparwasser T, Wu S, Vuthoori S, Ko K, Zavala F, Pamer EG, Littman DR, Lang RA (2002) In vivo depletion of CD11c+ dendritic cells abrogates priming of CD8+ T cells by exogenous cell-associated antigens. Immunity 17:211–220
45. Bennett CL, van Rijn E, Jung S, Inaba K, Steinman RM, Kapsenberg ML, Clausen BE (2005) Inducible ablation of mouse Langerhans cells diminishes but fails to abrogate contact hypersensitivity. J Cell Biol 169:569–576
46. Wu S, Wu Y, Capecchi MR (2006) Motoneurons and oligodendrocytes are sequentially generated from neural stem cells but do not appear to share common lineage-restricted progenitors in vivo. Development 133:581–590
47. Breitman ML, Rombola H, Maxwell IH, Klintworth GK, Bernstein, A (1990) Genetic ablation in transgenic mice with an attenuated diphtheria toxin A gene. Mol Cell Biol 10(2):474–479
48. Behringer RR, Mathews LS, Palmiter RD, Brinster RL (1988) Genetic ablation in transgenic mice with an attenuated diphtheria toxin A gene. Genes Dev 2:453–461
49. Quaife CJ, Hoyle GW, Froelick GJ, Findley SD, Baetge EE, Behringer RR, Hammang JP, Brinster RL, Palmiter RD (1994) Visualization and ablation of phenylethanolamine N-methyltransferase producing cells in transgenic mice. Transgenic Res 3:388–400
50. Garabedian EM, Roberts LJ, McNevin MS, Gordon JI (1997) Examining the role of Paneth cells in the small intestine by lineage ablation in transgenic mice. J Biol Chem 272:23729–23740
51. Soucy E, Wang Y, Nirenberg S, Nathans J, Meister M (1998) A novel signaling pathway from rod photoreceptors to ganglion cells in mammalian retina. Neuron 21:481–493
52. Chen J, Nathans J (2007) Genetic ablation of cone photoreceptors eliminates retinal folds in the retinal degeneration 7 (rd7) mouse. Invest Ophthalmol Vis Sci 48:2799–2805
53. Arase K, Saijo K, Watanabe H, Konno A, Arase H, Saito T (1999) Ablation of a specific cell population by the replacement of a uniquely expressed gene with a toxin gene. Proc Natl Acad Sci U S A 96:9264–9268
54. Itoh H, Beck PL, Inoue N, Xavier R, Podolsky DK (1999) A paradoxical reduction in susceptibility to colonic injury upon targeted transgenic ablation of goblet cells. J Clin Invest 104:1539–1547
55. Lee KJ, Dietrich P, Jessell TM (2000) Genetic ablation reveals that the roof plate is essential for dorsal interneuron specification. Nature 403:734–740
56. Bartell JG, Fantz DA, Davis T, Dewey MJ, Kistler MK, Kistler WS (2000) Elimination of male germ cells in transgenic mice by the diphtheria toxin A chain gene directed by the histone H1t promoter. Biol Reprod 63:409–416
57. Kaplan DH, Jenison MC, Saeland S, Shlomchik WD, Shlomchik MJ (2005) Epidermal langerhans cell-deficient mice develop enhanced contact hypersensitivity. Immunity 23:611–620
58. Zhang Y, Overbeek PA, Govindarajan V (2007) Perinatal ablation of the mouse lens causes multiple anterior chamber defects. Mol Vis 13:2289–2300
59. Althage MC, Ford EL, Wang S, Tso P, Polonsky KS, Wice BM (2008) Targeted ablation of glucose-dependent insulinotropic polypeptide-producing cells in transgenic mice reduces obesity and insulin resistance induced by a high fat diet. J Biol Chem 283:18365–18376
60. Mishra SK, Hoon MA (2010) Ablation of TrpV1 neurons reveals their selective role in thermal pain sensation. Mol Cell Neurosci 43(1):157–163
61. Tripathi P, Guo Q, Wang Y, Coussens M, Liapis H, Jain S, Kuehn MR, Capecchi MR, Chen F (2010) Ablation of TrpV1 neurons reveals their selective role in thermal pain sensation. Dev Biol 340:518–527
62. Longbottom R, Fruttiger M, Douglas RH, Martinez-Barbera JP, Greenwood J, Moss SE (2009) Genetic ablation of retinal pigment epithelial cells reveals the adaptive response of the epithelium and impact on photoreceptors. Proc Natl Acad Sci U S A 106:18728–18733
63. Gensch N, Borchardt T, Schneider A, Riethmacher D, Braun T (2008) Different autonomous myogenic cell populations revealed by ablation of Myf5-expressing cells during mouse embryogenesis. Development 135:1597–1604
64. Akazawa H, Komazaki S, Shimomura H, Terasaki F, Zou Y, Takano H, Nagai T, Komuro I (2004) Diphtheria toxin-induced autophagic cardiomyocyte death plays a pathogenic role in mouse model of heart failure. J Biol Chem 279:41095–41103
65. Kissenpfennig A, Henri S, Dubois B, Laplace-Builhe C, Perrin P, Romani N, Tripp CH, Douillard P, Leserman L, Kaiserlian D, Saeland S, Davoust J, Malissen B (2005) Dynamics and function of Langerhans cells in vivo: dermal dendritic cells colonize lymph node areas distinct from slower migrating Langerhans cells. Immunity 22:643–654

66. Stoecklinger A, Grieshuber I, Scheiblhofer S, Weiss R, Ritter U, Kissenpfennig A, Malissen B, Romani N, Koch F, Ferreira F, Thalhamer J, Hammerl P (2007) Epidermal langerhans cells are dispensable for humoral and cell-mediated immunity elicited by gene gun immunization. J Immunol 179:886–893
67. Luquet S, Perez FA, Hnasko TS, Palmiter RD (2005) NPY/AgRP neurons are essential for feeding in adult mice but can be ablated in neonates. Science 310:683–685
68. Chen H, Kohno K, Gong Q (2005) Conditional ablation of mature olfactory sensory neurons mediated by diphtheria toxin receptor. J Neurocytol 34:37–47
69. Clarke MC, Figg N, Maguire JJ, Davenport AP, Goddard M, Littlewood TD, Bennett MR (2006) Apoptosis of vascular smooth muscle cells induces features of plaque vulnerability in atherosclerosis. Nat Med 12:1075–1080
70. Cailhier JF, Partolina M, Vuthoori S, Wu S, Ko K, Watson S, Savill J, Hughes J, Lang RA (2005) Conditional macrophage ablation demonstrates that resident macrophages initiate acute peritoneal inflammation. J Immunol 174:2336–2342
71. Miyake Y, Kaise H, Isono K, Koseki H, Kohno K, Tanaka M (2007) Protective role of macrophages in noninflammatory lung injury caused by selective ablation of alveolar epithelial type II cells. J Immunol 178:5001–5009
72. Lahl K, Loddenkemper C, Drouin C, Freyer J, Arnason J, Eberl G, Hamann A, Wagner H, Huehn J, Sparwasser T (2007) Selective depletion of Foxp3+ regulatory T cells induces a scurfy-like disease. J Exp Med 204:57–63
73. Tatsumi S, Ishii K, Amizuka N, Li M, Kobayashi T, Kohno K, Ito M, Takeshita S, Ikeda K (2007) Targeted ablation of osteocytes induces osteoporosis with defective mechanotransduction. Cell Metab 5:464–475
74. Patsouris D, Li PP, Thapar D, Chapman J, Olefsky JM, Neels JG (2008) Ablation of CD11c-positive cells normalizes insulin sensitivity in obese insulin resistant animals. Cell Metab 8:301–309
75. Hatori M, Le H, Vollmers C, Keding SR, Tanaka N, Schmedt C, Jegla T, Panda S (2008) Inducible ablation of melanopsin-expressing retinal ganglion cells reveals their central role in non-image forming visual responses. PLoS One 3:e2451
76. Kamogawa Y, Minasi LA, Carding SR, Bottomly K, Flavell RA (1993) The relationship of IL-4- and IFN gamma-producing T cells studied by lineage ablation of IL-4-producing cells. Cell 75:985–995
77. Salomon B, Lores P, Pioche C, Racz P, Jami J, Klatzmann D (1994) Conditional ablation of dendritic cells in transgenic mice. J Immunol 152:537–548
78. Delaney CL, Brenner M, Messing A (1996) Conditional ablation of cerebellar astrocytes in postnatal transgenic mice. J Neurosci 16: 6908–6918
79. Bush TG, Savidge TC, Freeman TC, Cox HJ, Campbell EA, Mucke L, Johnson MH, Sofroniew MV (1998) Fulminant jejuno-ileitis following ablation of enteric glia in adult transgenic mice. Cell 93:189–201
80. Rindi G, Ratineau C, Ronco A, Candusso ME, Tsai M, Leiter AB (1999) Targeted ablation of secretin-producing cells in transgenic mice reveals a common differentiation pathway with multiple enteroendocrine cell lineages in the small intestine. Development 126:4149–4156
81. Visnjic D, Kalajzic I, Gronowicz G, Aguila HL, Clark SH, Lichtler AC, Rowe DW (2001) Conditional ablation of the osteoblast lineage in Col2.3deltatk transgenic mice. J Bone Miner Res 16:2222–2231
82. Bellier B, Thomas-Vaslin V, Saron MF, Klatzmann D (2003) Turning immunological memory into amnesia by depletion of dividing T cells. Proc Natl Acad Sci U S A 100:15017–15022
83. Ito M, Liu Y, Yang Z, Nguyen J, Liang F, Morris RJ, Cotsarelis G (2005) Stem cells in the hair follicle bulge contribute to wound repair but not to homeostasis of the epidermis. Nat Med 11:1351–1354
84. Cui W, Allen ND, Skynner M, Gusterson B, Clark AJ (2001) Inducible ablation of astrocytes shows that these cells are required for neuronal survival in the adult brain. Glia 34:272–282
85. Isles AR, Ma D, Milsom C, Skynner MJ, Cui W, Clark J, Keverne EB, Allen ND (2001) Conditional ablation of neurones in transgenic mice. J Neurobiol 47:183–193
86. Felmer R, Cui W, Clark AJ (2002) Inducible ablation of adipocytes in adult transgenic mice expressing the E. coli nitroreductase gene. J Endocrinol 175:487–498
87. Wang XD, Shou J, Wong P, French DM, Gao WQ (2004) Notch1-expressing cells are indispensable for prostatic branching morphogenesis during development and re-growth following castration and androgen replacement. J Biol Chem 279:24733–24744
88. Okuyama M, Kayama H, Atarashi K, Saiga H, Kimura T, Waisman A, Yamamoto M, Takeda K (2010) A novel in vivo inducible dendritic

cell ablation model in mice. Biochem Biophys Res Commun 397:559–563

89. Swiecki M, Gilfillan S, Vermi W, Wang Y, Colonna M (2010) Plasmacytoid dendritic cell ablation impacts early interferon responses and antiviral NK and CD8(+) T cell accrual. Immunity 33:955–966

90. Hanson J, Gille A, Zwykiel S, Lukasova M, Clausen BE, Ahmed K, Tunaru S, Wirth A, Offermanns S (2010) Nicotinic acid- and monomethyl fumarate-induced flushing involves GPR109A expressed by keratinocytes and COX-2-dependent prostanoid formation in mice. J Clin Invest 120:2910–2919

Chapter 26

Essentials of Recombinase-Based Genetic Fate Mapping in Mice

Patricia Jensen and Susan M. Dymecki

Abstract

Fate maps, by defining the relationship between embryonic tissue organization and postnatal tissue structure, are one of the most important tools on hand to developmental biologists. In the past, generating such maps in mice was hindered by their in utero development limiting the physical access required for traditional methods involving tracer injection or cell transplantation. No longer is physical access a requirement. Innovations over the past decade have led to genetic techniques that offer means to "deliver" cell lineage tracers noninvasively. Such "genetic fate mapping" approaches employ transgenic strategies to express genetically encoded site-specific recombinases in a cell type-specific manner to switch on expression of a cell-heritable reporter transgene as lineage tracer. The behaviors and fate of marked cells and their progeny can then be explored and their contributions to different tissues examined. Here, we review the basic concepts of genetic fate mapping and consider the strengths and limitations for their application. We also explore two refinements of this approach that lend improved spatial and temporal resolution: (1) Intersectional and subtractive genetic fate mapping and (2) Genetic inducible fate mapping.

Key words Fate map, Cre, Flp, Substractive genetic fate mapping, Intersectional genetic fate mapping, Genetic inducible fate mapping

1 Introduction

Knowing how embryonic tissue organization relates to postnatal tissue structure and function is fundamental for understanding development and disease. In studying the etiology of human disorders, especially informative are tissue fate maps generated in mammalian models. Yet mammalian models such as the mouse are poorly suited to traditional fate mapping techniques (reviewed in ref. 1). For example, in utero development makes it difficult to physically access embryonic cells for labeling by tracer injection. But no longer is physical access a fate mapping requirement. In 1998, technology was reported [2, 3] that allows, in effect, cell lineage tracers to be "delivered" to mouse embryonic cells in utero via noninvasive transgenic, rather than physical, means. The approach, called

Mark Lewandoski (ed.), *Mouse Molecular Embryology*, Methods in Molecular Biology, vol. 1092,
DOI 10.1007/978-1-60327-292-6_26, © Springer Science+Business Media New York 2014

"genetic fate mapping," exploits a type of molecule referred to as a site-specific recombinase (SSR), which, being genetically encoded, is amenable to in vivo delivery by transgenesis. The two recombinases most commonly used are Cre (named because it *c*auses *re*combination of the bacteriophage P1 genome) and Flp (named for its ability to invert, or "flip," a DNA segment in *Saccharomyces cerevisiae*). Through their capacity to produce precise DNA excisions, Cre and Flp each can be engineered to act as genetic on-switches, able to convert a silenced reporter transgene into a constitutively expressed one that has lineage tracing capability (Fig. 1).

In this fate mapping strategy, there are two basic elements, a recombinase-expressing transgene and a silenced reporter transgene that can be activated by recombination. For the first element, recombinase expression is directed in utero to a desired embryonic cell population. This is achieved by using gene regulatory elements, either incorporated into a transgene that drives recombinase expression selectively in certain cells or in the form of a gene knockin where recombinase-encoding sequence is inserted into a capable endogenous locus. Only in the recombinase-expressing cells is the second element, the silenced reporter transgene, switched, by recombinase action, to the "on," reporter-expressing configuration. This is achieved through SSR-mediated excision of an "off" or "stop" portion of the reporter transgene (Fig. 1). The consequence is that SSR-expressing embryonic cells are marked by reporter expression. Importantly, their descendant cells are also marked. This is because they inherit the "on" form of the reporter transgene and because the transgene has been engineered to remain on, that is to sustain constitutive reporter expression, regardless of subsequent cell differentiation and independent of any further recombinase expression. Thus, the reporter transgene has, in effect, been transformed into an indelible cell lineage tracer. The non-SSR-expressing embryonic cells (which typically is most of the embryo) are reporter-negative because they continue to harbor the unrecombined "silenced" form of the reporter transgene. So in this way, fate mapping no longer requires physical access to embryonic cells but rather genetic access via gene enhancer and promoter elements. Thanks to the extensive gene-expression data being generated presently and over the past decade [5–10] such driver elements are available and can be exploited to drive recombinase

Fig. 1 (continued) Genetic indicator transgene. Cells expressing the SSR will activate production of the reporter molecule (for example, β-gal). Activation of reporter molecule expression is permanent, and all cells descended from the SSR expressing (*gene A*-expressing) progenitors will continue expressing the reporter. Descendant cells are depicted here as *blue circles*. (Reproduced from ref. 4 with permission from Elsevier Science) (Color figure online)

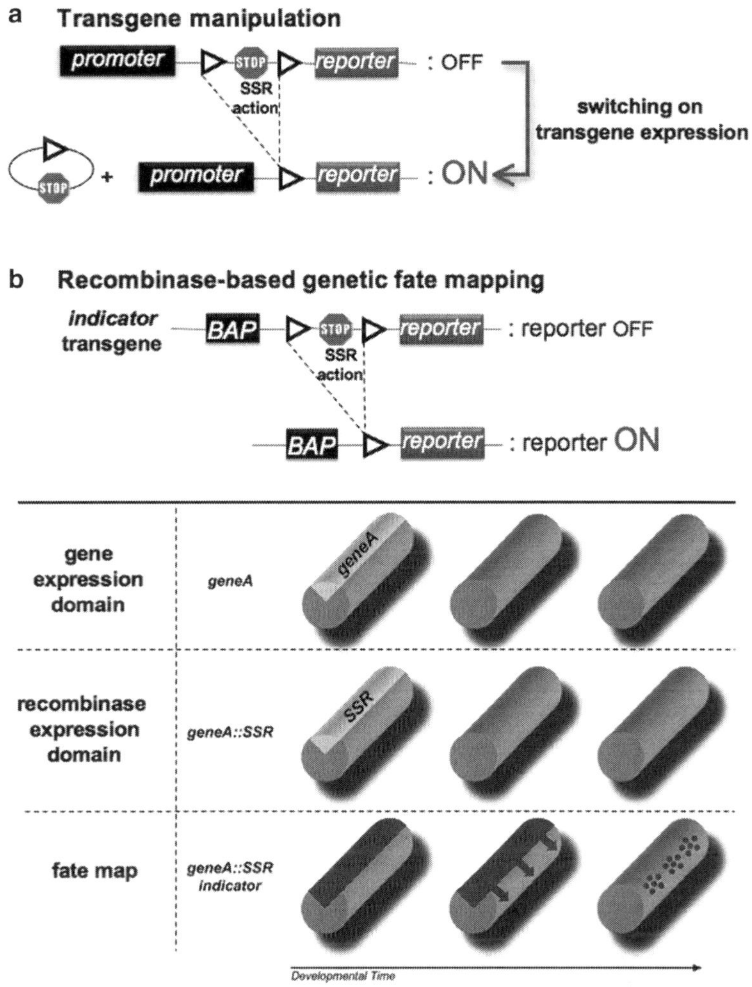

Fig. 1 Genetic fate mapping relies on site-specific recombinase-mediated DNA excision to "switch on" reporter expression. (**a**) Structure of a generic SSR-responsive transgene inserted as a single copy into the mouse genome. SSR-mediated recombination between two directly oriented SSR recognition sites (*triangles*) results in deletion of intervening transcriptional stop sequences (*red octagonal stop sign*) and consequent expression of a reporter molecule. (**b**) Illustration of how site-specific recombination can be used for genetic fate mapping. In the *top panel* the generic SSR-responsive transgene is modified by incorporation of a broadly active promoter (BAP) ideally capable of driving transgene expression in any cell type at any stage in development, such that after a recombination event in a given cell, that cell and all its progeny cells should be marked by reporter expression regardless of subsequent cell differentiation. The *lower panel* illustrates the strategy for SSR-based genetic fate mapping. *Cylinders* represent the neural tube at early (*left*) to late (*right*) developmental stages. *Top row* illustrates the transient expression of hypothetical *gene A* by progenitor cells located in the dorsal neural tube (*yellow domain*) at an early developmental stage. *Middle row* illustrates SSR-expressing transgene utilizing enhancer elements from *gene A*. *Bottom row* illustrates coupling of *gene A::SSR* with an

expression in most any embryonic cell type. In short, fate maps now can be generated in mice with relative ease.

Genetic fate mapping offers additional benefits. Perhaps most importantly, mouse genetics can be exploited concomitantly. Cells marked by genetic fate mapping can be studied in the context of mutant genes and other genetic alterations to reveal effects, for example, on their contributions to different tissues. Another benefit of genetic fate mapping is that the embryonic cells under study are selected based on their expression of a particular gene or by the activity of a particular enhancer element. As a consequence, the cell marking achievable from animal-to-animal is exceptionally precise and reproducible. Further, this molecular signature—used to select the embryonic cell population for labeling—can provide important clues about signaling pathways possibly relevant for acquiring particular fates. Further, it can reveal homologies in development among other cell populations that are otherwise disparate anatomically but at some point earlier during their development shared similar gene-expression signatures.

In this chapter, we review the basic concepts of recombinase-based genetic fate mapping and considerations for their application—as with all methods, there are strengths and limitations. We then describe two strategy variations that lend improved spatial and temporal resolution to genetic fate maps: the first variation is called intersectional and substractive genetic fate mapping [11, 12] and the second, genetic inducible fate mapping [13]. Early applications of these tools centered on studies of nervous system development (among examples are refs. 2, 3, 11, 12, 14–22) but today are being applied to study numerous cell types in a wide range of tissues and biological processes in the mouse.

2 Recombinase-Based Genetic Fate Mapping

Cre and Flp are the hinge-pins of genetic fate mapping technologies because of their capacity to produce precise, predetermined rearrangements of chromosomal DNA. DNA deletions, insertions, inversions, or exchanges are possible. Specifically which type of rearrangement depends on orientation of the DNA recognition sites (reviewed in ref. 23). Cre recombinase recognizes what are called *lox*P sites (locus of crossover (x) in P1) [24], and Flp, *FRT* sites (Flp recombinase recognition target) [25]. Both type of sites, while comprised of different DNA sequences, are 34-bp in length and not normally found in the mouse (or fly) genome, allowing for their creative insertion to engineer predetermined modifications. For example, when target sites are positioned on the same DNA molecule and oriented in the same direction, SSR action catalyzes deletion of the sequence lying in between, leaving behind a single target site. It is this DNA deletion reaction that has been exploited most for mouse fate mapping (Fig. 1). Of note, part of the foundation for SSR use in higher eukaryotes stems from work in *drosophila*, where SSRs are commonly used to catalyze exchange of chromosomal DNA that lies distal to target sites positioned on homologous chromosomes during the G2 phase of the cell cycle, with X segregation during mitosis allowing for the generation of mutant cell clones in an otherwise wild-type fly [26–29]. Thus, whether in mice or flies, once introduced into a cell, Cre or Flp can modify a *lox*P- or *FRT*-containing target locus, and depending on the configuration, turn it on, off (Fig. 1), or exchange distal sequence [26–29]. Turning on permanent expression of a reporter molecule, like β-galactosidase (β-gal) or green fluorescent protein (GFP), in select embryonic cells and their descendants is the basis of genetic fate mapping (Fig. 1).

As briefly introduced above, genetic fate mapping requires two types of transgenes or modified loci: (1) a recombinase "driver" transgene expressing the particular SSR in a gene- or enhancer-specific fashion to achieve specificity in the initially labeled cell type and (2) an "indicator" transgene designed to permanently express a reporter molecule following site-specific excisional recombination. When *SSR driver* and *indicator* transgenes are partnered in a double transgenic (*SSR driver, indicator*) animal, SSR-mediated recombination of the indicator transgene leads to constitutive reporter expression (Fig. 1). It is worth noting here that we use the term "indicator," as opposed to "reporter," to describe the target (reporter-encoding) transgene because it serves to indicate (by reporter expression) not only cells that are currently under the influence of SSR action but also cells that previously underwent SSR-recombination but are no longer expressing the recombinase. In other words, the presence of a recombined (reporter-expressing) target transgene indicates those cells with a history of recombinase expression, which means a history of expression of the particular driver gene or enhancer element. This feature is central to genetic fate mapping and worth emphasizing: once an indicator transgene has been "activated" by SSR-mediated recombination, the encoded reporter molecule is expressed constitutively by that lineage from that point onward, independent of any further recombinase expression. This feature ensures robust marking of descendant cells regardless of cell type or developmental stage. In most cases, it is possible to visualize lineage contributions to adult structures even though much time may elapse between the initial (embryonic) recombination event and the actual (adult) tissue analysis. Thus, long-term cell lineage tracing is enabled.

This "long-term" cell lineage tracing offered by SSR-based genetic fate mapping (with its recombinase and indicator transgenes) is worth contrasting with the often "short-term," more ephemeral, cell marking achievable when a nonconditional (non-SSR) *progenitor cell-specific promoter::reporter* transgene or knockin allele is used.

In this case, the term "reporter transgene," rather than indicator transgene, is used because reporter expression directly reflects the transient nature of the upstream cell type-specific driver sequences. Marked by the reporter molecule, as a consequence, are cells in which the driver gene enhancer elements are active at the time of tissue harvest. In some cases, later time points can be analyzed for marked descendant cells because of persistence of the reporter molecule itself beyond its immediate window of transcription, allowing for short-term lineage tracing. The "fate maps" resulting from this surrogate type of approach, though, may suffer loss in accuracy because some cell lineages may be missed because of their more rapid elimination of reporter molecules and/or because of lowering starting level of reporter expression. Erroneous exclusion of certain cells from a fate map may result.

The long-term cell lineage tracing enabled by SSR-based genetic fate mapping methods relies on constitutive reporter expression driven by the recombined indicator allele—that is, constitutive reporter expression in descendant cells regardless of the differentiated descendant cell type. This means that the indicator transgene must have the potential to be able to sustain robust reporter expression in virtually every cell type at all developmental and adult stages, or at least in every cell type of the tissue or organ under study, in order to ensure completeness of the fate map. This is a tall order. Various promoter/enhancer elements are being tested and used for this purpose. One that has been met with success in many tissue types is use of the endogenous mouse *Gt(ROSA)26Sor* (*R26*) locus [30], with [12, 31, 32, 33, 34] or without augmentation by enhancer elements from the chicken *β-actin* gene and cytomegalovirus genome (*CAG* sequences) [35]. Other exploited loci include *tau* [36], for nervous system studies, and *colA1* [37]. The actual range of cell types that can be marked by a given indicator transgene must be determined empirically. For example, an approximation of scope can be ascertained by analyzing tissue from an animal in which the target indicator transgene had been partnered with a transgene that drives broad SSR expression early in the development of the tissue of interest such that most constituent cells will harbor the recombined "active" form of the indicator allele. Cells can then be assessed for robustness of reporter expression. Any unmarked populations should be noted, and caution taken when interpreting the actual experimental (progenitor-specific) fate mapping results—exclusion of such cells from the fate map should not be concluded as they simply may not be capable of being marked by the indicator allele.

As introduced above, the choice of embryonic cell type for genetic fate mapping is determined by the cell type- or tissue-specific enhancer elements employed to drive SSR expression.

Such elements can be engineered to drive SSR expression by one of multiple ways: by conventional transgenic methods (*cell type-specific promoter::SSR* transgene of less than ~20 kb or so), bacterial artificial chromosome (BAC) transgenic strategies (constructs typically around ~200 kb), or by targeted gene knockin approach. The choice of method is determined by experimental need and availability of cell type- or tissue-specific enhancer elements.

Just as it is critical to determine the scope of cell types an indicator allele is capable of mapping, it is equally critical to determine the extent to which SSR expression matches that of the endogenous driver element profile. Ectopic SSR expression (expression outside the endogenous driver profile) would confound fate mapping studies by switching on the lineage tracer in cells outside of the lineage being studied. This would result in overestimation of the descendant population by erroneously including in the map unrelated cells. Reciprocally, if SSR levels permit recombination in only a subset of those cells in which the driver element is actually active, the resultant fate map will underestimate the descendant population. Thus, it is important to establish the extent to which initial expression of the activated reporter matches the initial SSR driver gene-expression profile. Later, expression of the activated reporter will diverge from that of the driver because the reporter is cumulatively and permanently tracking all cells that ever expressed the driver gene, whereas driver gene expression is transient (Fig. 1b). In other words, it is critical to determine whether all or only part of an initial gene-expression domain is being fate mapped. This is of particular concern if the driver gene exhibits a gradient of expression. It is possible that the lowest expressers in the gradient may be missed because levels of recombinase are inadequate to activate reporter expression. Such discrepancies are critical to determine.

Worth noting are the different variants of Flp recombinase that can be used in mice. They are referred to as wild-type Flp (Flp-wt), low-activity Flp (FlpL), enhanced Flp (Flpe), and optimized Flpe (referred to as Flpo). The enhanced Flpe contains four point mutations that together confer increased protein thermostability resulting in up to a tenfold increase in recombinase activity compared to Flp-wt [38, 39]. Recently, Flpe has been codon-optimized [40] with the goal of boosting Flp activity in mouse embryonic stem (ES) cells. By contrast to Flpe and Flpo, FlpL contains a single amino acid substitution that renders the recombinase thermolabile [41] resulting in at least a fivefold reduction in recombinase activity from Flp-wt. When expressed in mice, Flpe and Flpo have been shown to function with similar efficacy as Cre [11, 12, 20, 22, 42–47]. By contrast, FlpL achieves more modest recombination efficiency [16, 20]. In mouse embryonic stem cells, Flpo shows the greatest activity [37].

3 Dual Recombinase-Mediated Intersectional and Subtractive Genetic Fate Mapping

Despite the molecular precision of recombinase-based genetic fate mapping, many biological questions remain unanswerable because of the broad extent of cell types marked by the expression of a single gene. For example, embryonic gene-expression domains are often restricted along the dorsoventral (DV) or anteroposterior (AP) axis of a tissue or germinal zone while extending the length of the orthogonal axes (Fig. 2a). In this case, the gene-expression domain may actually contain multiple subpopulations of cells each distinguished by their expression of a second gene whose expression overlaps along this orthogonal axis (Fig. 2a) [11, 12, 42, 47]. The ability to fate map these subpopulations and their descendent lineages is not possible using single recombinase-based genetic fate mapping. To overcome this limitation, gene combinations need to be employed. Towards this goal, a method called "intersectional" genetic fate mapping in which the initial population to be traced is selected based on its expression of two genes (and consequently two recombinases) rather than one (Fig. 2) was developed [11]. The ability to resolve highly selective cell populations for fate mapping can be substantial, depending on the extent of overlap between the two chosen driver genes.

For intersectional genetic fate mapping, two SSRs, Cre and Flp, are needed to activate expression of a reporter molecule—the "intersectional reporter" [11, 12, 31, 42, 48, 49]; reviewed in refs. 4, 23, 50 (Fig. 2b). In other words, two excisional recombination events are required, one mediated by Cre (driven by *gene A*) and the other by Flp (driven by *gene B*) (Fig. 2). Only those cells lying at the intersection of the two gene-expression domains (*A* and *B*) will therefore activate reporter expression (Fig. 2) and their descendant cells marked. This intersectional approach can indeed be highly efficient for marking progenitor cells lying at the intersection of two gene-expression domains, as demonstrated in recent developmental studies of the brainstem [11, 12, 42, 47].

Of note, in this approach, the expression of the two genes (e.g., driver *gene A* and driver *gene B*) does not have to coincide temporally. They may be expressed at different times in a cell's developmental history with activation of the intersectional reporter occurring only after the second recombination event has been completed. This means that temporal as well as spatial resolution in lineage tracing can be improved.

In addition to fate mapping intersecting Cre/Flp cell subpopulations, this technology can be engineered to allow simultaneous tracing of Cre/non-Flp lineages, referred to as "subtractive" populations. The subtractive populations are what remain after Cre/Flp-intersecting cells are subtracted from the Cre-only expressing domain [12] (Fig. 2c). In order to visualize the subtractive population, the intersectional indicator allele needs to be configured such that a second reporter molecule is encoded (the "subtractive reporter") and its expression is dependent only upon excision by Cre of the *lox*P-flanked (floxed) stop cassette (Fig. 2c). Thus, two genetically distinct lineages can be tracked simultaneously, one by the intersectional reporter and the other by the subtractive reporter. The subtractive population can also be engineered to be the reciprocal Flp/non-Cre population; this is achieved simply by changing the order of the floxed and flrted (*FRT*-flanked) stop cassettes in the intersectional indicator allele (Fig. 2c) [11, 31, 42]. It is worth noting that use of a reporter molecule with a relatively short half-life is helpful as the subtractive reporter. This ensures that the subtractive reporter will not persist substantially and therefore be detected in intersectional descendents despite its coding sequence having been excised from the indicator allele. Of course, this can be rigorously discerned by reporter co-detection experiments, with the intersectional population being defined by constitutive expression of the intersectional reporter possibly along with (unwanted) persistence of the subtractive reporter; the subtractive population would be rigorously identified as harboring only the subtractive reporter molecule and not both.

Beyond improving fate map resolution, intersectional alleles offer another practical advantage. Three different mouse lines can be generated from one initial transgene construction and strain generation: one dual recombinase-responsive indicator allele and two derivative strains that are either responsive to Cre only or to Flp only. The latter are derived through germline deletion of either the floxed or flrted cassettes [12].

4 Genetic Inducible Fate Mapping

To better resolve temporal aspects of lineage allocation during mouse development, a method coined GIFM [51, 52], for genetic inducible fate mapping, has been developed. This approach relies on a driver transgene expressing a ligand-regulated form of Cre or Flp to temporally control SSR activity (Fig. 3a). In GIFM, the SSR is fused to an estrogen receptor (ER) ligand-binding domain (LBD) that has been mutated, rendering it insensitive to the natural ligand 17β-estradiol at physiological concentrations but responsive to the synthetic ligand 4-hydroxytamoxifen (4-OHT) [53–58] (Fig. 3a). Temporal control of recombinase activity relies on the ability of the ER-LBD to sequester the SSR into a cytoplasmic heat shock protein 90 (Hsp90) complex in the absence of 4-OHT. Upon 4-OHT binding, the ER-LBD undergoes a conformational change that frees the fused SSR, allowing it to enter the nucleus

Fig. 2 Intersectional and subtractive genetic fate mapping strategy. (**a**) Multiple uniquely coded molecular subdomains may comprise a single gene-expression domain. Shown are schematics of the neural tube (*gray cylinder*), with different gene-expression domains depicted in different colors. The expression domain for hypothetical gene A (*yellow*) restricts along the dorsoventral (DV) axis but extends along the anteroposterior (AP) axis; by contrast, the expression domains for genes B (*pink*) and C (*red*) restrict along the AP axis but extend along the DV axis. Thus, the gene A expression domain (*yellow*) is subdivided into three molecularly distinct subdomains: one in which genes A and B are coexpressed (*tan domain*), another in which genes A and C are coexpressed (*orange domain*), and finally, that territory (*yellow*) marked by gene A expression, but not B or C. Similarly, both the gene B and C expression domains are each subdivided. (**b**) Structure of a prototypical dual recombinase (Cre and Flpe)-responsive indicator allele. In contrast to a single recombinase-responsive indicator allele (Fig. 1b), a dual recombinase-responsive indicator allele has two stop cassettes, one flanked by directly oriented *loxP* sites (*triangles*) and the other by *FRT* sites (*vertically oriented rectangles*). Cre-mediated stop cassette removal results in expression of nβ-gal, while the remaining *FRT*-flanked stop cassette prevents GFP expression. Following removal of both stop cassettes, requiring Cre- and Flpe-mediated excisions, GFP expression is turned on and nβ-gal expression off. (**c**) Illustration of intersectional and subtractive populations and the latter dependency on stop-cassette order. In the "PF" configured allele, the loxP-flanked stop cassette precedes the FRT-flanked cassette (*left panel*), while the reciprocal order characterizes the "FP" configuration (*right panel*). Shown are schematics of the neural tube (*gray cylinder*), with the expression domain for hypothetical gene A and Flpe recombinase (*yellow*) restricting along the DV axis but extending along the AP axis (*top row*); in contrast, the expression domain for gene B (*pink*) restricts along the AP axis but extends along the DV axis (*middle row*). When gene A::Flpe and gene B::cre are coupled with a PF dual recombinase-responsive indicator allele (*bottom row, left*), cells expressing cre and Flpe activate production of GFP (*green domain*, intersectional population), while cells expressing only cre activate production of nβ-gal (*blue domain*, subtractive population). When gene A::Flpe and gene B::cre are coupled with an FP-configured allele (*bottom row, right*), cells expressing cre and Flpe still activate production of GFP in the same intersectional population (*green domain*), but now cells expressing only Flpe (rather than cre) activate production of nβ-gal (*blue domain*, new subtractive population). (**d**) Illustration of the selective fate mapping achievable using an intersectional and subtractive approach. *Cylinders* represent the neural tube at early (*left*) to late (*right*) developmental stages. *Top row* illustrates transient Flpe expression, driven by *gene A*, in progenitor cells located in the dorsal neural tube (*yellow domain*) at an early developmental stage. *Middle row* illustrates transient Cre expression, driven by *gene B*, in progenitor cells located at a particular anteroposterior level of the neural tube at an early developmental stage (*pink domain*). *Bottom row* illustrates coupling *gene A::Flpe* and *gene B::cre* are coupled with a dual recombinase-responsive indicator allele (FP configuration), cells expressing Flpe and Cre activate production of GFP, while cells expressing only Flpe activate production of nβ-gal. Activation of reporter molecule expression is permanent, and all cells descended from Flpe-expressing or Flpe- and cre-expressing progenitors will continue expressing the blue or green marker, respectively. Descendant cells from the intersectional domain are denoted by *green circles*, those from the subtractive population by *blue circles*. (Reproduced from ref. 4 with permission from Elsevier Science) (Color figure online)

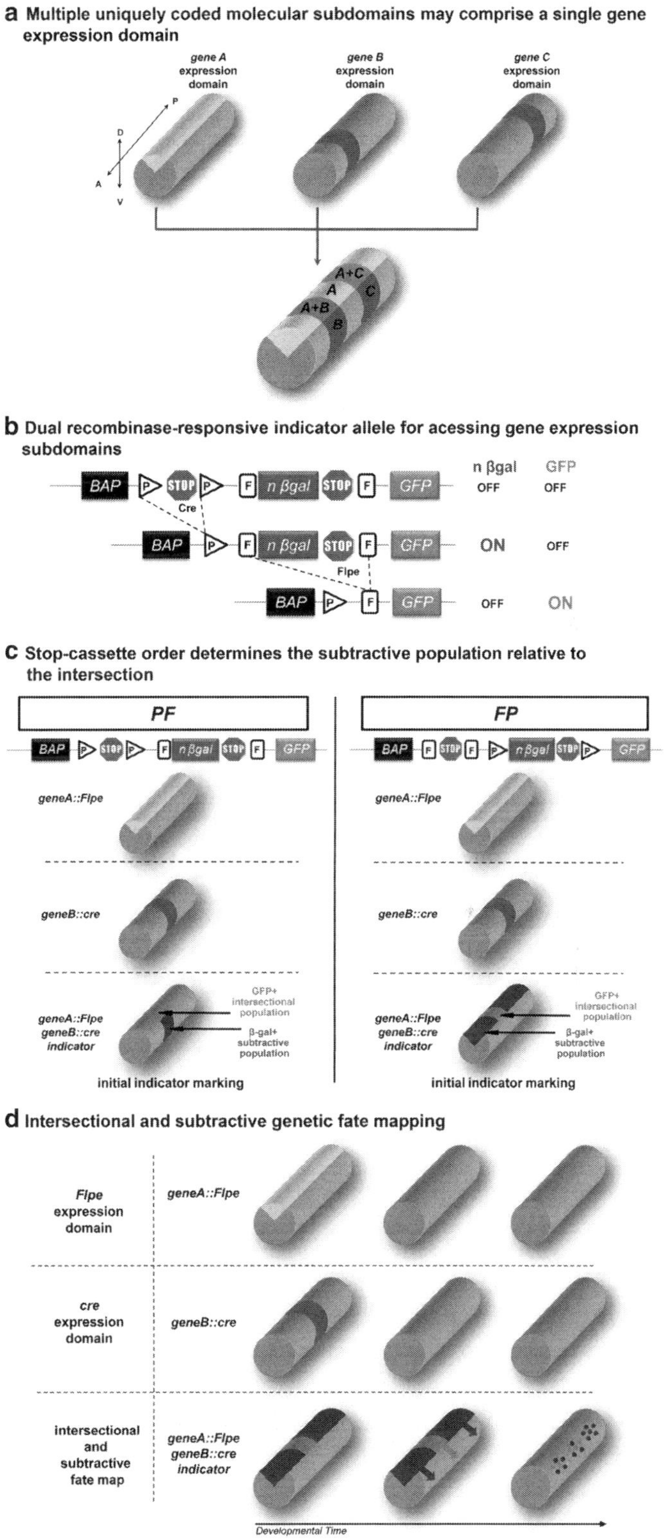

a Multiple uniquely coded molecular subdomains may comprise a single gene expression domain
gene A expression domain
gene B expression domain
gene C expression domain
D
P
A
V
A+C
A
C
A+B
B
b Dual recombinase-responsive indicator allele for acessing gene expression subdomains
n βgal
GFP
BAP
STOP
n βgal
GFP
OFF
OFF
Cre
ON
OFF
Flpe
OFF
ON
c Stop-cassette order determines the subtractive population relative to the intersection
PF
FP
geneA::Flpe
geneB::cre
geneA::Flpe
geneB::cre
indicator
GFP+ intersectional population
β-gal+ subtractive population
initial indicator marking
d Intersectional and subtractive genetic fate mapping
Flpe expression domain
cre expression domain
intersectional and subtractive fate map
Developmental Time

where it can mediate recombination of a target site-containing locus (Fig. 3b) (For a detailed review *see* ref. 52). There are at least three different ER-LBDs available—two human ER variants, ER^{T} [53–56, 58, 59] and ER^{T2} [13, 60–63], and one from the mouse, ER^{TM} [17, 57, 54, 65]. ER^{T2} appears to be the most sensitive form for both nuclear [13, 60–63] translocation and recombinase activity [13, 60–63]. $CreER^{T2}$ [13, 61–63], $FlpeER^{T2}$ [66], and $FlpoER^{T2}$ [42] fusions have been generated, and all are effective in vivo. Worth noting, $CreER^{T2}$ may outperform $FLPeER^{T2}$ in certain cell types when expressed at low levels (N.L. Hunter and S.M.D., unpublished observations), despite the fact that the constitutive forms, Cre and Flpe, show comparably robust activity in vivo.

Tamoxifen, the precursor to 4-OHT, is typically the reagent used in GIFM. It is both easier to work with (it is more soluble) and less costly than 4-OHT. Conversion of tamoxifen to 4-OHT takes approximately 6–12 h in vivo, resulting in an initial lag between administration of tamoxifen and onset of recombinase-mediated target gene recombination. The 6–12 h lag is followed by an approximately 24 h window during which recombination is catalyzed and the initial cell population is marked [13, 17, 18, 64, 66]. Given such kinetics, it is important to establish the extent of recombinase activity following tamoxifen administration; in other words, it is critical to establish the extent to which expression of the indicator transgene-encoded reporter molecule matches the expected driver gene-expression profile between 24 and 48 h after tamoxifen administration. The degree of concordance indicates whether most or only a subset of the highest driver gene expressors can be tracked. Once these parameters are set, it should be possible to visualize the fate of these cells at any later time point. The resulting fate map will mark just those cells that have emerged from a gene-expression domain during a particular 24 h window. Duration of the recombination window will of course vary depending on the tamoxifen dosing regimen, with higher amounts and repeated administrations lengthening that window.

Fig. 3 (continued) rows, much as done previously in Fig. 1b. *Top row* illustrates transient, midgestation cre recombinase expression, driven by *gene A*, in progenitor cells of the dorsal neural tube. *Second row* illustrates activation of nβ-gal, for example, as a lineage tracer in all cells that ever in their history expressed *gene A::cre*. Inducible genetic fate mapping is schematized in the bottom two rows. *Third row* illustrates transient, midgestation cre recombinase expression, driven by *gene A*, in progenitor cells of the dorsal neural tube. *Bottom row* illustrates selective activation of nβ-gal in late-emerging cohorts (*blue triangles*) following administration of 4-OHT and subsequent induction of recombinase-mediated recombination of the indicator transgene. (Reproduced from ref. 4 with permission from Elsevier Science) (Color figure online)

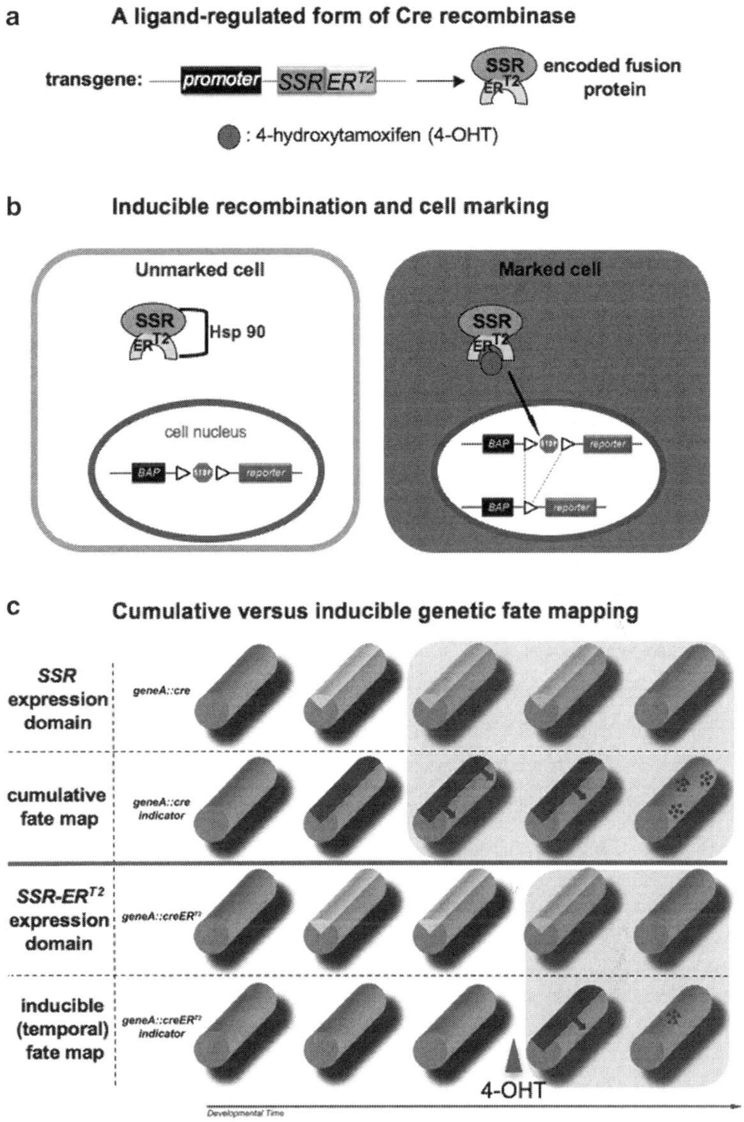

Fig. 3 Genetic inducible fate mapping strategy. (**a**) Schematic of a transgene encoding the site-specific recombinase-steroid fusion protein, SSR-ERT2, whose activity is regulated posttranslationally by the ligand 4-hydroxy tamoxifen (4-OHT, *orange circle*). (**b**) Inducible recombination and cell marking using SSR-ERT2. In the absence of 4-OHT, SSR-ERT2 is inactive due to sequestration of the fusion protein into an Hsp90 complex. Binding of 4-OHT to SSR-ERT2 results in a conformational change that disrupts the Hsp90 interaction, freeing the recombinase to enter the cell nucleus and mediate recombination at its target sites (*triangles*) positioned within an indicator transgene. Excisional recombination renders cells positive for reporter expression (for example, cytoplasmic β-gal as indicated in *dark blue*). (**c**) Cumulative versus inducible genetic fate mapping. Development of the neural tube is again rendered as simple cylinders progressing left to right in each row. Cumulative genetic fate mapping is schematized in the top two

It is important to note, however, that tamoxifen, when administered at high or repeated doses, can be lethal to the developing embryo. Therefore, it is critical to establish the administration regimen of tamoxifen that maximizes recombination at the desired embryonic stage but minimizes unwanted side effects. It appears that inbred mouse strains are more sensitive to higher doses of tamoxifen than outbred strains such as Swiss Webster [51, 52]. The coadministration of progesterone with high tamoxifen doses also may improve litter viability [51, 52]. Another consideration is gestational age. Later-stage embryos are better able to tolerate higher doses of tamoxifen. The level of expressed SSR-ER^{T2} protein must also be considered. While higher levels of SSR-ER^{T2} protein are better able to induce recombination at lower tamoxifen levels, if too high, the capacity for tight induction may be compromised resulting in unwanted recombination even in the absence of tamoxifen.

By contrast to both standard and intersectional genetic fate mapping where populations of progenitor cells and their descendants are marked, GIFM has the potential to allow for clonal analyses because low doses of tamoxifen can permit, in certain cases, the marking of a single progenitor cell in a region and therefore just its daughter cells [67]; reviewed in ref. 51. Another SSR-based approach, distinct from GIFM, that permits clonal analyses and is therefore important to include in this review is a technique called mosaic analysis with double markers (MADM). This approach, like the FRT-mediated clonal analyses in *drosophila* mentioned above, relies on rare SSR-mediated translocation events between two homologous chromosomes during the G2 phase of the cell cycle; X segregation of the recombined chromosomes during mitosis then results in two daughter cells each expressing one or the other marker [32]; reviewed in ref. 68.

While the majority of Cre and CreER transgenics function well with few side effects, it is important to note that toxicity, seen as upregulated apoptosis and cell cycle arrest for example, has been reported in a number of cases [69]. DNA damage secondary to abortive recombination attempts on endogenous noncanonical *lox*P sites may be the cause [69]—a situation most likely to occur when Cre activity is quite high. Especially susceptible may be CreER transgenics because there is no selection against extremely high CreER expressors. This is because Cre activity is held in check by heat shock complex sequestration, allowing founder lines of a range of expression levels to be generated and propagated, even when expression levels are high such that noncanonical recombination might occur under tamoxifen induction. By contrast, for constitutive *cre* lines, there may be selection against high expressors in the founder generation, thus weeding out the most problematic lines. Thus critical for sussing out Cre toxicity is a comparison between tamoxifen-induced and uninduced CreER-positive animals, in the absence of any target transgene [69]. Toxicity associated with Flp recombinase (and its Flpe, Flpo, and ER variants) has not been reported, but may be lurking and simply not observed yet. Thus, single recombinase controls, regardless of which recombinase, are essential.

5 Conclusion

Genetic fate mapping, and its variations, have enabled powerful science: (1) it is now possible to more easily generate fate maps in a mammalian system; (2) using mice, with its genetics, the behaviors of marked cells and their contributions to different tissues can now be studied in the context of mutant genes and other genetic alterations, thereby informing on how gene products affect development or lead to disease; and (3) progenitor cells can be selected for tracking based on an expressed gene—a feature that enables exceptionally precise and reproducible cell marking from animal to animal. Moreover, this latter feature can serve to illuminate potential roles of gene products in development, as well as developmental homologies between unexpected and otherwise disparate anatomical structures.

Elaborations on these SSR basics are ongoing in many laboratories, with the field advancing in numerous directions. One example is to use these paradigms of conditional gene "delivery" to express various genetically encoded effector molecules capable of revealing functional properties of a genetic lineage or of the processes it participates in refs. 70, 71. The possibilities indeed seem endless; even further so when combined with classical and complementary methodologies, such as the clonal information afforded by retroviral lineage tracing or the neuroanatomical information provided upon combining axon tract tracing techniques with genetic fate mapping, just to name a few exciting examples.

References

1. Stern CD, Fraser SE (2001) Tracing the lineage of tracing cell lineages. Nat Cell Biol 3:E216–E218
2. Dymecki SM, Tomasiewicz H (1998) Using Flp-recombinase to characterize expansion of Wnt1-expressing neural progenitors in the mouse. Dev Biol 201:57–65
3. Zinyk DL, Mercer EH et al (1998) Fate mapping of the mouse midbrain-hindbrain constriction using a site-specific recombination system. Curr Biol 8:665–668
4. Dymecki SM, Kim JC (2007) Molecular neuroanatomy's "Three Gs": a primer. Neuron 54:17–34
5. Portales-Casamar E, Swanson DJ et al (2010) A regulatory toolbox of MiniPromoters to drive selective expression in the brain. Proc Natl Acad Sci U S A 107:16589–16594
6. Gong S, Zheng C et al (2003) A gene expression atlas of the central nervous system based on bacterial artificial chromosomes. Nature 425:917–925
7. Gray PA, Fu H et al (2004) Mouse brain organization revealed through direct genome-scale TF expression analysis. Science 306:2255–2257
8. Visel A, Thaller C et al (2004) GenePaint.org: an atlas of gene expression patterns in the mouse embryo. Nucleic Acids Res 32:D552–D556

9. Magdaleno S, Jensen P et al (2006) BGEM: an in situ hybridization database of gene expression in the embryonic and adult mouse nervous system. PLoS Biol 4:e86
10. Lein ES, Hawrylycz MJ et al (2007) Genome-wide atlas of gene expression in the adult mouse brain. Nature 445:168–176
11. Awatramani R, Soriano P et al (2003) Cryptic boundaries in roof plate and choroid plexus identified by intersectional gene activation. Nat Genet 35:70–75
12. Farago AF, Awatramani RB et al (2006) Assembly of the brainstem cochlear nuclear complex is revealed by intersectional and subtractive genetic fate maps. Neuron 50:205–218
13. Kimmel RA, Turnbull DH et al (2000) Two lineage boundaries coordinate vertebrate apical ectodermal ridge formation. Genes Dev 14:1377–1389
14. Chai Y, Jiang X et al (2000) Fate of the mammalian cranial neural crest during tooth and mandibular morphogenesis. Development 127:1671–1679
15. Jiang X, Rowitch DH et al (2000) Fate of the mammalian cardiac neural crest. Development 127:1607–1616
16. Rodriguez CI, Dymecki SM (2000) Origin of the precerebellar system. Neuron 27:475–486
17. Zirlinger M, Lo L et al (2002) Transient expression of the bHLH factor neurogenin-2 marks a subpopulation of neural crest cells biased for a sensory but not a neuronal fate. Proc Natl Acad Sci U S A 99:8084–8089
18. Zervas M, Millet S et al (2004) Cell behaviors and genetic lineages of the mesencephalon and rhombomere 1. Neuron 43:345–357
19. Ahn S, Joyner AL (2005) In vivo analysis of quiescent adult neural stem cells responding to Sonic hedgehog. Nature 437:894–897
20. Landsberg RL, Awatramani RB et al (2005) Hindbrain rhombic lip is comprised of discrete progenitor cell populations allocated by Pax6. Neuron 48:933–947
21. Sgaier SK, Millet S et al (2005) Morphogenetic and cellular movements that shape the mouse cerebellum; insights from genetic fate mapping. Neuron 45:27–40
22. Hunter NL, Dymecki SM (2007) Molecularly and temporally separable lineages form the hindbrain roof plate and contribute differentially to the choroid plexus. Development 134:3449–3460
23. Branda CS, Dymecki SM (2004) Talking about a revolution: the impact of site-specific recombinases on genetic analyses in mice. Dev Cell 6:7–28
24. Hoess R, Abremski K et al (1984) The nature of the interaction of the P1 recombinase Cre with the recombining site loxP. Cold Spring Harb Symp Quant Biol 49:761–768
25. McLeod M, Craft S et al (1986) Identification of the crossover site during FLP-mediated recombination in the Saccharomyces cerevisiae plasmid 2 microns circle. Mol Cell Biol 6:3357–3367
26. Golic KG, Lindquist S (1989) The FLP recombinase of yeast catalyzes site-specific recombination in the Drosophila genome. Cell 59:499–509
27. Dang DT, Perrimon N (1992) Use of a yeast site-specific recombinase to generate embryonic mosaics in Drosophila. Dev Genet 13:367–375
28. Xu T, Rubin GM (1993) Analysis of genetic mosaics in developing and adult Drosophila tissues. Development 117:1223–1237
29. Lee T, Luo L (1999) Mosaic analysis with a repressible cell marker for studies of gene function in neuronal morphogenesis. Neuron 22:451–461
30. Zambrowicz BP, Imamoto A et al (1997) Disruption of overlapping transcripts in the ROSA beta geo 26 gene trap strain leads to widespread expression of beta-galactosidase in mouse embryos and hematopoietic cells. Proc Natl Acad Sci U S A 94:3789–3794
31. Engleka KA, Manderfield LJ, Brust RD, Li L, Cohen A, Dymecki SM, Epstein JA (2012) Islet1 derivatives in the heart are of both neural crest and second heart field origin. Circ Res 110:922–926
32. Zong H, Espinosa JS et al (2005) Mosaic analysis with double markers in mice. Cell 121:479–492
33. Madisen L, Zwingman TA et al (2010) A robust and high-throughput Cre reporting and characterization system for the whole mouse brain. Nat Neurosci 13:133–140
34. Yamamoto M, Shook NA et al (2009) A multifunctional reporter mouse line for Cre- and FLP-dependent lineage analysis. Genesis 47:107–114
35. Niwa H, Yamamura K et al (1991) Efficient selection for high-expression transfectants with a novel eukaryotic vector. Gene 108:193–199
36. Hashimoto Y, Muramatsu K et al (2008) Neuron-specific and inducible recombination by Cre recombinase in the mouse. Neuroreport 19:621–624
37. Beard C, Hochedlinger K et al (2006) Efficient method to generate single-copy transgenic mice by site-specific integration in embryonic stem cells. Genesis 44:23–28
38. Buchholz F, Angrand PO et al (1998) Improved properties of FLP recombinase evolved by cycling mutagenesis. Nat Biotechnol 16:657–662

39. Rodriguez CI, Buchholz F et al (2000) High-efficiency deleter mice show that FLPe is an alternative to Cre-loxP. Nat Genet 25:139–140
40. Raymond CS, Soriano P (2007) High-efficiency FLP and PhiC31 site-specific recombination in mammalian cells. PLoS One 2:e162
41. Buchholz F, Ringrose L et al (1996) Different thermostabilities of FLP and Cre recombinases: implications for applied site-specific recombination. Nucleic Acids Res 24: 4256–4262
42. Jensen P, Farago AF et al (2008) Redefining the serotonergic system by genetic lineage. Nat Neurosci 11:417–419
43. Raymond CS, Soriano P (2010) ROSA26Flpo deleter mice promote efficient inversion of conditional gene traps in vivo. Genesis 48:603–606
44. Kranz A, Fu J, Duerschke K, Weidlich S, Naumann R, Stewart AF, Anastassiadis K (2010) An improved Flp deleter mouse in C57Bl/6 based on Flpo recombinase. Genesis 48:512–520
45. Lao Z, Raju GP, Bai CB, Joyner AL (2012) MASTR: a technique for mosaic mutant analysis with spatial and temporal control of recombination using conditional floxed alleles in mice. Cell Rep 2:386–396
46. Hirsch MR, d'Autreaux F, Dymecki SM, Brunet JF, Goridis C (2013) A Phox2b::FLPo transgenic mouse line suitable for intersectional genetics. Genesis 51(7):506–514. doi:10.1002/dvg.22393
47. Robertson SD, Plummer NW, de Marchena J, Jensen P (2013) Developmental origins of central norepinephrine neuron diversity. Nat Neurosci. doi:10.1038/nn.3458
48. Cocas LA, Miyoshi G, Carney RS, Sousa VH, Hirata T, Jones KR, Fishell G, Huntsman MM, Corbin JG (2009) Emx1-lineage progenitors differentially contribute to neural diversity in the striatum and amygdala. J Neurosci 29:15933–15946
49. Bang SJ, Jensen P, Dymecki SM, Commons KG (2012) Projections and interconnections of genetically defined serotonin neurons in mice. Eur J Neurosci 35:85–96
50. Dymecki SM, Ray RS et al (2010) Mapping cell fate and function using recombinase-based intersectional strategies. Methods Enzymol 477:183–213
51. Legue E, Joyner AL (2010) Genetic fate mapping using site-specific recombinases. Methods Enzymol 477:153–181
52. Joyner AL, Zervas M (2006) Genetic inducible fate mapping in mouse: establishing genetic lineages and defining genetic neuroanatomy in the nervous system. Dev Dyn 235:2376–2385
53. Logie C, Stewart AF (1995) Ligand-regulated site-specific recombination. Proc Natl Acad Sci U S A 92:5940–5944
54. Metzger D, Clifford J et al (1995) Conditional site-specific recombination in mammalian cells using a ligand-dependent chimeric Cre recombinase. Proc Natl Acad Sci U S A 92:6991–6995
55. Feil R, Brocard J et al (1996) Ligand-activated site-specific recombination in mice. Proc Natl Acad Sci U S A 93:10887–10890
56. Brocard J, Warot X et al (1997) Spatio-temporally controlled site-specific somatic mutagenesis in the mouse. Proc Natl Acad Sci U S A 94(26):14559–14563
57. Danielian PS, Muccino D et al (1998) Modification of gene activity in mouse embryos in utero by a tamoxifen-inducible form of Cre recombinase. Curr Biol 8(24):1323–1326
58. Schwenk F, Kuhn R et al (1998) Temporally and spatially regulated somatic mutagenesis in mice. Nucleic Acids Res 26:1427–1432
59. Vooijs M, Jonkers J et al (2001) A highly efficient ligand-regulated Cre recombinase mouse line shows that LoxP recombination is position dependent. EMBO Rep 2:292–297
60. Feil R, Wagner J et al (1997) Regulation of Cre recombinase activity by mutated estrogen receptor ligand-binding domains. Biochem Biophys Res Commun 237:752–757
61. Indra AK, Warot X et al (1999) Temporally-controlled site-specific mutagenesis in the basal layer of the epidermis: comparison of the recombinase activity of the tamoxifen-inducible Cre-ER(T) and Cre-ER(T2) recombinases. Nucleic Acids Res 27:4324–4327
62. Imai T, Jiang M et al (2001) Impaired adipogenesis and lipolysis in the mouse upon selective ablation of the retinoid X receptor alpha mediated by a tamoxifen-inducible chimeric Cre recombinase (Cre-ERT2) in adipocytes. Proc Natl Acad Sci U S A 98:224–228
63. Seibler J, Zevnik B et al (2003) Rapid generation of inducible mouse mutants. Nucleic Acids Res 31:e12
64. Hayashi S, McMahon AP (2002) Efficient recombination in diverse tissues by a tamoxifen-inducible form of Cre: a tool for temporally regulated gene activation/inactivation in the mouse. Dev Biol 244:305–318
65. Guo Q, Loomis C et al (2003) Fate map of mouse ventral limb ectoderm and the apical ectodermal ridge. Dev Biol 264:166–178
66. Hunter NL, Awatramani RB et al (2005) Ligand-activated Flpe for temporally regulated gene modifications. Genesis 41:99–109

67. Legue E, Nicolas JF (2005) Hair follicle renewal: organization of stem cells in the matrix and the role of stereotyped lineages and behaviors. Development 132:4143–4154
68. Luo L (2007) Fly MARCM and mouse MADM: genetic methods of labeling and manipulating single neurons. Brain Res Rev 55:220–227
69. Naiche LA, Papaioannou VE (2007) Cre activity causes widespread apoptosis and lethal anemia during embryonic development. Genesis 45:768–775
70. Kim JC, Cook MN et al (2009) Linking genetically defined neurons to behavior through a broadly applicable silencing allele. Neuron 63:305–315
71. Ray RS, Corcoran AE, Brust RD, Kim JC, Richerson GB, Nattie E, Dymecki SM (2011) Impaired respiratory and body temperature control upon acute serotonergic neuron inhibition. Science 333:637–642

Index

A

B

Mark Lewandoski (ed.), *Mouse Molecular Embryology*, Methods in Molecular Biology, vol. 1092, DOI 10.1007/978-1-60327-292-6, © Springer Science+Business Media New York 2014

F

G

H

I

J

K

L

M

N

T

U

V

W

X

Y

Z

Printed by Publishers' Graphics LLC
LMO131224.15.18.78 20131224